CARBON NANOTUBES - CURRENT PROGRESS OF THEIR POLYMER COMPOSITES

Edited by **Mohamed Reda Berber**
and **Inas Hazzaa Hafez**

Carbon Nanotubes - Current Progress of their Polymer Composites
http://dx.doi.org/10.5772/61428
Edited by Mohamed Reda Berber and Inas Hazzaa Hafez

Contributors
Samy Yousef, Hai Duong, Lixing Dai, Jun Sun, Chuankun Jia, Du Yuan, Yun Guang Zhu, Mohammad Jellur Rahman, Tetsu Mieno, Saeed Bastani, Masoume Kaviani Darani, Xiaohua Zhang, Xueping Yu, Jingna Zhao, Qingwen Li, Zekerya Dursun, Şükriye Ulubay Karabiberoğlu, Çağrı Ceylan Koçak, Abdul Hakim Shah, Chenzhong Yao, Yexiang Tong, Peter Hammer, Samarah Vargas Harb, Celso Valentim Santilli, Fábio Cesar Dos Santos, Sandra Hellena Pulcinelli, Kevin Knowles, Javier Pozuelo, Marta González, Guillermo Mokry, María De Nicolás, Juan Baselga, Parveen Saini, Steve Francis Albert Acquah, Lara Al-Hariri, Branden Leonhardt, Mesopotamia Nowotarski, James Magi, Kaelynn Chambliss, Thaís Venzel, Sagar Delekar, Nikolai Kouklin, Go Yamamoto, Keiichi Shirasu, Toshiyuki Hashida, Gülşen Evingur, Önder Pekcan, Yuming Zhao

CBS, Edition **2017**

Published by InTech
Janeza Trdine 9, 51000 Rijeka, Croatia

Publishing Process Manager Edi Lipovic
Technical Editor SPi Global
Cover InTech Design team

Additional hard copies can be obtained from orders@intechopen.com

Carbon Nanotubes - Current Progress of their Polymer Composites , Edited by Mohamed Reda Berber and Inas Hazzaa Hafez

p. cm.

Print ISBN 978-953-51-2469-6
Online ISBN 978-953-51-2470-2

INTECH

open science | open minds

Contents

Preface

This book shows the recent advances of carbon nanotubesapplications (CNTs), in particular, the polymer functionalized carbon nanotubes. It also includes a comprehensive description of carbon nanotubes' preparation, properties, and characterization. Therefore, in this book, we have attempted to provide detailed information about the polymer-carbon nanotube composites. With regard to the unique structure and properties of carbon nanotubes, a series of important findings have been reported. The unique properties of carbon nanotubes, including thermal, mechanical, and electrical properties, have been documented in detail in this book.

This book comprises 18 chapters. The chapter of Parveen Saini and co-workers covers the designing, development, properties, and applications of carbon nanotube–loaded polymer composites. The chapter of Hai M. Duong et al. presents an overview of the synthesis approaches of aligned CNT composites. To enhance the load transfer within CNT fibers, polymer infiltration of CNT fibers is reviewed. The chapter of Xiaohua Zhang et al. shows the bio-inspired design and fabrication of super-strong and multifunctional carbon nanotube composites. The chapter of T. Hosseini and N. Kouklin highlights the device properties and photovoltaic applications of carbon nanotube-polymer composites. The chapters of Gülşen Akın Evingür et al. and Lixing Dai and Jun Sun show the optical, mechanical, and electrical properties of polymer composites doped by multiwalled CNTs. The chapter of Peter Hammer et al. highlights the protective coatings based on PMMA-silica nanocomposites reinforced with CNTs. The chapter of Steve F. A. Acquah et al. describes the use of carbon nanotubes as additives in three-dimensional (3D) printing. The chapter of Samy Yousef discusses the use of polymer-CNT nanocomposite in the artificial joints. The chapter of S. Bastani and M. K. Darani shows the use of carbon nanotube as UV-curable nanocomposite coatings. The chapter of Javier Pozuelo et al. shows the use of the carbon nanotube-polymer composites as electromagnetic shielding materials in the GHz range. The chapter of M. J. Rahman and T. Mieno highlights the safer production of water dispersible carbon nanotubes and nanotube/cotton composite materials. The chapter of Li Wang and Yuming Zhao covers the design, reversible dispersion, and applications of stimuli-responsive carbon nanotube composites. The chapter of Keiichi Shirasu et al. shows the application of aligned carbon nanotube-reinforced polymer composite to electrothermal actuator. The chapter of Zekerya Dursun et al. shows the use of carbon nanotube-conducting polymer composites as electrode material in electroanalytical applications. The chapter of Yexiang Tong et al. shows the use of CNT-polymer composites for supercapacitor. The chapter of Chuankun Jia shows the applications of carbon nanotube-polymer composites for energy storage. The chapter of Abdul Hakim Shah covers the applications of carbon nanotubes and their polymer nanocomposites for gas sensors.

We would like to take this opportunity to thank all the researchers who have made direct contributions to the writing of this book. Also, we would like to thank all the editorial members of InTech publisher, in particular, Mr. Edi Lipović, the publishing process manager for his effective editing and support during the different stages of the production of this book. Finally, I would like to thank all my family members, Ayten, Lama, Perla, and Yuki for their continued patience, understanding, and support throughout my editing process of this book.

Mohamed Reda Berber
Chemistry department, Faculty of Science, Tanta University, Egypt
Department of Applied Chemistry, Kyushu University, Japan

Inas Hazzaa Hafez
Department of Water Resources and Agriculture Engineering,
Faculty of Agriculture, Damanhour University, Egypt
Department of Applied Chemistry, Kyushu University, Japan

Chapter 1

Carbon Nanotube-Based Polymer Composites: Synthesis, Properties and Applications

Waseem Khan, Rahul Sharma and Parveen Saini

Additional information is available at the end of the chapter

http://dx.doi.org/10.5772/62497

Abstract

The present chapter covers the designing, development, properties and applications of carbon nanotube-loaded polymer composites. The first section will provide a brief overview of carbon nanotubes (CNTs), their synthesis, properties and functionalization routes. The second section will shed light on the CNT/polymer composites, their types, synthesis routes and characterization. The last section will illustrate the various applications of CNT/polymer composites; important properties, parameters and performance indices backed by comprehensive literature account of the same. The chapter concludes with the current challenges and future aspects.

Keywords: carbon nanotubes (CNTs), conducting polymer, nanocomposites, EMI shielding and supercapacitors, theroelectrics and photovoltaics

1. Introduction

Since their discovery in 1991 by Prof. Iijima [1], carbon nanotubes (CNTs) have been a subject of global research focus, owing to their remarkable properties and related fascinating applications [2–14]. It is worth mentioning here that except in few cases, CNTs cannot be used in its bulk form (i.e. powder, aligned stacks, films/papers, etc.) due to the poor translation of outstanding inherent properties of individual CNTs into its macroscopic forms. Therefore, the most applications of CNTs involve their strategic combination with other materials in the form of alloys, blends, composites or hybrid materials [15–23]. In particular, idea of incorporating CNTs as filler inside various polymer-based matrices (e.g. conventional polymers such as thermoplastics, thermosets or elastomers as well as conjugated polymers) to form CNTs/polymer nanocomposites [13, 14] has revolutionized the materials science and technology. This facilitates the synergistic combination of flexibility, low density and facile processing of conventional polymers with outstanding mechanical, thermal and electrical properties of CNTs, or even introduction of additional electrical/thermal/electromagnetic attributes, thereby extending their field of applicability. In this context, attempts have also been made to form CNTs-filled conjugated polymer (CP) composites, to club the specialties of CNTs with good electroactivity, interesting doping-dependent properties and solution processability of CPs Therefore, inspired by their scientific and technological potential, over two decades, a lot of research work has been done on CNTs/polymer nanocomposite [19–21, 24–28] and the area is still growing stronger.

2. Carbon nanotubes

2.1. Structure of carbon nanotubes

The electronic configuration of carbon atom is $1s^2\ 2s^2\ 2p^2$ indicating that it has two strongly bound electrons in the 1s orbital and four relatively weakly bound electrons in 2s and 2p orbitals known as valence electrons. The small energy difference between 2s and 2p levels allows the carbon atom to exist in several hybridization states; sp, sp^2 and sp^3 in different materials (**Figure 1**).

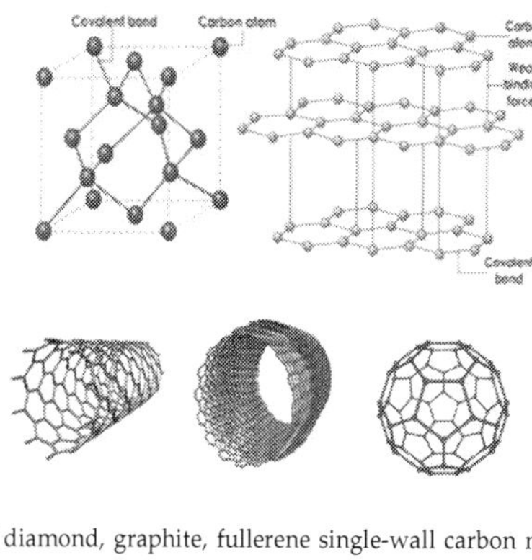

Figure 1. Schematic diagrams of diamond, graphite, fullerene single-wall carbon nanotube (SWCNT) and multi-wall carbon nanotube (MWCNT).

The hybridization flexibility enables the atomic orbitals of carbon to arrange themselves in structures of different dimensionalities ranging from diamond (3D), graphite (2D), carbon nanotubes (1D) and fullerene (0D). CNTs are hollow cylinders of graphene with extraordinary electronic and mechanical properties. They exist in two varieties viz. CNTs composed of a single graphene sheet called single-wall carbon nanotubes (SWCNTs) and an array of coaxial nanotubes known as multiwall carbon nanotubes (MWCNTs) [29, 30].

In the honeycomb lattice (**Figure 2**), the vector $\mathbf{C_h}$ is called chiral vector, while **T** is called the translational vector. The rectangle generated by the chiral vector $\mathbf{C_h}$ and the translational vector **T** is the unit cell of the SWCNT in real space. The chiral angle is defined as the angle between the chiral vector $\mathbf{C_h}$ and the unit vector $\mathbf{a_1}$. The chiral vector can be expressed in terms of the unit vectors $\mathbf{a_1}$ and $\mathbf{a_2}$ by means of integers (n, m) (which are called chiral indices):

$$C_h = n\, a_1 + m\, a_2$$

The wrapping of graphene sheet is governed by the different orientations of the chiral vector leading to the different CNT geometries. When the chiral indices are equal (n = m), the SWCNT is called armchair, and the chiral angle is 30°. If one of the chiral index is zero (n, 0) or (0, m), the SWCNT is named zig-zag, and in this case, the chiral angle is 0° (achiral nanotubes). In the other cases (n ≠ m), the nanotube is called chiral and its chiral angle is $0° < \theta < 30°$.

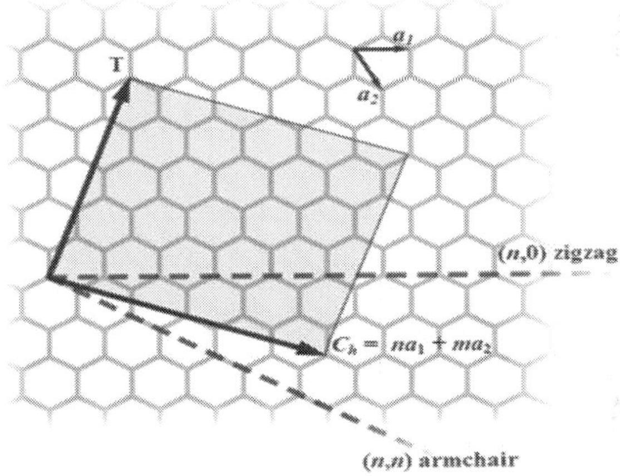

Figure 2. Chiral vector C_h and chiral angle θ definition for a nanotube on graphene sheet. a_1 and a_2 are the unit cell vectors of the two-dimensional hexagonal graphene sheet.

The expressions for the main parameters of a tube as a function of the chiral indices (n, m) are 29, 30:

$$\text{nanotube diameter } D = a\sqrt{(n^2 + m^2 + nm)}$$

$$\text{with, } a = a_{C\text{-}C}\sqrt{3} = 1.44\,\text{Å}$$

$$\text{Chiral angle } \theta,\ \cos\theta = \frac{(2n + m)}{2\sqrt{(n^2 + m^2 + nm)}}$$

In addition to above structural details, important physical attributes of CNTs that decide their properties are described below:

2.1.1. Length, diameter and aspect ratio

The actual length of CNTs is expected to influence their physical status as well as acquired configuration inside polymer matrix. In general, long-length CNTs show a flexible fibre-like character with tube buckling and entangled configuration. Therefore, its processing inside polymer matrix is a bit difficult in terms of de-entanglements. The long length ensures lower electrical percolation threshold, good mechanical properties due to the large-distance stress transfer and crack propagation prevention ability. In contrast, the short-length CNTs display rod-like character, provide more stiffness (compared to long-length CNTs of similar loading) to composite, and has relatively large percolation threshold, inferior stress transfer, poor crack countering property, low tensile/flexural strength and elongation of composites. However, it can be easily aligned and tends to show better interfacial polarization. Like the length of CNTs, diameters also affect the properties, for example SWCNTs, large diameter offers large interfacial surface area per unit length but less stiffness. Similarly, for MWCNTs, large diameter means more number of cylindrical graphitic tubular shells (i.e. thick walls), which tends to improve inherent mechanical properties of tubes, its stiffness, electrical properties and ability to undergo functionalization without much harm in tube's properties. Another important parameter is aspect ratio (i.e. length/diameter ratio), which affects the properties via interplay of length and diameter. However, it could be bit elusive and care has to be taken while talking about aspect ratio because a short-length and small-diameter CNT can have same aspect ratio as a long-length and large-diameter CNT, but their properties and translated effect on composites may be altogether different. Nevertheless, CNTs are widely used as very small, high aspect ratio conductive additives for plastics of all types. The high aspect ratio (up 1000 or more) gives CNTs an edge over other conductive fillers (e.g. carbon black, chopped carbon fibre, carbon nanofibres, stainless steel fibres or whiskers) as a lower loading of CNTs is needed to achieve the same electrical conductivity. This low loading preserves polymer resin's toughness, especially at low temperatures and maintains other key properties of the matrix resin.

2.1.2. Defects

A perfect single-wall CNT is only a theoretical construction and even perfect hexagonal sp^2 structure possesses different types of defects. Generally, a defective site has a high reactivity meaning that at those points, the chemical reactions are favoured. A simple example of a defect is the presence of non-hexagonal-shaped carbon rings such as a pentagon and heptagon pair (also known as Stone/Wales defect or 5/7 defect) [31–34]. They are localized mainly at tube ends and near tube bending locations. A Stone/Wales defect has very significant role to play in the modification of density of states of nanotubes which has implications for possible nanodevice applications [35–39]. Defects are very important because they can modify the electronic properties of the nanotubes thereby influencing their applications. The distribution of the 5/7 defects in the capping of the CNTs can induce the presence of new localized states in the valence and/or the conduction band, and so to modify the field emission properties [40, 41]. The introduction of a 5/7 defect in the hexagonal structure can induce a local deformation of the diameter of the tube, a change of the chirality or the formation of a CNT intramolecular junction. The defects can also be present along the sidewall, and after acid treatment, it is possible to open nanotube ends and attach chemical functional groups facilitating the interaction of CNTs with other moieties. In addition to being important for the potential applications of nanotubes, defects also play a key role in the functionalization process.

2.2. Synthesis methods of carbon nanotubes

2.2.1. Arc discharge method

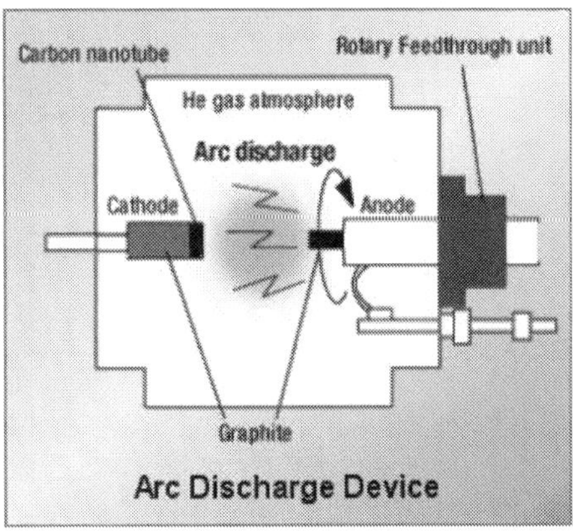

Figure 3. Schematic diagram of Arc-discharge setup.

The carbon arc-discharge method was first brought to light by Krätschmer et al. [42] who utilized it to achieve the production of fullerenes in macroscopic quantities. As mentioned in the introduction of this chapter, Iijima discovered the catalyst-free formation of multiwall carbon nanotubes (MWCNTs), while investigating the other carbon nanostructures formed along with the fullerenes and more particularly, the solid carbon deposit forming onto the cathode. **Figure 3** shows the schematic diagram of the arc-discharge method. This method creates CNTs through arc-induced vapourization of two carbon rods placed end to end, separated by approximately 1 mm, in an enclosure usually filled with inert gas (argon or helium) at low pressure. A direct current of 50–100 A, driven by a potential difference of approximately 20 V, creates a high-temperature discharge between the two electrodes, vapourizes the surface of one of the carbon electrodes, and forms a small rod-shaped deposit on the other electrode. This technique produces a complex mixture of components including non-tubular forms of the carbon such as nanoparticles, fullerene-like structures including C_{60}, multiwall shells, amorphous carbon, etc. [43–45]. Therefore, this requires further purification to separate the CNTs from the soot and the residual catalytic metals present in the crude product. In this technique, the high yield production of CNTs depends on the uniformity of the plasma arc, and the temperature of the deposit forming on the carbon electrode.

2.2.2. Laser ablation method

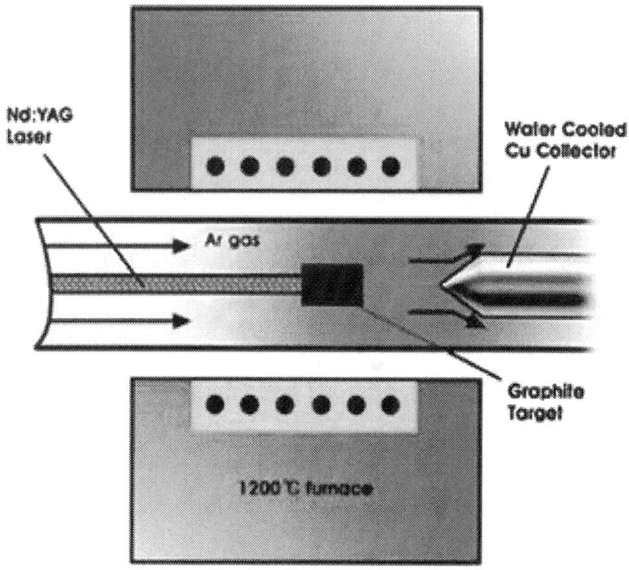

Figure 4. Schematic diagram of laser ablation setup.

The laser ablation technique was used successfully to synthesize fullerene for the first time in 1985 by Kroto et al. [46]. However, the synthesis of carbon nanotubes by this technique could be possible only 10 years later in 1995 by Guo et al. [47].

Figure 4 shows the schematic diagram of laser ablation method. Samples are prepared by laser vapourization of graphite rods with a 50:50 catalyst mixture of cobalt and nickel at 1200°C in flowing argon, followed by heat treatment in a vacuum at 1000°C to remove the C_{60} and other fullerenes. The target is vapourized more uniformly by a second laser pulse. The use of two successive laser pulses minimizes the amount of carbon deposited as soot. The second laser pulse breaks up the larger particles ablated by the first one and feeds them into the growing nanotube structure.

The material produced by this method appears as a 'mat' of ropes, 10–20 nm in diameter and up to 100 μm or more in length. Each rope primarily consists of a bundle of SWCNTs, aligned along a common axis. It is possible to vary the average nanotube diameter and size distribution by varying the growth temperature, the catalyst composition and other process parameters.

Both these solid carbon source-based synthesis methods have some drawbacks. One issue is the scaling up of the process to the industrial level. Secondly, the synthesized CNTs have impurities of metal catalyst particles and unwanted carbon forms such as fullerenes, amorphous carbon, multiwall shells, single-wall nanocapsule and need further purification.

2.2.3. Chemical vapour deposition method

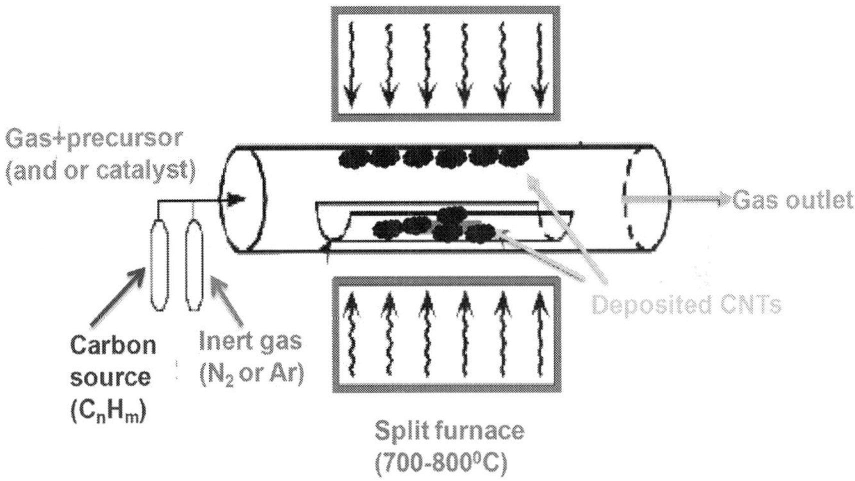

Figure 5. Schematic diagram of chemical vapour deposition setup.

Chemical vapour deposition (CVD) of gaseous carbon source (hydrocarbons, CO) over a metal catalyst is a classical method that has been used to produce various carbon materials such as carbon fibres and filaments for a long time [48]. However, CVD technique was first reported to produce MWCNTs by Endo and his research group [49]. Three years later, Dai in Smalley's group successfully adapted CO-based CVD to produce SWCNTs [50].

Figure 5 shows the schematic diagram of CVD method. In this technique, a carbon source is taken in the gas phase and an energy source such as plasma or a resistively heated coil is employed to transfer energy to a gaseous carbon molecule. The CVD technique uses hydrocarbons such as methane, carbon monoxide or acetylene as carbon source. During CVD, a substrate covered with metal catalysts (such as nickel, cobalt, iron or their combination) is heated to approximately 700°C. The growth starts after two gases are passed through the chamber, that is hydrocarbon gas and the other a carrier gas such as nitrogen, hydrogen or argon. The advantages of the CVD process are low power input, low-temperature range, relatively high purity and most importantly, the possibility to scale up the process. This technique can produce both MWCNTs and SWCNTs depending on the temperature, wherein production of SWCNTs occurs at a higher temperature than MWCNTs.

2.3. Properties of carbon nanotubes

2.3.1. Electrical properties

Depending on their chirality and diameter, CNTs can be either metallic or semiconducting in their electrical behaviour [20, 27, 51, 52]. In terms of the chiral index, a CNT will be metallic if $|n - m| = 3q$ otherwise, it will be semiconducting [53]. Theoretically, metallic nanotubes can carry an electric current density of 4×10^9 A/cm^2, which is more than 1000 times greater than those of metals such as copper [3]. Because of its nanoscale cross-section, electrons propagate only along the tube axis. As a result, carbon nanotubes are frequently referred to as one dimensional conductor. The maximum electrical conductance of a single-wall carbon nanotube is $2G_0$, where $G_0 = 2e^2/h$ is the conductance of a single ballistic quantum channel [54].

2.3.2. Thermal properties

All nanotubes are expected to be very good thermal conductors along the tube axis, exhibiting a property known as ballistic conduction, but good insulators laterally to the tube axis [55–58]. A SWCNT has been shown to have a thermal conductivity along its axis of about 3500 Wm^{-1} K^{-1} at room temperature [59]. This value is almost 10 times higher than that of copper, a metal well known for its good thermal conductivity which is 385 Wm^{-1} K^{-1}. A SWCNT has a room-temperature thermal conductivity in the radial direction of about 1.52 W/m/K^2 which matches very well with the thermal conductivity of soil. The temperature stability of CNTs is estimated to be upto 2800°C in vacuum and about 750°C in air [20].

2.3.3. Mechanical properties

CNTs are the strongest and the stiffest materials yet discovered in terms of tensile strength and elastic modulus, respectively [20, 27, 60–62]. This strength is a direct consequence of covalent sp^2 bonds formed between the individual carbon atoms. It has been shown that CNTs are very strong in the axial direction. Young's modulus of the order of 270–950 GPa and tensile strength of 11–63 GPa were obtained [63].

On the other hand, there was evidence that in the radial direction, they are rather soft. The first transmission electron microscope observation of radial elasticity suggested that even the van der Waals forces can deform two adjacent nanotubes [64]. Later, nanoindentations with atomic force microscope were performed by several groups to quantitatively measure radial elasticity of MWCNTs [65, 66] and tapping/contact mode atomic force microscopy was also performed on SWCNTs [65]. The results show that MWCNTs are radially deformable to a large extent notwithstanding their axial rigidness and strength under tensile load.

2.3.4. Magnetic properties

Although pure CNTs are non-magnetic in nature, but the most synthesized CNTs (both SWCNTs and MWCNTs) are known to possess magnetic properties that can be attributed to the presence of entrapped catalyst nanoparticles in their inner cavity. Such catalytic particles are formed either in situ during the growth of CNTs (e.g. from ferrocene during MWCNTs synthesis by CVD) or from previously added metallic catalyst particles (e.g. from Fe-, Co- or Ni-filled electrodes during MWCNTs/SWCNTs synthesis via arc discharge). In general, SWCNTs tend to have very high amount of catalytic residues and posses stronger magnetic character than MWCNTs. As the magnetism is due to the metallic impurities, CNTs tend to lose magnetic character upon treatment under harsh conditions, for example acidic treatment such as functionalization [27, 67] or high-temperature annealing [68].

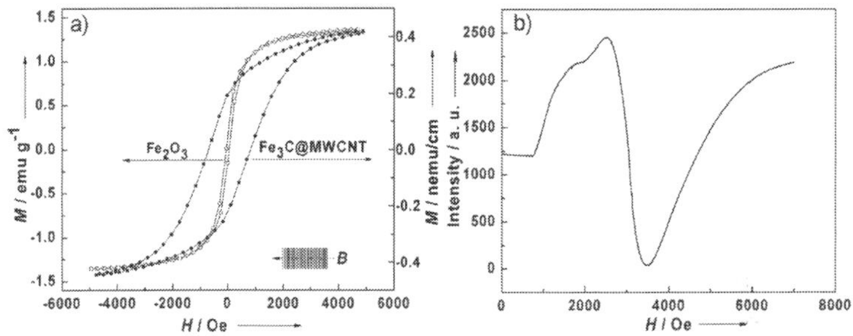

Figure 6. (a) M versus H curves for Fe_2O_3 nanoparticles and Fe_3C@MWCNT (B indicates magnetic field, lines the nanotubes, and arrow the direction of the magnetic field). (b) EPR spectrum of Fe_3C@MWCNT. Reproduced from [69] with permission from Wiley.

Nevertheless, the CNTs display ferromagnetic behaviour (**Figure 6a**) marked by the presence of hysteresis loop (i.e. nonzero corecivity and retentivity values) [69]. The ferromagnetic nature of CNTs is also supported by the electron paramagnetic resonance (EPR) spectra (**Figure 6b**), showing the presence of a high density of delocalized π electrons corresponding to g values of 2.1365 and 2.2132 for the first and second peaks, respectively. The obtained g values are very near to the characteristic g values of ferromagnetic materials. It is suggested that presence of iron catalyst inside CNTs may be helpful in their magnetic manipulation, for example for magnetic alignment of CNTs in composites [20, 70] or for targeted drug delivery [69]. Further, the magnetic properties also contribute towards electromagnetic energy absorption and play a role in improving its contribution towards total microwave shielding effectiveness [21, 71].

2.3.5. *Surface properties/wettability characteristics*

The pure CNTs are made up of graphitic cylindrical walls made up of carbon atoms. Therefore, being of non-polar in nature, the surfaces of CNTs (especially MWCNTs) are highly hydrophobic in nature with good affinity towards non-polar materials such as hydrocarbons, parrafins, oils or organic solvents. As CNTs also display relatively high-specific surface area, when 3D shape architectured, due to the inherent hydrophobicity they can be useful for water purification applications, for example for separation of non-polar pollutants such as oils, solvents or even organic dyes.

Owing to their chemical inertness, CNTs are difficult to disperse in water and in organic media, and they pose high resistance to wetting. Historically, unlike fullerenes, their chemistry was considered very poor for a long time. Difficulties also arise in making composites of such inert nanotubes with other materials which is important for many device applications. A suitable functionalization of the nanotubes (i.e. the attachment of 'chemical functionalities') represents a strategy for overcoming these barriers and has thus become an attractive field for synthetic chemists and materials scientists.

3. Functionalization of carbon nanotubes

Functionalization enhances the solubility and processibility, and allows combining the unique properties of nanotubes with those of other materials [20, 26, 27, 72]. It also improves the interaction of the nanotube with other entities, such as a solvent, polymer and other organic molecules and also with other nanotubes. A functionalized nanotube displays different mechanical and electrical properties as compared to pristine nanotube and thus may be utilized for several applications.

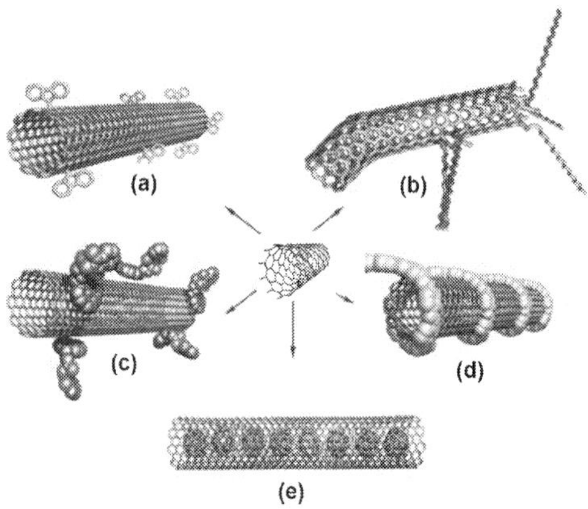

Figure 7. Different possibilities of the functionalization of SWCNTs (a) Sidewall functionalization. (b) Defect-group functionalization. (c) Non-covalent exohedral functionalization with molecules through π-stacking. (d) Non-covalent exohedral functionalization with polymers. (e) Endohedral functionalization, in this case C60@SWCNT. Reproduced from [73] with permission from Springer-Verlag Berlin.

The process of functionalization can conveniently be divided into three major types depending upon the chemistry involved (**Figure 7**) [73].

3.1. Covalent functionalization

Covalent functionalization utilizes the covalent linkage of functional entities onto the nanotube's carbon scaffold. Depending upon the site of interaction, it can be of two types—covalent sidewall functionalization and defect functionalization. Direct covalent sidewall functionalization involves a change of hybridization from sp^2 to sp^3 and the loss of conjugation. Defect functionalization is based on the transformations of defect sites already present. Defect sites can be the open ends and holes in the sidewalls, terminated, for example by functional groups and Stone/Wales defects (5–7 defects) in hexagonal graphene framework. In addition to these, oxidative purification generated oxygenated sites are also considered as defects. SWCNTs demonstrate low dispersability, and they occur in the form of bundles. This situation warrants the use of a highly reactive reagent for the covalent bond formation at the sidewalls. It is not possible to tell beforehand that whether these addition reactions are more likely to take place at defect sites or intact hexagonal regions of sidewall. Several covalent routes have been taken for covalent functionalization such as oxidative purification [74, 75], amidation [76], esterification [77], thiolation [78], halogenation [79–81], hydrogenation [82], cycloadditions [81, 83–87], electrochemical functionalization [88, 89].

3.2. Non-covalent functionalization

Covalent functionalization suffers from the drawback of damaging the structure of CNTs. Therefore, with a view to retain the structural integrity and π network of CNTs, non-covalent functionalization is particularly attractive. The adsorption forces, such as van der Waals' and π-stacking interactions, are responsible for the non-covalent interaction between surface active reagents and CNTs. Broadly, surfactants and polymers are utilized for this non-destructive functionalization of CNTs. Surface active molecules such as sodium dodecylsulfate (SDS) or benzylalkonium chloride have been successfully used for the formation of non-covalent aggregates [90–92]. On the other side, polymer wrapping around CNTs can be accomplished through the use of polymers such as poly (*m*-phenylene-*co*-2, 5-dioctoxy-*p*-phenylenevinylene) (PmPV, 2) in organic solvents such as chloroform. The stable solution of the SWNT/PmPV complex exhibits conductivity eight-times higher than that of PmPV, without any compromise on its optical properties. The polar side-chain polymers such as polyvinylpyrrolidone (PVP) or polystyrenesulfonate (PSS) give stable solutions of the SWNT/polymer complexes in water [93].

The covalent and non-covalent functionalizations are essentially exohedral derivatizations.

3.3. Endohedral functionalization

The hollow inner cavity of SWCNTs serves as a capillary for the storage of nanoparticles and fullerenes, etc. The rich endohedral chemistry of SWCNTs is amply illustrated by the incorporation of fullerenes and metallofullerenes in their cavity [94, 95].

4. CNTs-based polymer composites

The spectacular properties of CNTs such as their high strength and stiffness make them ideal candidates for structural applications. At present, polymer nanocomposite is one of the biggest application areas for CNTs. The extraordinary properties of CNTs coupled with easily tailorable characteristics of polymers give rise to truly versatile CNT-polymer nanocomposites [20, 21, 26, 27, 96–106]. The emergence of CNTs as filler materials has contributed in the realization of CNT-polymer nanocomposites as next generation advanced structural material. Keeping in view, the ubiquitous need for the creation of materials with tailored properties for various applications, it is hardly an exaggeration to call the present age as the 'age of composites'.

By definition, a composite is a multiphase material formed from a combination of materials which differ in composition or form, retain their own chemical and physical properties, and maintain an interface between components which act in concert to provide improved specific or synergistic characteristics not obtainable by any of the original components acting alone [20, 107, 108]. Any composite is composed of two categories of materials: the reinforcement or filler and the matrix. The reinforcements contribute useful properties (mechanical, electrical, thermal, optical, etc.) to enhance the matrix properties. In the past decades, we have reached the technological design limits of optimizing composites with traditional micro-meter scale fillers/reinforcements. The limitations of traditional micrometer-scale polymer composites have prompted a considerable research effort to focus on polymer nanocomposites. This new hopes-laden advancement over the traditional polymer composites is characterized by the presence of at least one phase which is less than 100 nm in at least one dimension. Historically, nanocomposites are not entirely new as some nanocomposites such as carbon black and fumed silica-filled polymers [109, 110] have been used a long time ago. However, the discovery of CNTs added an additional impetus in the polymer nanocomposite research promising hitherto unknown potential for various applications. **Table 1** shows a comparison between various fibre reinforcements and carbon nanotubes in terms of various parameters.

Fibre	Diameter (μm)	Density (g/cm^3)	Tensile strength (GPa)	Modulus (GPa)
Carbon	7	1.66	2.4–3.1	120–170
S-glass	7	2.50	3.4–4.6	90
Aramid	12	1.44	2.8	70–170
Boron	100–140	2.50	3.5	400
Quartz	9	2.2	3.4	70
SiC fibres	10–20	2.3	2.8	190
SiC whiskers	0.002	2.3	6.9	–
Carbon NTs	0.001–0.1	~1.33	Up to ~50	Up to ~1000

Credit: Fisher/ Northwestern University.

Table 1. Comparison and contrast of advanced fibre reinforcements vs. carbon nanotubes in terms of diameter, density, tensile strength and modulus.

4.1. Nanocomposite fabrication methods

A variety of synthesis techniques are in practice for incorporation of CNTs into various polymeric host matrices [14, 20, 21, 26, 27, 97, 108, 111–114]. The main motive is to deagglomerate the CNTs and realize their uniform dispersion inside polymer matrix. Currently, there is no single technique which is universally applicable to all the situations. Depending upon the thermal or chemical properties of matrix polymer, ease of its synthesis from suitable monomer, desired performance indices of composites and cost constraints, with some trade-off in properties, one can choose the suitable processing method for a particular case. This section briefly describes the important processing techniques for synthesis of CNT-based polymer nanocomposites.

4.1.1. Solution processing

The solution processing is the most common technique to form CNT-based polymer nano-composites which exploits intensive agitation (e.g. refluxing, mechanical/magnetic stirring, vigorous shaking, high shear homogenization, bath/probe sonication) aided rigorous and thorough mixing of CNTs with polymer in a solvent, so as to facilitate nanotube de-bundling and their dispersion inside host polymer matrix [20, 21, 27, 108, 114]. It is important to note that this technique is limited to the polymers which are soluble in solvent(s).

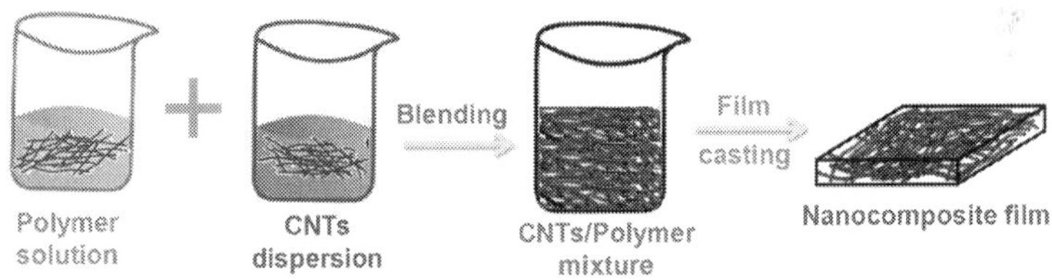

Figure 8. Schematic representation of solution processing method.

Typical process involves dispersion of nanotubes in a suitable solvent and mixing with the polymer solution (**Figure 8**), followed by film casting and solvent evaporation leaving behind nanocomposite film/sheet. The choice of solvent is mainly governed by solubility of matrix polymer. Further, the solvent for CNTs and polymer dispersion may be same or different, but should be of good missibility to realize intimate mixing between phases. In many cases, CNTs are not separately dispersed rather they are directly added to polymer solution followed by intensive mixing before film casting. The major shortcoming associated with the usage of high-power ultrasonication or shear mixing for a long time is that it can lead to the shortening of tube lengths, thereby deteriorating the composite properties. Plenty of literature [14, 21, 97, 106, 114–119] is available on the formation of CNT-based nanocomposites by this method using both organic as well as aqueous media and variety of polymer matrices [14, 21, 120–123]. To enable better dispersion and to solve the problem of tube shortening upon high-power agitation, some efforts have been made to use surfactants for tube dispersion or to use physically/chemically functionalized CNTs [120, 124, 125]. However, the use of surfactant carries the unavoidable limitation of retaining surfactant in the nanocomposites which hinders the thermal/electrical transport properties of nanocomposites [120]. In contrast, the function-alized CNTs often provide positive results in terms of CNTs deagglomeration, dispersion and their improved interfacial adhesion with matrix polymer, which gets reflected in terms of superior electrical, thermal, mechanical and dielectric properties of resultant nanocomposites [14, 27, 106, 114]. The boiling point of dispersion medium or solvent (s) is found to have a profound impact on the properties of formed nanocomposites [14, 27, 114, 121]. In general, low-boiling point (i.e. fast evaporating or drying) solvents are preferred, due to their ease of removal from the solution-casted nanocomposites mass. In contrast, high-boiling point solvents are difficult to be removed and tend to get trapped in the solidifying/curing mass. Such a trapped solvent may interfere with the curing reaction (in thermosets) or can act as softener (in thermoplastics), thereby adversely affecting electrical, thermal or mechanical properties. In addition, efforts to remove the solvent and to prevent evaporation generated voids formation add towards complexity and cost in terms of requirement of systems for controlled heating, vacuum/pressure [14]. Another, limitation encountered quite often with solution processing is that the slow evaporation of solvent provides sufficient time for CNTs re-agglomeration and differential settling, resulting in inhomogenous CNTs dispersion in matrix (e.g. CNTs content lowest at the casted film/sheet's surface, shows a uniform/random gradient across thickness and maximum at both surfaces due to the extensive tube settling) and observation of non-uniform and inferior properties. The solvent evaporation rate-related limitations can be resolved by gently pouring CNT/polymer nanocomposite dispersion on a rotating substrate (spin coating) [122] or over a heated substrate (drop-casting) [126]. However, use of spin coating is limited only to thin films (few nanometres thick) which cannot be peeled off from the substrate, whereas drop casting has issues in terms of uniform drying across thickness and high possibility of void formation. Another versatile method exploits coagulation [114, 123, 127] of CNT/polymer dispersion by pouring into an excess of non-solvent, thereby achieving rapid precipitation of polymer chains which immediately entrap CNTs (without providing sufficient time for CNTs diffusion and settling). Nevertheless, solution processing is still widely used and is one of the important steps in the processing of thermosetting matrices-based nanocomposites.

4.1.2. Melt processing

The melt processing is considered as viable option for making thermoplastic matrices-based CNT/polymer composites due to its low cost and amenability towards large scale synthesis for industrial applications. Here, elevated temperatures are exploited to melt the receptor thermoplastic matrix polymers, which form a viscous liquid and made to flow [21, 26, 108, 114, 128]. The molten polymer flow induces high shear forces which assist in partial de-agglomeration of CNTs bundles and their dispersion inside matrix. The melt mixing can be carried out in batch or continuous operation using high shear mixer (e.g. Sigma mixer) and extruder, respectively. Sigma mixer is often used to prepare highly concentrated nanocomposites called masterbatches, which maybe used to synthesize desired CNTs-loading composites by their mixing with proportionate amount of neat matrix polymer using extruder. An extruder may consist of single or twin screws with twin screw version being more effective in terms of mixing uniformity and properties. A typical twin screw extruder (**Figure 9**) consists of two co- or counter-rotating screws inside barrel housing. The polymer granules are caught by the rotating screws and pushed forward; they become melted inside heated melting (feed) zone due to the externally provided heat and shearing of the material between screw and barrel. The CNTs are loaded into the extruder via separate hopper, such that melt-phase mixing takes place due to the combination of shearing and kneading action, and by the time molten-mixture reaches the homogenization zone, it has already achieved significant degree of mixing. Finally, mixture passes to die before coming out as semisolid strands, which may be cooled (via air drying or by passing through water bath) and chopped into granules for further use, for example for compression moulding. The extruded output may also be diverted to injection moulding machine to form desired shape samples.

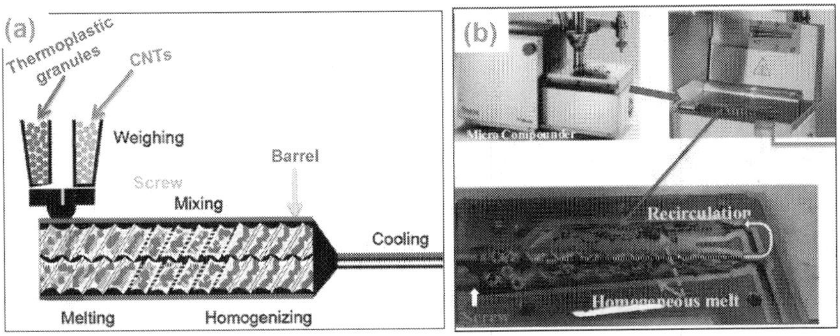

Figure 9. Schematic representation of twin screw extruder for melt phase mixing of CNTs with thermoplastic matrices. Reprinted from [108] with permission from Wiley. (b) Twin screw extruder with melt recirculation provision. Reprinted from [129] with permission from Elsevier.

In the past, melt mixing/blending is successfully exploited for dispersion of CNTs inside various thermoplastic matrices, for example polystyrene [130], polypropylene [130, 131] and acrylonitrile-butadiene-styrene (ABS) [22, 130], polyamide-6 [22], polyethylene [132]. Although melt-mixing technique is simple but the issues of high shear force and elevated temperature need to be properly addressed in order to avoid the deterioration of nanocomposites. While high shear force facilitates CNT dispersion, it can also lead to the undesirable CNT fragmentation or even polymer chain scission. Therefore, the shear force needs to be optimized to achieve desired dispersion without compromising the structural integrity of CNTs. Similar situation prevails with high temperature which promotes CNTs dispersion but can degrade the intrinsic properties of polymer making the need of optimizing the temperature indispensable. Further, it is found that melt-mixing technique is not much effective in breaking of agglomeration of CNTs compared to solution processing [131, 133]. Besides, the melt-mixing fails to realize dispersion of high loading (>5 wt%, which are required for thermal/electrical/electromagnetic applications) inside polymeric matrices, due to the viscosity buildup and screw rpm/reside time limitations [129]. Recently, the above issues are resolved by using a twin screw extruded equipped with melt recirculation provision (**Figure 9b**) [128, 129], such that upto 10 wt% CNTs can be loaded inside polypropylene copolymer (PPCP) matrix without any deleterious effect on mechanical properties.

4.1.3. In situ polymerization

In situ polymerization [14, 21, 108] remains the only viable option for the preparation of composites based on insoluble or thermally unstable matrix polymers, which cannot be processed by solution or melt processing routes. However, many a times, it is used in other cases too (where above limitations are not applicable), due to the distinguished advantages of in situ polymerization in terms of ability to allow formation of high CNTs-loading nanocomposites, facilitating good CNTs dispersion within polymer matrix and ensuring excellent intimacy between CNTs and matrix polymers.

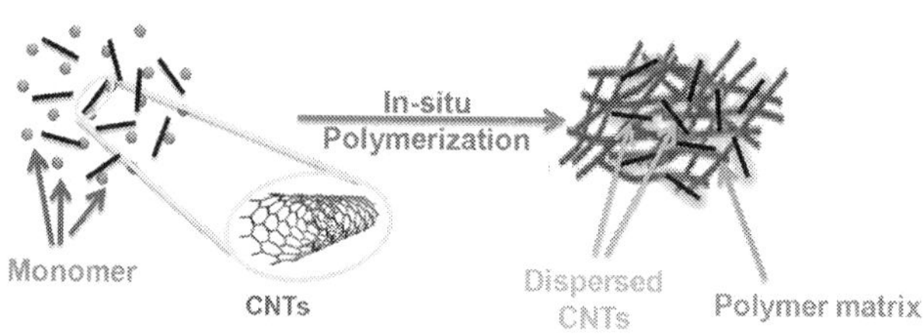

Figure 10. Schematic representation of in situ polymerization process.

This strategy involves dispersion of CNTs in monomer (**Figure 10**) followed by in situ polymerization leading to the formation of CNT/polymer nanocomposites. Exploitation of functionalized CNTs or use of monomer-grafted CNTs are known to improve the initial dispersion of the nanotubes in the monomer and consequently in the formed nanocomposites. This results in a stronger and more active interface between nanotube and polymer which is pivotal to the nanocomposite performance for structural, electronic, electromagnetic or electrochemical applications. This method has been used to synthesize CNTs-filled composites with various polymers, for example thermoplastics, thermosets or conjugated polymers-based matrices [14, 21, 26, 27, 111, 112]. In addition to composite formation, in situ polymerization is also used for physical functionalization of CNTs (via surface polymer wrapping), for their onward use as hybrid filler for nanocomposites.

4.1.4. Miscellaneous routes

In addition to aforementioned methods, some other less popular methods are also available for CNT-polymer nanocomposites preparation, for example wet spinning [134], batch mixing inside banbury mixer [129], twin screw pulverization [135], electrophoretic deposition [136] and latex processing [137], twin screw extrusion complemented by melt-recirculation provision [129]. The issue of the synthesis of CNT-based polymer nanocomposites is still an open arena as no single technique is entirely satisfactory on all grounds. Therefore, some efforts have also been made to use combination of techniques, for example solution processing with melt mixing [133]; in situ polymerization with solvent processing [23]; or in situ polymerization with melt processing [114].

http://dx.doi.org/10.5772/62497

5. Properties of CNT/polymer nanocomposites

With the advent of CNTs on technological landscape, there are now myriad possibilities for tailoring the properties of CNT/polymer nanocomposites. The electrical, thermal, dielectric, rheological and mechanical properties of CNTs filled nanocomposites are significantly enhanced compared to neat polymers. However, the issue of property enhancement is a challenging one as the improvement in one property might come at the cost of other. A number of factors such as the nature of matrix polymer, aspect ratio or actual length/diameter of CNTs, its pretreatment (e.g. covalent functionalization, surface coating of polymer or surfactant), loading level, exploited processing technique and the presence of tertiary phase (e.g. compatibilizer) are known to exert decisive influence on the properties of formed nanocomposites. The next section briefly describes the important properties of CNT/polymer nanocomposites.

5.1. Electrical properties

The very high intrinsic conductivity and low aspect ratio (length/diameter ratio) of CNTs compared to other carbon-based fillers (e.g. carbon black, graphite, carbon fibre), inspired researchers to synthesize CNTs filled electrically conducting nanocomposites. The incorporation of CNTs leads to onset of electrical conductivity within otherwise insulating matrix. This can be attributed to formation of 3D electrically conductive networks within host thermoset matrix so that electrons can easily hop/tunnel between dispersed filler particles [23]. The minimum filler loading where first continuous network of filler particles were formed within matrix polymer is known as percolation threshold. At this point, electrical conductivity of composites displays a sharp rise, that is several orders of magnitude (**Figure 11a**).

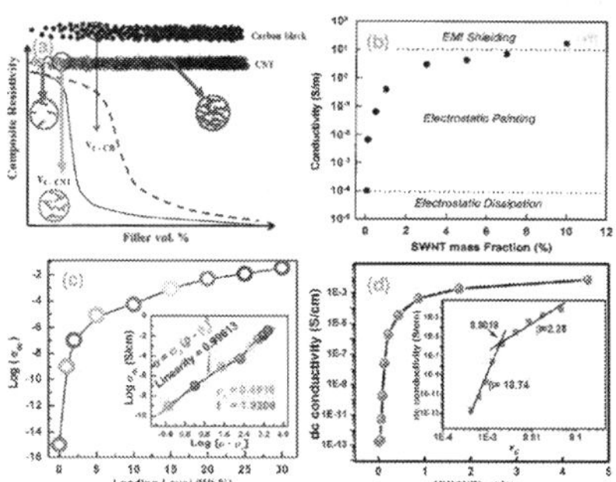

Figure 11. (a) Schematic representation of difference between percolation behaviour of CNTs compared to carbon black. (b) Electrical conductivity of SWNT/polycarbonate nanocomposites as a function of nanotube loading, showing a typical percolation behaviour. Dashed lines represent the lower limits of electrical conductivity required for the specified applications. Reproduced from [138] with permission from AIP. (c) Variation of conductivity (σ_{dc}) of PANI-MWCNT nanofiller-loaded polystyrene solution blends. Inset shows the percolation and scaling details. Reproduced from [23] with permission from Elsevier (d) Plot of electrical conductivity (σ) versus MWCNT (vol%) for PPCP/MWCNTs composites. Inset shows the log–log plot of r as a function of (v_c). Reproduced from [129] with permission from Elsevier.

Depending upon the level of achieved electrical conductivity, these conductive nanocomposites may have a multitude of applications (**Figure 11b**) including electromagnetic interference (EMI) shielding, transparent conductive coating, electrostatic painting and electrostatic dissipation [138, 139]. Especially, electrically conductive CNT-polymer composites are used in anti-static packaging applications, as well as in specialized components in the electronics, automotive and aerospace sectors.

The critical conducting filler loading (i.e. percolation threshold) for a given matrix-filler combination can be calculated by plotting the electrical conductivity as a function of the reduced volume fraction of filler (**Figure 11c**) and performing data fitting with a power law function [23]:

$$\sigma = \sigma_o (v - v_c)^t \quad (1)$$

where σ is the electrical conductivity of the composite, σ_o is characteristic conductivity, v is the volume fraction of filler, v_c is volume fraction at the percolation threshold, and t is the critical exponent. The log (σ) versus log ($v - v_c$) plot (inset, **Figure 11c**) gives a straight line according to Eq. 10. In practical situations, where the densities of polymer matrix and filled inclusion are same (e.g. for organic fillers like ICPs, CNTs or graphene), the mass fraction, p, and volume fraction (v) of the filler can be assumed same. The percolation threshold can also be determined by plotting the log (σ) versus log (v) plot and finding the point of intersection of lines (**Figure 11d**) corresponding to different (β) values.

The CNT/polymer composites show very low percolation threshold for electrical conductivity [14, 27, 114]. For SWCNT/polymer composites, the percolation thresholds ranging from 0.005 vol% to several vol% have been reported [140]. On the other hand, in case of MWCNT/polymer composites percolation threshold up to 0.002 vol% has been reported [141]. It is found that compared to common conductive fillers, for example metallic or graphitic particles in any shape (spherical, platelet-like or fibrous) and size, CNTs display much lower percolation threshold (**Figure 11**). This can be ascribed to combination of their high inherent conductivity and very high aspect ratio.

It is found that the percolation threshold in CNT/polymer nanocomposites depends on several parameters viz. dispersion [14, 20, 27, 114, 124], alignment [14, 21, 106, 123, 129, 133, 142], aspect ratio [140, 143]. Higher aspect ratio is obtained for well-dispersed nanotubes relative to nanotube bundles making percolation threshold decrease with better dispersion. Bryning et al. [140] reported a smaller percolation threshold with the higher aspect ratio nanotubes for SWCNT/epoxy composites. The percolation threshold is also significantly affected by alignment of the nanotubes in the polymer matrix. In the alignment condition, there are fewer contacts between the tubes which results in a reduced electrical conductivity and a higher percolation threshold as compared to random orientation of nanotubes. It is a common notion that chemical functionalization reduces the electrical conductivity of nanotubes through disruption of the extended π-conjugation of nanotubes. However, it is also reported that functionalization can improve the electrical properties of the composites [144–147]. Valentini et al. [146] concluded that the amine-functionalized SWCNT in epoxy matrix allows migration of intrinsic charges raising the conductivity of the composite. Similar result is also obtained by Tamburri et al. [147] for the functionalization of SWCNT with hydroxyl and carboxylic groups in 1,8-diaminophthalene. Therefore, it is established that negative effects of functionalization for SWCNTs conductivity are often counterbalanced by the improved dispersion caused by functionalization, with overall effect leading to positive outcome.

5.2. Thermal properties

Due to the excellent thermal conductivity of CNTs, some efforts have been made to incorporate them into various polymer matrices to improve thermal conductivity of formed composites [148–153]. It is found that, besides CNTs content, the thermal conductivity of CNT/polymer nanocomposites also depends on state of their dispersion and alignment as well as aspect ratio and the presence of metal impurities. Biercuk et al. [150] synthesized an epoxy composite with 1 wt% raw laser-oven SWCNTs that showed a 125% increase in thermal conductivity at room temperature. Choi et al. [154] reported a 300% increase in thermal conductivity at room temperature with 3 wt% SWCNTs in epoxy. They also obtained an additional 10% increment in case of magnetic alignment. Du et al. [155] reported an infiltration method with an epoxy and a nanotube-rich phase showing a 220% increase in thermal conductivity at 2.3 wt% SWCNTs loading. Among various CNTs variants (SWCNTs, MWCNTs, DWCNTs) in epoxy composites, the MWCNTs are found to most significantly improve the thermal conductivity of polymer composites. This is due to their relatively low interfacial area (therefore, less phonon scattering at the interface) and the existence of shielded internal layers which promote the conduction of phonons and minimizes the matrix coupling losses [149]. Besides, thermal conductivity improvement, CNTs are also known to improve the thermal stability of the polymer composites. This may be attributed to better heat dissipation characteristics. Nevertheless, these good thermal conductivity nanocomposites are considered very promising candidates for a number of applications such as thermal interface materials, heat sinks, printed circuit boards, connectors and other high-performance thermal management systems.

5.3. Mechanical properties

The outstanding intrinsic mechanical properties of CNTs (including ultra-high-specific strength and modulus) make them especially lucrative as fillers in CNT-polymer nanocomposite materials. Due to these properties, only they were talked about as filler material for composite cables for NASA's fascinating project of space elevator [156]. Therefore, several efforts have been made in the past, to translate the fraction of exceptional mechanical properties of CNTs into formed nanocomposites.

In general, the tensile modulus and strength of CNT-polymer nanocomposites increase with nanotube loading (**Figure 12a–c**), dispersion, and alignment in the polymer matrix [14, 20, 21, 27, 114, 128, 129] However, unlike modulus (**Figure 12c**), the tensile strength (**Figure 12b**) does not follow a monotonic increase [14, 20] with CNTs loading, due to CNTs re-aggrega-

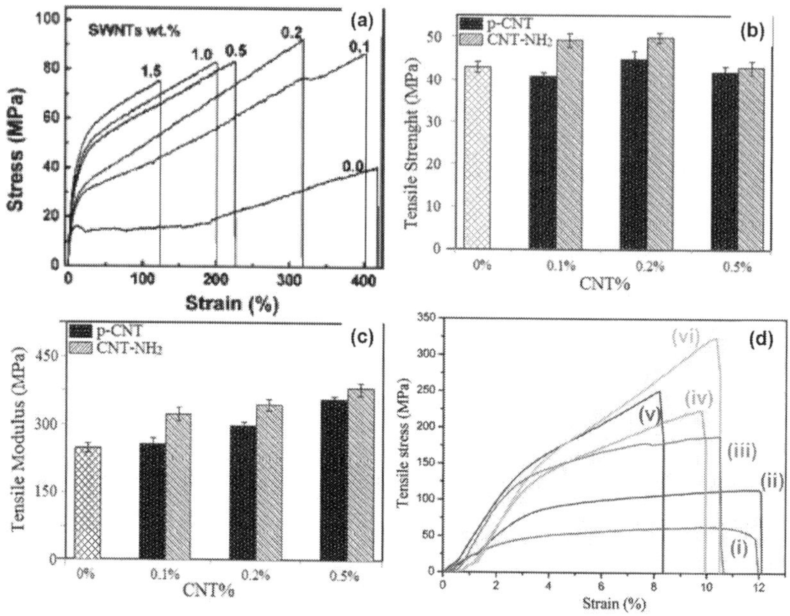

Figure 12. (a) Stress–strain profiles of SWNT-nylon-6 composite fibres at different SWCNT loadings. Reproduced from [157] with permission from ACS. (b) Tensile strength and (c) tensile modulus of epoxy nanocomposites containing various % loadings of pure CNTs (p-CNT) and amino functionalized (CNT-NH_2). Reproduced from [158] with permission from ACS. (d) Typical stress–strain curves for (i) electrospun pure PAN terpolymer nanofibre sheets; (ii) electrospun PAN terpolymer nanofibre sheets with 2 wt% loading of grafted MWCNTs; (iii) hot-stretched pure PAN terpolymer nanofibre sheets; (iv–vi) hot-stretched PAN terpolymer nanofibre sheets with 1, 2 and 3 wt% loadings of grafted MWCNTs, respectively. Reproduced from [159] with permission from ACS.

tion, viscosity buildup issues, incomplete CNTs wetting by polymer and their poor dispersion. Therefore, optimum CNTs loading and means to overcome above negative factors are keys towards realizing composites with good mechanical properties. It is observed that the theoretical predictions of mechanical properties and experimental findings often differ due to the number of issues including the presence of voids, lack of perfect orientation, poor filler dispersion and insufficient load transfer due to the lack of interfacial adhesion. Moreover, nanotube agglomeration decreases the modulus of the nanotubes compared to that of isolated nanotubes because the nanotubes involve only weak dispersive forces between them. According to the molecular simulation studies and elasticity calculations done by Liao et al. [160], when the atomic bonding between the nanotubes and the matrix is not present, there are basically two sources of nanotube/matrix adhesion (i) electrostatic and van der Waals interactions and (ii) stress/deformation resulting from the difference in the coefficients of thermal expansion between the filler and matrix. The functional groups on the nanotube sidewalls are known to improve the compatibility with the polymer matrix. This facilitates interfacial load transfer via CNT/polymer bonding leading to improved mechanical properties [14, 20, 158]. The stress–strain profile of these composites give a 153 and 103% increase in Young's modulus and tensile strength, respectively (**Figure 12a**). The importance of CNTs alignment has also been highlighted [20, 114, 159], and it is shown that the aligned CNTs-based nanocomposites (**Figure 12d**) display better strength and modulus in orientation direction compared to those based on random CNTs alignment. In spite of so many pros and cons, these lightweight yet structurally strong CNTs-based nanocomposites are considered as promising material for structural applications, for example bullet proof garments or body armours, aircraft or automobile parts, industrial components.

5.4. Dielectric properties

It has been demonstrated that the incorporation of CNTs within insulating or conducting polymer matrices leads to improvement of dielectric properties [21, 23]. This can be attributed to the localization of charges at the CNTs/polymer interfaces, resulting in Maxwell–Wagner interfacial polarization. As a result, both real and imaginary permittivity of the polymer composites scales with CNTs loading (Figure 13).

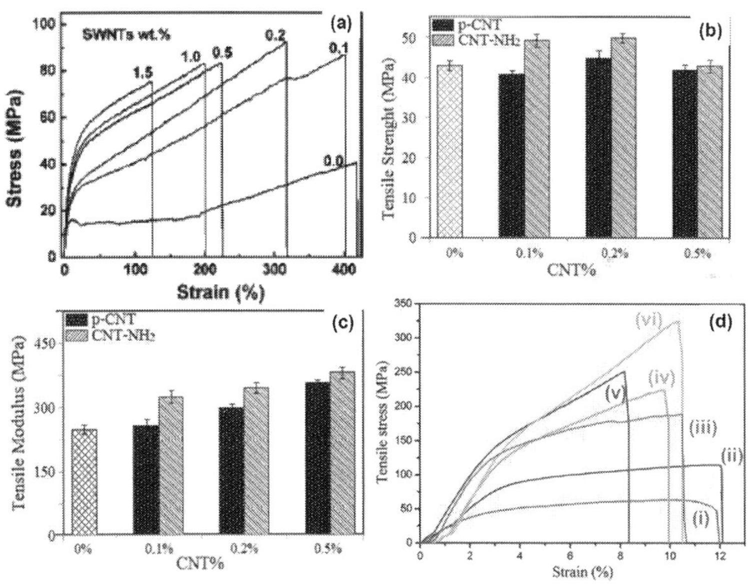

Figure 13. Complex permittivity spectra of the composites using "long-SWCNTs" with loading from 0.01 to 15 wt%. Reproduced from [161] with permission from Elsevier.

The parameter ε' (real permittivity) represents the charge storage (or dielectric constant), whereas ε'' (imaginary permittivity) is a measure of dielectric dissipation or losses. It is found that dielectric properties are dependent on the CNTs' aspect ratio, actual length, functionalization/doping status and loading level as well as nature of polymeric matrix and employed processing technique [21, 161]. Such properties are best interpreted in terms of effective medium theory and have direct influence on electromagnetic wave-blocking characteristics of the CNTs/polymer nanocomposites.

5.5. Rheological properties

The rheological properties of CNT/polymer nanocomposites have significance for composite processing as well as a probe of the composite dynamics and microstructure. At high frequencies, the response is almost independent of the filler concentration, indicating that the short-range polymer dynamics are not influenced by the nanotubes. The glass transition temperatures of the composites remain constant in the absence of strong interfacial bonds and for low nanotube loadings. At low frequencies, with the increase of nanotube concentration, the rheological behaviour gradually shifts from a liquid-like behaviour to a solid-like behaviour. This is in accordance with earlier findings in silicate nanocomposites [162]. Applying a power law function to the data provides a rheological percolation threshold associated with the onset of solid-like response. As observed with respect to electrical percolation, the rheological percolation also depends on aspect ratio, nanotube dispersion and alignment. In a study by Mitchell et al. [163], the effect of dispersion was demonstrated by functionalizing SWCNT such that the rheological percolation threshold decreased from 3 wt% when using pristine SWCNT to 1.5 wt% in functionalized SWCNT/polystyrene composites.

Pötschke et al. [164] also showed the temperature dependence of the rheological percolation threshold. In SWCNT/ PC composite, the percolation threshold decreases from ~5 to ~0.5 wt % MWCNT when the temperature rises from 170 to 280°C. Besides, elucidating structural information, rheological properties also provide useful information about the processability of the nanocomposites.

6. Application of CNT/polymer nanocomposites

As already discussed, incorporation of CNTs inside various polymer matrices enables the formation of advanced nanocomposites with improved or even novel set of electrical, thermal, mechanical, electrochemical or electromagnetic properties. Accordingly, they are known to display a number of dependent applications, which are described in details in the following section.

6.1. Electromagnetic interference (EMI) shielding

The excellent electrical conductivity and high aspect ratio of CNTs compared to other carbon-based fillers (e.g. carbon black, graphite, carbon fibre) and their excellent corrosion resistance, low density along with ultra-high-specific strength compared to metals, inspired the designing of CNTs-based polymer composites. Both SWCNTs- and MWCNTs-based composites have been prepared by their incorporation in various polymer matrices (e.g. insulating thermoplastic or thermosetting polymers, conjugated polymers), and their EMI SE performance was measured [14, 21, 23, 97, 106, 128, 129, 165, 166]. The introduction of CNTs inside host polymeric matrices leads to improvement of electrical conductivity as well as real- and imaginary-permittivity values [14, 21]. These are direct manifestations of increase in number of conducting links and interfacial polarization phenomenon and lead to improvement in SE (**Figure 14**).

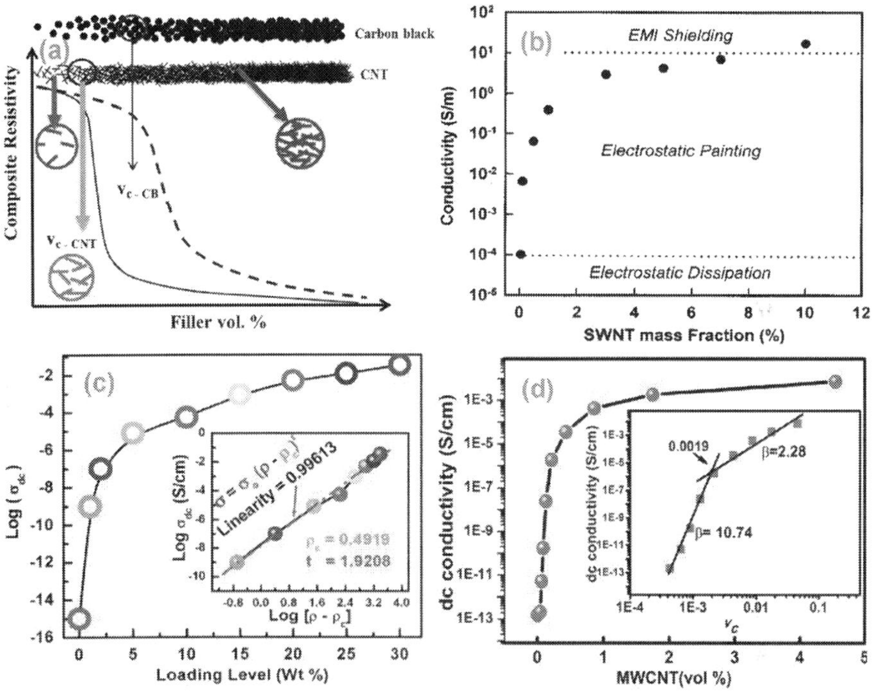

Figure 14. EMI shielding effectiveness (plots labelled A–D) for SWNT-polymer materials (wt% 3–15) studied in this work (10 MHz–1.5 GHz). Plots labelled (E–H) are higher-frequency data on MWNT-based material presented for comparison: (E) MWCNTs in PS; (F) MWCNTs in PMMA; (G) MWCNTs in epoxy resin and the value of the y axis for G is the reflection loss; (H) MWNTs in silica. Impact of wall integrity and aspect ratio on the EMI shielding effectiveness of the composites containing 10 wt% SWCNTs. Reprinted from [166] with permission from ACS.

The SE value also depends upon the nature of polymer matrix (**Figure 14a**), CNT-loading level and state of CNTs de-agglomeration/dispersion. Besides, CNT aspect ratio and wall defects (**Figure 14b**) also play crucial role in deterring shielding performance [14, 166]. In general, composites based on higher aspect ratio CNTs tend to display higher conductivities and better real and imaginary permittivity values compared to low aspect ratio CNTs, which gets reflected in terms of better shielding performance of the former. Further, annealed CNTs (containing very low defects)-based composites display superior shielding performance compared to unannealed CNT-filled composites.

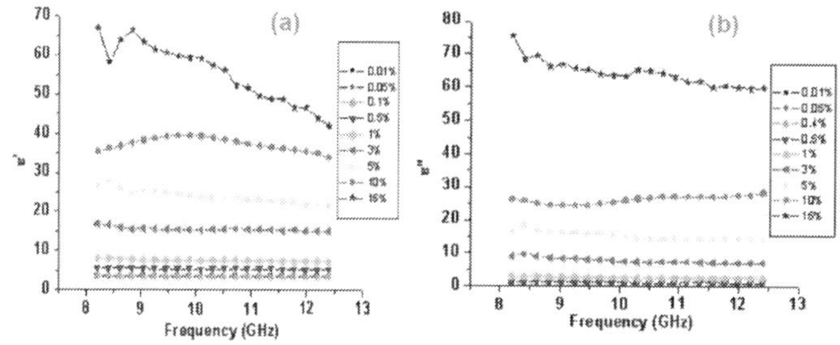

Figure 15. (a) Variation of reflection (SE_R) and absorption (SE_A) losses with MWCNT loading and magnified SEM image (inset) showing dispersed CNTs and voids, (b) Schematic representation of radiation shield interaction and involved multiple internal reflection (MIR) phenomenon. Reprinted from [106] with permission from Springer.

It has also been demonstrated that increased interfacial polarization and improvement of CNT dispersion via surface coating of conducting polymers [23, 106] lead to improvement of shielding response in the composites. It is suggested that CNTs networks inside composites trigger multiple reflections (**Figure 15b**), which are considered beneficial for improvement of absorption and overall shielding (**Figure 15a**). As the high CNT loading is required (where mechanical properties often degrade due to agglomeration and inhomogeneous filler dispersion issues) for achieving high shielding performance, some efforts have also been made to prepare high CNT loading (>10 wt%) yet mechanically strong composites via melt recirculation aided extrusive mixing [128, 129]. These composites display good shielding effectiveness. Recently, it is shown that at relatively low CNTs loading (<4 wt%), high aspect ratio and long-length CNTs-based composites display better shielding performance, whereas at higher (>4 wt %) CNTs loading, low aspect ratio and low-length CNTs-based composite show superior SE, probably due to the complex interplay between, impedance matching, interfacial polarization and multiple reflection phenomenon [128].

6.2. Supercapacitor electrodes

CNTs, due to their superior electrical properties, good mechanical and thermal stability, readily accessible surface area and unique pore structure are an attractive candidate for supercapacitor electrode applications [167, 168]. The composite formation strategy, involving incorporation of CNTs into various conjugated polymers (e.g. PANI, PPY) matrix, is considered an effective solution for improving the mechanical and electrochemical properties of electrodes [102, 103, 168–171]. For example, the sulfonated multiwall carbon nanotube (MWCNTs) in the composites [102] can greatly improve the cycle stability of PANI (**Figure 16**, left image), only showing 5.4% loss from their initial specific capacitance even after 1000 cycles. This was attributed to the use of MWCNTs with exceptional mechanical properties as a support and the formation of the charge-transfer complex, which could reduce the cycle degradation problems of PANI caused by volume changes or mechanical problems [102, 168]. Similarly, sulfonated-MWCNT/polypyrrole nanocomposite (MWCNTs/C-SO_3H/PPy) [170] exhibited good rate ability, high-specific capacitance (357 F/g), and high-specific capacitance retention rate (specific capacitance loss was only 3% even after the 1000 cycles). Apart from preserving the conducting polymer active material from mechanical changes (such as shrinkage and breaking) during long cycling, CNTs also improved the charge transfer characteristics, thereby facilitating the realization of high charge/discharge rates [103, 171].

It has also been realized that due to their excellent electrical and mechanical properties and open tubular mesoporous network structure, CNTs can as act good support materials for pseuocapacitive materials like conjugated polymers [167, 168]. Zhang and co-workers [167] have reported the use of a carbon nanotube array directly connected to the current collector (Ta foil) as the support to make composite electrodes with hierarchical porous structures. The electrochemical studies have shown that these PANI/CNT composite electrode (**Figure 16**, right image) with nanosize hierarchical porous structure, large surface area, and superior conductivity had high-specific capacitance (1030 F/g), superior rate capability (95% capacity retention at current density of 118 A/g), and high stability (i.e. only 5.5% capacity loss after 5000 cycles) between potential window of −0.2 to +0.7 V (vs. SCE) in 1.0 M H_2SO_4 electrolyte. These results are considered as manifestations of efficient charge transport (in the composite electrode); high-specific capacity (due to the efficient utilization of electrode materials); improved ionic conductivity; and countering of mechanical stability problems or volume changes by CNTs. Some efforts have also been made to deposit conjugated polymer over CNTs-based membranes to be directly used in supercapacitor with metal backing. For example, pulsed electrochemically deposited PPy over MWCNT membrane-based electrode [172] displayed remarkable specific capacitance of 427 F/g. Similarly, attempts have also been made to combine composite and paper electrode strategy, for example Oh et al. [173] prepared the highly porous sheets comprised of SWCNT/PPy composite by vacuum filtration of SWCNT/PPy methanol dispersions. The nanotube network provided mechanical strength and improved electrochemical performance of the nanocomposites. The highest specific capacitance of 131 F/g was obtained for nanocomposite with 1:1 SWCNT: PPy ratio.

6.3. Photovoltaics

The excellent electron transport properties of carbon nanotubes (CNTs) and their larger electron affinity compared to available polymer-based light-harvesting donors prompted to use them in bulk heterojunction (BHJ)-type hybrid solar cells. Their acceptor characteristics have been demonstrated by their ability to rapidly quench (at very low CNTs loading) the photoluminescence (PL) of polymer-based donor [100, 101, 174–176] which is a direct evidence of charge transfer from polymer (donor) to CNTs (acceptor). Further, CNTs also act as good transporter for exciton dissociation generated electrons. The initial effort in the direction [39] demonstrated that incorporation of 1% SWCNTs inside poly(3-octylthiophene) (P3OT) donor lead to efficient exciton dissociation and improved open-circuit voltage (V_{OC}). Another study has highlighted the influence of SWCNTs on the crystallinity enhancement and morphology improvement (continuous interpenetrating network formation) in poly(3-hexylthiophene) (P3HT)-based solar cells [176]. It is also outlined that workfunction of P3HT-modified SWCNTs increases due to the shifting of Fermi level towards vacuum level, leading to the improvement of V_{OC}. Another work [174] showed that the solar cells BHJ active layer based

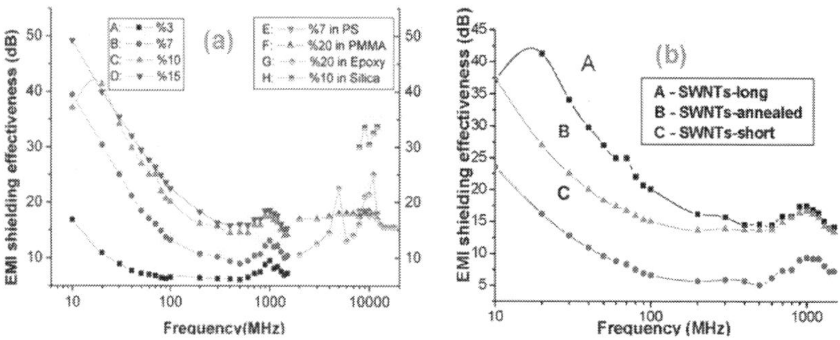

Figure 16 Cycle stability of (a) PANI nanorods, (b) PANI/sMWCNT-2, and (c) PANI/sMWCNT-4 electrodes in the voltage range of −0.2 to 0.75 V at a current density of 1 A/g (left image). Reproduced from [102] with permission from Elsevier. Schematic representation of the microstructure and energy storage characteristics of a polyaniline/carbon nanotube (PANI/CNT) composite electrode (right image). Reproduced from [168] with permission from RSC.

on semiconducting SWCNTS (s-SWCNTs)/P3HT blend (**Figure 17a**) form nanofilaments (**Figure 17b**) which helps in preventing bundling of s-SWCNT via formation of core shell structures (**Figure 17c** and inset).

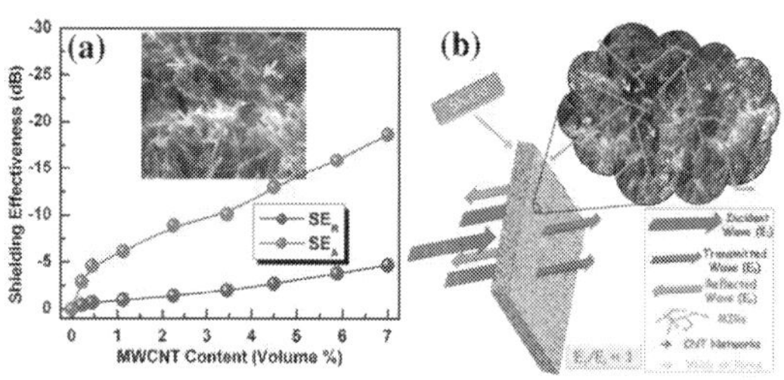

Figure 17. (a) Device structure of SWCNT/P3HT nanofilament BHJ layer-based solar cells. (b) AFM phase image of P3HT/s-SWCNT nanofilaments in a sample with 3 wt% SWCNT showing the worm-like morphology of the active layer. (c) Bright field TEM image of P3HT/s-SWCNTs blend having 3% s-SWCNT and schematic (inset) of P3HT coating over SWCNTs. (d) Band diagram for P3HT/s-SWCNTs interfaces and Voc. (e) Dependence of Jsc and Voc on the loading concentration of SWCNTs. Reproduced from [175] and [174] with permission from AIP and ACS respectively.

Such ordered configuration containing synergistic combination of s-SWCNTs and P3HT lead to improvement of charge separation, Voc improvement (**Figure 17d**) and efficient carrier transport, which collectively contribute towards light-to-electricity conversion efficiency enhancement (**Figure 17e**). The role of CNT content has also been investigated, and it was pointed that short-circuit current (Jsc) and Voc of devices critically rely on the SWCNT loading (**Figure 17e**). For example, in the present system, both the Jsc and Voc display proportionate increases up to 3 wt% (V_{oc} of 1.04 V and Jsc of 1.99 mA/cm^2) CNT content and decreases afterwards. Duly supported by morphological and electrical studies that was ascribed to the formation of optimal co-continuous interpenetrating-type morphology only around 3 wt% s-

SWCNT loading. Above this value, phase segregation takes place, thereby decreasing the exciton breaking and charge transport efficiencies. It has also been shown that P3HT/s-SWCNT BHJs can generate photocurrent from photons absorbed both in the P3HT and in the s-SWCNT and can achieve an IQE of 26% in the near-infrared region. Another work demonstrated that incorporation of MWCNTs into the P3HT/C60 BHJ layer increased the fill factor by 20% with a corresponding improvement of efficiency compared to the polythiophene/C_{60} bilayer device containing no MWCNTs [175].

6.4. Thermoelectrics

Recently, several attempts have been made to combine the excellent electrical conductivity, tunability of Seebeck coefficient, outstanding mechanical properties and good thermal stability of CNTs with the solution processabilty, low thermal conductivity, cost advantages and facile and scalable synthesis of conducting polymers. In particular, nanoscale heterostructuring is considered as good approach to regulate thermopower and electrical/thermal conductivity. In early attempts, conducting polymer/CNTs nanocomposites [105, 177, 178] (e.g. PANI/CNTs and PEDOT/CNTs) were synthesized by in situ polymerization (**Figure 18a**) to introduce heterojunctions and to see the effects of constituents on thermoelectric properties.

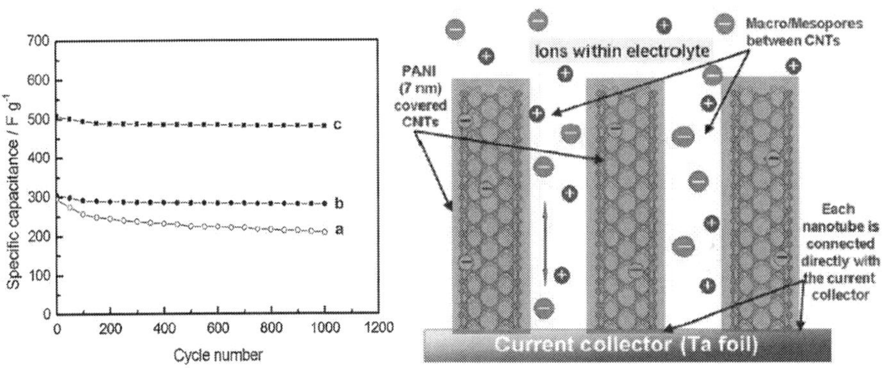

Figure 18. (a) Schematic representation of the formation of CNT/conjugated polymer by in situ polymerization leading to coating of nanotubes by polymer and formation of heat/electricity flow regulating nanojunctions. (b) Electrical conductivities and thermopower (i.e. seebeck coefficient) and (c) Power factor for the CNT/PEDOT-PSS composites. Reproduced from [179] with permission from ACS.

The presence of the heterojunctions is expected to improve the thermoelectric transport properties, that is obstructing the heat flow and favouring the electronic conduction. Indeed, it has been observed that poly(3,4-ethylenedioxy-thiophene)-poly(styrenesulfonate) (PEDOT:PSS)/CNTs composites display improved electrical conductivity without significantly altering or decreasing thermopower (i.e. seebeck coefficient) (**Figure 18b**), ultimately resulting in improved power factor (PF) (**Figure 18c**). This behaviour results from thermally disconnected, but electrically connected junctions in the nanotube network, which makes it feasible to tune the properties in favour of a higher thermoelectric figure of merit [178, 180]. Similarly, when bundle of CNTs were coated and bounded by PANI, the thermally insulating PANI interfacial layer can act as energy filters, which allow the high-energy carriers to pass and scatter the low-energy carriers, thereby increasing the Seebeck coefficient [177]. Furthermore, the growth of PANI over CNTs forms an ordered chain structure [105] which reduced the $\pi - \pi$ conjugation defects along the PANI backbone causing increase in carrier mobility (i.e. increased electrical conductivity). As a result, the composite possesses similar thermal conductivity as that of pure PANI (though orders of magnitude lower than pure CNTs) but significantly improved PF compared to pure PANI or CNTs (**Figure 19**)

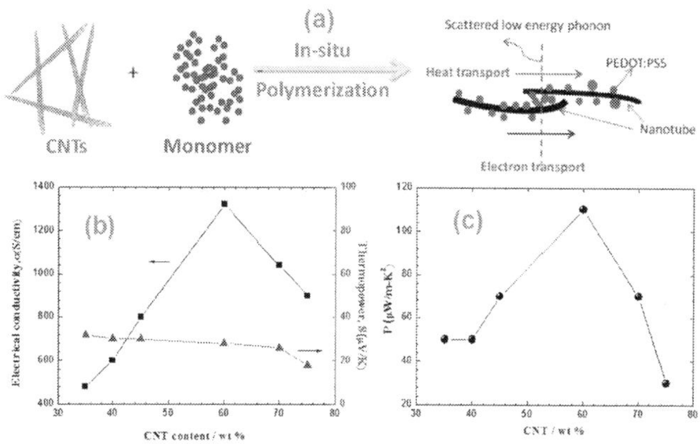

Figure 19. Seebeck coefficient and electrical conductivity (a), power factor and thermal conductivity (b) of SWCNT/PANI composites with different SWCNT content. The dashed line is the calculated electrical conductivity and thermal conductivity based on the particle mixture rule. Reprinted from [105] with permission from ACS.

Besides ordering-induced improvement in carrier mobility, such systems also show anisotropic thermoelectric properties [181] with more than a doubled improvement of power factor in the orientation direction. This provides a novel and effective way of improving the thermoelectric properties of conducting polymers.

In addition to conducting polymers, efforts have also been made to combine CNTs with conventional insulating polymers to improve thermoelectric PF. For example, in polyvinylidene fluoride (PVDF)/SWCNT composite [182] thin films-based systems, due to the decrease in CNTs content (from 100 to 5 wt%), the beneficial effects of increasing Seebeck coefficient and decreasing thermal conductivity were outweighed by negative effect of decreasing electrical conductivity, resulting in an increase in a thermoelectric figure of merit (ZT). Similarly, CNT/poly(vinyl acetate) (PVAc) composites [104] show highest thermoelectric performance at 20 wt% CNTs loading, with an electrical conductivity of 48 S/cm, thermal conductivity of 0.34 W/mK and room temperature ZT value larger than 6×10^{-3}. Besides, in addition to improvement in thermoelectric properties, due to their exceptional mechanical

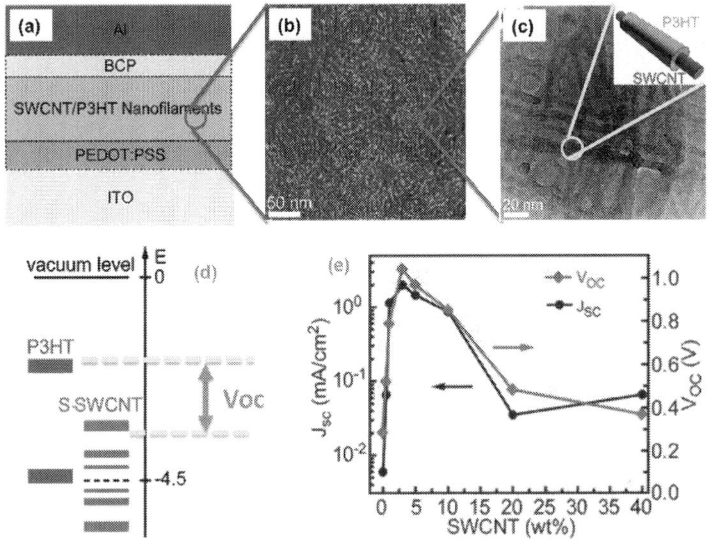

Figure 20. Effect of MWCNTs contents on the stress–strain curves for PA membranes. Reprinted from [183] with permission from Elsevier.

properties CNTs are also expected to improve the mechanical properties of their composite-based thermoelectric materials (both bulk materials and thin films).

6.5. Water purification

In recent past, membrane-based filtration techniques have emerged as potential alternatives for waste water purification applications. Owing to their cost-effective facile synthesis, good thermal stability, high mechanical strength and biocompatibility, polymers such as polysulfone, polyamides, cellulose nitrate, polyethersulfone, membranes are the most promising/ preferred. However, due to very small pore size, bacteriological contamination and pore blockings by adsorption of inorganic/organic impurities; low throughput and fouling are the common limitations of these membranes. In this context, CNTs owing to their strong antimicrobial activity, tunable surface chemistry and high mechanical strength have emerged as promising filler candidate for making composite membranes with improved antifouling characteristics and mechanical strength. Shawky et. al. [183] has synthesized a MWCNT/ polyamide nanocomposite membrane which exhibited excellent mechanical strength (**Figure 20**) and very good salt rejection ability with high permeability (**Table 2**).

The continuous network formation between structurally compact aromatic polyamide matrix and CNTs is solely responsible for high mechanical strength, good salt rejection ability though at the expense of slightly lower permeability.

MWCNTs loading (mg/g)	Permeability (L/m²/h bar)	Flux (L/m² h)	Salt rejection (%)
0	0.76 ± 0.08	32 ± 0.7	24 ± 1.1
2.5	0.75 ± 0.09	32 ± 0.4	28 ± 1.0
5	0.73 ± 0.07	31 ± 1.1	35 ± 0.7
10	0.72 ± 0.10	30 ± 0.9	69 ± 0.9
15	0.71 ± 0.11	28 ± 0.8	76 ± 1.1

Reprinted from [183] with permission from Elsevier.

Table 2. Membrane performance as a function of different MWCNTs loading at constant PA concentration (10%).

However, these membranes suffer from fouling tendency as the incorporation of hydrophobic CNTs in polyamide membranes leads to facile irreversible adsorption of organic/inorganic/ biological impurities. It is shown that the fouling problems can be circumvented by altering the surface morphology of the membrane. Since CNTs can be easily transformed from hydrophobic to hydrophilic by acid treatment; therefore, incorporation of CNTs functionalized with hydrophilic/amphiphilic groups is considered very advantageous for long-term uninterrupted performance of theses membranes [184–188]. In this direction, Choi et. al. [187] have fabricated a polysulfone/MWCNT hybrid membrane by phase inversion process. The surface-modified MWCNTs provide hydrophilicity and conductivity to the membrane. The authors reported that the pore size of the membrane increased with increase in CNT loading up to 1.5 wt% but further loading of CNTs increased viscosity of blend solution which led to decrease in pore size. The membrane with 4 wt% of MWCNT has pore size just smaller than pure polysulfone membranes and exhibited higher flux and good salt rejection ability. Nevertheless, the CNTs reinforced polymeric membranes are still a new concept and more efforts in the direction are necessary to further improve their performance and address the negative influence on permeability.

6.6. Gas and chemical vapour sensors

The major drawbacks such as high operating temperature, elevated costs and complex fabrication protocols associated with the conventional metal oxide-based sensors have prompted researchers to look for new avenues to come up with new materials not plagued by these shortcomings. This insatiable quest has unfolded the arena of CNTs-based polymer nanocomposites. The synergistic combination of the remarkable electrical transport and mechanical properties of CNTs and easily tailorable electroactive nature of conducting polymers has the potential to give one of the best-sensing platforms for the efficient gas/ chemical vapour detection [99, 189]. Therefore, dedicated efforts have been made to combine CNTs with various conducting polymers such as polyaniline [190–192], poly(3,4-ethylenedioxythiophene)/poly(styrenesulfonate) (PEDOT/PSS)[193], polypyrrole (PPy) [194], hexafluoroisopropanol substituted polythiophene (HFIP-PT), poly(3-hexylthiophene) (P3HT) [195], poly 3-methyl thiophene [98], to fabricate portable, more stable, highly sensitive, cost-effective and energy efficient chemiresistors especially for the detection of extremely minute quantities of environmentally hazardous analytes. In general, incorporation of CNTs tends to improve the mechanical strength, charge transport properties, porosity and specific interactions (due to high-specific surface area available for physico-chemical adsorption) with analytes (gas or chemical vapour), whereas conducting polymer contributes towards electroactivity and improved CNTs dispersion (thus better film formability compared to pristine CNTs). Therefore, the CNTs/conducting polymer nanocomposites often display improved sensitivity, response or selectivity.

The chemical functionalization of CNTs can also help to improve the affinity for one specific analyte species over another during the sensing of analyte mixture. For example, chemically polymerized 3-methylthiophene in the presence of COOH-functionalized MWCNTs, mixed with polyethylene oxide (used as a binder) deposited between two palladium electrodes was sensitive to chloromethanes (CH_3Cl, CH_2Cl_2, $CHCl_3$, CCl_4, CH_4) with fast response times [98]. However, it was not sensitive to acetone, acetaldehyde, benzaldehyde, tetrahydrofuran, methanol, and ethanol vapours providing high selectivity. The functionalized carbon nanotubes (CNT) with acidic groups (e.g.–COOH) have capability to doped PANI and can improve the sensing performances, for example good selectivity towards chloroform vapour over the other chlorinated methane vapour [192]. The targeted functionalization of CNTs by conducting polymer via electrochemical route is demonstrated to facilitate creation of high-density individually addressable nanosensor arrays, with improved sensitivity, detection limit and reproducibility [191].

It is also shown that oxygen plasma treatment of CNTs and their alignment in the nanocomposites film can improve the sensing performance in terms of sensitivity, selectivity, rapidity of response, good reversibility and stability [193]. Such effects are considered as manifestations of improved morphology and charge transport properties of thin film. For example, oxygen plasma functionalized and dielectrophoretically aligned CNTs-loaded PEDOT film-based sensors display (**Figure 21**) excellent responses for detection of 2–300 ppm NH_3 and 6–1000 ppb trimethylamine gases at room temperature. The presence of CNTs is conjugated polymer matrix is also known to improve sensitivity due to increase in degree of interactions during adsorption or desorption of the analyte [194]. For example, the sensitivity for NO_2 gas of polypyrrole—single-walled CNT nanocomposite is about ten times higher than that of pristine polypyrrole due to increase in the specific surface area by uniform polypyrrole coating on the single-wall CNTs. Nevertheless, though CNTs/polymer composites demonstrated to display improved sensing performance, the detailed sensing mechanism is still unclear, the cross-selectivity is poor and delayed recovery, stability and performance drift are the major issues to be resolved.

7. Conclusion and future directions

In conclusion, it is notable that CNTs-based polymer nanocomposites present an array of possibilities for their use in various technology developments. The electrical properties and dependent applications of CNTs-based nanocomposites (e.g. EMI shielding, antistatic) are

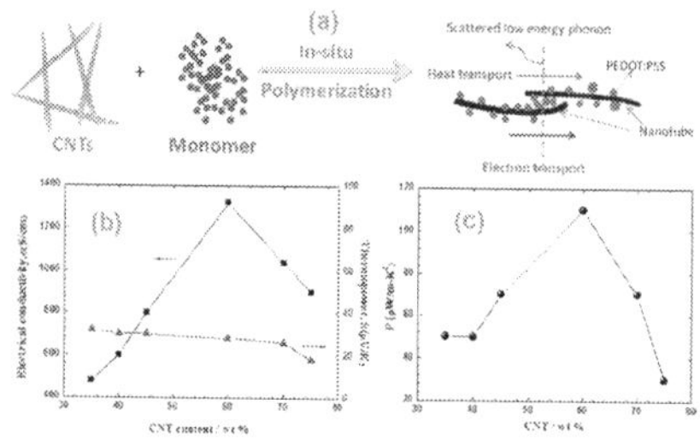

Figure 21. Selective response of the drop coated and AC-DEP assembled PEDOT/PSS-SWCNTs nanocomposite films to various vapours of 10 ppm. Reproduced from [193], with permission from Elsevier.

already satisfactory, though the scope for improvement cannot be ruled out especially in terms of cost reduction. In the context of thermal properties, CNTs-polymer nanocomposites with good thermal stability and appreciable thermal conductivity have already been exploited for heat sink and thermal-interfacing applications. However, for mechanical properties in particular, some key challenges need to be addressed and resolved so that the full potential of CNTs can be realized. Out of these challenges, the ability to acquire homogeneous dispersion of CNTs and their alignment in the polymer matrix remain major bottlenecks. The limitations faced due to this adversely affect the available filler surface area and thus the load transfer between the filler and the matrix leading to a compromise in terms of mechanical properties. With the increased understanding of functionalization chemistry in the recent years, the issue of dispersion has been partly circumvented. However, to compete with the existing carbon fibre-based nanocomposites, further efforts are needed in this direction. Novel functionalization routes are needed which enable maximum possible homogeneity in CNTs dispersion with least sacrifice on the part of mechanical properties of nanocomposites. Due to the large costs involved in the synthesis and processing of these nanocomposites, their commercial viability also needs mention in terms of key challenges. Furthermore, the high thermal conductivity of CNTs can be capitalized only when the high interfacial thermal resistance of nanotube networks can be minimized. In a nutshell, the future of CNTs-based polymer nanocomposites decisively hinges on the success achieved in the address of these key challenges.

Author details

Waseem Khan, Rahul Sharma and Parveen Saini*

*Address all correspondence to: pksaini@nplindia.org; parveensaini580@gmail.com

Polymeric and Soft Materials Section, Materials Physics and Engineering Division, CSIR-National Physical Laboratory, New Delhi, India

References

[1] Iijima, S. Helical microtubules of graphitic carbon. *Nature* 354, 56–58 (1991).

[2] Sinha, S., Barjami, S., Iannacchione, G., Schwab, A. & Muench, G. Off-axis Thermal Properties of Carbon Nanotube Films. *J. Nanoparticle Res.* 7, 651–657 (2005).

[3] Hong, S. & Myung, S. Nanotube Electronics: A flexible approach to mobility. *Nat. Nanotechnol.* 2, 207–208 (2007).

[4] Zeng, Y. *et al.* Enhanced adsorption of malachite green onto carbon nanotube/polyaniline composites. *J. Appl. Polym. Sci.* 127, 2475–2482 (2013).

[5] Shao, D. *et al.* Polyaniline Multiwalled Carbon Nanotube Magnetic Composite Prepared by Plasma-Induced Graft Technique and Its Application for Removal of Aniline and Phenol. *J. Phys. Chem. C* 114, 21524–21530 (2010).

[6] Ravindran, S., Chaudhary, S., Colburn, B., Ozkan, M. & Ozkan, C. S. Covalent Coupling of Quantum Dots to Multiwalled Carbon Nanotubes for Electronic Device Applications. *Nano Lett.* 3, 447–453 (2003).

[7] Novak, J. P. *et al.* Nerve agent detection using networks of single-walled carbon nanotubes. *Appl. Phys. Lett.* 83, 4026 (2003).

[8] Niyogi, S. *et al.* Chemistry of Single-Walled Carbon Nanotubes. *Accounts Chem. Res.* 35, 1105–1113 (2002).

[9] Li, J. *et al.* Carbon Nanotube Sensors for Gas and Organic Vapor Detection. *Nano Lett.* 3, 929–933 (2003).

[10] Kardimi, K. *et al.* Synthesis and characterization of carbon nanotubes decorated with Pt and PtRu nanoparticles and assessment of their electrocatalytic performance. *Int. J. Hydrog. Energy* 37, 1243–1253 (2012).

[11] Bryning, M. B. *et al.* Carbon Nanotube Aerogels. *Adv. Mater.* 19, 661–664 (2007).

[12] Bekyarova, E. *et al.* Chemically Functionalized Single-Walled Carbon Nanotubes as Ammonia Sensors †. *J. Phys. Chem. B* 108, 19717–19720 (2004).

[13] Saini, P. in *Fundamentals of Conjugated Polymer Blends, Copolymers and Composites* (ed. Saini, P.) 449–518 (John Wiley & Sons, Inc., 2015). at <http://doi.wiley.com/10.1002/9781119137160.ch9>

[14] Saini, P. in *Thermoset Nanocomposites* (ed. Mittal, V.) 211–237 (Wiley-VCH Verlag GmbH & Co. KGaA, 2013). at <http://doi.wiley.com/10.1002/9783527659647.ch10>

[15] Xia, Z. *et al.* Direct observation of toughening mechanisms in carbon nanotube ceramic matrix composites. *Acta Mater.* 52, 931–944 (2004).

[16] Flahaut, E. *et al.* Carbon nanotube-metal-oxide nanocomposites: microstructure, electrical conductivity and mechanical properties. *Acta Mater.* 48, 3803–3812 (2000).

[17] Li, Q., Viereckl, A., Rottmair, C. A. & Singer, R. F. Improved processing of carbon nanotube/magnesium alloy composites. *Compos. Sci. Technol.* 69, 1193–1199 (2009).

[18] Johnson, R. R., Johnson, A. T. C. & Klein, M. L. Probing the Structure of DNA-Carbon Nanotube Hybrids with Molecular Dynamics. *Nano Lett.* 8, 69–75 (2008).

[19] Peng, C., Zhang, S., Jewell, D. & Chen, G. Z. Carbon nanotube and conducting polymer composites for supercapacitors. *Prog. Nat. Sci.* 18, 777–788 (2008).

[20] Thostenson, E., Li, C. & Chou, T. Nanocomposites in context. *Compos. Sci. Technol.* 65, 491–516 (2005).

[21] Saini, P. & Aror, M. in *New Polymers for Special Applications* (ed. De Souza Gomes, A.) (InTech, 2012). at <http://www.intechopen.com/books/new-polymers-for-special-applications/microwave-absorption-and-emi-shielding-behavior-of-nanocomposites-based-on-intrinsically-conducting->

[22] Meincke, O. *et al.* Mechanical properties and electrical conductivity of carbon-nanotube filled polyamide-6 and its blends with acrylonitrile/butadiene/styrene. *Polymer* 45, 739–748 (2004).

[23] Saini, P., Choudhary, V., Singh, B. P., Mathur, R. B. & Dhawan, S. K. Enhanced microwave absorption behavior of polyaniline-CNT/polystyrene blend in 12.4-18.0GHz range. *Synth. Met.* 161, 1522–1526 (2011).

[24] Cai, L., Tabata, H. & Kawai, T. Self-assembled DNA networks and their electrical conductivity. *Appl. Phys. Lett.* 77, 3105 (2000).

[25] Hersam, M. C., Hoole, A. C. F., O'Shea, S. J. & Welland, M. E. Potentiometry and repair of electrically stressed nanowires using atomic force microscopy. *Appl. Phys. Lett.* 72, 915 (1998).

[26] Moniruzzaman, M. & Winey, K. I. Polymer Nanocomposites Containing Carbon Nanotubes. *Macromolecules* 39, 5194–5205 (2006).

[27] Ma, P.-C., Siddiqui, N. A., Marom, G. & Kim, J.-K. Dispersion and functionalization of carbon nanotubes for polymer-based nanocomposites: A review. *Compos. Part Appl. Sci. Manuf.* 41, 1345–1367 (2010).

[28] Liu, Y. & Kumar, S. Polymer/Carbon Nanotube Nano Composite Fibers-A Review. *Acs Appl. Mater. Interfaces* 6, 6069–6087 (2014).

[29] Saito, Y., Dresselhaus, G. & Dresselhaus, M. S. *Physical properties of carbon nanotubes.* (London and Imperial College Press, 1998).

[30] Dresselhaus, M. S., Dresselhaus, G. & Avouris, P. *Carbon Nanotubes: synthesis and structure and properties and applications.* (Berlin Springer-Verlag, 2001).

[31] Meunier, V. & Lambin, P. Scanning tunneling microscopy and spectroscopy of topological defects in carbon nanotubes. *Carbon* 38, 1729–1733 (2000).

[32] Charlier, J.-C., Ebbesen, T. W. & Lambin, P. Structural and electronic properties of pentagon-heptagon pair defects in carbon nanotubes. *Phys. Rev. B* 53, 11108–11113 (1996).

[33] Lambin, P., Márk, G. I. & Biró, L. P. Structural and electronic properties of coiled and curled carbon nanotubes having a large number of pentagon-heptagon pairs. *Phys. Rev. B* 67, (2003).

[34] Miyamoto, Y., Rubio, A., Berber, S., Yoon, M. & Tománek, D. Spectroscopic characterization of Stone-Wales defects in nanotubes. *Phys. Rev. B* 69, (2004).

[35] Kim, P., Odom, T. W., Huang, J.-L. & Lieber, C. M. Electronic Density of States of Atomically Resolved Single-Walled Carbon Nanotubes: Van Hove Singularities and End States. *Phys. Rev. Lett.* 82, 1225–1228 (1999).

[36] Ouyang, M., Huang, J. L., Cheung, C. L. & Lieber, C. M. Atomically resolved single-walled carbon nanotube intramolecular junctions. *Science* 291, 97 (2001).

[37] Kim, H. *et al.* Scanning Tunneling Spectroscopy of a Semiconducting Heterojunction Nanotube on the Au(111) Surface. *Surf Sci* 581, 241 (2005).

[38] Kim, H. *et al.* Direct Observation of Localized Defect States in Semiconductor Nanotube Junctions. *Phys. Rev. Lett.* 90, (2003).

[39] Ishigami, M. *et al.* Identifying Defects in Nanoscale Materials. *Phys. Rev. Lett.* 93, (2004).

[40] Carroll, D. L. *et al.* Electronic Structure and Localized States at Carbon Nanotube Tips. *Phys. Rev. Lett.* 78, 2811–2814 (1997).

[41] Dean, K. A. & Chalamala, B. R. Experimental studies of the cap structure of single-walled carbon nanotubes. *J. Vac. Sci. Technol. B Microelectron. Nanometer Struct.* 21, 868 (2003).

[42] Krätschmer, W., Lamb, L. D., Fostiropoulos, K. & Huffman, D. R. Solid C60: a new form of carbon. *Nature* 347, 354–358 (1990).

[43] Allouche, H. *et al.* Physical characteristics of the graphite electrode electric- arc as parameters for the formation of single-wall carbon nanotubes. in 1053–1054 (2000).

[44] Razafinimanana, M. *et al.* Influence of doped graphite electrode in electric arc for the formation of single wall carbon nanotubes. in 649–654

[45] Pacheco, M., Allouche, H., Monthioux, M., Razafinimanana, A. & Gleizes, A. Correlation between the plasma characteristics and the morphology and structure of the carbon phases synthesised by electric arc discharge. in (2001).

[46] Kroto, H. W., Heath, J. R., O'Brien, S. C., Curl, R. F. & Smalley, R. E. C60: Buckminsterfullerene. *Nature* 318, 162–163 (1985).

[47] Guo, T. *et al.* Self-Assembly of Tubular Fullerenes. *J. Phys. Chem.* 99, 10694–10697 (1995).

[48] Baker, R. T. K. Catalytic growth of carbon filaments. *Carbon* 27, 315–323 (1989).

[49] Endo, M. *et al.* The production and structure of pyrolytic carbon nanotubes (PCNTs). *J. Phys. Chem. Solids* 54, 1841–1848 (1993).

[50] Dai, H. *et al.* Single-wall nanotubes produced by metal-catalyzed disproportionation of carbon monoxide. *Chem. Phys. Lett.* 260, 471–475 (1996).

[51] Dai, H., Wong, E. W. & Lieber, C. M. Probing Electrical Transport in Nanomaterials: Conductivity of Individual Carbon Nanotubes. *Science* 272, 523–526 (1996).

[52] Odom, T. W., Huang, J. L., Kim, P. & Lieber, C. M. Atomic structure and electronic properties of single-walled carbonnanotubes. *Nature* 391, 62–64 (1998).

[53] Ouyang, M. Energy Gaps in 'Metallic' Single-Walled Carbon Nanotubes. *Science* 292, 702–705 (2001).

[54] Charlier, J.-C., Blase, X. & Roche, S. Electronic and transport properties of nanotubes. *Rev. Mod. Phys.* 79, 677–732 (2007).

[55] Hone, J., Whitney, M., Piskoti, C. & Zettl, A. Thermal conductivity of single-walled carbon nanotubes. *Phys. Rev. B* 59, R2514-R2516 (1999).

[56] Hone, J. *et al.* Electrical and thermal transport properties of magnetically aligned single wall carbon nanotube films. *Appl. Phys. Lett.* 77, 666 (2000).

[57] Berber, S., Kwon, Y.-K. & Tománek, D. Unusually High Thermal Conductivity of Carbon Nanotubes. *Phys. Rev. Lett.* 84, 4613–4616 (2000).

[58] Che, J., C¸a˘gın, T. & Goddard III, W. A. Thermal conductivity of carbon nanotubes. *Nanotechnology* 11, 65–69 (2000).

[59] Pop, E., Mann, D., Wang, Q., Goodson, K. & Dai, H. Thermal Conductance of an Individual Single-Wall Carbon Nanotube above Room Temperature. *Nano Lett.* 6, 96–100 (2006).

[60] Yakobson, B. I. & Avouris, P. Mechanical Properties of Carbon Nanotubes. *Top. Appl Phys* 80, 287–327 (2001).

[61] Qian, D., Wagner, G. J., Liu, W. K., Yu, M.-F. & Ruoff, R. S. Mechanics of carbon nanotubes. *Appl. Mech. Rev.* 55, 495 (2002).

[62] Reich, S., Thomsen, C. & Maultzsch, J. *Carbon Nanotubes: Basic Concepts and Physical Properties*. (New York: Wiley-VCH, 2004).

[63] Yu, M. Strength and Breaking Mechanism of Multiwalled Carbon Nanotubes Under Tensile Load. *Science* 287, 637–640 (2000).

[64] Ruoff, R. S., Tersoff, J., Lorents, D. C., Subramoney, S. & Chan, B. Radial deformation of carbon nanotubes by van der Waals forces. *Nature* 364, 514–516 (1993).

[65] Yang, Y. H. & Li, W. Z. Radial elasticity of single-walled carbon nanotube measured by atomic force microscopy. *Appl. Phys. Lett.* 98, 041901 (2011).

[66] Yu, M.-F., Kowalewski, T. & Ruoff, R. S. Investigation of the Radial Deformability of Individual Carbon Nanotubes under Controlled Indentation Force. *Phys. Rev. Lett.* 85, 1456–1459 (2000).

[67] Kuryliszyn-Kudelska, I., Małolepszy, A., Mazurkiewicz, M., Stobinski, L. & Dobrowolski, W. Magnetic Properties of 'As-Prepared' and Chemically Modified Multiwalled Carbon Nanotubes. *Acta Phys. Pol.* 119, 597–599 (2011).

[68] Maheshwari, P. H., Singh, R. & Mathur, R. B. Effect of heat treatment on the structure and stability of multiwalled carbon nanotubes produced by catalytic chemical vapor deposition technique. *Mater. Chem. Phys.* 134, 412–416 (2012).

[69] Gupta, V. & Kotnala, R. K. Multifunctional Ferromagnetic Carbon-Nanotube Arrays Prepared by Pulse-Injection Chemical Vapor Deposition. *Angew. Chem. Int. Ed.* 51, 2916–2919 (2012).

[70] Walters, D. A. *et al.* In-plane-aligned membranes of carbon nanotubes. *Chem. Phys. Lett.* 338, 14–20 (2001).

[71] Kim, H. M. *et al.* Electrical conductivity and electromagnetic interference shielding of multiwalled carbon nanotube composites containing Fe catalyst. *Appl. Phys. Lett.* 84, 589 (2004).

[72] Bahr, J. L., Mickelson, E. T., Bronikowski, M. J., Smalley, R. E. & Tour, J. M. Dissolution of small diameter single-wall carbon nanotubes in organic solvents? *Chem. Commun.* 193–194 (2001). doi:10.1039/b008042j

[73] Hirsch, A. Functionalization of single-walled carbon nanotubes. *Angew Chem Int Ed* 41, 1853–1859 (2002).

[74] Rinzler, A. G. *et al.* Large-scale purification of single-wall carbon nanotubes: process, product, and characterization. *Appl. Phys. Mater. Sci. Process.* 67, 29–37 (1998).

[75] Liu, J. *et al.* Fullerene pipes. *Science* 280, 1253 (1998).

[76] Holzinger, M. Functionalization of single-walled carbon nanotubes. (University Erlangen-Nürnberg, 2002).

[77] Riggs, J. E., Guo, Z., Carroll, D. L. & Sun, Y.-P. Strong Luminescence of Solubilized Carbon Nanotubes. *J. Am. Chem. Soc.* 122, 5879–5880 (2000).

[78] Lim, J. K. *et al.* Selective thiolation of single-walled carbon nanotubes. *Synth. Met.* 139, 521–527 (2003).

[79] Nakajima, T., Kasamatsu, S. & Matsuo, Y. Synthesis and characterization of fluorinated carbon nanotubes. *Eur J Solid State Inorg Chem* 33, 831 (1996).

[80] Unger, E., Graham, A., Kreupl, F., Liebau, M. & Hoenlein, W. Electrochemical functionalization of multi-walled carbon nanotubes for solvation and purification. *Curr. Appl. Phys.* 2, 107–111 (2002).

[81] Chen, Y. K. *et al.* Purification and opening of carbon nanotubes via bromination. *Adv. Mater.* 8, 1012–1015 (1996).

[82] Owens, F. J. & Iqbal, Z. Electrochemical functionalization of carbon nanotubes with hydrogen. in (2002).

[83] Holzinger, M. *et al.* Sidewall Functionalization of Carbon Nanotubes This work was supported by the European Union under the 5th Framework Research Training Network 1999, HPRNT 1999-00011 FUNCARS. *Angew. Chem. Int. Ed.* 40, 4002 (2001).

[84] Coleman, K. S., Bailey, S. R., Fogden, S. & Green, M. L. H. Functionalization of Single-Walled Carbon Nanotubes via the Bingel Reaction. *J. Am. Chem. Soc.* 125, 8722–8723 (2003).

[85] Georgakilas, V. *et al.* Organic Functionalization of Carbon Nanotubes. *J. Am. Chem. Soc.* 124, 760–761 (2002).

[86] Cui, Burghard, M. & Kern, K. Reversible Sidewall Osmylation of Individual Carbon Nanotubes. *Nano Lett.* 3, 613–615 (2003).

[87] Lu, X., Tian, F., Wang, N. & Zhang, Q. Organic Functionalization of the Sidewalls of Carbon Nanotubes by Diels-Alder Reactions: A Theoretical Prediction. *Org. Lett.* 4, 4313–4315 (2002).

[88] Bahr, J. L. *et al.* Functionalization of Carbon Nanotubes by Electrochemical Reduction of Aryl Diazonium Salts: A Bucky Paper Electrode. *J. Am. Chem. Soc.* 123, 6536–6542 (2001).

[89] Kooi, S. E., Schlecht, U., Burghard, M. & Kern, K. Electrochemical Modification of Single Carbon Nanotubes. *Angew. Chem. Int. Ed.* 41, 1353–1355 (2002).

[90] Bandow, S. *et al.* Purification of Single-Wall Carbon Nanotubes by Microfiltration. *J. Phys. Chem. B* 101, 8839–8842 (1997).

[91] Duesberg, G. S., Burghard, M., Muster, J. & Philipp, G. Separation of carbon nanotubes by size exclusion chromatography. *Chem. Commun.* 435–436 (1998). doi:10.1039/a707465d

[92] Krstic, V., Duesberg, G. S., Muster, J., Burghard, M. & Roth, S. Langmuir-Blodgett Films of Matrix-Diluted Single-Walled Carbon Nanotubes. *Chem. Mater.* 10, 2338–2340 (1998).

[93] O'Connell, M. J. *et al.* Reversible water-solubilization of single-walled carbon nanotubes by polymer wrapping. *Chem. Phys. Lett.* 342, 265–271 (2001).

[94] Smith, B. W. & Luzzi, D. E. Formation mechanism of fullerene peapods and coaxial tubes: a path to large scale synthesis. *Chem. Phys. Lett.* 321, 169–174 (2000).

[95] Smith, B. W., Monthioux, M. & Luzzi, D. E. Carbon nanotube encapsulated fullerenes: a unique class of hybrid materials. *Chem. Phys. Lett.* 315, 31–36 (1999).

[96] Jeon, I.-Y., Wook, D., Ashok, N. & Baek, J.-B. in *Carbon Nanotubes - Polymer Nanocomposites* (ed. Yellampalli, S.) (InTech, 2011). at <http://www.intechopen.com/books/carbon-nanotubes-polymer-nanocomposites/functionalization-of-carbon-nanotubes>

[97] Saini, P., Choudhary, V., Singh, B. P., Mathur, R. B. & Dhawan, S. K. Polyaniline-MWCNT nanocomposites for microwave absorption and EMI shielding. *Mater. Chem. Phys.* 113, 919–926 (2009).

[98] A chemical sensor for chloromethanes using a nanocomposite of multiwalled carbon nanotubes with poly(3-methylthiophene). *Sensors Actuators B Chem.* 106, 766–771 (2005).

[99] Kar, P., Choudhury, A. & Verma, S. K. in *Fundamentals of Conjugated Polymer Blends, Copolymers and Composites* (ed. Saini, P.) 619–686 (John Wiley & Sons, Inc., 2015). at <http://doi.wiley.com/10.1002/9781119137160.ch12>

[100] Rud, J. A., Lovell, L. S., Senn, J. W., Qiao, Q. & Mcleskey, J. T. Water soluble polymer/carbon nanotube bulk heterojunction solar cells. *J. Mater. Sci.* 40, 1455–1458 (2005).

[101] Kanai, Y. & Grossman, J. C. Role of Semiconducting and Metallic Tubes in P3HT/Carbon-Nanotube Photovoltaic Heterojunctions: Density Functional Theory Calculations. *Nano Lett.* 8, 908–912 (2008).

[102] Zhu, Z., Wang, G., Sun, M., Li, X. & Li, C. Fabrication and electrochemical characterization of polyaniline nanorods modified with sulfonated carbon nanotubes for supercapacitor applications. *Electrochimica Acta* 56, 1366–1372 (2011).

[103] Liu, P. in *Fundamentals of Conjugated Polymer Blends, Copolymers and Composites* (ed. Saini, P.) 419–447 (John Wiley & Sons, Inc., 2015). at <http://doi.wiley.com/10.1002/9781119137160.ch8>

[104] Yu, C., Kim, Y. S., Kim, D. & Grunlan, J. C. Thermoelectric Behavior of Segregated-Network Polymer Nanocomposites. *Nano Lett.* 9, 1283–1283 (2009).

[105] Yao, Q., Chen, L., Zhang, W., Liufu, S. & Chen, X. Enhanced Thermoelectric Performance of Single-Walled Carbon Nanotubes/Polyaniline Hybrid Nanocomposites. *Acs Nano* 4, 2445–2451 (2010).

[106] Saini, P. & Choudhary, V. Enhanced electromagnetic interference shielding effectiveness of polyaniline functionalized carbon nanotubes filled polystyrene composites. *J. Nanoparticle Res.* 15, (2013).

[107] Daniel, I. M. *Engineering mechanics of composite materials.* (Oxford University Press, 2006).

[108] Saini, P. in *Fundamentals of Conjugated Polymer Blends, Copolymers and Composites* (ed. Saini, P.) 1–118 (John Wiley & Sons, Inc., 2015). at <http://doi.wiley.com/10.1002/9781119137160.ch1>

[109] Berrod, G., Vidal, A., Papirer, E. & Donnet, J. B. Reinforcement of siloxane elastomers by silica. Interactions between an oligomer of poly(dimethylsiloxane) and a fumed silica. *J. Appl. Polym. Sci.* 23, 2579–2590 (1979).

[110] Johnson, B. L. *Industrial and Engineering Chemistry*. (1951).

[111] Choudhary, V. & Gupt, A. in *Carbon Nanotubes - Polymer Nanocomposites* (ed. Yellampalli, S.) (InTech, 2011). at <http://www.intechopen.com/books/carbon-nanotubes-polymer-nanocomposites/polymer-carbon-nanotube-nanocomposites>

[112] Choudhary, V., Singh, B. P. & Mathur, R. B. in *Syntheses and Applications of Carbon Nanotubes and Their Composites* (ed. Suzuki, S.) (InTech, 2013). at <http://www.intechopen.com/books/syntheses-and-applications-of-carbon-nanotubes-and-their-composites/carbon-nanotubes-and-their-composites>

[113] Coleman, J. N., Khan, U., Blau, W. J. & Gun'ko, Y. K. Small but strong: A review of the mechanical properties of carbon nanotube-polymer composites. *Carbon* 44, 1624–1652 (2006).

[114] Spitalsky, Z., Tasis, D., Papagelis, K. & Galiotis, C. Carbon nanotube-polymer composites: Chemistry, processing, mechanical and electrical properties. *Prog. Polym. Sci.* 35, 357–401 (2010).

[115] Badaire, S., Poulin, P., Maugey, M. & Zakri, C. In Situ Measurements of Nanotube Dimensions in Suspensions by Depolarized Dynamic Light Scattering. *Langmuir* 20, 10367–10370 (2004).

[116] Bandyopadhyaya, R., Nativ-Roth, E., Regev, O. & Yerushalmi-Rozen, R. Stabilization of Individual Carbon Nanotubes in Aqueous Solutions. *Nano Lett.* 2, 25–28 (2002).

[117] Wu, H.-X. *et al.* Polymer-wrapped multiwalled carbon nanotubes synthesized via microwave-assisted in situ emulsion polymerization and their optical limiting properties. *Carbon* 45, 2866–2872 (2007).

[118] Pei, X., Hu, L., Liu, W. & Hao, J. Synthesis of water-soluble carbon nanotubes via surface initiated redox polymerization and their tribological properties as water-based lubricant additive. *Eur. Polym. J.* 44, 2458–2464 (2008).

[119] Cheng, F., Imin, P., Maunders, C., Botton, G. & Adronov, A. Soluble, Discrete Supramolecular Complexes of Single-Walled Carbon Nanotubes with Fluorene-Based Conjugated Polymers. *Macromolecules* 41, 2304–2308 (2008).

[120] Bryning, M. B., Milkie, D. E., Islam, M. F., Kikkawa, J. M. & Yodh, A. G. Thermal conductivity and interfacial resistance in single-wall carbon nanotube epoxy composites. *Appl. Phys. Lett.* 87, 161909 (2005).

[121] Lau, K., Lu, M., Cheung, H., Sheng, F. & Li, H. Thermal and mechanical properties of single-walled carbon nanotube bundle-reinforced epoxy nanocomposites: the role of solvent for nanotube dispersion. *Compos. Sci. Technol.* 65, 719–725 (2005).

[122] De la Chapelle, M. L. *et al.* Raman characterization of singlewalled carbon nanotubes and PMMA-nanotubes composites. *Synth. Met.* 103, 2510–2512 (1999).

[123] Du, F., Fischer, J. E. & Winey, K. I. Coagulation method for preparing single-walled carbon nanotube/poly(methyl methacrylate) composites and their modulus, electrical conductivity, and thermal stability. *J. Polym. Sci. Part B Polym. Phys.* 41, 3333–3338 (2003).

[124] Barrau, S. *et al.* Effect of Palmitic Acid on the Electrical Conductivity of Carbon Nanotubes-Epoxy Resin Composites. *Macromolecules* 36, 9678–9680 (2003).

[125] Islam, M. F., Rojas, E., Bergey, D. M., Johnson, A. T. & Yodh, A. G. High Weight Fraction Surfactant Solubilization of Single-Wall Carbon Nanotubes in Water. *Nano Lett.* 3, 269–273 (2003).

[126] Benoit, J.-M. *et al.* Transport properties of PMMA-Carbon Nanotubes composites. *Synth. Met.* 121, 1215–1216 (2001).

[127] Dalton, A. B. *et al.* Super-tough carbon-nanotube fibres. *Nature* 423, 703–703 (2003).

[128] Verma, P., Saini, P. & Choudhary, V. Designing of carbon nanotube/polymer composites using melt recirculation approach: Effect of aspect ratio on mechanical, electrical and EMI shielding response. *Mater. Des.* 88, 269–277 (2015).

[129] Verma, P., Saini, P., Malik, R. S. & Choudhary, V. Excellent electromagnetic interference shielding and mechanical properties of high loading carbon-nanotubes/polymer composites designed using melt recirculation equipped twin-screw extruder. *Carbon* 89, 308–317 (2015).

[130] Andrews, R., Jacques, D., Qian, D. & Rantell, T. Multiwall Carbon Nanotubes: Synthesis and Application. *Accounts Chem. Res.* 35, 1008–1017 (2002).

[131] Bhattacharyya, A. R. *et al.* Crystallization and orientation studies in polypropylene/single wall carbon nanotube composite. *Polymer* 44, 2373–2377 (2003).

[132] Tang, W., Santare, M. H. & Advani, S. G. Melt processing and mechanical property characterization of multi-walled carbon nanotube/high density polyethylene (MWNT/HDPE) composite films. *Carbon* 41, 2779–2785 (2003).

[133] Haggenmueller, R., Gommans, H. H., Rinzler, A. G., Fischer, J. E. & Winey, K. I. Aligned single-wall carbon nanotubes in composites by melt processing methods. *Chem. Phys. Lett.* 330, 219–225 (2000).

[134] Vigolo, B. Macroscopic Fibers and Ribbons of Oriented Carbon Nanotubes. *Science* 290, 1331–1334 (2000).

[135] Xia, H., Wang, Q., Li, K. & Hu, G.-H. Preparation of polypropylene/carbon nanotube composite powder with a solid-state mechanochemical pulverization process. *J. Appl. Polym. Sci.* 93, 378–386 (2004).

[136] Dhand, C. *et al.* Preparation of polyaniline/multiwalled carbon nanotube composite by novel electrophoretic route. *Carbon* 46, 1727–1735 (2008).

[137] Regev, O., ElKati, P. N. B., Loos, J. & Koning, C. E. Preparation of Conductive Nanotube-Polymer Composites Using Latex Technology. *Adv. Mater.* 16, 248–251 (2004).

[138] Ramasubramaniam, R., Chen, J. & Liu, H. Homogeneous carbon nanotube/polymer composites for electrical applications. *Appl. Phys. Lett.* 83, 2928 (2003).

[139] Baughman, R. H. Carbon Nanotubes--the Route Toward Applications. *Science* 297, 787–792 (2002).

[140] Bryning, M. B., Islam, M. F., Kikkawa, J. M. & Yodh, A. G. Very Low Conductivity Threshold in Bulk Isotropic Single-Walled Carbon Nanotube-Epoxy Composites. *Adv. Mater.* 17, 1186–1191 (2005).

[141] Sandler, J. K. W., Kirk, J. E., Kinloch, I. A., Shaffer, M. S. P. & Windle, A. H. Ultra-low electrical percolation threshold in carbon-nanotube-epoxy composites. *Polymer* 44, 5893–5899 (2003).

[142] Du, F., Fischer, J. E. & Winey, K. I. Effect of nanotube alignment on percolation conductivity in carbon nanotube/polymer composites. *Phys. Rev. B* 72, (2005).

[143] Bai, J. B. & Allaoui, A. Effect of the length and the aggregate size of MWNTs on the improvement efficiency of the mechanical and electrical properties of nanocomposites-experimental investigation. *Compos. Part Appl. Sci. Manuf.* 34, 689–694 (2003).

[144] Li, X. *et al.* Fabrication and characterization of well-dispersed single-walled carbon nanotube/polyaniline composites. *Carbon* 41, 1670–1673 (2003).

[145] Ramanathan, T., Liu, H. & Brinson, L. C. Functionalized SWNT/polymer nanocomposites for dramatic property improvement. *J. Polym. Sci. Part B Polym. Phys.* 43, 2269–2279 (2005).

[146] Valentini, L., Armentano, I., Puglia, D. & Kenny, J. M. Dynamics of amine functionalized nanotubes/epoxy composites by dielectric relaxation spectroscopy. *Carbon* 42, 323–329 (2004).

[147] Tamburri, E. *et al.* Modulation of electrical properties in single-walled carbon nanotube/conducting polymer composites. *Carbon* 43, 1213–1221 (2005).

[148] Yuen, S.-M. *et al.* Silane-modified MWCNT/PMMA composites - Preparation, electrical resistivity, thermal conductivity and thermal stability. *Compos. Part Appl. Sci. Manuf.* 38, 2527–2535 (2007).

[149] Gojny, F. H. *et al.* Evaluation and identification of electrical and thermal conduction mechanisms in carbon nanotube/epoxy composites. *Polymer* 47, 2036–2045 (2006).

[150] Biercuk, M. J. *et al.* Carbon nanotube composites for thermal management. *Appl. Phys. Lett.* 80, 2767 (2002).

[151] Evseeva, L. E. & Tanaeva, S. A. Thermal conductivity of micro-and nanostructural epoxy composites at low temperatures. *Mech. Compos. Mater.* 44, 87–92 (2008).

[152] Guthy, C., Du, F., Brand, S., Winey, K. I. & Fischer, J. E. Thermal Conductivity of Single-Walled Carbon Nanotube/PMMA Nanocomposites. *J. Heat Transf.* 129, 1096 (2007).

[153] Velasco-Santos, C., Martínez-Hernández, A. L., Fisher, F. T., Ruoff, R. & Castaño, V. M. Improvement of Thermal and Mechanical Properties of Carbon Nanotube Composites through Chemical Functionalization. *Chem. Mater.* 15, 4470–4475 (2003).

[154] Choi, E. S. *et al.* Enhancement of thermal and electrical properties of carbon nanotube polymer composites by magnetic field processing. *J. Appl. Phys.* 94, 6034 (2003).

[155] Du, F., Guthy, C., Kashiwagi, T., Fischer, J. E. & Winey, K. I. An infiltration method for preparing single-wall nanotube/epoxy composites with improved thermal conductivity. *J. Polym. Sci. Part B Polym. Phys.* 44, 1513–1519 (2006).

[156] at <http://science.nasa.gov/science-news/science-at-nasa/2005/27jul_nanotech/>

[157] Gao, J. *et al.* Continuous Spinning of a Single-Walled Carbon Nanotube-Nylon Composite Fiber. *J. Am. Chem. Soc.* 127, 3847–3854 (2005).

[158] Zabihi, O., Ahmadi, M., Akhlaghi bagherjeri, M. & Naebe, M. One-pot synthesis of aminated multi-walled carbon nanotube using thiol-ene click chemistry for improvement of epoxy nanocomposites properties. *Rsc Adv* 5, 98692–98699 (2015).

[159] Ji, J. *et al.* Significant Improvement of Mechanical Properties Observed in Highly Aligned Carbon-Nanotube-Reinforced Nanofibers. *J. Phys. Chem.* C 113, 4779–4785 (2009).

[160] Liao, K. & Li, S. Interfacial characteristics of a carbon nanotube-polystyrene composite system. *Appl. Phys. Lett.* 79, 4225 (2001).

[161] Huang, Y. *et al.* The influence of single-walled carbon nanotube structure on the electromagnetic interference shielding efficiency of its epoxy composites. *Carbon* 45, 1614–1621 (2007).

[162] Krishnamoorti, R. & Giannelis, E. P. Rheology of End-Tethered Polymer Layered Silicate Nanocomposites. *Macromolecules* 30, 4097–4102 (1997).

[163] Mitchell, C. A., Bahr, J. L., Arepalli, S., Tour, J. M. & Krishnamoorti, R. Dispersion of Functionalized Carbon Nanotubes in Polystyrene. *Macromolecules* 35, 8825–8830 (2002).

[164] Pötschke, P., Abdel-Goad, M., Alig, I., Dudkin, S. & Lellinger, D. Rheological and dielectrical characterization of melt mixed polycarbonate-multiwalled carbon nanotube composites. *Polymer* 45, 8863–8870 (2004).

[165] Yang, Y., Gupta, M. C., Dudley, K. L. & Lawrence, R. W. Novel Carbon Nanotube-Polystyrene Foam Composites for Electromagnetic Interference Shielding. *Nano Lett.* 5, 2131–2134 (2005).

[166] Li, N. *et al.* Electromagnetic Interference (EMI) Shielding of Single-Walled Carbon Nanotube Epoxy Composites. *Nano Lett.* 6, 1141–1145 (2006).

[167] Zhang, L. L., Zhou, R. & Zhao, X. S. Graphene-based materials as supercapacitor electrodes. *J. Mater. Chem.* 20, 5983 (2010).

[168] Wang, G., Zhang, L. & Zhang, J. A review of electrode materials for electrochemical supercapacitors. *Chem Soc Rev* 41, 797–828 (2012).

[169] Zhang, L. L. & Zhao, X. S. Carbon-based materials as supercapacitor electrodes. *Chem. Soc. Rev.* 38, 2520 (2009).

[170] Fu, Q. *et al.* Novel non-covalent sulfonated multiwalled carbon nanotubes from p-toluenesulfonic acid/glucose doped polypyrrole for electrochemical capacitors. *Synth. Met.* 161, 373–378 (2011).

[171] Frackowiak, E., Khomenko, V., Jurewicz, K., Lota, K. & Béguin, F. Supercapacitors based on conducting polymers/nanotubes composites. *J. Power Sources* 153, 413–418 (2006).

[172] Fang, Y. *et al.* Self-supported supercapacitor membranes: Polypyrrole-coated carbon nanotube networks enabled by pulsed electrodeposition. *J. Power Sources* 195, 674–679 (2010).

[173] Oh, J. *et al.* Preparation and electrochemical characterization of porous SWNT-PPy nanocomposite sheets for supercapacitor applications. *Synth. Met.* 158, 638–641 (2008).

[174] Ren, S. *et al.* Toward Efficient Carbon Nanotube/P3HT Solar Cells: Active Layer Morphology, Electrical, and Optical Properties. *Nano Lett.* 11, 5316–5321 (2011).

[175] Miller, A. J., Hatton, R. A. & Silva, S. R. P. Water-soluble multiwall-carbon-nanotube-polythiophene composite for bilayer photovoltaics. *Appl. Phys. Lett.* 89, 123115 (2006).

[176] Geng, J. & Zeng, T. Influence of Single-Walled Carbon Nanotubes Induced Crystallinity Enhancement and Morphology Change on Polymer Photovoltaic Devices. *J. Am. Chem. Soc.* 128, 16827–16833 (2006).

[177] Meng, C., Liu, C. & Fan, S. A Promising Approach to Enhanced Thermoelectric Properties Using Carbon Nanotube Networks. *Adv. Mater.* 22, 535–539 (2010).

[178] Yu, C., Choi, K., Yin, L. & Grunlan, J. C. Light-Weight Flexible Carbon Nanotube Based Organic Composites with Large Thermoelectric Power Factors. *Acs Nano* 5, 7885–7892 (2011).

[179] Kim, D., Kim, Y., Choi, K., Grunlan, J. C. & Yu, C. Improved Thermoelectric Behavior of Nanotube-Filled Polymer Composites with Poly(3,4-ethylenedioxythiophene) Poly(styrenesulfonate). *Acs Nano* 4, 513–523 (2010).

[180] Yu, C., Kim, Y. S., Kim, D. & Grunlan, J. C. Thermoelectric Behavior of Segregated-Network Polymer Nanocomposites. *Nano Lett.* 8, 4428–4432 (2008).

[181] Wang, Q., Yao, Q., Chang, J. & Chen, L. Enhanced thermoelectric properties of CNT/PANI composite nanofibers by highly orienting the arrangement of polymer chains. *J. Mater. Chem.* 22, 17612 (2012).

[182] Hewitt, C. A. *et al.* Varying the concentration of single walled carbon nanotubes in thin film polymer composites, and its effect on thermoelectric power. *Appl. Phys. Lett.* 98, 183110 (2011).

[183] Shawky, H. A., Chae, S.-R., Lin, S. & Wiesner, M. R. Synthesis and characterization of a carbon nanotube/polymer nanocomposite membrane for water treatment. *Desalination* 272, 46–50 (2011).

[184] Liu, Y.-L., Chang, Y., Chang, Y.-H. & Shih, Y.-J. Preparation of Amphiphilic Polymer-Functionalized Carbon Nanotubes for Low-Protein-Adsorption Surfaces and Protein-Resistant Membranes. *Acs Appl. Mater. Interfaces* 2, 3642–3647 (2010).

[185] Celik, E., Park, H., Choi, H. & Choi, H. Carbon nanotube blended polyethersulfone membranes for fouling control in water treatment. *Water Res.* 45, 274–282 (2011).

[186] Qiu, S. *et al.* Preparation and properties of functionalized carbon nanotube/PSF blend ultrafiltration membranes. *J. Membr. Sci.* 342, 165–172 (2009).

[187] Choi, J.-H., Jegal, J. & Kim, W.-N. Fabrication and characterization of multi-walled carbon nanotubes/polymer blend membranes. *J. Membr. Sci.* 284, 406–415 (2006).

[188] Brunet, L. *et al.* Properties of Membranes Containing Semi-dispersed Carbon Nanotubes. *Environ. Eng. Sci.* 25, 565–576 (2008).

[189] Hatchett, D. W. & Josowicz, M. Composites of Intrinsically Conducting Polymers as Sensing Nanomaterials. *Chem. Rev.* 108, 746–769 (2008).

[190] Choi, H. H., Lee, J., Dong, K.-Y., Ju, B.-K. & Lee, W. Gas Sensing performance of composite materials using conducting polymer/single-walled carbon nanotubes. *Macromol. Res.* 20, 143–146 (2012).

[191] Zhang, T., Nix, M. B., Yoo, B.-Y., Deshusses, M. A. & Myung, N. V. Electrochemically Functionalized Single-Walled Carbon Nanotube Gas Sensor. *Electroanalysis* 18, 1153–1158 (2006).

[192] Kar, P. & Choudhury, A. Carboxylic acid functionalized multi-walled carbon nanotube doped polyaniline for chloroform sensors. *Sensors Actuators B Chem.* 183, 25–33 (2013).

[193] Jian, J. *et al.* Gas-sensing characteristics of dielectrophoretically assembled composite film of oxygen plasma-treated SWCNTs and PEDOT/PSS polymer. *Sensors Actuators B Chem.* 178, 279–288 (2013).

[194] An, K. H., Jeong, S. Y., Hwang, H. R. & Lee, Y. H. Enhanced Sensitivity of a Gas Sensor Incorporating Single-Walled Carbon Nanotube-Polypyrrole Nanocomposites. *Adv. Mater.* 16, 1005–1009 (2004).

[195] Wang, F., Gu, H. & Swager, T. M. Carbon Nanotube/Polythiophene Chemiresistive Sensors for Chemical Warfare Agents. *J. Am. Chem. Soc.* 130, 5392–5393 (2008).

Chapter 2

Advanced Fabrication and Properties of Aligned Carbon Nanotube Composites: Experiments and Modeling

Hai M. Duong, Feng Gong, Peng Liu and
Thang Q. Tran

Additional information is available at the end of the chapter

http://dx.doi.org/10.5772/62510

Abstract

Aligned carbon nanotube (CNT) composites have attracted a lot of interest due to their superb mechanical and physical properties. This article presents a brief overview of the synthesis approaches of aligned CNT composites. The three major methods for fabricating aligned CNT fibers are first reviewed, including wet-spinning, dry-spinning and floating catalysts. The obtained CNT fibers, however, have limited mechanical and physical properties due to their porous structure and poor CNT alignment within the fibers. Appropriate treatments are required to densify the fibers to enhance their properties. The main approaches for the densification of CNT fibers are then discussed. To further enhance load transfer within CNT fibers, polymer infiltration is always used. Typical studies on polymer infiltration of CNT fibers are reviewed, and the properties of the obtained composites indicate the superiority of this composite fabrication method over the conventional dispersion method. Since aligned CNT composites are usually obtained in structures of long fiber or thin film, it is difficult to measure the thermal conductivity of these composites. An off-lattice Monte Carlo model is developed to accurately predict the thermal conductivity of aligned CNT composites.

Keywords: Aligned CNT composite, CNT fiber, fiber densification, polymer infiltration, Monte Carlo model

1. Introduction

Both single-walled nanotubes (SWNTs) and multi-walled nanotubes (MWNTs), which possess remarkable multifunctional properties such as high Young's modulus of 1 TPa [1], ultrahigh thermal conductivity of 2000–3500 W/mK [2,3], and outstanding electrical conductivity of 3×10^4

S/cm [4], have attracted much attention over the past decades [5]. Carbon nanotubes (CNTs) have been considered promising effective additives for developing high-performance composites. Currently, in the widely-used approaches for fabrication of CNT-based composites, CNTs are randomly dispersed into the matrix. This approach, however, typically results in low volume fraction, poor dispersion and random orientation of CNTs in matrices, inducing very limited enhancements and much lower properties than expected. To overcome these limitations, various CNT Structures such as buckypapers [6], CNT arrays [7–9], and CNT yarns [10] have been developed to pre-arrange the CNTs prior to fabricating composites.

Among those CNT post-treatments, assembling CNTs into CNT fibers has attracted tremendous attention. In general, there are three major methods for production of CNT fibers: (1) wet-spinning from CNT/acid or polymer solutions [11–13]; (2) dry-spinning from vertically aligned CNT arrays [10,14–17]; and (3) direct-assembling from CNT aerogels formed in chemical vapor deposition (CVD) [5,18–24]. The first method is also known as the wet-spinning method, while the others are known as dry-spinning methods.

The obtained CNT fibers commonly possess satisfactory mechanical and electrical properties [5], and even higher strength and better flexibility than commercial carbon fibers and polymer fibers [25]. In order to further improve their properties (mechanical strength in particular), polymer infiltration is usually performed on the CNT fibers to obtain CNT fiber/polymer composites. The polymer can greatly enhance the inter-tube load transfer, inducing high mechanical strength of the composites.

2. Methods for assembling CNTs into CNT fibers

2.1. Spinning from CNT Solution

In 2000, Vigolo et al. [11] first fabricated CNT ribbons and fibers via the coagulation spinning approach that was widely used to synthesize polymer fibers. **Figure 1**(**a**) shows the schematic of the experimental setup used to make nanotube ribbons. In this method, SWNTs are homogeneously dispersed in a solution of sodium dodecyl sulfate (SDS), which helps prevent CNTs from agglomeration. The CNT dispersion is then injected into the co-flowing stream of a polymer solution that contains 5.0 wt.% of polyvinylalcohol (PVA) to form CNT ribbons, as shown in **Figure 1**(**b**). **Figure 1**(**e**) shows a typical SEM image of the as-obtained CNT ribbon, revealing a preferential orientation of the nanotubes along the ribbon's main axis. After the ribbons are washed and dried, most of the surfactants and polymers are removed. The ribbons are collapsed into fibers due to capillary force, as shown in **Figure 1**(**c**). These fibers are more flexible than traditional carbon fibers, as shown in **Figure 1**(**d**). The diameter of CNT fibers varies from 10 to 100 μm depending on fabrication conditions. The tensile strength, modulus and electrical conductivity of the obtained fibers are 300 MPa, 40 GPa and 10 S/cm, respectively.

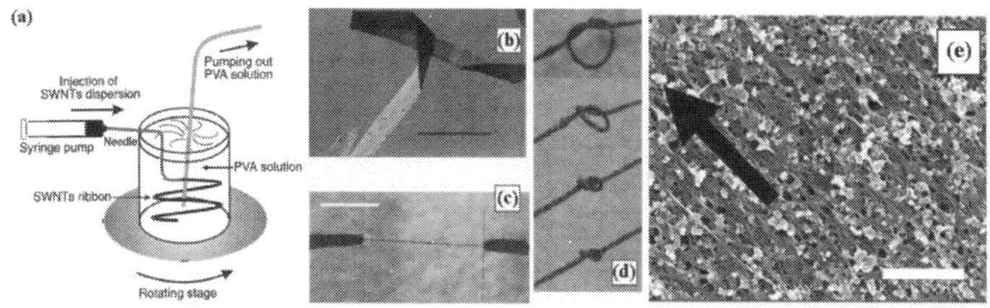

Figure 1. (a) Schematic of the experimental setup used to make nanotube ribbons. (b–d) Optical micrographs of nanotube ribbons and fibers. (b) A single folded ribbon between horizontal and vertical crossed polarizers (scale bar = 1.5 mm); (c) A freestanding nanotube fiber between two glass substrates (scale bar = 1 mm); and (d) Tying knots reveals the high flexibility and resistance to torsion of the nanotube microfibers. (e) Scanning electron micrograph shows SWNT bundles are preferentially oriented along the main axis of the ribbon (scale bar = 667 nm) [11].

Vigolo's technique is remarkable for further studies on fabricating continuous CNT fibers on a large scale, although this technique has some disadvantages. For example, the drawing of these as-spun gel fibers is slow (~1 cm/min), and the solid fibers are short (~10 cm) [26]. In addition, the fibers' mechanical properties are low compared with those of component individual nanotubes. Moreover, mechanical performance is improved, mostly because PVA chains in a CNT fiber enhance load transfer efficiency between CNTs. However, the existence of the non-conductive PVA leads to lower electrical and thermal conductivity of the obtained CNT fibers than pure CNT sheets [27]. Thus, fibers composed solely of CNTs are desirable. In 2004, Ericson et al. [12] developed a method to fabricate well-aligned macroscopic fibers composed solely of SWNTs. In this method, purified SWNTs are dispersed in 102% sulfuric acid, which charges SWNTs and promotes their ordering into an aligned phase of individual mobile SWNTs surrounded by acid anions. This ordered dispersion is extruded into continuous lengths of macroscopic neat SWNT fibers. The obtained fibers possess a Young's modulus of 120 ±10 GPa and a tensile strength of 116 ±10 MPa. Because these pure CNT fibers do not contain polymers, they demonstrate better electrical and thermal properties than fibers containing polymers, with an electrical conductivity of 500 S/cm and thermal conductivity of ~21 W/m K. In another method proposed by Behabtu et al. [13], high-quality CNTs are dissolved in chlorosulfonic acid and extruded into a coagulant (acetone or water) to remove the acid. The forming filament is further stretched and tensioned to ensure high CNT alignment in the structure. The resulting fibers possess a Young's modulus of 120 ± 50 GPa and strength of 1.0 ± 0.2 GPa. The tensile strength shows a tenfold improvement over wet-spun fibers fabricated using the method developed by Ericson et al. [12]. At the same time, they display outstanding electrical conductivity (~29000 ± 3000 S/cm) and thermal conductivity (~380 ± 15 W/m K).

2.2. Spinning from vertically aligned CNT arrays

Just like drawing a thread from a silk cocoon, CNT fibers can be synthesized from a vertically aligned CNT array. In 2002, Jiang et al. [10] spun a 30-cm-long CNT fiber from a CNT array

(~100 μm in height). In 2004, Zhang et al. [14] modified this technique by introducing twist during spinning. In this method, the nanotube arrays (~30 μm in height) are grown on an iron catalyst–coated substrate by CVD. Afterward, yarns are drawn from the array and twisted with a variable-speed motor. **Figure 2**(**a**) and (**b**) clearly show the SEM images of the structures formed during the spinning process. The obtained fibers have a tensile strength of ~460 MPa, modulus of ~30 GPa and electrical conductivity of 300 S/cm.

Since then, many efforts have been made to optimize spinning processes and to improve the performance of CNT fibers. Spinning fibers from higher CNT arrays can effectively improve fiber performance. For example, Zhang et al. [15] reported the spinning of CNT fibers from relatively long CNT arrays (0.65 mm), which resulted in the strength and Young's modulus of the CNT fibers reaching 1.91 GPa and 330 GPa, respectively. Furthermore, Li et al. [16] spun CNT fibers from 1 mm CNT arrays. Their tensile strength reached up to 3.3 GPa, which is much higher than that of CNT fibers from the 0.65 mm array. In aiming to achieve the goal of providing a continuous process for the solid-state fabrication of CNT yarns from CNT forests, Lepro et al. [17] spun fibers from CNT forests grown on both sides of highly flexible stainless steel sheets, instead of the conventionally used silicon wafers, as shown in **Figure 2**(**c**), (**d**) and (**e**). They reported that the catalyst layer is shown to be re-usable, decreasing the need for catalyst renewal during a proposed continuous process.

Figure 2. (a) and (b) SEM images of a carbon nanotube yarn in the process of being simultaneously drawn and twisted during spinning from a nanotube forest outside the SEM [14]; (c–e) Spinnable CNT forest grown on flexible stainless steel substrate [28].

Studies have found that not all CNT arrays can be spun into fibers, and the degree of spinnability of CNTs is closely related to the morphology of CNT arrays [29,30]. Several research groups have made efforts to study the mechanism of spinning fibers from CNT arrays. Kuznetsov et al. [28] developed a structural model for the drawing of sheets and fibers from CNT arrays. Huynh et al. [30] studied the roles of catalyst, substrate, temperature, gas flow rates, reaction time with acetylene, etc. to identify and understand the key parameters and develop a robust, scalable process. More recently, Zhu et al. [31] pointed out that the entangled structures at the ends of CNT bundles are critical for the continuous drawing process. Further fundamental studies of this mechanism are critical for fabricating spinnable CNT arrays and improving the properties of CNT fibers.

2.3. Spinning from CNT aerogels

In both of the above-mentioned methods, individual CNTs are first produced in the form of CNT powders or arrays. In this method, fibers are achieved through the post-process of spinning. Unlike the in-direct methods, CNT fibers can be assembled directly in a CVD process in which individual CNTs are synthesized. In 2000, Zhu et al. [18] first reported the direct synthesis of 20cm long ordered SWNTs with a diameter of approximately 0.3mm using a floating catalyst CVD method in a vertical furnace. In 2004, Li et al. [19] reported a method for the direct spinning of long CNT fibers from aerogels formed during CVD. **Figure 3**(a) is a schematic of this direct spinning process. In this method, reaction precursors are mixed and introduced into a tube furnace operated at 1200°C. In a reducing hydrogen atmosphere, the

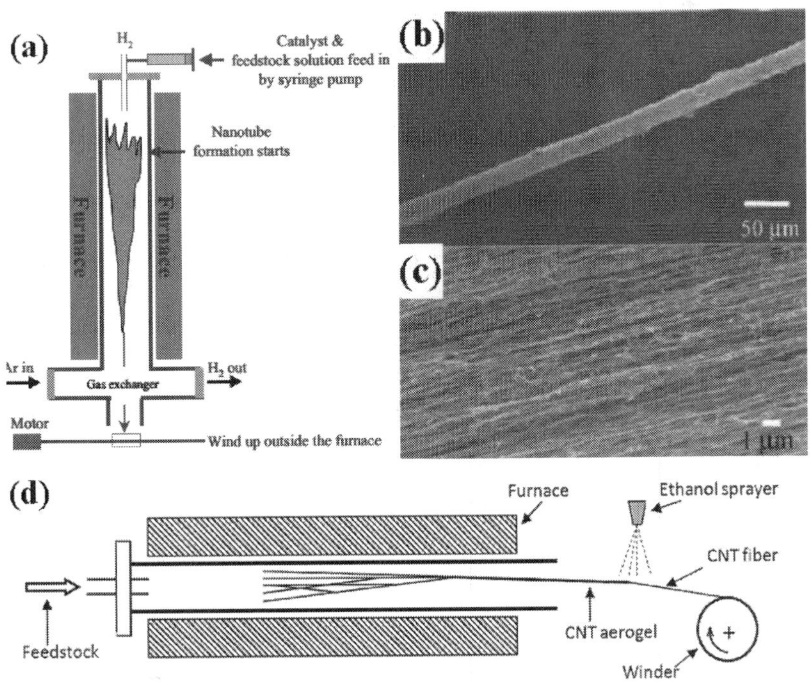

Figure 3. (a) Schematic diagram of the direct spinning process for CNT fibers [21]; (b) and (c)SEM micrographs of a fiber that consists of well-aligned MWNTs [19]; (d) Schematic diagram of the CNT fiber-spinning process using a horizontal furnace [24].

nanotubes form an aerogel in the hot zone of the furnace and are stretched into cylindrical hollow socks which are then pulled and collected continuously out of the furnace as fibers. **Figure 3(b)** and **(c)** show SEM images of the aligned CNT fibers after condensation by acetone vapor. CNT fibers, spun directly and continuously from aerogels, demonstrate both high strength (up to 357 GPa) and high stiffness (8.8 GPa), which are comparable to those of commercial fibers [20].

3. Methods for enhancing CNT alignment

The as-spun CNT fibers fabricated using the above methods usually have a porous structure, and the CNTs within the fibers have poor alignment [14,19,32]. Hence, the CNT fibers should be further densified to obtain a more closely-packed structure and better alignment of the CNTs. Since the van der Waals interaction strongly depends on the contact area between CNTs, much space and many pores between CNT bundles could lower the degree of this interaction. By applying the densification process, the densified CNT fibers can have reduced interspace between CNTs and an improved contact area, leading to an increased van der Waals interaction. As a result, these highly dense structures have a stronger van der Waals interaction between CNT bundles, hence improving fiber performance [32–34].

3.1. Enhancing CNT alignment for fibers spun from the wet-spinning technique

While classical composite fibers consist of CNTs embedded in a polymeric matrix, fibers fabricated by the wet-spinning technique consist of an interconnected network of polymers and CNTs. The spinning conditions, such as the flow velocity of the polymer solution and the injection rate of the CNT solution, has no measurable effect on the CNT orientation in the resulting fibers. However, when the fibers are immersed in an appropriate solvent or heated, the network of polymers and CNTs can be loosened and stretched, resulting in a significant improvement in CNT alignment. For example, Vigolo et al. [35] enhanced the CNT alignment of their SWNT/PVA fibers by re-wetting, swelling and re-drying the fibers vertically under a tensile load with a weight attached to the end of the fiber. The solvent used in the study was comprised of water, acetone and acetonitrile. Once re-wetted and swollen by the solvent, the fibers could be stretched up to 160% with significantly improved alignment, as shown in **Figure 4**. This indicates that the networks of CNTs and absorbed polymers form cross linked assemblies that can be elastically deformed. As a result, their strength and stiffness increased from 10 GPa and 125 MPa, to 40 GPa and 230 MPa, respectively, after the stretching.

Separately, Miaudet et al. [36] reported an increase in tensile strength of CNT/PVA fibers from 1.4 GPa to 1.8 GPa after the hot-stretched treatment. The reduction of PVA chain alignment from ±27° to as low as 4.3°, and the nanotube alignment to as low as 9°, suggested that better alignment of the CNT and PVA chains was the main reason for the improved mechanical performance of the hot-stretched fibers.

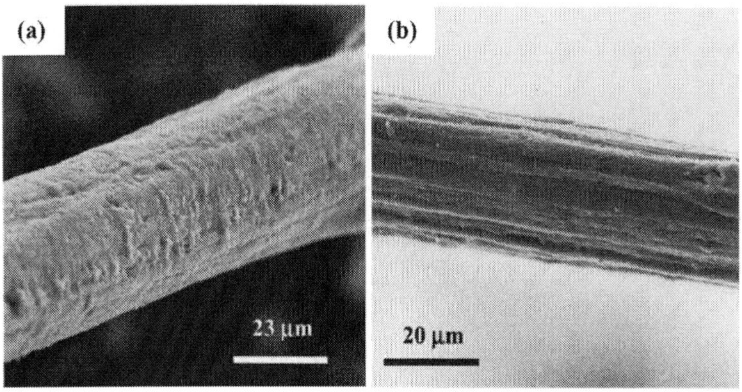

Figure 4. SEM images of an as-spun fiber (a), and (b) a stretched CNT fiber [35].

3.2. Enhancing CNT alignment for fibers spun from dry-spinning and floating catalyst techniques

CNT fibers spun from dry-spinning and floating catalyst techniques can be densified by applying a mechanical force in their lateral direction. The densification methods can be classified into two categories: indirect approaches (such as liquid densification, twisting [32], drawing through dies [37], or an aligning and tension system [38]); and direct approaches (such as rubbing/false twisting [39] and mechanical compression [40]).

3.2.1. Indirect approaches

3.2.1.1. Liquid densification

The alignment and mechanical performance of as-spun CNT fibers can be improved through liquid densification. In this method, a liquid such as acetone or ethanol is absorbed into the fibers and subsequently evaporated, resulting in a dense CNT structure. The fibers are densified due to the surface tension of the solvent and the fiber diameter is reduced accordingly. The densification process slightly improves nanotube alignment (**Figure 5**) and enhances load transfer between nanotubes, ensuring that most of them are fully load-bearing. Liu et al. [41] studied the mechanical properties of twisted fibers with and without acetone densification. The diameter of the yarn changed from 11.5 to 9.7 μm after shrinking. Although the maximum strain of the yarn remained unchanged (~2.3%), the Young's modulus (~56 GPa) of the shrunk fiber was slightly greater than before shrinking (~48 GPa). Liu et al. [41] also reported a change in diameter and maximum load of twisted fibers before and after shrinking. After acetone shrinking, the diameter reduction ranged from 15% to 24%, and the load increase ranged from 15 to 40%, indicating that tensile strength was enhanced. Among common solvents (water, ethanol and acetone) used to shrink CNT yarns, acetone showed the best shrinking effect.

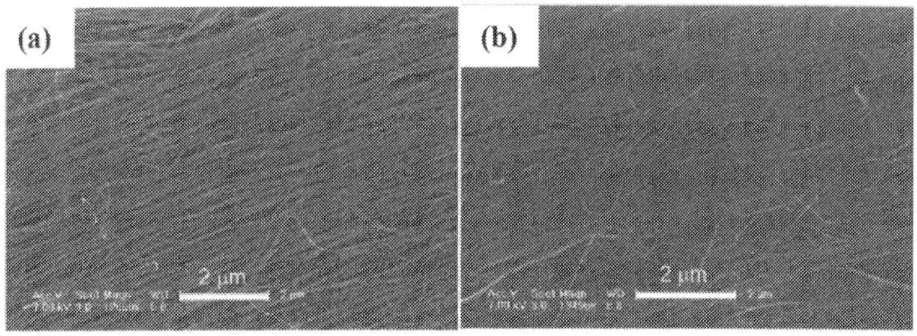

Figure 5. (a) SEM images of a twisted fiber before, and after shrinking (b) [41].

3.2.1.2. Twisting

The as-spun fiber is relatively loose with noticeable spaces between CNT bundles. Increasing the twist angle is an effective method for densifying CNT fibers. Since it brings CNTs into closer contact with each other, twisting improves the friction coefficient μ between CNTs, therefore contributing positively to fiber strength [15]. Zhang et al. [15] compared the tensile behaviors of twisted and untwisted CNT fibers spun from the same 650 mm height array. After twisting, the diameter of the fiber decreased from 4 to 3 μm (**Figure 6**), while tensile strength increased from 0.85 GPa to 1.9 GPa [15].

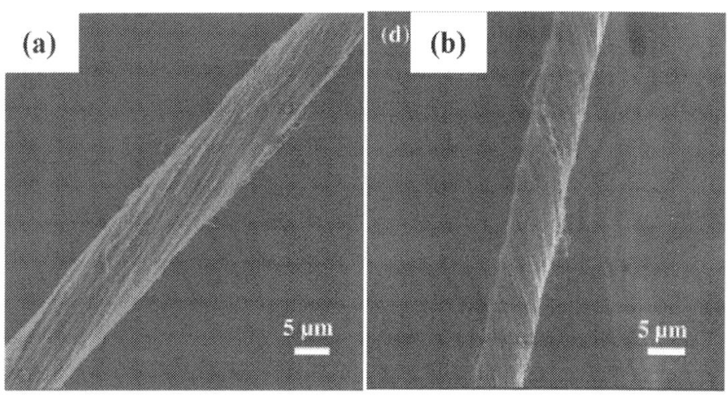

Figure 6. SEM images of as-spun CNT fiber (a), and the same fiber after post-spin twisting (b) [15].

3.2.1.3. Drawing through dies

The as-spun CNT fibers can be densified by being drawn through dies of different diameters. The average measured fiber diameter was determined by die size. Sugano et al. [37] densified CNT fibers spun from CNT arrays by drawing them through densifying dies with different diameters (d = 30, 35, 55, 75 μm) (**Figure 7(a)**). The fibers were deformed elastically after drawing CNTs through the die, and their density increased with decreasing die diameter. As CNT fibers are held together by van der Waals forces between MWNTs, these forces increased due to higher apparent density with decreasing distance between MWNTs, as shown in **Figure 7(b)**. As a result, their strength was significantly enhanced after treatment.

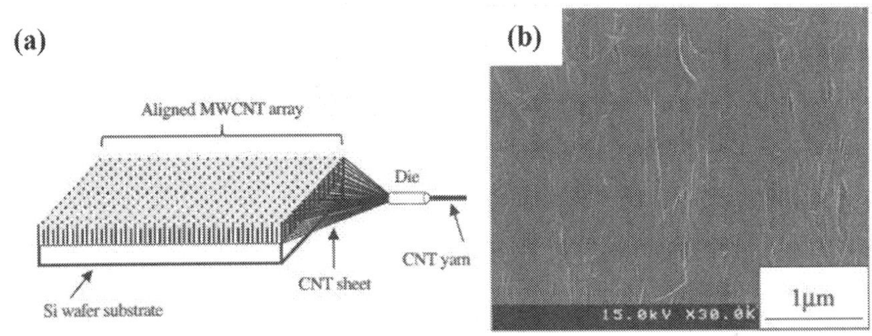

Figure 7. (a) Schematic view of untwisted CNT yarn in the process of being drawn from the aligned MWNT array and past the sheet through a die; (b) SEM image of surface morphologies of the resulting CNT fiber.

3.2.1.4. Aligning and tension system

The aligning and tension system is one of the most effective methods of enhancing CNT alignment and performance of CNT fibers. Tran et al. [38] first modified the traditional dry-spinning process to improve CNT alignment of their CNT fibers (**Figure 8(a)**). In this modified system, a capstan effect rod system (CERS) is added to a dry-spinning system to regulate tension and torque to the fibers. As the fibers pass through a CERS, the increased tension extends and aligns the bundles (**Figure 8(b)** and **8(c)**). This process has two effects: (i) aligning CNTs in the fibers during the initial tensioning; and (ii) condensing the CNT bundles. The first effect increases contact length between bundles, and the second effect reduces the distance between CNTs. The significant increase in fiber strength from 0.45 to 1.2 GPa after the treatment is due to better alignment of the fiber bundles and higher fiber compaction.

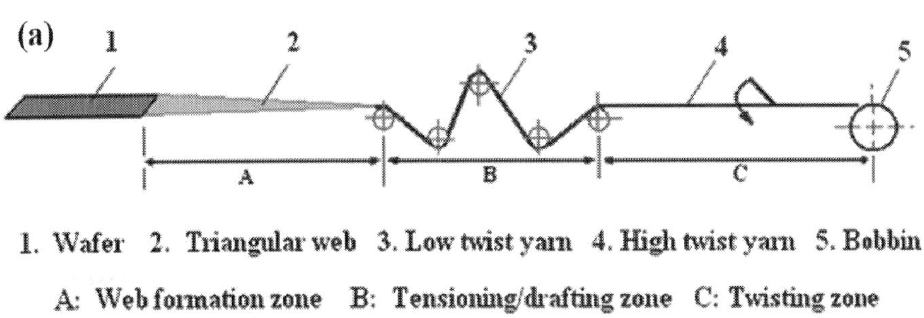

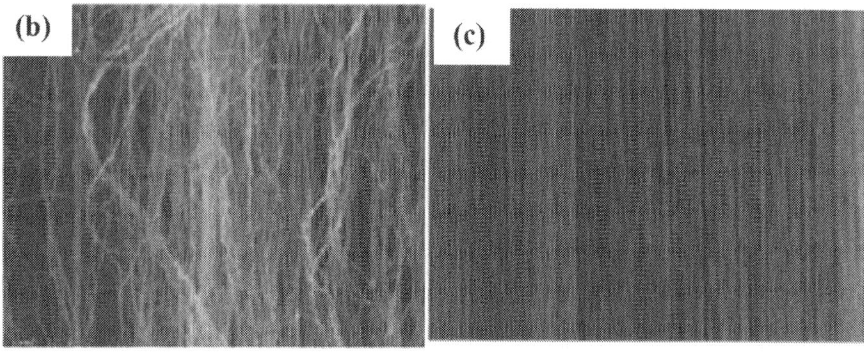

Figure 8. (a) The schematic of modified CNT yarn spinning; (b) SEM images of surface morphologies of CNT fiber spun from traditional process; and modified process (c) [38].

Generally, the drawback of the indirect approaches is their low densifying forces. The liquid densification method, for example, employs the surface tension of volatile solvents such as acetone or ethanol to densify the CNT fibers. Its densifying force is therefore limited by the low surface tension of the solvents used [32]. Similarly, the compressive force produced by drawing CNT fibers through a die results from the drawing forces and the die size used. These drawing forces are limited by fiber strength, while a significantly smaller die could damage the fiber structure, resulting in poor strength [37]. Therefore, CNT fibers cannot be adequately densified with these methods, and their performance remains unsatisfactory.

3.2.2. *Direct approaches*

Direct approaches are considered the best solution to overcome the above limitations. As the densifying forces are applied directly to CNT fibers, the forces can condense the fibers into a much denser structure [40].

3.2.2.1. *Rubbing/false twisting*

CNT fibers can be densified using several traditional textile twistless methods such as rubbing. Miao et al. [39] used a rubbing roller system (**Figure 9(a)**) to densify CNT web drawn from a vertically aligned CNT forest into a compact twistless yarn. As the system used a constant rate false twisting process, there is only a temporary twist on the incoming side of the yarn, and the yarn on the outgoing side (and thus the final yarn) will be twistless. As can be seen in **Figure 9(b)**, the resulting yarn consists of a high packing density sheath with CNTs lying straight and parallel to the yarn axis, and a low density core with many microscopic voids. With an increased contact length between CNT bundles in the high packing density sheath, the mechanical performance of the core-sheath structured, twistless carbon nanotube yarns are significantly higher than that of their corresponding twist-densified yarns.

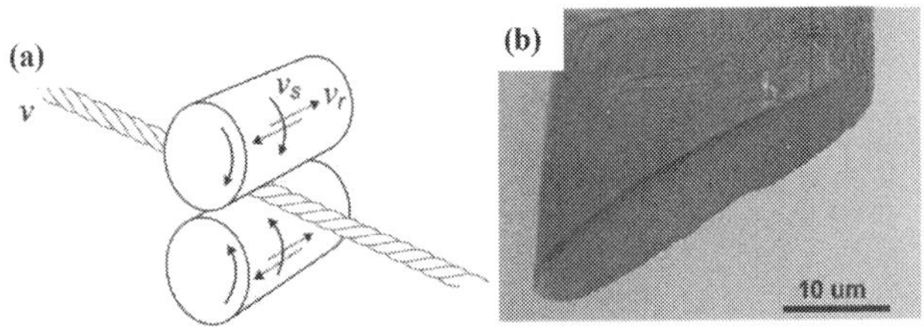

Figure 9. (a) The schematic of the core-sheath, twistless CNT yarn fabricated by a rubbing roller system; and (b) SEM image of the resulting yarn.

3.2.2.2. Mechanical compression

Wang et al. [40] reported that CNT fibers densified by the pressurized rolling system showed highly packed structures with a densification factor of up to 10 (**Figure 10(a)**). Moreover, the densified fibers can reach an impressive average strength of 4.34 GPa, which is the highest extrinsic CNT fiber strength reported to date [40]. In addition, Tran et al. [24] presented a modified densification method to produce a highly packed CNT structure. As shown in **Figure 10(b)**, CNT fibers were sandwiched between two sheets of A4 paper and pressed by a stainless-steel spatula with an applied force of approximately 100 *N*, at 45° to the fiber axis. The spatula was subsequently slid across the A4 paper, along the fiber axis, to compress and mechanically densify the fibers into a ribbon shape while the compressive force was maintained. The CNT ribbon in this study also showed a densification factor of up to 10, while the strength and electrical conductivity of the densified fibers approached impressive values of 2.81 GPa and 12,000 S/cm, respectively. The study showed that the mechanical densification treatment may have increased the CNT bundle size and inter-CNT contact, and induced better alignment (**Figure 10(c)** and **(d)**), resulting in improved properties of the densified CNT fiber.

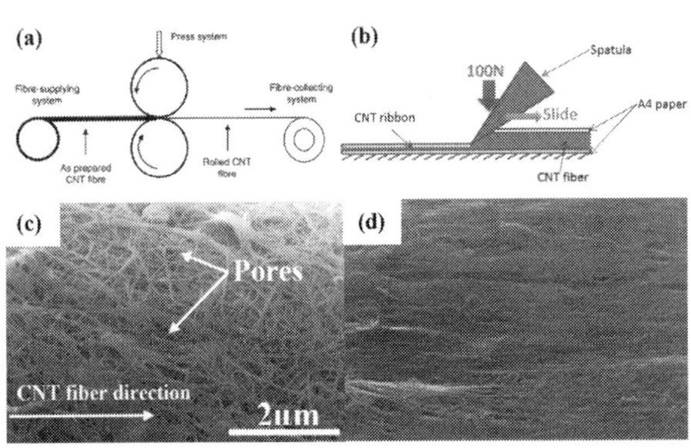

Figure 10. The schematic of mechanical densification methods using (a) pressurized rolling system, and (b) spatula; SEM images of the surface morphology of the (c) as-spun CNT fiber, and (d) densified CNT ribbons [24].

4. Mechanical and physical properties of aligned CNT polymer composites

The outstanding physical and mechanical properties of aligned CNT assemblies make them promising for the research and development of high-performance composites. The final properties of the composites are affected by many factors, such as, morphology of individual nanotubes and impregnating method.

4.1. Impregnating method

Pure CNT assemblies including fibers and films have an ineffective load transfer between CNTs as the CNTs interact with each other via weak van der Waals forces. Among all methods used to enhance load transfer between CNTs, polymer impregnation is one of the most effective treatments in enhancing the mechanical properties of CNT assemblies. This reinforcement mainly stems from the enhanced inter-tube load transfer and the crystallinity of the impregnated polymer. Several impregnating methods are used to fabricate aligned CNT polymer composites.

4.1.1. Dip coating/soaking

Dip coating or soaking is widely used to impregnate polymer into CNT assemblies. In this method, CNT assembles are immerged into polymer solution for sufficient infiltration and then taken out for curing. Liu et al. [32] reports the mechanical properties of PVA impregnated fibers spun from CNT arrays. **Figure 11** shows the schematic of the CNT/polymer fiber manufacturing process and compares the tensile properties of a CNT/PVA fiber with two types of pure CNT fibers. As can be seen, the CNT/PVA composite fiber with 19 wt.% PVA possesses a tensile strength of 1.95 GPa. This result is 255% higher than that of simply twisting the CNT fiber and 103% higher than that of a CNT fiber subjected to twisting and shrinking by acetone. The greater strength of the CNT/PVA fiber stems from the decrease in fiber diameter due to the high wettability between dimethyl sulfoxide (DMSO) and CNTs, and the increase in tensile load due to improved load transfer efficiency between CNTs after PVA impregnation (**Figure 11(b)**).

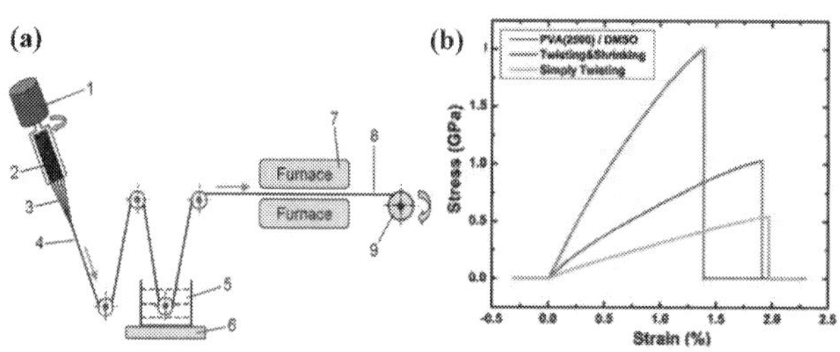

Figure 11. (a) Schematic of the CNT/polymer fiber manufacturing process; and (b) stress-strain curves of a typical SWNT/PVA yarn and two types of pure SWNT yarns. [32]

Similarly, Tran et al. [24] reported an outstanding enhancement of the electrical and mechanical performances of MWNT fibers through the combined treatments of mechanical densification and epoxy infiltration. Compared to the mechanical performances of CNT fibers produced by different post-treatments, the combined post-treatments employed in their study showed better effects, with enhancement factors of more than 13.5 for tensile strength and 63 for stiffness. After the first mechanical treatment, their condensed CNT ribbons achieved a tensile strength much greater than that of the best CNT fibers spun with wet-spinning and array-spinning methods, as shown in **Figure 12**. When further combined with epoxy infiltration, the CNT/epoxy ribbons reached significantly greater strength (up to 5.2 GPa) and stiffness (up to 444 GPa), which are very comparable to those of commercial PAN carbon fibers as shown in **Figure 12**. Furthermore, while the strength of their CNT/epoxy ribbons was comparable to that of the best double-walled CNT (DWNT) ribbons produced by the floating-catalyst method [40], their stiffness was much higher. The results suggest that by using a polymer infiltration treatment, the performance of MWNT fibers with low electrical and mechanical properties could achieve the performance of many other high-strength fibers.

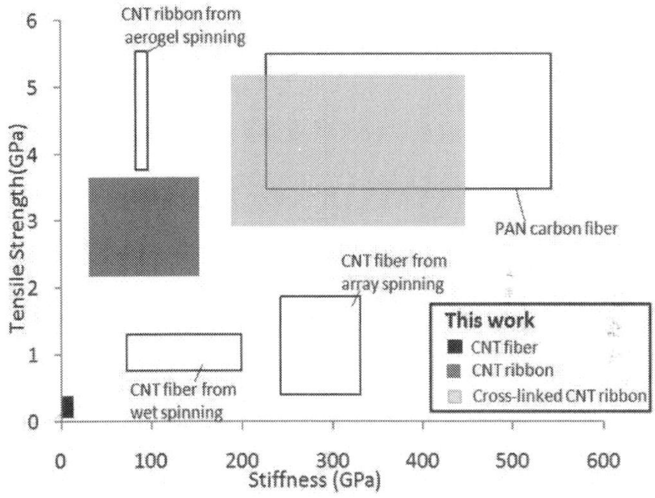

Figure 12. Comparisons of mechanical properties of the best CNT fibers from array-spinning [15] and wet-spinning [13], ribbons from aerogel spinning, and PAN carbon fibers [40].

Liu et al. [23] fabricated CNT/polyimide aerogel (CNT/PIA) composite fibers by dip-coating the CNT fibers in a sol solution, and then drying them using the supercritical CO_2 drying process. In the CNT/PIA composite fibers, CNT fibers are tightly wrapped by porous polyimide aerogel, showing a core-shell structure. This core-shell structure resulted in the light weight, low density and high surface area of the composite fiber. Owing to the superior properties of the CNT fibers (stiffness of ~450 MPa, tensile strength of ~83 MPa and electrical conductivity of ~419 S/cm), the CNT/PIA composite fibers achieve significant enhancements in mechanical and electrical properties (stiffness of ~68.1 MPa, strength of ~11.6 MPa and electrical conductivity of ~418 S/cm), compared with the pure PIA and other CNT/PIA composite [42–46]. It was also found that the mechanical and electrical properties of CNT/PIA composite fibers decline with an increase in the diameter of CNT fibers.

4.1.2. Resin transfer molding

Resin transfer molding (RTM) is a very common and cost-effective method to fabricate composites in industries, in which the liquid resins are first injected to the preforms and then cured to be solid. Given its capability of making composites with large sizes and complex shapes, RTM is expected to be appropriate for preparing CNT/epoxy composites in large scale. Liu et al. [5] developed aligned CNT/epoxy composite films by combining layer-by-layer and vacuum-assisted RTM (VA-RTM) method using direct chemical vapor deposition (CVD)-spun CNT plies. The CNTs in the plies are well-condensed during the VA-RTM process (**Figure 13(a)**), leading to much higher mass fractions of CNTs (up to 24.4 wt.%) compared with those obtained from the conventional dispersion methods. Due to good alignment of the condensed CNTs in the plies, the CNT/epoxy composite with 24.4 wt.% CNTs achieves ~5 and ~10 times enhancements in their Young's modulus and strength, respectively. A high tensile toughness of up to 6.39×10^3 kJ/m^3 was also obtained in the composite films, making them promising candidates for protective materials, as shown in **Figure 13(b)**. The electrical conductivity of the aligned CNT/epoxy composites reaches as high as 252.8 S/cm for the composite with 24.4 wt.% CNTs, which is ~20 times greater than that of the CNT/epoxy composites obtained using dispersion methods [47–49].

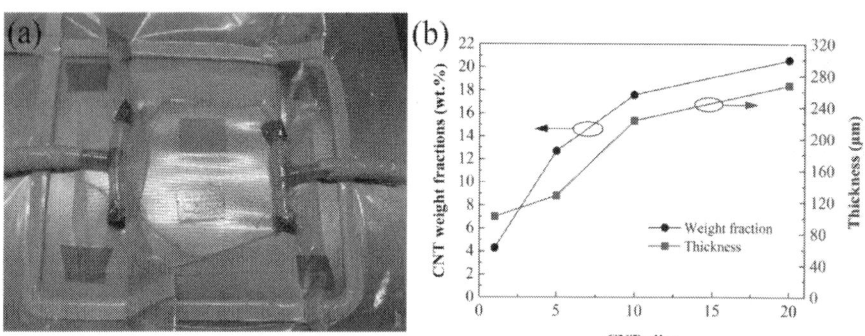

Figure 13. (a) Experimental set-up of the RTM process for preparing CNT/epoxy films; and (b) The CNT weight fractions and the thickness of CNT/epoxy composite films as a function of CNT plies. [5].

4.1.3. Spray winding and layer-by-layer deposition

The aforementioned methods are mostly regarded as off-line methods in which highly packed CNT assembles are used as preforms. Because the preforms are already tightly packed,

http://dx.doi.org/10.5772/62510

however, these methods often have the problem of the uniformity of infiltration. As a result, the un-infiltrated part may become defects and limit the overall performance of the composites [50]. In order to control the uniformity of infiltration and avoid over-infiltration, Liu et al. [50, 51] developed a one-step approach of "spray winding" to fabricate high-performance CNT composites. In this on-line infiltration method, CNT sheets, drawn out from CNT arrays, were continuously collected (wound) onto a rotating mandrel under the spray of a polymer solution, as shown in **Figure 14(a)**. The spray-wound CNT/PVA composite films, the CNT fraction is tunable, and could be as high as 65 wt.% to reach the best mechanical properties. The best film had the tensile strength, stiffness and toughness up to 1.8 GPa, 45 GPa, and 100 J/g, respectively, much better than the fibers made by the same CNT and PVA and many other CNT/polymer composites. The high performance can be attributed to the long CNTs, highly-aligned tube morphology, and good interfacial bonding between CNT and PVA, which were obtained simultaneously. In order to control the exact layers of the composite films in large scale, Zhang et al. [52] reported a "layer-by-layer (LBL) deposition" method to produce CNT polymer composites, as shown in **Figure 12(b)**. This on-line deposition method allowed PVA to infiltrate into the CNT film efficiently, resulting in a remarkable improvement in the mechanical property of CNT/PVA composite. The composite film possessed a tensile strength of 1.7 GPa, which is almost one order of magnitude and 20 times higher than those of the pure CNT and PVA films, respectively.

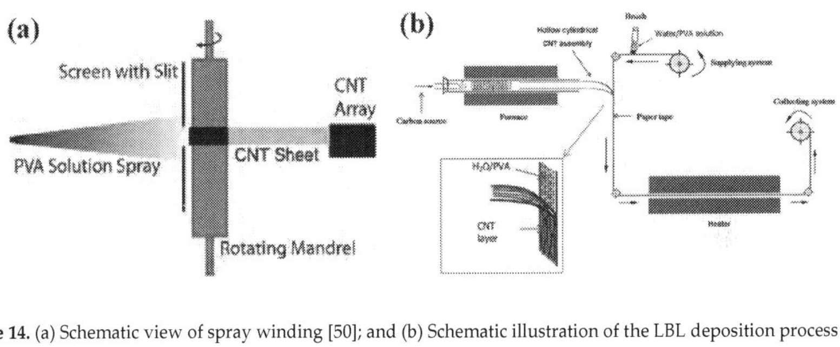

Figure 14. (a) Schematic view of spray winding [50]; and (b) Schematic illustration of the LBL deposition process [52].

4.2. Effect of CNT morphology

Individual nanotube morphologies, such as length and alignment, have great influence on mechanical and physical properties of CNT polymer composites. Wang et al. [53] reported an ultrastrong and stiff CNT/composite using a stretch-winding process. The unstretched composites exhibited strength of 2.0 GPa, Young's modulus of 130 GPa and electrical conductivity of 820 S/cm. After stretching the strength, Young's modulus and electrical conductivity were increased to as high as 3.8GPa, 293 GPa and 1230 S/cm, respectively. These remarkable improvements can be ascribed to the enhancement of CNT alignment and decreasing of waviness. The alignment of the CNTs was characterized by polarized Raman. Specifically, the shift of the intensity ratio ($IG_{||}/IG_{\perp}$) of G band peaks was measured. Following the stretch-winding process, the intensity ratio for the 12%-stretched sheet is increased to 7.6 from 1.6, indicating that alignment of the CNTs in the nanocomposites was significantly improve via the stretching process. Therefore, the improved CNT alignment is correlated with the observed improvements in mechanical and electrical properties of the composites.

Park et al. [54] studied the effects of nanotube length and alignment on thermal conductivity of MWNT/epoxy composites. It was found that the long-MWCNT composites exhibited higher thermal conductivity than the short-MWCNT composites with the same weight percentages. At room temperature, 10 wt.% short-MWCNT/epoxy composite showed thermal conductivity of 0.35 W/mK, while the long-MWCNT composites showed 2.6 W/mK even at lower concentration of 6.38 wt.%. To improve the in-plane thermal conductivity, CNT sheets (60 wt.%) were stretched mechanically. The thermal conductivity increased up to 83 W/mK (25% stretched) and 103 W/mK (40% stretched) along the alignment direction compared to 55 W/mK of the random sample.

5. Mesoscopic modeling of thermal conduction in aligned CNT composites

Due to the unique morphology of aligned CNT composites, it is difficult to directly measure their thermal conductivity, especially for composites in thin film and long fiber structures. Proper computational modeling is required to accurately predict the thermal conductivity of aligned CNT composites. The widely-used effective medium theories (EMTs) can well predict the thermal conductivity of CNT composites obtained using dispersion methods. However, the EMTs generally fail to predict the thermal conductivity of aligned CNT composites, since they cannot take into account the complex morphology of CNTs and the thermal boundary resistances (TBRs) at both CNT-CNT and CNT-matrix interfaces [55]. The TBRs are the resistances to the heat flow at interfaces, which have been regarded as the bottleneck of thermal conduction in CNT composites [56].

5.1. Thermal conduction model for two-phase aligned CNT composites

In order to accurately predict the thermal conductivity of aligned CNT composites, Duong et al. [57] developed an off-lattice Monte Carlo (MC) approach by quantifying thermal energy through a large quantity of random thermal walkers. Thermal walkers have a random Brownian motion in the polymer matrix, which is described by the position changes of thermal walkers in each direction. The position changes take values from a normal distribution with a zero mean and a standard deviation, as expressed as below:

$$\sigma = \sqrt{2D_{\mathrm{m}}\Delta t} \tag{1}$$

where D_{m} is the thermal diffusivity of polymer matrices, and Δt is the time increment of the simulation [58]. When a walker jumps to the CNT-polymer matrix interface, it may jump into the CNT with a probability $f_{\text{m-CNT}}$, or remain within the polymer matrix with a probability (1-

$f_{m\text{-}CNT}$). The probability is a function of the TBR between polymer and CNT, $R_{m\text{-}CNT}$, obtained using the acoustic mismatch theory (AMT) [59]:

$$f_{m-CNT} = 4/(\rho C_p v R_{m\text{-}CNT}) \tag{2}$$

where ρ is the density of polymer, C_p is the specific heat of polymer, and v is the sound velocity in the polymer matrix. Due to the ballistic phonon transport and ultrahigh thermal conductivity in the CNT [60], thermal walkers are assumed to travel at an infinite speed inside the SWNT [61]. The walker is allowed to exit from a CNT to the matrix by using another probability $f_{CNT\text{-}m}$, which is related to $f_{m\text{-}CNT}$ in a way that satisfies the second thermodynamic theorem [62, 63]:

$$V_{CNT} f_{CNT-m} = C_{f-CNT} \sigma_m A_{CNT} f_{m-CNT} \tag{3}$$

wherein V_{CNT} and A_{CNT} are the volume and surface area of a CNT, and σ_m is the standard deviation of Brownian motion in the polymer matrix. $C_{f\text{-}CNT}$ is the thermal equilibrium factor at the polymer-CNT interface, which is dependent on the geometry of the CNTs, and the interfacial area between the CNT and the matrix.

By using the developed MC approach, Duong et al modeled SWNT-epoxy and SWNT-polymethyl methacrylate (PMMA) composites [64]. The thermal conductivity of SWNT-epoxy and SWNT-PMMA composites were accurately predicted. The effects of the SWNT orientation, weight fraction and TBRs on the thermal conductivity of composites were quantified. The quantitative findings showed that in SWNT-PMMA composites with 1.0wt.% of SWNT loading, aligned SWNTs achieved enhanced thermal conductivity 15 times higher than that of PMMA, whereas, the randomly dispersed SWNTs only resulted in thermal conductivity ~5 times higher than that of PMMA [64]. This indicated the superiority of aligned CNT composites.

Since CNTs are normally grown into forests or spun into fibers, the contacts between CNTs may play a significant role to modify the thermal conductivity of composites. Duong et al. then modified their model to study the effect of CNT-CNT contacts on the thermal conductivity of both SWNT-epoxy and MWNT-epoxy composites [65,66]. A representative volume element (RVE) was built based on the real CNT-epoxy composites, as shown in **Figure 15**. In MWNT-epoxy composites with 20 vol % of MWNT loading, aligned MWNTs without contacts achieved a thermal conductivity nearly 40 times higher than that of epoxy, while, aligned MWNTs with contacts induced a thermal conductivity only 20 times higher than that of epoxy. This indicated that CNT-CNT contacts in aligned CNTs may reduce the thermal conductivity of composites. The anisotropic thermal conductivity of aligned CNT composites was also quantified. In both SWNT-epoxy and MWNT-epoxy composites, the thermal conductivity parallel to the aligned CNTs was much higher than that perpendicular to the aligned CNTs. The SWNT-epoxy composites had more significantly anisotropic thermal conduction than MWNT-epoxy composites.

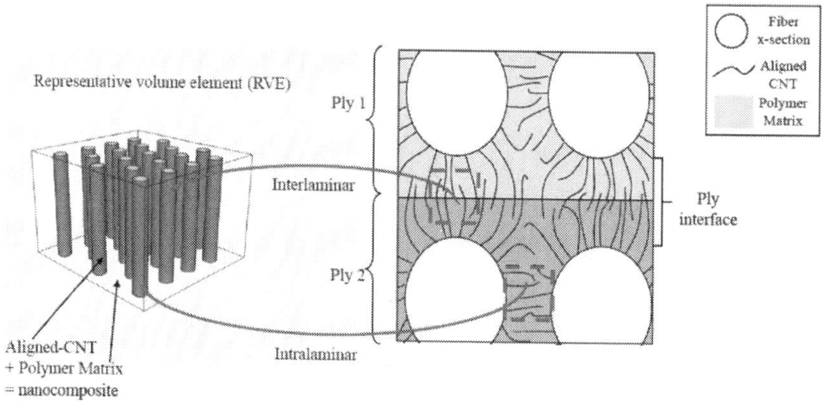

Figure 15. Schematic drawing of CNT reinforced composites, and aligned CNTs in a polymer as a representative volume element [65].

Bui et al. modified Duong's approach to investigate the thermal behavior of the SWNT-polystyrene (PS) composites at different volume fractions and at various temperature [67]. It was found that the thermal conductivity of SWNT-PS composites increased with the temperature rise. By validating with experimental data [68], the TBRs at both SWNT-PS and SWNT-SWNT interfaces were estimated by using the MC approach. The calculations at various temperature showed that the TBR between SWNT and PS increased with the rise of temperature. A TBR value of SWNT-SWNT was estimated to be 12.15×10^{-8} m^2K/W at 300K, which was higher than that between SWNT and PS (2.21×10^{-8} m^2K/W). Bui et al. also conducted the comparable study between graphene-polymer composites and SWNT-polymer composites [69]. The quantitative results showed that graphene nanosheets were more effective than SWNTs to enhance the thermal conduction in polymer composites.

5.2. Thermal conduction model for multi-phase aligned CNT composites

Recently, multiphase polymer composites, which contain more than one type of additive in the matrix, have attracted much attention [70,71]. The multiphase composites can combine the merits of all the components, inducing advanced multifunctional properties. Diverse multiphase polymer composites have been developed, such as CNT/nanoparticle/polymer composites [70,71], CNT/graphene/polymer composites [72,73], CNT/fiber/polymer composites [74,75] and CNT-stabilized polymer blends [76]. Since there is no effective approach to study thermal properties of CNT multiphase composites, Gong et al. [77] extended Duong's MC approach to investigate heat transfer phenomena in CNT multiphase composites. In Gong's model, a three-phase poly(ether ether ketone) (PEEK) composite containing SWNTs and tungsten disulfide (WS_2) nanoparticles was chosen as a case study, as shown in **Figure 16(a)** [78]. The TBRs at all interfaces (i.e. SWNT-PEEK, WS_2-PEEK and SWNT-SWNT) were taken into account in their model. The results showed that the thermal conductivity of multiphase composites increased when the CNT concentration increased, and when the TBRs of CNT-PEEK and WS_2-PEEK interfaces decreased. The thermal conductivity of composites with CNTs aligned parallel to the heat flux was enhanced ~2.7 times relative to that of composites with randomly-dispersed CNTs.

The model could also quantitatively study the effect of the complex morphology and dispersion of SWNTs, (e.g., individual and bundled SWNTs, the number of SWNT bundles, and the number of SWNTs per bundle), on the thermal conductivity of multiphase composites. It was found that the TBR at the SWNT-SWNT interface played a significant role in the thermal conduction of the composite with SWNT bundles. As presented in **Figure 16(b)**, the thermal conductivity of the three-phase composite decreased with the rise of SWNT-SWNT TBR. A critical SWNT-SWNT TBR was found to be 0.155×10^{-8} m^2K/W, which dominated the heat transfer mechanism in the three-phase composite. Proper treatment may be used to reduce the SWNT-SWNT TBR, such as the condensation of SWNT fibers, to enhance the thermal conductivity of multiphase composites with aligned SWNTs [78]. Besides the CNT multiphase composites, Gong et al. also modified the MC model to study the thermal conduction mechanisms in graphene composites [79] and CNT aerogels [80], as well as the multiphase biological systems containing CNTs [77,81,82], which indicated that the MC approach may be applicable for modeling heat transfer in diverse aligned CNT composite systems.

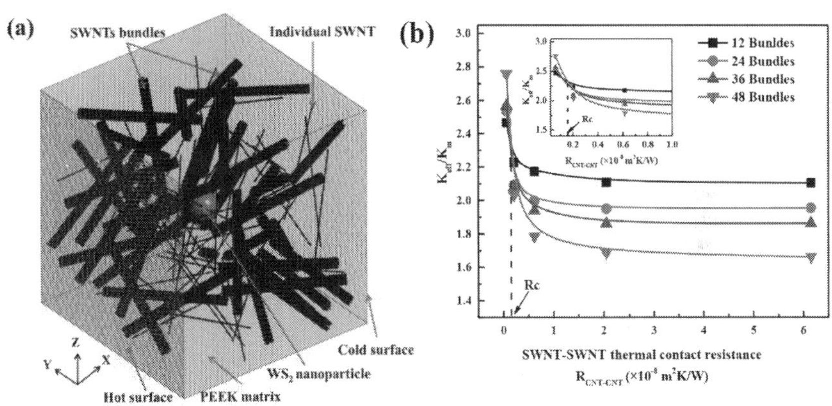

Figure 16. (a) Schematic plot of the computational model of the SWNT/WS_2/PEEK composites with SWNT bundles; (b) Effect of the SWNT-SWNT TBR on the thermal conductivity of the SWNT/WS_2/PEEK composites [78].

6. Conclusions and outlook

Both the CNT fibers and their polymer composites fabricated using the methods outlined in this article attain superior mechanical, electrical and thermal properties compared with CNT composites fabricated using the conventional methods. Their advanced properties make them promising candidates for diverse applications, such as protective materials in airplanes and electrode materials in energy storage devices. For the diverse industrial applications of the aligned CNT composites, more studies should be carried out to fabricate the composites on a large scale and at low cost. New synthesis approaches can be developed to control the diameter of composite fibers and the size of composite films. To enhance their mechanical properties, cross linking should be created within CNT fibers through proper post-treatments. Chemical compositions and fabrication conditions require optimization for better polymer infiltration into the aligned CNT fibers, to achieve enhanced properties of the aligned CNT composites.

Author details

Hai M. Duong[1*], Feng Gong[1,2], Peng Liu[1] and Thang Q. Tran[1]

*Address all correspondence to: mepdhm@nus.edu.sg

1 Department of Mechanical Engineering, National University of Singapore, Singapore

2 School of Chemical, Biological and Materials Engineering, University of Oklahoma, USA

References

[1] Demczyk BG, Wang YM, Cumings J, Hetman M, Han W, Zettl A, et al. Direct mechanical measurement of the tensile strength and elastic modulus of multiwalled carbon nanotubes. Mater Sci Eng A. 2002;334(1–2):173-8.

[2] Pop E, Mann D, Wang Q, Goodson KE, Dai HJ. Thermal conductance of an individual single-wall carbon nanotube above room temperature. Nano Lett. 2006;6(1):96-100.

[3] Kim P, Shi L, Majumdar A, McEuen PL. Thermal transport measurements of individual multiwalled nanotubes. Phys Rev Lett. 2001;87(21):215502.

[4] Li QW, Li Y, Zhang XF, Chikkannanavar SB, Zhao YH, Dangelewicz AM, et al. Structure-dependent electrical properties of carbon nanotube fibers. Adv Mater. 2007;19(20):3358-63.

[5] Liu P, Lam A, Fan Z, Tran TQ, Duong HM. Advanced multifunctional properties of aligned carbon nanotube-epoxy thin film composites. Mater Design. 2015;87:600-5.

[6] Vennerberg D, Kessler MR. Anisotropic buckypaper through shear-induced mechanical alignment of carbon nanotubes in water. Carbon. 2014;80:433-9.

[7] Wardle BL, Saito DS, García EJ, Hart AJ, de Villoria RG, Verploegen EA. Fabrication and characterization of ultrahigh-volume- fraction aligned carbon nanotube–polymer composites. Adv Mater. 2008;20(14):2707-14.

[8] Roberto Guzmán de V, Namiko Y, Antonio M, Brian LW. Multi-physics damage sensing in nano-engineered structural composites. Nanotechnology. 2011;22(18): 185502.

[9] García EJ, Hart AJ, Wardle BL, Slocum AH. Fabrication and nanocompression testing of aligned carbon-nanotube–polymer nanocomposites. Adv Mater. 2007;19(16):2151-6.

[10] Jiang K, Li Q, Fan S. Nanotechnology: spinning continuous carbon nanotube yarns. Nature. 2002;419(6909):801.

[11] Vigolo B, Pénicaud A, Coulon C, Sauder C, Pailler R, Journet C, et al. Macroscopic fibers and ribbons of oriented carbon nanotubes. Science. 2000;290(5495):1331-4.

[12] Ericson LM, Fan H, Peng H, Davis VA, Zhou W, Sulpizio J, et al. Macroscopic, neat, single-walled carbon nanotube fibers. Science. 2004;305(5689):1447-50.

[13] Behabtu N, Young CC, Tsentalovich DE, Kleinerman O, Wang X, Ma AW, et al. Strong, light, multifunctional fibers of carbon nanotubes with ultrahigh conductivity. Science. 2013;339(6116):182-6.

[14] Zhang M, Atkinson KR, Baughman RH. Multifunctional carbon nanotube yarns by downsizing an ancient technology. Science. 2004;306(5700):1358-61.

[15] Zhang X, Li Q, Tu Y, Li Y, Coulter JY, Zheng L, et al. Strong carbon-nanotube fibers spun from long carbon-nanotube arrays. Small. 2007;3(2):244-8.

[16] Li Q, Zhang X, DePaula RF, Zheng L, Zhao Y, Stan L, et al. Sustained growth of ultralong carbon nanotube arrays for fiber spinning. Adv Mater- Deerfield Beach Then Weinheim. 2006;18(23):3160.

[17] Lepro X, Lima MD, Baughman RH. Spinnable carbon nanotube forests grown on thin, flexible metallic substrates. Carbon. 2010;48(12):3621-7.

[18] Zhu H, Xu C, Wu D, Wei B, Vajtai R, Ajayan P. Direct synthesis of long single-walled carbon nanotube strands. Science. 2002;296(5569):884-6.

[19] Li Y-L, Kinloch IA, Windle AH. Direct spinning of carbon nanotube fibers from chemical vapor deposition synthesis. Science. 2004;304(5668):276-8.

[20] Vilatela JJ, Windle AH. Yarn-like carbon nanotube fibers. Adv Mater. 2010;22(44): 4959-63.

[21] Motta M, Moisala A, Kinloch IA, Windle AH. High performance fibres from 'dog bone'carbon nanotubes. Adv Mater. 2007;19(21):3721-6.

[22] Wang J, Luo X, Wu T, Chen Y. High-strength carbon nanotube fibre-like ribbon with high ductility and high electrical conductivity. Nature communications. 2014;5. 3848. (DOI:10.1038/ncomms4848)

[23] Liu P, Tran TQ, Fan Z, Duong HM. Formation mechanisms and morphological effects on multi-properties of carbon nanotube fibers and their polyimide aerogel-coated composites. Compos Sci Technol. 2015;117:114-20.

[24] Tran TQ, Fan Z, Liu P, Myint SM, Duong HM. Super-strong and highly conductive carbon nanotube ribbons from post-treatment methods. Carbon. 2016;99:407-15.

[25] Koziol K, Vilatela J, Moisala A, Motta M, Cunniff P, Sennett M, et al. High-performance carbon nanotube fiber. Science. 2007;318(5858):1892-5.

[26] HowardáEbron V. Continuous carbon nanotube composite fibers: properties, potential applications, and problems. J Mater Chem. 2004;14(1):1-3.

[27] Kozlov ME, Capps RC, Sampson WM, Ebron VH, Ferraris JP, Baughman RH. Spinning solid and hollow polymer-free carbon nanotube fibers. Adv Mater. 2005;17(5):614-7.

[28] Kuznetsov AA, Fonseca AF, Baughman RH, Zakhidov AA. Structural model for dry-drawing of sheets and yarns from carbon nanotube forests. ACS Nano. 2011;5(2):985-93.

[29] Zhang Y, Zou G, Doorn SK, Htoon H, Stan L, Hawley ME, et al. Tailoring the morphology of carbon nanotube arrays: from spinnable forests to undulating foams. ACS Nano. 2009;3(8):2157-62.

[30] Huynh CP, Hawkins SC. Understanding the synthesis of directly spinnable carbon nanotube forests. Carbon. 2010;48(4):1105-15.

[31] Zhu C, Cheng C, He Y, Wang L, Wong T, Fung K, et al. A self-entanglement mechanism for continuous pulling of carbon nanotube yarns. Carbon. 2011;49(15):4996-5001.

[32] Liu K, Sun Y, Lin X, Zhou R, Wang J, Fan S, et al. Scratch-resistant, highly conductive, and high-strength carbon nanotube-based composite yarns. ACS Nano. 2010;4(10): 5827-34.

[33] Hill FA, Havel TF, Hart AJ, Livermore C. Enhancing the tensile properties of continuous millimeter-scale carbon nanotube fibers by densification. ACS Appl Mater Interf. 2013;5(15):7198-207.

[34] Natarajan B, Lachman N, Lam T, Jacobs D, Long C, Zhao M, et al. The evolution of carbon nanotube network structure in unidirectional nanocomposites resolved by quantitative electron tomography. ACS Nano. 2015;9(6):6050-8.

[35] Vigolo B, Poulin P, Lucas M, Launois P, Bernier P. Improved structure and properties of single-wall carbon nanotube spun fibers. Appl Phys Lett. 2002;81(7):1210-2.

[36] Miaudet P, Badaire S, Maugey M, Derré A, Pichot V, Launois P, et al. Hot-drawing of single and multiwall carbon nanotube fibers for high toughness and alignment. Nano Lett. 2005;5(11):2212-5.

[37] Sugano K, Kurata M, Kawada H. Evaluation of mechanical properties of untwisted carbon nanotube yarn for application to composite materials. Carbon. 2014;78:356-65.

[38] Tran CD, Humphries W, Smith SM, Huynh C, Lucas S. Improving the tensile strength of carbon nanotube spun yarns using a modified spinning process. Carbon. 2009;47(11): 2662-70.

[39] Miao M. Production, structure and properties of twistless carbon nanotube yarns with a high density sheath. Carbon. 2012;50(13):4973-83.

[40] Wang JN, Luo XG, Wu T, Chen Y. High-strength carbon nanotube fibre-like ribbon with high ductility and high electrical conductivity. Nature Commun. 2014;5.

[41] Liu K, Sun Y, Zhou R, Zhu H, Wang J, Liu L, et al. Carbon nanotube yarns with high tensile strength made by a twisting and shrinking method. Nanotechnology. 2010;21(4). 045708. (DOI: 10.1088/0957-4484/21/4/045708)

[42] Ounaies Z, Park C, Wise KE, Siochi EJ, Harrison JS. Electrical properties of single wall carbon nanotube reinforced polyimide composites. Compos Sci Technol. 2003;63(11): 1637-46.

[43] Jiang X, Bin Y, Matsuo M. Electrical and mechanical properties of polyimide–carbon nanotubes composites fabricated by in situ polymerization. Polymer. 2005;46(18): 7418-24.

[44] Zhu B-K, Xie S-H, Xu Z-K, Xu Y-Y. Preparation and properties of the polyimide/multi-walled carbon nanotubes (MWNTs) nanocomposites. Compos Sci Technol. 2006;66(3–4):548-54.

[45] So HH, Cho JW, Sahoo NG. Effect of carbon nanotubes on mechanical and electrical properties of polyimide/carbon nanotubes nanocomposites. Eur Polym J. 2007;43(9): 3750-6.

[46] Jiang Q, Wang X, Zhu Y, Hui D, Qiu Y. Mechanical, electrical and thermal properties of aligned carbon nanotube/polyimide composites. Compos Part B Eng. 2014;56:408-12.

[47] Sandler J, Shaffer MSP, Prasse T, Bauhofer W, Schulte K, Windle AH. Development of a dispersion process for carbon nanotubes in an epoxy matrix and the resulting electrical properties. Polymer. 1999;40(21):5967-71.

[48] Allaoui A, Bai S, Cheng HM, Bai JB. Mechanical and electrical properties of a MWNT/ epoxy composite. Compos Sci Technol. 2002;62(15):1993-8.

[49] Moisala A, Li Q, Kinloch IA, Windle AH. Thermal and electrical conductivity of single- and multi-walled carbon nanotube-epoxy composites. Compos Sci Technol. 2006;66(10):1285-8.

[50] Liu W, Zhang X, Xu G, Bradford PD, Wang X, Zhao H, et al. Producing superior composites by winding carbon nanotubes onto a mandrel under a poly (vinyl alcohol) spray. Carbon. 2011;49(14):4786-91.

[51] Liu W, Zhao H, Inoue Y, Wang X, Bradford PD, Kim H, et al. Poly (vinyl alcohol) reinforced with large-diameter carbon nanotubes via spray winding. Compos Part A Appl Sci Manuf. 2012;43(4):587-92.

[52] Zhang L, Xu W, Luo XG, Wang JN. High performance carbon nanotube based composite film from layer-by-layer deposition. Carbon. 2015;90:215-21.

[53] Wang X, Yong Z, Li Q, Bradford PD, Liu W, Tucker DS, et al. Ultrastrong, stiff and multifunctional carbon nanotube composites. Mater Res Lett. 2013;1(1):19-25.

[54] Park JG, Cheng Q, Lu J, Bao J, Li S, Tian Y, et al. Thermal conductivity of MWCNT/epoxy composites: the effects of length, alignment and functionalization. Carbon. 2012;50(6):2083-90.

[55] Gong F, Bui K, Papavassiliou DV, Duong HM. Thermal transport phenomena and limitations in heterogeneous polymer composites containing carbon nanotubes and inorganic nanoparticles. Carbon. 2014;78(0):305-16.

[56] Han ZD, Fina A. Thermal conductivity of carbon nanotubes and their polymer nanocomposites: a review. Prog Polym Sci. 2011;36(7):914-44.

[57] Duong HM, Papavassiliou DV, Lee LL, Mullen KJ. Random walks in nanotube composites: improved algorithms and the role of thermal boundary resistance. Appl Phys Lett. 2005;87(1):013101.

[58] Einstein A. The electrodynamics of moving bodies. Ann Phys. 1905;17:891-921.

[59] Swartz ET, Pohl RO. Thermal-boundary resistance. Rev Mod Phys. 1989;61(3):605-68.

[60] Balandin AA. Thermal properties of graphene and nanostructured carbon materials. Nat Mater. 2011;10(8):569-81.

[61] Bui K, Papavassiliou DV. Numerical calculation of the effective thermal conductivity of nanocomposites. Num Heat Transfer Part A Appl. 2013;63(8):590-603.

[62] Duong HM, Papavassiliou DV, Mullen KJ, Wardle BL, Maruyama S. Calculated thermal properties of single-walled carbon nanotube suspensions. J Phys Chem C. 2008;112(50):19860-5.

[63] Duong HM, Papavassiliou DV, Mullen KJ, Wardle BL, Maruyama S. A numerical study on the effective thermal conductivity of biological fluids containing single-walled carbon nanotubes. Int J Heat Mass Transfer. 2009;52(23-24):5591-7.

[64] Duong HM, Papavassiliou DV, Mullen KJ, Maruyama S. Computational modeling of the thermal conductivity of single-walled carbon nanotube - polymer composites. Nanotechnology. 2008;19(6):065702.

[65] Duong HM, Yamamoto N, Papavassiliou DV, Maruyama S, Wardle BL. Inter-carbon nanotube contact in thermal transport of controlled-morphology polymer nanocomposites. Nanotechnology. 2009;20(15):155702.

[66] Duong HM, Yamamoto N, Bui K, Papavassiliou DV, Maruyama S, Wardle BL. Morphology effects on nonisotropic thermal conduction of aligned single-walled and multi-walled carbon nanotubes in polymer nanocomposites. J Phys Chem C. 2010;114(19): 8851-60.

[67] Bui K, Grady BP, Papavassiliou DV. Heat transfer in high volume fraction CNT nanocomposites: Effects of inter-nanotube thermal resistance. Chem Phys Lett. 2011;508(4-6):248-51.

[68] Peters JE, Papavassiliou DV, Grady BP. Unique thermal conductivity behavior of single-walled carbon nanotube–polystyrene composites. Macromolecules. 2008;41(20): 7274-7.

[69] Bui K, Duong HM, Striolo A, Papavassiliou DV. Effective heat transfer properties of graphene sheet nanocomposites and comparison to carbon nanotube nanocomposites. J Phys Chem C. 2011;115(10):3872-80.

[70] Naffakh M, Diez-Pascual AM, Gomez-Fatou MA. New hybrid nanocomposites containing carbon nanotubes, inorganic fullerene-like WS2 nanoparticles and poly(ether ether ketone) (PEEK). J Mater Chem. 2011;21(20):7425-33.

[71] Naffakh M, Diez-Pascual AM, Marco C, Ellis G. Morphology and thermal properties of novel poly(phenylene sulfide) hybrid nanocomposites based on single-walled carbon nanotubes and inorganic fullerene-like WS2 nanoparticles. J Mater Chem. 2012;22(4):1418-25.

[72] Yu AP, Ramesh P, Sun XB, Bekyarova E, Itkis ME, Haddon RC. Enhanced thermal conductivity in a hybrid graphite nanoplatelet - carbon nanotube filler for epoxy composites. Adv Mater (Weinheim, Germany). 2008;20(24):4740.

[73] Gupta TK, Singh BP, Mathur RB, Dhakate SR. Multi-walled carbon nanotube–graphene–polyaniline multiphase nanocomposite with superior electromagnetic shielding

[74] [77] Liao WH, Tien HW, Hsiao ST, Li SM, Wang YS, Huang YL, et al. Effects of multiwalled carbon nanotubes functionalization on the morphology and mechanical and thermal properties of carbon fiber/vinyl ester composites. ACS Appl Mater Interf. 2013;5(9):3975-82.

[75] Diez-Pascual AM, Ashrafi B, Naffakh M, Gonzalez-Dominguez JM, Johnston A, Simard B, et al. Influence of carbon nanotubes on the thermal, electrical and mechanical properties of poly(ether ether ketone)/glass fiber laminates. Carbon. 2011;49(8):2817-33.

[76] Cheng HKF, Basu T, Sahoo NG, Li L, Chan SH. Current advances in the carbon nanotube/thermotropic main-chain liquid crystalline polymer nanocomposites and their blends. Polymers. 2012;4(2):889-912.

[77] Gong F, Papavassiliou DV, Duong HM. Off-Lattice Monte Carlo simulation of heat transfer through carbon nanotube multiphase systems taking into account thermal boundary resistances. Num Heat Transfer Part A Appl. 2014;65(11):1023-43.

[78] [81] Gong F, Duong HM, Papavassiliou DV. Inter-carbon nanotube contact and thermal resistances in heat transport of three-phase composites. J Phys Chem C. 2015;119(14): 7614-20.

[79] Fan Z, Gong F, Nguyen ST, Duong HM. Advanced multifunctional graphene aerogel – Poly (methyl methacrylate) composites: experiments and modeling. Carbon. 2015;81(0):396-404.

[80] Gong F, Tam YS, Nguyen ST, Duong HM. Prediction of thermal resistances and heat conduction of carbon nanotube aerogels in various permeated gases. Chem Phys Lett. 2015;627:116-20.

[81] Gong F, Hongyan Z, Papavassiliou DV, Bui K, Lim C, Duong HM. Mesoscopic modeling of cancer photothermal therapy using single-walled carbon nanotubes and near infrared radiation: insights through an off-lattice Monte Carlo approach. Nanotechnology. 2014;25(20):205101.

[82] Fang Y, Lv Y, Gong F, Wu Z, Li X, Zhu H, et al. Interface tension-induced synthesis of monodispersed mesoporous carbon hemispheres. J Am Chem Soc. 2015;137(8):2808-11.

Chapter 3

Bio-inspired Design and Fabrication of Super-Strong and Multifunctional Carbon Nanotube Composites

Xiaohua Zhang, Xueping Yu, Jingna Zhao and Qingwen Li

Additional information is available at the end of the chapter

http://dx.doi.org/10.5772/62810

Abstract

Carbon nanotubes (CNTs) are ideal scaffolds to design and architect high-performance composites at high CNT volume fractions. In these composites, the CNT alignment determines the level of aggregation and the structure morphology, and thus the load transfer efficiency between neighboring CNTs. Here, we discuss two major solutions to produce high-volume fraction CNT composites, namely the layer-by-layer stacking of aligned CNT sheets and the stretching of entangled CNT webs (networks). As inspired by the growth procedure of natural composites, the aggregation of CNTs can be well controlled during the assembling process. As a result, the CNTs can be highly packed, aligned, and importantly unaggregated, with the impregnated polymers acting as interfacial adhesion or mortars to build up the composite structure. The CNT/bismaleimide composites can yield a super-high tensile strength up to 6.27–6.94 GPa and a modulus up to 315 GPa.

Keywords: carbon nanotube composite, high volume fraction, super-strong, multifunctionality, bio-inspired

1. Introduction

Since their discovery in 1991 [1], carbon nanotubes (CNTs) have generated huge activity in most areas of science and engineering due to their unprecedented mechanical, electrical, and thermal properties. For example, lightweight multifunctional composites with enhanced properties can be produced by effectively incorporating individual CNTs into polymer matrices [2, 3]. The first polymer nanocomposites using CNTs as a filler were reported in 1994, where the CNTs were aligned within the epoxy matrix by the shear forces induced by cutting with a diamond knife [4]. In the following decades, tremendous research has been done to develop CNT-reinforced composites with a high strength and modulus. However, as compared to individual CNTs, the composites produced by these "conventional" fabrication methods usually do not exhibit significantly improved mechanical and electrical performances. This is mainly due to the limited content of CNT (usually under 5 wt%). Furthermore, it is also important to introduce a uniform CNT dispersion in polymer matrix and adhesion between different constituents in improving the composite performances.

Recently, a new type of CNT composite material has been developed using macroscopic CNT assemblies as raw materials, namely high-volume fraction CNT composites, where the volume fraction of CNT is usually much larger than 50%. To obtain these composites, the macroscopic forms of entangled or aligned CNTs, that is, fibers [5–7], forests [8–12], and membranes [13–16] are used as scaffolds, and the polymer is impregnated into the free pores of these CNT network. Of great importance, the super-aligned CNT sheets drawn out from spinnable CNT forests [17] and the entangled CNT webs (networks) grown with an injection chemical vapor deposition method [18] have superb advantages in obtaining super-strong high-volume fraction CNT composites.

Furthermore, to design the structure for high-performance composites at high CNT fractions, nature can offer us with scientific and technological clues from the formation process of biological composites. The natural composite materials are usually built up by common organic components via the naturally mild approaches [19], such as super-tough spider fibers [20], strong hard nut skins [21], and wear-resistant molluscan shells [22]. In these composites, the major components such as proteins, cellulose molecules, and nanometer-sized crystals of carbonated calcium phosphates or calcium carbonates are homogeneously distributed and orientated along with other co-existing components [20, 23]. This provides us new approaches to obtain super-strong CNT composites [24]. For CNT composites, due to the strong tendency to agglomerate between CNTs [25–29], it is still difficult to uniformly disperse CNTs within polymer matrix at a high-volume fraction and thus to mimic the natural composites. Fortunately, CNTs can be treated as linear macromolecules, and thus, the processing on them can be dealt with in a biomimic way. Therefore, it becomes possible to mimic the formation process of biological composites to design new type of CNT composites.

This chapter thus aims at the structural design of super-strong and multifunctional CNT composites at high-volume fractions. Various processing methods are presented in the following sections, namely the layer-by-layer stacking of aligned CNT sheets, stretching on entangled CNT webs, and, most importantly, bio-inspired aggregation control to optimize the composite structure.

2. Conventional composite processing

Solution processing and melt processing have been widely used to prepare composites, where CNTs and polymer matrix can be well mixed together. For the solution treatment, CNT and polymer are mixed in a suitable solvent before evaporating the solvent to form a composite film. One benefit of such method is that agitation of the CNT powder in a solvent facilitates CNT de-aggregation and dispersion. In general, agitation of CNT or of CNT/polymer mixture is provided by magnetic stirring, shear mixing, reflux, and, as widely used, ultrasonication.

In an early study [30], multiwalled CNTs were dispersed in chloroform by sonication. Then, polyhydroxyaminoether (PHAE) was dissolved in the CNT/chloroform dispersion and mixed by additional sonication. The suspension was poured into a Teflon mould and dried in ambient conditions to obtain the CNT composites. In the following decade, various similar methods have been reported. For example, CNTs were first chemically modified and then dispersed in water [31]. The dispersion was blended with poly(vinyl alcohol) (PVA)/water solution to give composite solutions which were used to prepare the composites.

As pristine CNTs cannot be well-dispersed in most solvents, surfactants such as sodium dodecyl sulfate (SDS), sodium dodecyl sulfonate (SDSA), polyvinylpyrrolidone (PVP), and dodecyl tri-methyl ammonium bromide (DTAB) were also used to assist the CNT dispersion before mixing with the polymer solution [32–38]. This technique results in excellent dispersion with no derogatory effects on film properties observed. However, the existence of surfactant could affect the interfacial strength between CNT and polymer.

For those polymers which are insoluble in the designed solvent, melt processing has become a common alternative. For example, amorphous polymers can be processed above their glass transition temperature, and semi-crystalline polymers can be heated above the melt temperature to induce sufficient softening [39–43]. In general, melt processing involves the melting of polymer pellets to form a viscous liquid. CNTs can be mixed into the melt by shear mixing. Bulk samples can then be fabricated by techniques such as compression molding, injection molding, or extrusion. In such approach, the optimized processing conditions depend on not only the types of CNT, but also the whole range of polymer–nanotube combinations. This is because nanotubes can effect melt properties such as viscosity, resulting in unexpected polymer degradation under conditions of high shear rates [39].

The dispersion and melting treatments have been widely used to obtain the composite structure by well mixing polymer and CNTs. Besides the structure design, interfacial covalent bonding is often applied in these methods in order to increase the composite performances. For example, in situ polymerization enables grafting of polymer molecules onto the wall of CNT and thus allows the preparation of composites with high nanotube loading. It is also particularly important for the preparation of insoluble and thermally unstable polymers, which cannot be processed by solution or melt processing. For example, due to their π-bonds, CNTs can participate in the polymerization of poly(methyl methacrylate) (PMMA), which resulted in a strong interface between the CNT and the PMMA matrix [44–46]. Poly(p-phenylene benzobisoxazole) (PBO)/CNT composites were also obtained with in situ polymerization by introducing CNTs in poly(phosphoric acid) (PPA) [47]. For CNT composites using polyamide, in situ polymerization can be realized between carboxylated CNTs and ε-caprolactam monomer [48].

In situ polymerization has been applied for the preparation of composites with CNT and various types of polymer. In this approach, covalent functionalization of CNT surface plays a special role for CNT processing and applications [49–55]. "Grafting from" and "grafting to" are the two main strategies to introduce covalent bonding between polymers and CNTs [2]. In the former approach, the initiators are immobilized onto CNT surfaces and then, the monomers are in situ polymerized with the formation of covalent bonding between polymer and CNT. The latter approach is based on attachment of already preformed end-functionalized polymer molecules to functional groups on CNT surface via different chemical reactions. There have been many reports on the grafting techniques, where the grafted polymers include PMMA [56–60], polyethylene (PE) [61–64], polystyrene [65–67], chitosan [68, 69], and so on.

There have been tremendous development on CNT composites using these conventional processing methods over the past 20 years. However, the dispersion of long CNTs is still hindered by their entanglement and aggregation, and the CNTs are limited to a low fraction and randomly orientated. Consequently, the final composite strength is usually below 400 MPa [70, 71]. Further, in situ polymerization or hot stretching is also not very effective in improving the mechanical properties [2, 28, 72]. Therefore, further development is still of great necessity, especially on the design of composite structure.

3. Layer-by-layer stacking of aligned CNTs

The large mass scale production of CNT has pushed the rapid development of CNT in various applications [73, 74]. However, the random orientation and entanglement hinder the structure design of high-volume fraction and aligned CNT composites. The development on spinnable CNT forests opened a new way towards such composite structure. CNT forests (also called CNT carpets) superficially resemble bamboo forests, except that the CNT "trees" in these forests can be over 50,000 times longer than their diameter, and this very high aspect ratio is useful for optimizing the mechanical and electrical properties. As a unique type of CNT forests, known as spinnable forests, one can continuously transform the vertically aligned CNTs into a horizontally aligned CNT sheet by a simple dry drawing or spinning method [17, 75–79]. Thus, the spinnability (or processability/drawability) is defined by the stable width and continuous length of the spun-out CNT sheets and the available spinning rates [80].

3.1. CNT/polymer composite films

The spinnable CNT forests have been successfully used to fabricate high-performance composite films. At first, the spun-out aligned CNTs were stacked together at different orientations to make a CNT preform [81]. After a resin transfer molding process, homogeneously dispersed CNT/epoxy composites with a CNT loading up to 16.5 wt% were obtained. The Young's modulus and tensile strength of the composites reached 20.4 and 231.5 MPa, corresponding to 716 and 160% improvement compared to pure epoxy.

To further improve the CNT volume fraction, the CNT sheets were stacked in a continuous winding way with the aid of solution spraying [12]. **Figure 1A** shows the schematic of spray winding, where a CNT sheet is continuously wound onto a rotating mandrel on which micrometer-sized polymer solutions are deposited. With the long-chain polymer PVA, the aligned composite film exhibited a tensile strength up to 1.8 GPa at a CNT content of 65 wt%

[12]. Due to the alignment, the composite film was also stiff (modulus of 40–96 GPa), tough (energy absorbed before fracturing of 38–100 J g^{-1}), and electrically conductive (conductivity of 780 S cm^{-1}). By following this method, high-performance CNT composites with epoxy, polyimide (PI), and bismaleimide (BMI) were produced [82–85].

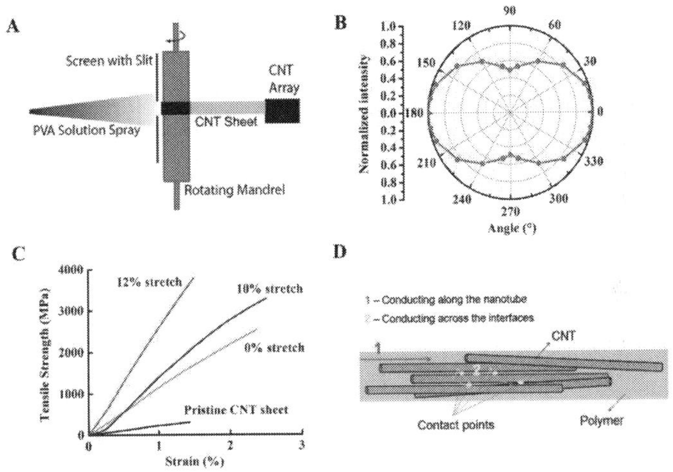

Figure 1. (A) Schematic view of spray winding. A CNT sheet is drawn out from a spinnable forest and continuously wound onto a rotating mandrel on which micrometer-sized droplets of polymer solution are deposited [12]. (B) Normalized intensity of G′-band peak as a function of the angle between the sample's longitudinal direction and the polarization axis of the incident laser beam. (C) Typical stress–strain curves of pristine CNT sheet, unstretched and stretched composites, demonstrating a significant improvement on mechanical properties through aligning and straightening of CNTs [82]. (D) Schematic presentation of heat conduction mechanism in high-volume fraction CNT composites [83].

The layer-by-layer stacking allowed a high level of CNT alignment in the final composites. The alignment can be quantitatively measured with polarized Raman spectroscopy [8, 86]. **Figure 1B** shows the normalized intensity of Raman G′-band peak as a function of the angle between the sample's longitudinal direction and the polarization axis of the incident laser beam; according to the results, we have reported previously [12]. The normalized G′ peak was strongest at 0° (along the CNT orientation) and monotonically decreased from 1 to 0.493 at 90° (perpendicular to the orientation).

To further improve the CNT alignment, a pair of stationary rods was placed between the CNT forest and the rotating mandrel. Due to the surface friction, the rods induced a certain shear stretching on the CNT sheet [82]. As a result, the mechanical, thermal, and electrical properties of the composites were simultaneously improved; the film exhibited a strength of 3.8 GPa, Young's modulus of 293 GPa, electrical conductivity of 1230 S cm^{-1}, and thermal conductivity of 41 Wm^{-1} K^{-1}. **Figure 1C** shows the stress–strain curves of pristine CNT sheet without matrix and CNT composite sheets stretched to various stretch ratios, where the effect of CNT alignment and straightness can be clearly seen.

In these composite films, the longer the CNTs the higher the thermal and electrical conductivities are. However, the mechanical properties including the strength and modulus exhibited no CNT length dependency [83]. (A) Setup for layer-by-layer stacking of aligned CNT sheets. (B) As-prepared CNT film on a mandrel. (C) Flexible CNT film strip. (D,E) Photographs of a shining CNT strip and a strip coated with a 12 nm gold layer. Inset in (D) is the shape of a water droplet on the strip. (F) SEM image of a woven fabric consisting of CNT strips. **Figure 1D** shows a schematic of thermal (electrical) conducting mechanism for the high-volume fraction CNT composites. Due to the long tube length, thermal conduction across the interfaces is as not dominant as that along the CNTs. On the contrary, the thermally insulating polymer restricts the phonon mobility at the interface in short-CNT-reinforced composites. This means that the high interfacial thermal resistance can strongly limit the total thermal conductivity. However, with increasing the tube length, several issues might affect the mechanical performance. First, the average tube diameter and wall number increased with the tube length, leaving the aspect ratio remaining nearly unchanged. Second, the increased tube length usually causes more aggregation between the CNTs. These two issues both hinder the efficiency of interfacial stress transfer.

3.2. Super-strong CNT assembly film

As inspired by the forest-based fiber spinning where macroscopic one-dimensional assembly fiber with pure (neat) CNTs can be produced [75, 76], the layer-by-layer stacking can be also used to fabricate pure CNT films with high mechanical performances. Although no polymer molecules are impregnated, various solvents such as ethanol, acetone, and *N,N*-dimethylmethanamide (DMF) are used to densify the CNTs. The solvent can be either sprayed onto the mandrel during the winding process or dip onto the CNT film after the winding [87]. **Figure 2**

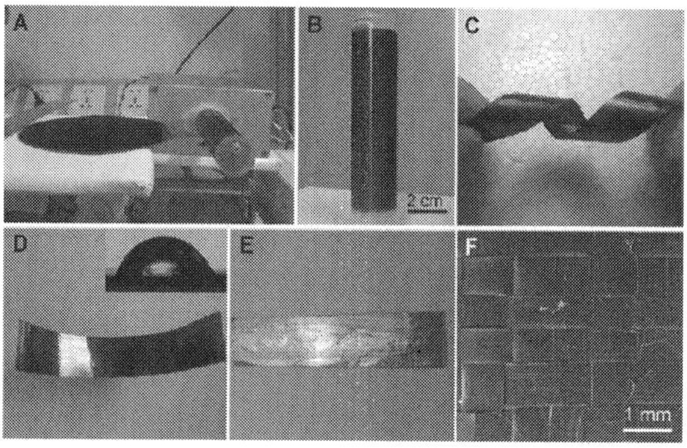

Figure 2. Pure CNT assembly films can be produced by layer-by-layer stacking of as-drawn CNT sheets with the aid of solvent densification [87] (A) Setup for layer-by-layer stacking of aligned CNT sheets. (B) As-prepared CNT film on a mandrel. (C) Flexible CNT film strip. (D,E) Photographs of a shining CNT strip and a strip coated with a 12 nm gold layer. Inset in (D) is the shape of a water droplet on the strip. (F) SEM image of a woven fabric consisting of CNT strips.

shows the CNT films obtained using 2–3-walled 4–6-nm-diameter CNTs. The CNT film had a smooth and shinning surface, was flexible, and exhibited tensile strengths of 1.1–1.9 GPa and Young's modulus of 40–90 GPa. Obviously, the alignment allowed a high degree of utilization of the intrinsic properties of CNT.

The CNT alignment was further improved when a stretch-dip-drying approach [88] was applied to the as-spun CNT films (**Figure 3A**). After such treatment, the CNT film was observed to significantly shrink in width by 22–39%, corresponding to an increased packing density. As compared to the simple solvent densification, the mass density of the film increased from 0.5–0.85 to 1.21–1.35 g cm^{-1}. Therefore, with the improved CNT alignment, the obtained CNT films exhibited significantly improved strength, up to 2.58–3.19 GPa.

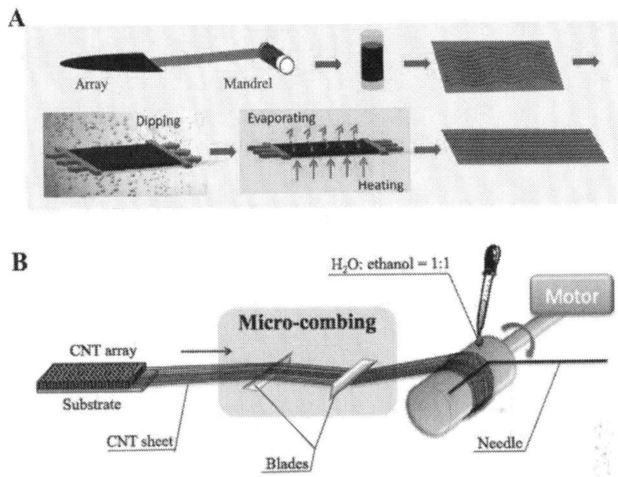

Figure 3. Two solutions to further improve the CNT alignment. (A) CNT alignment can be remarkably improved by the stretch-dip-drying method [88]. (B) Schematic view of the microcombing process which can mitigate the CNT waviness [89].

In another study, the as-drawn CNT sheets were "combed" to become straighter and aligned before being layer-by-layer stacked [89]. **Figure 3B** shows the schematic view of such microcombing. As the microcombing mitigates the CNT waviness and thus reduces the sheet defects, the final CNT film exhibited very high Young's modulus of 172 GPa and the similar tensile strength of 3.2 GPa. Surprisingly, the electrical conductivity was as high as 1800 S cm^{-1}, higher than other CNT films fabricated using the aligned CNT sheets.

All these studies show clearly that the CNT alignment plays the most important role in determining the mechanical and electrical properties. Nevertheless, solvents still have interesting influences on the microstructure of aggregated CNTs. For example, the acetone-densified pure CNT films usually had a higher strain at break than the ethanol-densified CNT films [89]. Such phenomenon is due to the different volatility of solvent [90]. When a high-volatile solvent was used, the CNT films exhibited a certain level of networking, which can absorb additional energy during the tensile stretching. As a result, the film's plasticity and toughness can be improved. This means that, even for the aligned CNT films, it is still possible to tune the internal microstructure of the highly packed CNTs, and thus to influence the film's mechanical and electrical properties.

4. Stretching on entangled CNT webs

As the CNT length can only reach several hundred micrometers for the forest-based spinning, the aspect ratio of the CNTs is thus limited to be in the range of 10^4. On solution to increase the aspect ratio is the decrease in diameter, like the spinnable few-walled CNT forests [87, 91]. A different way is to grow super-long CNTs, up to millimeter-long. For example, the millimeter-long and small-diameter (~3–8 nm) 2–5-walled CNTs provided a aspect ratio up to 10^5 [92]. These CNTs substantially entangled with each other due to the floating catalyst synthesis and aero-gel condense method [18]. The films with entangled CNTs (CNT webs) can reach up to a meter long and are commercially available, which makes them practical for manufacturing bulk composites [93].

4.1. Direct stretching

A simple mechanical-stretch method can be used to align the CNTs in the entangled webs [93]. For example, for a 40%-stretched CNT film (i.e., the post-stretch film was 40% longer than the pre-stretch one), the degree of CNT alignment can be dramatically improved (**Figure 4A**). Polarized Raman scattering tests were conducted to calculate the alignment degree [94]. From the trend of the best fitting, it was predicted that the near-perfect alignment (more than 95% CNTs aligned along stretch direction) at an approximate 50% stretch ratio [93]. This means that it is still necessary to further improve the plasticity of the CNT webs.

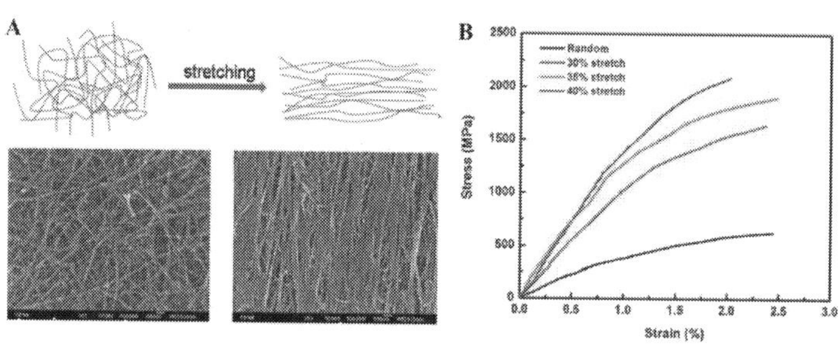

Figure 4. Direct stretching on entangled CNT webs resulted in the remarkably improved mechanical performances [93]. (A) Schematic illustration of mechanical stretching to align CNTs in the entangled webs. (B) Typical tensile stress–strain curves of produced CNT/BMI composites at different stretch ratios.

The mechanical properties of the pure CNT webs were improved with the stretch ratio. The film's tensile strength and Young's modulus were just about 205 MPa and 1.10 GPa for the unstretched films. After the stretching, the CNT packing density also increased with the alignment. For the 30, 35, and 40% stretch ratios, the strength increased up to 390, 508, and 668 MPa, respectively, and the modulus along the alignment direction showed even more dramatic improvements, up to 11.93, 18.21, and 25.45 GPa, respectively.

When BMI was introduced into the CNT webs, either as-produced or stretched, there were further improvements on the mechanical properties. The tensile strength of the randomly dispersed CNT/BMI composite (CNT loading of about 60 wt%) was approximately 620 MPa, and the Young's modulus was 47 GPa. After stretching by 30, 35, and 40%, the strength and modulus became 1600 MPa and 122 GPa, 1800 MPa and 150 GPa, and 2088 MPa and 169 GPa, respectively (**Figure 4B**).

4.2. CNT functionalization

Although the improvement in CNT alignment benefited the mechanical properties, substantial CNT pull-outs were observed at the fracture of the composites [93], corresponding to weak interfacial bonding. To overcome this problem, a follow-up effort was performed by improving the bonding with epoxidation functionalization [95]. To realize the interfacial reaction between CNT and BMI, the stretched CNT films were placed in peroxide acid (*m*-chloroperoxybenzoic acid or *m*-CPBA)/dichloromethane solution for functionalization. The functionalized CNT films were placed in a vacuum oven at 80°C for 30 min to evaporate the residual dichloromethane. Then, the CNT films were impregnated with BMI resin solution to prepare prepregs with approximately 60 wt% CNT loading. During the curing process, the reaction between the functionalized CNTs and BMI resin formed interfacial covalent bonds, whose mechanism is proposed as shown in **Figure 5A**. First, *o,o'*-diallyl bisphenol A reacted with the epoxide groups of the pre-functionalized CNTs, according to the epoxy-phenol reaction mechanism [96]. Then, in ene and Diels–Alder reactions, three-dimensional cross-linked structures were formed

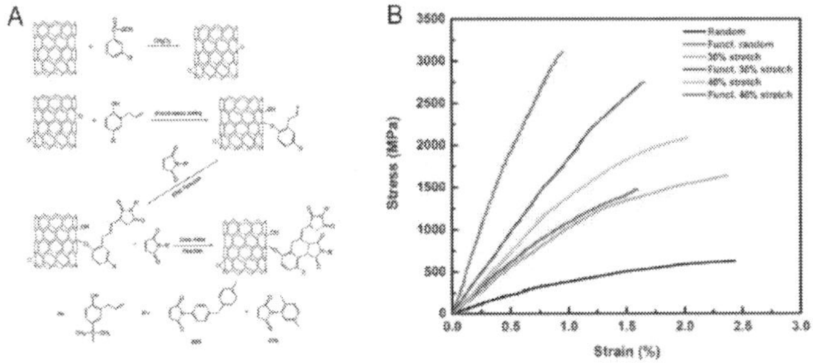

Figure 5. CNT functionalization improved the interfacial bonding strength and thus increased the tensile properties of the CNT/BMI composites [95]. (A) Proposed reaction mechanism of functionalized CNTs and BMI resin. (B) Typical tensile stress–strain curves of random and stretch-aligned CNT/BMI composites with and without functionalization.

between the derivative and the other two BMI components [97, 98]. The covalent bonding between CNT and BMI dramatically enhanced the interfacial adhesion, and thus, the improved load transfer was achieved by the functionalization.

Figure 5B shows typical stress–strain curves of CNT/BMI composite films along the CNT alignment direction. As compared to the pristine random CNT/BMI composites, the tensile strength and Young's modulus dramatically increased with the increase of alignment. After the functionalization to for interfacial covalent bonds, the mechanical properties of the resultant film were further improved. The strength and modulus of functionalized random CNT/BMI composites were 1437 MPa and 124 GPa, respectively. For the functionalized and 30%-stretched composites, the strength and modulus increased up to 2843 MPa and 198 GPa, respectively. The 40% stretch alignment together with interfacial functionalization resulted in the highest mechanical performances, where the strength and modulus were surprisingly as high as 3081 MPa and 350 GPa, respectively.

5. Bio-inspired aggregation control

The high-volume fraction CNT composites have exhibited exciting advantages in achieving high mechanical and electrical performances. However, due to the aggregation of nanometer-sized components, there is still a severe problem in paving the way to stronger materials [99]. The aggregation control is very necessary and can be demonstrated by a comparison of the structures of carbon fiber-reinforced polymer, aggregated and unaggregated CNTs in composites, as shown in **Figure 6** [24]. The large specific surface area is one important advantage of CNT. When the aggregated CNTs were used to replace solid carbon fibers, as observed in many reports [12, 82, 87, 93, 95], larger interfacial contact area was formed, and tensile strengths of the CNT composites range from 2.08 to 3.8 GPa [82, 93, 95]. However, in the aggregation phase, the load transfer is not as efficient as at the interface. Thus, such aggregation phase might become the weak parts in the composites and hinders further reinforcement. In an ideal structure, the nanometer-sized components should be uniformly distributed in the matrix

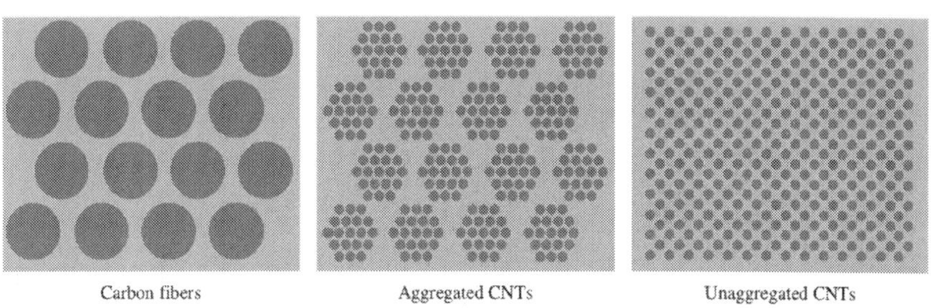

Figure 6. Schematics of carbon fiber-reinforced polymer, composite structure with aggregated CNTs, and ideal structure containing unaggregated CNTs, respectively [24].

without formation of any aggregation phases. Therefore, all the interfaces can play roles in shear load transfer.

The CNT aggregation arises from van der Waals (vdW) attraction and can be enhanced in wet environment. The situation becomes very severe in the layer-by-layer stacking of aligned CNT sheets with the aid of solution spray [12, 82, 87], where the CNTs (or more commonly, small-sized CNT bundles) usually aggregate first into large-sized bundles and then are surrounded by polymer matrix. **Figure 7A** and **B** show the aggregation phase of CNT, which were results of the layer-by-layer stacked aligned CNTs and the highly stretched CNT webs, respectively. Usually, the CNT aggregation was observed in a scale of hundreds of nanometer. Nevertheless, due to the entanglement, the preformed CNT webs were found to be more optimal than the forest-based CNT sheets in realizing the aggregation control.

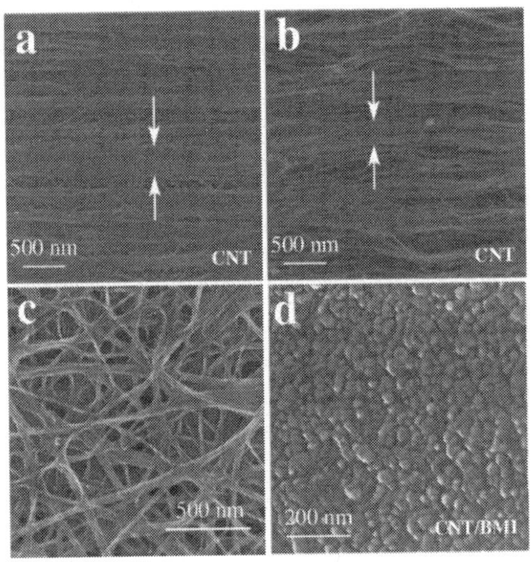

Figure 7. SEM characterization of different CNT assembly structures [24]. (A) CNT aggregation in the layer-by-layer stacking of as-spun aligned CNT sheets. (B) CNT aggregation in the stretched dry films composed by entangled CNTs. (C) Acetone-densified CNT webs maintained the feature of entanglement. (D) Cross section of the optimal CNT/BMI composite structure using focused ion beam treatment, where the aggregation level of CNTs was limited within a dimension of 20–50 nm.

5.1. Impregnation without introducing aggregation

As-inspired by the natural composite structures, the entanglement was utilized to produce the optimal composite structure where the CNTs and polymer matrix co-existed and uniformly distributed among each other. This is because that the CNT webs can maintain the CNT entanglement after liquid densification. For example, after being densified with acetone, the pore sizes of the CNT webs decreased from >500 nm to ~100–200 nm, while the CNTs were still unaggregated and randomly distributed (**Figure 7C**). This is reminiscent of the formation process of biological composites. During the formation, the matrix and the major components are simultaneously grown from stem cells at the optimized fractions to maximize the interfacial stress transfer. Therefore, we should introduce the polymer matrix into the CNT network prior to any other treatments, just to "grow" an initial composite configuration without CNT aggregations. Here, the BMI resins were dissolved in acetone and impregnated into the CNT webs. After the stretching, densification, and thermal curing, the CNTs were still found to be unaggregated (at least no larger than 50–100 nm), as shown in **Figure 7D**.

5.2. Multi-step stretching processes

The entangled CNTs can be re-assembled and aligned by a high level of stretching. This requires the CNT film to possess high plasticity. The raw films can be stretched by 10–15% in length. After the stretching, the tensile strength was improved from 180–198 to 500–600 MPa owing to the improved alignment. For the "wet" films which were infiltrated with 1 wt% BMI resin/acetone solutions, their strain at break were up to 20–25%, corresponding to the improved plasticity. This means that the impregnation prior to stretching also resulted in improved processability.

After the hot-pressing to cure BMI resins, the unstretched CNT/BMI films finally exhibited a tensile strength of 478–501 MPa. On the contrast, by first stretching the "wet" film by 20% and

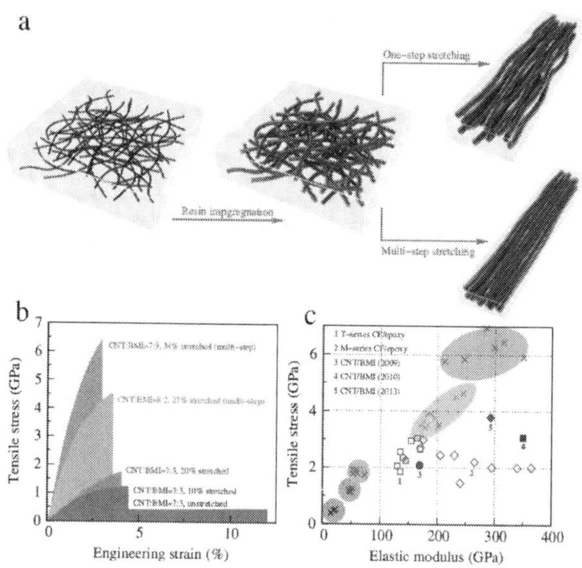

Figure 8. Bio-inspired aggregation control showed great advantages in realizing super-strong CNT composites [24]. (A) The schematic of preparation and stretching processes of CNT/polymer films. (B) The mechanical performances of CNT/BMI composite films under different stretching strategies. (C) A comparison of tensile strength and elastic modulus for CNT/BMI and carbon fiber/epoxy composites.

then curing the film, the strength increased up to 1.74–1.92 GPa. To further improve the mechanical properties, the stretching method was modified to a multi-step way (**Figure 8A**) and the "wet" films could be stretched by 27–34% after 8 to 10 steps. Notice that, in each step, 3% additional stretching, according to the immediate film length, was applied to the film, and there was always 5–10 min between steps to allow sufficient structure relaxation. After 8 or 10 steps, the total stretching magnitude was $1.03^8-1 = 0.267$ or $1.03^{10}-1 = 0.344$, respectively. In this multi-step way, the CNTs were fully aligned and the packing density during the hot-pressing was also improved. At this stage, the small-sized CNT bundles were well surrounded by the BMI resin molecules and maintained unaggregated phases. After being cured, the CNT/BMI composite films stably exhibited an extremely high tensile strength up to 4.5–6.94 GPa, depending on the CNT-to-resin mass ratio and the total stretching magnitude. At the same time, the elastic modulus was up to 232–315 GPa. Figure 8B shows the typical stress–strain curves for various CNT/BMI composite films, and Figure 8C provides the comparison with carbon fiber/epoxy composites.

5.3. Structural characterization

Based on the aggregation control and high CNT alignment, a big step has been realized towards the ideal composite structure. The tensile strength of >6 GPa is obviously much larger than those of carbon fiber/epoxy composites, in good agreement with their different composite structures (Figure 6). Figure 7D shows the cross-sectional picture of the CNT/BMI composite structure obtained by focused ion beam. Although individual CNTs were difficult to distinguish, CNT bundles with small size were found perpendicular to the cross section and their surfaces were all surrounded by BMI resins. Obviously, the maximized interface contacts can allow the most efficient interfacial stress transfer. Notice that the level of CNT aggregation was within 20–50 nm. This means that there is still a big challenge to obtain the ideal composite structure where individual CNTs are aligned, highly packed, and unaggregated.

5.4. Electrical properties

The ability to conduct electricity of a thin film is usually characterized by sheet resistance or square resistance, in units of "ohms per square". The square resistance of the as-produced CNT web was 1.194 Ω sq^{-1}. After being impregnated with BMI resins, the resistance decreased to 0.926 Ω sq^{-1}, as a result of the enhanced densification. After being stretched, the CNTs became aligned and the connection between CNT bundles were separated by the matrix. As a result, the resistance increased to 1.461 and 1.931 Ω sq^{-1} before and safer the curing process. The electrical conductivity of the final film was ~1700 S cm^{-1}, about 0.3 or 12% of copper's or stainless steel's electrical conductivity.

6. Conclusion

High-volume fraction CNT composites have exhibited exciting mechanical and electrical performances due to the high utilization of the CNT's intrinsic properties. The layer-by-layer stacking of aligned CNT sheets and the stretching of entangled CNT webs have provided two different solutions to realize the high-strength CNT composite films. Considering the easy aggregation between CNTs, the bio-inspired aggregation control makes a big step to approach the ideal composite structure where the CNTs are highly packed, aligned, and unaggregated. The highest tensile strength of the CNT/BMI composite film can be up to 6.94 GPa, much higher than the strength of carbon fiber-reinforced polymers. We anticipate that the present fabrication method can be generalized for developing multifunctional and smart nanocomposites.

Acknowledgements

The authors thank financial supports from the National Natural Science Foundation of China (11302241, 11404371, 21273269, 21473238, 21503267, 51561145008) and the Youth Innovation Promotion Association of the Chinese Academy of Sciences (2015256, Grant to X.Z.).

Author details

Xiaohua Zhang*, Xueping Yu, Jingna Zhao and Qingwen Li

*Address all correspondence to: xhzhang2009@sinano.ac.cn

Key Laboratory of Nano-Devices Applications, Suzhou Institute of Nano-Tech Nano-Bionics, Chinese Academy of Sciences, Suzhou, China

References

[1] S. Iijima. Helical microtubules of graphitic carbon. Nature 354, 56 (1991).

[2] J. N. Coleman, U. Khan, W. J. Blau, and Y. K. Gun'ko. Small but strong: a review of the mechanical properties of carbon nanotube–polymer composites. Carbon 44, 1624 (2006).

[3] M. Moniruzzaman and K. I. Winey. Polymer nanocomposites containing carbon nanotubes. Macromolecules 39, 5194 (2006).

[4] P. M. Ajayan, O. Stephan, C. Colliex, and D. Trauth. Aligned carbon nanotube arrays formed by cutting a polymer resin–nanotube composite. Science 265, 1212 (1994).

[5] A. B. Dalton, S. Collins, E. Muñoz, J. M. Razal, V. H. Ebron, J. P. Ferraris, J. N. Coleman, B. G. Kim, and R. H. Baughman. Super-tough carbon-nanotube fibres. Nature 423, 703 (2003).

[6] C. Fang, J. Zhao, J. Jia, Z. Zhang, X. Zhang, and Q. Li. Enhanced carbon nanotube fibers by polyimide. Appl. Phys. Lett. 97, 181906 (2010).

[7] S. Li, X. Zhang, J. Zhao, F. Meng, G. Xu, Z. Yong, J. Jia, Z. Zhang, and Q. Li. Enhancement of carbon nanotube fibres using different solvents and polymers. Compos. Sci. Technol. 72, 1402 (2012).

[8] N. R. Raravikar, L. S. Schadler, A. Vijayaraghavan, Y. Zhao, B. Wei, and P. M. Ajayan. Synthesis and characterization of thickness-aligned carbon nanotube–polymer composite films. Chem. Mater. 17, 974 (2005).

[9] H. Cebeci, R. G. de Villoria, A. J. Hart, and B. L. Wardle. Multifunctional properties of high volume fraction aligned carbon nanotube polymer composites with controlled morphology. Compos. Sci. Technol. 69, 2649 (2009).

[10] T.-W. Chou, L. Gao, E. T. Thostenson, Z. Zhang, and J.-H. Byun. An assessment of the science and technology of carbon nanotube-based fibers and composites. Compos. Sci. Technol. 70, 1 (2010).

[11] P. D. Bradford, X. Wang, H. Zhao, J.-P. Maria, Q. Jia, and Y. T. Zhu. A novel approach to fabricate high volume fraction nanocomposites with long aligned carbon nanotubes. Compos. Sci. Technol. 70, 1980 (2010).

[12] W. Liu, X. Zhang, G. Xu, P. D. Bradford, X. Wang, H. Zhao, Y. Zhang, Q. Jia, F.-G. Yuan, Q. Li, Y. Qiu, and Y. Zhu. Producing superior composites by winding carbon nanotubes onto a mandrel under a poly(vinyl alcohol) spray. Carbon 49, 4786 (2011).

[13] J. N. Coleman, W. J. Blau, A. B. Dalton, E. Muñoz, S. Collins, B. G. Kim, J. Razal, M. Selvidge, G. Vieiro, and R. H. Baughman. Improving the mechanical properties of single-walled carbon nanotube sheets by intercalation of polymeric adhesives. Appl. Phys. Lett. 82, 1682 (2003).

[14] Z. Wang, Z. Liang, B. Wang, C. Zhang, and L. Kramer. Processing and property investigation of single-walled carbon nanotube (SWNT) buckypaper/epoxy resin matrix nanocomposites. Compos. Part A 35, 1225 (2004).

[15] P. Gonnet, Z. Liang, E. S. Choi, R. S. Kadambala, C. Zhang, J. S. Brooks, B. Wang, and L. Kramer. Thermal conductivity of magnetically aligned carbon nanotube buckypapers and nanocomposites. Curr. Appl. Phys. 6, 119 (2006).

[16] Z. Špitalský, C. Aggelopoulos, G. Tsoukleri, C. Tsakiroglou, J. Parthenios, S. Georga, C. Krontiras, D. Tasis, K. Papagelis, and C. Galiotis. The effect of oxidation treatment on the properties of multi-walled carbon nanotube thin films. Mater. Sci. Eng. B 165, 135 (2009).

[17] M. Zhang, S. Fang, A. A. Zakhidov, S. B. Lee, A. E. Aliev, C. D. Williams, K. R. Atkinson, and R. H. Baughman. Strong, transparent, multifunctional, carbon nanotube sheets. Science 309, 1215 (2005).

[18] Y.-L. Li, I. A. Kinloch, and A. H. Windle. Direct spinning of carbon nanotube fibers from chemical vapor deposition synthesis. Science 304, 276 (2004).

[19] T. L. B. Ha, T. M. Quan, D. N. Vu, and D. M. Si. Naturally derived biomaterials: preparation and application. "Regenerative Medicine and Tissue Engineering" (Edited by J. A. Andrades, InTech - Open Access, Rijeka, Croatia, 2013) Chapter 11, pp. 247–274.

[20] T. Giesa, M. Arslan, N. M. Pugno, and M. J. Buehler. Nanoconfinement of spider silk fibrils begets superior strength, extensibility, and toughness. Nano Lett. 11, 5038 (2011).

[21] C. M. Preston and B. G. Sayer. What's in a nutshell: an investigation of structure by carbon-13 cross-polarization magic-angle spinning nuclear magnetic resonance spectroscopy. J. Agric. Food Chem. 40, 206 (1992).

[22] S. M. Porter. Seawater chemistry and early carbonate biomineralization. Science 316, 1302 (2007).

[23] Q. Cheng, L. Jiang, and Z. Tang. Bioinspired layered materials with superior mechanical performance. Acc. Chem. Res. 47, 1256 (2014).

[24] Y. Han, X. Zhang, X. Yu, J. Zhao, S. Li, F. Liu, P. Gao, Y. Zhang, T. Zhao, and Q. Li. Bio-inspired aggregation control of carbon nanotubes for ultra-strong composites. Sci. Rep. 5, 11533 (2015).

[25] J. N. Coleman, U. Khan, and Y. K. Gun'ko. Mechanical reinforcement of polymers using carbon nanotubes. Adv. Mater. 18, 689 (2006).

[26] B. Fiedler, F. H. Gojny, M. H. G. Wichmann, M. C. M. Nolte, and K. Schulte. Fundamental aspects of nano-reinforced composites. Compos. Sci. Technol. 66, 3115 (2006).

[27] X.-L. Xie, Y.-W. Mai, and X.-P. Zhou. Dispersion and alignment of carbon nanotubes in polymer matrix: a review. Mater. Sci. Eng. R 49, 89 (2005).

[28] Z. Spitalsky, D. Tasis, K. Papagelis, and C. Galiotis. Carbon nanotube–polymer composites: chemistry, processing, mechanical and electrical properties. Prog. Polym. Sci. 35, 357 (2010).

[29] M. Rahmat and P. Hubert. Carbon nanotube–polymer interactions in nanocomposites: a review. Compos. Sci. Technol. 72, 72 (2011).

[30] L. Jin, C. Bower, and O. Zhou. Alignment of carbon nanotubes in a polymer matrix by mechanical stretching. Appl. Phys. Lett. 73, 1197 (1998).

[31] M. S. P. Shaffer and A. H. Windle. Fabrication and characterization of carbon nanotube/poly(vinyl alcohol) composites. Adv. Mater. 11, 937 (1999).

[32] X. Gong, J. Liu, S. Baskaran, R. D. Voise, and J. S. Young. Surfactant-assisted processing of carbon nanotube/polymer composites. Chem. Mater. 12, 1049 (2000).

[33] A. Dufresne, M. Paillet, J. Putaux, R. Canet, F. Carmona, P. Delhaes, and S. Cui. Processing and characterization of carbon nanotube/poly (styrene-co-butyl acrylate) nanocomposites. J. Mater. Sci. 37, 3915 (2002).

[34] O. Probst, E. M. Moore, D. E. Resasco, and B. P. Grady. Nucleation of polyvinyl alcohol crystallization by single-walled carbon nanotubes. Polymer 45, 4437 (2004).

[35] F. Dalmas, L. Chazeau, C. Gauthier, K. Masenelli-Varlot, R. Dendievel, J.-Y. Cavaille, and L. Forro. Multiwalled carbon nanotube/polymer nanocomposites: processing and properties. J. Polym. Sci. B: Polym. Phys. 43, 1186 (2005).

[36] M. Zhang, L. Su, and L. Mao. Surfactant functionalization of carbon nanotubes (CNTs) for layer-by-layer assembling of CNT multi-layer films and fabrication of gold nanoparticle/CNT nanohybrid. Carbon 44, 276 (2006).

[37] S. Bose, R. A. Khare, and P. Moldenaers. Assessing the strengths and weaknesses of various types of pre-treatments of carbon nanotubes on the properties of polymer/carbon nanotubes composites: a critical review. Polymer 51, 975 (2010).

[38] Y. Y. Huang and E. M. Terentjev. Dispersion of carbon nanotubes: mixing, sonication, stabilization, and composite properties. Polymers 4, 275 (2012).

[39] P. Pötschke, A. R. Bhattacharyya, A. Janke, and H. Goering. Melt mixing of polycarbonate/multi-wall carbon nanotube composites. Compos. Interfaces 10, 389 (2003).

[40] O. Breuer and U. Sundararaj. Big returns from small fibers: a review of polymer/carbon nanotube composites. Polym. Compos. 25, 630 (2004).

[41] E. J. Siochi, D. C. Working, C. Park, P. T. Lillehei, J. H. Rouse, C. C. Topping, A. R. Bhattacharyya, and S. Kumar. Melt processing of SWCNT–polyimide nanocomposite fibers. Compos. Part B 35, 439 (2004).

[42] W. D. Zhang, L. Shen, I. Y. Phang, and T. Liu. Carbon nanotubes reinforced nylon-6 composite prepared by simple melt-compounding. Macromolecules 37, 256 (2004).

[43] B. Lin, U. Sundararaj, and P. Pötschke. Melt mixing of polycarbonate with multi-walled carbon nanotubes in miniature mixers. Macromol. Mater. Eng. 291, 227 (2006).

[44] Z. Jia, Z. Wang, C. Xu, J. Liang, B. Wei, D. Wu, and S. Zhu. Study on poly (methyl methacrylate)/carbon nanotube composites. Mater. Sci. Eng. A 271, 395 (1999).

[45] C. Velasco-Santos, A. L. Martínez-Hernández, F. T. Fisher, R. Ruoff, and V. M. Castano. Improvement of thermal and mechanical properties of carbon nanotube composites through chemical functionalization. Chem. Mater. 15, 4470 (2003).

[46] K. W. Putz, C. A. Mitchell, R. Krishnamoorti, and P. F. Green. Elastic modulus of single-walled carbon nanotube/poly (methyl methacrylate) nanocomposites. J. Polym. Sci. B: Polym. Phys. 42, 2286 (2004).

[47] S. Kumar, T. D. Dang, F. E. Arnold, A. R. Bhattacharyya, B. G. Min, X. Zhang, R. A. Vaia, C. Park, W. W. Adams, R. H. Hauge, R. E. Smalley, S. Ramesh, and P. A. Willis. Synthesis, structure, and properties of PBO/SWNT composites. Macromolecules 35, 9039 (2002).

[48] C. Zhao, G. Hu, R. Justice, D. W. Schaefer, S. Zhang, M. Yang, and C. C. Han. Synthesis and characterization of multi-walled carbon nanotubes reinforced polyamide 6 via in situ polymerization. Polymer 46, 5125 (2005).

[49] Y. Lin, B. Zhou, K. A. S. Fernando, P. Liu, L. F. Allard, and Y.-P. Sun. Polymeric carbon nanocomposites from carbon nanotubes functionalized with matrix polymer. Macromolecules 36, 7199 (2003).

[50] C. A. Dyke and J. M. Tour. Covalent functionalization of single-walled carbon nanotubes for materials applications. J. Phys. Chem. A 108, 11151 (2004).

[51] K. Balasubramanian and M. Burghard. Chemically functionalized carbon nanotubes. Small 1, 180 (2005).

[52] L. Xie, F. Xu, F. Qiu, H. Lu, and Y. Yang. Single-walled carbon nanotubes functionalized with high bonding density of polymer layers and enhanced mechanical properties of composites. Macromolecules 40, 3296 (2007).

[53] C. Bartholome, P. Miaudet, A. Derré, M. Maugey, O. Roubeau, C. Zakri, and P. Poulin. Influence of surface functionalization on the thermal and electrical properties of nanotube–PVA composites. Compos. Sci. Technol. 68, 2568 (2008).

[54] Y. Hou, J. Tang, H. Zhang, C. Qian, Y. Feng, and J. Liu. Functionalized few-walled carbon nanotubes for mechanical reinforcement of polymeric composites. ACS Nano 3, 1057 (2009).

[55] P.-C. Ma, N. A. Siddiqui, G. Marom, and J.-K. Kim. Dispersion and functionalization of carbon nanotubes for polymer-based nanocomposites: a review. Compos. Part A 41, 1345 (2010).

[56] H. Xia, Q. Wang, and G. Qiu. Polymer-encapsulated carbon nanotubes prepared through ultrasonically initiated in situ emulsion polymerization. Chem. Mater. 15, 3879 (2003).

[57] S. Qin, D. Qin, W. T. Ford, D. E. Resasco, and J. E. Herrera. Functionalization of single-walled carbon nanotubes with polystyrene via grafting to and grafting from methods. Macromolecules 37, 752 (2004).

[58] G. L. Hwang, Y.-T. Shieh, and K. C. Hwang. Efficient load transfer to polymer-grafted multiwalled carbon nanotubes in polymer composites. Adv. Funct. Mater. 14, 487 (2004).

[59] H.-J. Jin, H. J. Choi, S. H. Yoon, S. J. Myung, and S. E. Shim. Carbon nanotube-adsorbed polystyrene and poly (methyl methacrylate) microspheres. Chem. Mater. 17, 4034 (2005).

[60] M. Salami-Kalajahi, V. Haddadi-Asl, F. Behboodi-Sadabad, S. Rahimi-Razin, and H. Roghani-Mamaqani. Properties of PMMA/carbon nanotubes nanocomposites prepared by "grafting through" method. Polym. Compos. 33, 215 (2012).

[61] X. Tong, C. Liu, H.-M. Cheng, H. Zhao, F. Yang, and X. Zhang. Surface modification of single-walled carbon nanotubes with polyethylene via in situ Ziegler-Natta polymerization. J. Appl. Polym. Sci. 92, 3697 (2004).

[62] M. L. Shofner, V. N. Khabashesku, and E. V. Barrera. Processing and mechanical properties of fluorinated single-wall carbon nanotube–polyethylene composites. Chem. Mater. 18, 906 (2006).

[63] B.-X. Yang, K. P. Pramoda, G. Q. Xu, and S. H. Goh. Mechanical reinforcement of polyethylene using polyethylene-grafted multiwalled carbon nanotubes. Adv. Funct. Mater. 17, 2062 (2007).

[64] J. F. Vega, Y. da Silva, E. Vicente-Alique, R. Núñez-Ramírez, M. Trujillo, M. L. Arnal, A. J. Müller, P. Dubois, and J. Martínez-Salazar. Influence of chain branching and molecular weight on melt rheology and crystallization of polyethylene/carbon nanotube nanocomposites. Macromolecules 47, 5668 (2014).

[65] H. Li, F. Cheng, A. M. Duft, and A. Adronov. Functionalization of single-walled carbon nanotubes with well-defined polystyrene by "click" coupling. J. Am. Chem. Soc. 127, 14518 (2005).

[66] B. Fragneaud, K. Masenelli-Varlot, A. Gonzalez-Montiel, M. Terrones, and J.-Y. Cavaillé. Mechanical behavior of polystyrene grafted carbon nanotubes/polystyrene nanocomposites. Compos. Sci. Technol. 68, 3265 (2008).

[67] B. P. Grady, A. Paul, J. E. Peters, and W. T. Ford. Glass transition behavior of single-walled carbon nanotube–polystyrene composites. Macromolecules 42, 6152 (2009).

[68] Z. Wu, W. Feng, Y. Feng, Q. Liu, X. Xu, T. Sekino, A. Fujii, and M. Ozaki. Preparation and characterization of chitosan-grafted multiwalled carbon nanotubes and their electrochemical properties. Carbon 45, 1212 (2007).

[69] J. Venkatesan, Z.-J. Qian, B. Ryu, N. A. Kumar, and S.-K. Kim. Preparation and characterization of carbon nanotube-grafted-chitosan–natural hydroxyapatite composite for bone tissue engineering. Carbohydr. Polym. 83, 569 (2011).

[70] J. N. Coleman, M. Cadek, R. Blake, V. Nicolosi, K. P. Ryan, C. Belton, A. Fonseca, J. B. Nagy, Y. K. Gun'ko, and W. J. Blau. High performance nanotube-reinforced plastics: understanding the mechanism of strength increase. Adv. Funct. Mater. 14, 791 (2004).

[71] H. Deng, R. Zhang, C. T. Reynolds, E. Bilotti, and T. Peijs. A novel concept for highly oriented carbon nanotube composite tapes or fibres with high strength and electrical conductivity. Macromol. Mater. Eng. 294, 749 (2009).

[72] B. S. Shim, J. Zhu, E. Jan, K. Critchley, S. Ho, P. Podsiadlo, K. Sun, and N. A. Kotov. Multiparameter structural optimization of single-walled carbon nanotube composites: toward record strength, stiffness, and toughness. ACS Nano 3, 1711 (2009).

[73] Y. Wang, F. Wei, G. Luo, H. Yu, and G. Gu. The large-scale production of carbon nanotubes in a nano-agglomerate fluidized-bed reactor. Chem. Phys. Lett. 364, 568 (2002).

[74] Q. Zhang, J.-Q. Huang, M.-Q. Zhao, W.-Z. Qian, and F. Wei. Carbon nanotube mass production: principles and processes. ChemSusChem 4, 864 (2011).

[75] K. Jiang, Q. Li, and S. Fan. Spinning continuous carbon nanotube yarns. Nature 419, 801 (2002).

[76] M. Zhang, K. R. Atkinson, and R. H. Baughman. Multifunctional carbon nanotube yarns by downsizing an ancient technology. Science 306, 1358 (2004).

[77] X. Zhang, K. Jiang, C. Feng, P. Liu, L. Zhang, J. Kong, T. Zhang, Q. Li, and S. Fan. Spinning and processing continuous yarns from 4-inch wafer scale super-aligned carbon nanotube arrays. Adv. Mater. 18, 1505 (2006).

[78] Q. Li, X. Zhang, R. F. DePaula, L. Zheng, Y. Zhao, L. Stan, T. G. Holesinger, P. N. Arendt, D. E. Peterson, and Y. T. Zhu. Sustained growth of ultralong carbon nanotube arrays for fiber spinning. Adv. Mater. 18, 3160 (2006).

[79] K. Liu, Y. Sun, L. Chen, C. Feng, X. Feng, K. Jiang, Y. Zhao, and S. Fan. Controlled growth of super-aligned carbon nanotube arrays for spinning continuous unidirectional sheets with tunable physical properties. Nano Lett. 8, 700 (2008).

[80] L. Zheng, G. Sun, and Z. Zhan. Tuning array morphology for high-strength carbon-nanotube fibers. Small 6, 132 (2010).

[81] Q. F. Cheng, J. P. Wang, J. J. Wen, C. H. Liu, K. L. Jiang, Q. Q. Li, and S. S. Fan. Carbon nanotube/epoxy composites fabricated by resin transfer molding. Carbon 48, 260 (2010).

[82] X. Wang, Z. Z. Yong, Q. W. Li, P. D. Bradford, W. Liu, D. S. Tucker, W. Cai, H. Wang, F. G. Yuan, and Y. T. Zhu. Ultrastrong, stiff and multifunctional carbon nanotube composites. Mater. Res. Lett. 1, 19 (2013).

[83] X. Wang, Q. Jiang, W. Xu, W. Cai, Y. Inoue, and Y. Zhu. Effect of carbon nanotube length on thermal, electrical and mechanical properties of CNT/bismaleimide composites. Carbon 53, 145 (2013).

[84] Y.-N. Liu, M. Li, Y. Gu, K. Wang, D. Hu, Q. Li, and Z. Zhang. A modified spray-winding approach to enhance the tensile performance of array-based carbon nanotube composite films. Carbon 65, 187 (2013).

[85] Q. Jiang, X. Wang, Y. Zhu, D. Hui, and Y. Qiu. Mechanical, electrical and thermal properties of aligned carbon nanotube/polyimide composites. Compos. Part B 56, 408 (2014).

[86] M. D. Frogley, Q. Zhao, and H. D. Wagner. Polarized resonance Raman spectroscopy of single-wall carbon nanotubes within a polymer under strain. Phys. Rev. B 65, 113413 (2002).

[87] J. Di, D. Hu, H. Chen, Z. Yong, M. Chen, Z. Feng, Y. Zhu, and Q. Li. Ultrastrong, foldable, and highly conductive carbon nanotube film. ACS Nano 6, 5457 (2012).

[88] Y. Wang, M. Li, Y. Gu, X. Zhang, S. Wang, Q. Li, and Z. Z. Zhang. Tuning carbon nanotube assembly for flexible, strong and conductive films. Nanoscale 7, 3060 (2015).

[89] L. Zhang, X. Wang, W. Xu, Y. Zhang, Q. Li, P. D. Bradford, and Y. Zhu. Strong and conductive dry carbon nanotube films by microcombing. Small 11, 3830 (2015).

[90] X. Yu, J. Zou, Z. Lan, C. Jiang, J. Zhao, D. Zhang, X. Zhang, M. Miao, and Q. Li. Solvent-tunable microstructures of aligned carbon nanotube films. (Unpulished results).

[91] J. Jia, J. Zhao, G. Xu, J. Di, Z. Yong, Y. Tao, C. Fang, Z. Zhang, X. Zhang, L. Zheng, and Q. Li. A comparison of the mechanical properties of fibers spun from different carbon nanotubes. Carbon 49, 1333 (2011).

[92] D. Lashmore, Supercapacitors and methods of manufacturing same, (2008), US Patent No. 20080225464.

[93] Q. Cheng, J. Bao, J. Park, Z. Liang, C. Zhang, and B. Wang. High mechanical performance composite conductor: multi-walled carbon nanotube sheet/bismaleimide nanocomposites. Adv. Funct. Mater. 19, 3219 (2009).

[94] J. E. Fischer, W. Zhou, J. Vavro, M. C. Llaguno, C. Guthy, R. Haggenmueller, M. J. Casavant, D. E. Walters, and R. E. Smalley. Magnetically aligned single wall carbon nanotube films: preferred orientation and anisotropic transport properties. J. Appl. Phys. 93, 2157 (2003).

[95] Q. Cheng, B. Wang, C. Zhang, and Z. Liang. Functionalized carbon-nanotube sheet/bismaleimide nanocomposites: mechanical and electrical performance beyond carbon-fiber composites. Small 6, 763 (2010).

[96] D. Gagnebien, P. J. Madec, and E. Maréchal. Synthesis of poly(sulphone-*b*-siloxane)s—I. Model study of the epoxy-phenol reaction in the melt. Eur. Polym. J. 21, 273 (1985).

[97] G. Wei and H. J. Sue. Fracture mechanisms in preformed polyphenylene oxide particle-modified bismaleimide resins. J. Appl. Polym. Sci. 74, 2539 (1999).

[98] Y. Li, J. Miranda, and H.-J. Sue. Hygrothermal diffusion behavior in bismaleimide resin. Polymer 42, 7791 (2001).

[99] H.D. Wagner. Nanocomposites: paving the way to stronger materials. Nat. Nanotechnol. 2, 742 (2007).

Chapter 4

Carbon Nanotube–Polymer Composites: Device Properties and Photovoltaic Applications

T. Hosseini and N. Kouklin

Additional information is available at the end of the chapter

http://dx.doi.org/10.5772/62692

Abstract

This chapter provides an in-depth coverage of recent advances in the areas of the development and characterization of electro-optically active, device-grade carbon nanotube (CNT)–polymer blends. These new organic–inorganic multifunctional nanocomposites share many advanced characteristics which make them ideally suited for industrial scale, high-throughput manufacturing of lightweight, flexible electronic, light switching and emitting as well as energy harvesting devices of extremely low cost. The fundamental aspects and the physical mechanisms controlling light–matter interaction, photo-conversion, and photo-generated charge-carrier transport in these nanotube–polymer composites as well as the influence of the processing conditions on the electronic properties and device-related performances are further reviewed and discussed.

Keywords: Carbon nanotube, Polymer, Composite, Exciton, Photovoltaic, Solar cell, Diode, Detector

1. Introduction

Blends of conjugated polymers and high performance carbon-based nanosemiconductors are an emerging class of easy-to-fabricate organic–inorganic nanocomposite materials with the potential to profoundly influence many electronic device market segments, including optoelectronics. Extraordinary characteristics of carbon nanotubes (CNTs) and prevalence of interfacial regions and the nanoscopic phase are a source of drastic change and gain in the opto-electrical response of the polymer matrix that typically falls outside of classical scaling behavior of conventional polymer composites. These novel nanocomposites and their based devices can be fabricated using roll-to-roll techniques that makes them ideally suited to industrial scale, high-throughput manufacturing of lightweight, flexible electronic, light switching and emitting as well as energy harvesting devices at extremely low cost [1–6].

Conjugated polymers exhibit electronic and light emission properties that are similar to those of crystalline semiconductors and have been already implemented in organic optoelectronic devices such as organic light-emitting diodes (OLEDs), switches, and organic photovoltaic (OPV) cells [7]. Incorporating n-type dopants in the form of metallic CNTs into p-type polymer matrix has been shown to greatly enhance performance of such OPV cells by increasing the rates of non-radiative dissociation of excitons as well as charge-carrier collection efficiency. The formation of optimally loaded networks of electrically conductive nanotube network in turn entails detailed the consideration of the influence of the process parameters on the physical characteristics and interaction of the polymer with the nanotubes in a liquid phase. As the absorption coefficient of photosensitive polymers remains large, light is typically absorbed within a very thin layer, which drastically benefits the efficiency-to-cost ratio for these cells [8, 9]. The π-conjugation in polymers results in an energy separation of ~1–3 eV between the lowest unoccupied molecular orbital (LUMO) and the highest occupied molecular orbital (HOMO). As a result, the light absorption–emission spectrum falls in the visible near-infrared (NIR) spectral range that complements that of single-walled carbon nanotubes (SWNTs), that is, the near IR–UV [2, 10–12]. An abrupt, type-II band alignment between polymer matrix and carbon phase is required and can be realized for many nanotube–polymer composites to achieve sufficiently fast interfacial charge separation and pronounced photovoltaic effect [13, 14].

Among different classes of nanomaterials including semiconductor quantum dots and fullerenes, SWNTs have been proven particularly suitable for uses in OPV, photodetector, and light-emitting diode applications based on conjugated polymers because of their large aspect ratio and remarkable optoelectronic properties including bandgap tunability, strong optical absorptivity, ballistic transport, solution-processability, and excellent chemical stability [15, 16]. Owing to their quasi-one-dimensional structure and improved transport characteristics, a class of SWNTs has been confirmed to exhibit many favorable device functionalities which make them attractive for application in a variety nano-electronic and mechanical devices and systems, among which are interconnects, rectifies, field-effect transistors, analyte, and light sensors. Compared with other nanostructures, SWNTs are also known to exhibit strong multi-range absorption in part associated with resonance-type interband electronic transitions (e.g., S_{11}, M_{11}, S_{22}) as well as free carrier and plasmonic excitations. Recent experiments further confirmed on the presence of a strong photoconduction response in the infrared (IR) which can in turn afford many new opportunities in engineering nanophotovoltaic and optoelectronic organic polymer–SWNT-based devices operating over multiple spectral ranges, including IR [17–19].

High conversion efficiencies of ~5 and 9% were extracted in case of polymer-based OPV cells featuring C_{60} molecules and CNTs, respectively. Yet, unlike to C_{60}, polymers incorporating aligned CNTs demonstrate much larger intrinsic charge mobility at lower percolation threshold/limit. At the same time, the increased photo-generated charge transport and in turn

http://dx.doi.org/10.5772/62692

collection efficiency facilitate the development of OPV cells featuring larger light absorption (thicker active device layer) and electrical power output, which translates into overall higher efficiency-to-cost ratio for these cells. Combining SWNTs with electrooptically active polymers thus provides an attractive route to creating a new generation of multifunctional device-grade organic–inorganic electronic materials for uses as sensors, OLEDs, PV cells, electromagnetic absorbers, and other electronic devices [20–22].

In this chapter, we review the progress while focusing on the fundamental aspects behind the light–matter interaction, photo-conversion, and photo-carrier generation as well charge-carrier transport in SWNT–polymer composites. The fabrication, structural–mechanical, and transport characteristics of various nanotubes–polymer-based composites are reviewed in Section 2. Key photo-physical processes that take place at the interface between SWNT and polymer molecules including energy transfer, exciton dissociation, charge transfer, and related effects are reviewed in Section 3. Section 4 discusses the electronic and optoelectronic devices built based on SWNT/polymer composites including OPV cells, light-emitting diodes, and IR sensors.

2. Carbon nanotubes/polymer composites: synthesis and properties

Recent studies involving fabrication and characterization of structural and underlying device characterizations have identified several processing-related challenges pertaining to producing polymer/nanotubes composites of high purity, structural anisotropy/alignment, and uniform dispersion [23, 24]. Because of the π-orbitals of the sp^2-hybridized C atoms, CNTs show a tendency for strong intermolecular interaction and spontaneous aggregation (van der Waals interaction) into large diameter bundles that are not readily dispersible in organic solvents or polymer matrix. To address the dispersion-related and mixing challenges, the use of surfactants [25–31], performing shear mixing [32–34], sidewall chemical modification [35, 36], and in situ polymerization [37–41] were proposed. Among all these strategies, covalent chemical functionalization and introduction of defects into SWNT surfaces have been proven highly effective in achieving stable SWNT suspensions in polar solvents as discussed below.

2.1. Defect functionalization

In defect functionalization, nanotubes are treated by oxidative methods that also help remove metal particles and amorphous carbon deposits, that is, raise purity. The resultant SWNTs oftentimes gain in localized surface defect density most of which are in the form as carboxyl, that is, –COOH attachments. Mawhinney et al. [42] studied surface defect site density of oxidatively treated SWNTs by probing the amounts of CO_2(g) and CO(g) released during heating to up to 1273 K. The results indicated that as much as ~5% of the carbon atoms in such SWNTs can be defect-associated. Acid–base titration method [43] yielded similar results, that is, 1–3% of acidic sites in purified SWNTs. The density of defective sites created at the surfaces by this method is viewed generally insufficient for good nanotubes dispersion in the polymer matrix. However, the strategy can be used for covalent attachment of organic groups by first converting them into acid chlorides that can be next linked to amines to form amides. Such modified CNTs show significantly higher solubility in organic solvents as compared with unprocessed nanotubes [44].

2.2. Non-covalent functionalization

Non-covalent functionalization routes are of further interest because they do not compromise the integrity of the nanotube backbone while helping improve their solubility and processability. This type of functionalization is primarily done with the aid of surfactants, bio-macromolecules, or non-covalent surface attachments, such as wrapping sidewalls with polymers. Successful implementation of both cationic and anionic surfactants as well as nonionic surfactants has been demonstrated in several studies involving non-covalent functionalization of nanotubes via microemulsion [45–47], ultrasonication [48], sonication [49], and emulsion polymerization [50–54]. Biological macromolecules such as protein/DNA and glucose [44, 55] have been also linked to CNTs via dialysis [56], electro-active interaction [57], and ultrasonication [58]. Poly (4-vinyl pyridine), poly (phenyl acetylene), poly(styrene)–poly(methacrylic acid) [59], poly(m-phenylenevinylene-co-2,5-dioctoxy-p-phenylenevinylene) [60], and poly[2-methoxy-5-(20-ethylhexyloxy)21,4-phenylene vinylene] [61], have been reported to non-covalently wrap nanotubes using sol–gel chemistry, solution mixing, and immobilization methods, **Figure 1**.

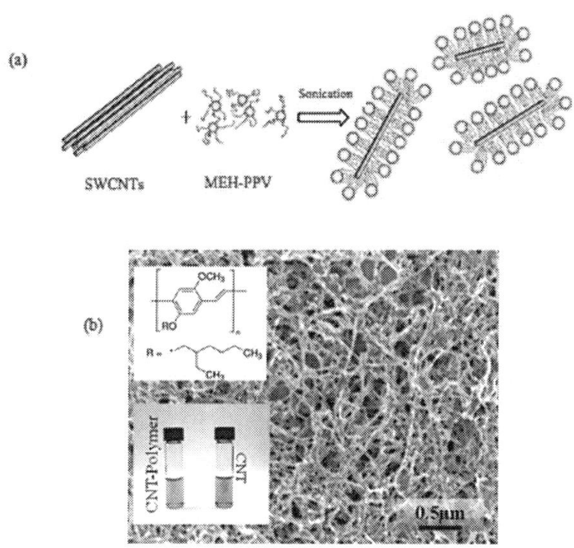

Figure 1. (a) Scheme for the formation of MEH-PPV/carbon nanotubes composites, (b) top-view SEM image of the MEH-PPV encased nanotubes films fabricated by immobilization method. The lower inset shows a digital photo of the suspensions of the SWNTs and MEH-PPV–SWNTs.

2.3. Covalent functionalization

Despite the fact that sp^2-hybridized C atoms form a chemically stable backbone, a number of strategies were developed to covalently link chemical groups to CNTs [62–64]. In the case of covalent functionalization, the translational symmetry of nanotubes is disrupted by changing sp^2 carbon atoms to sp^3 carbon atoms that were reported to affect the electronic and transport properties of nanotubes [65, 66]. This route is highly effective in increasing solubility as well as dispersion of nanotubes in many organic solvents as well as polymers. Covalent functionalization can be accomplished either by the modification of surface-bound carboxyl groups on the nanotubes or by the direct elemental reaction with carbon atoms such as in the case of CH_x-modified nanotubes. Poly (ε-caprolactone) [67, 68], poly(L-lactide) [69, 70], poly(methyl methacrylate) [71–73], Polystyrene [74–76], poly(N-isopropyl acrylamide) [77–80], polyimide [81, 82], polyvinyl (acetate-co-vinyl alcohol) [83], have been used to covalently attach to CNTs.

From the standpoint of device application, non-covalent functionalization remains preferred over the covalent approach, as the latter has the propensity to induce strong structural damage [24, 84]. The dispersion of nanotubes in polymer matrices is one of the most critical bottlenecks in the preparation of CNTs/polymer composites. Additional strategies to enhance dispersion of nanotubes included melt mixing and in situ polymerization, whereas Ni et al. confirmed considerable improvement in the dispersion of multi-walled CNTs in poly(vinyl alcohol) (PVA) matrix through gum Arabic treatment [24].

3. Photo-physical properties of carbon nanotube–polymer composites

3.1. Energy transfer in carbon nanotube–polymer composites

Absorption of a photon by aromatic polymers leads to a formation of bound electron–hole pair known as exciton, which can dissociate radiatively by emitting a lower energy photon. The presence of semiconducting SWNTs has been shown to strongly affect the rate of radiative recombination by inducing the transfer of either holes or electrons to the nanotubes which depends on the electronic band alignment between SWNTs and polymer [85]. Alternatively, resonant energy transfer from polymers to SWNTs has been confirmed experimentally [86, 87]. In Umeyama et al. [86] study, a conjugated polymer, poly [(p-phenylene-1,2-vinylene)-co-(p-phenylene-1,1-vinylidene)] (coPPV), was synthesized and used to study the influence of SWNTs on the light emission characteristics of the former. UV–vis–NIR absorption and AFM measurements revealed that SWNTs were dispersed well in organic solvents likely via π–π interaction. The composite solution of coPPV–SWNTs exhibited a strong NIR emission originating from SWNT when the polymer was subject to a direct optical excitation with the light source operating at ~400–500 nm.

The efficiency and rate of the energy transfer from polymers to SWNTs have been shown to be strongly dependent on the polymer concentration/aggregation on SWNTs [22, 88]. Further studies point to the polymer π-conjugation chain that governs the energy transfer in the polymer–SWNT system to remain more extended compared with that of the pure polymer system [85]. Massuyeau et al. [89] studied energy transfer between the polymer and nanotubes by examining steady state PL spectra of a series of composite films containing both metallic and semiconducting nanotubes. The results of these studies show that there is a substantial spectral overlap of PL and optical absorption of SWNTs, which favors the Förster energy transfer between polymer chains and CNTs.

3.2. Charge transfer in carbon nanotube–polymer composites

Combining CNTs with polymers offers an attractive route not only to mechanically reinforcing polymer films but also to enhancing polymers' charge transport properties and modifying electronic properties through a morphological modification or electronic coupling between the two [90]. The effect of nanotube doping has been systematically investigated by embedding nanotube powders in the emission, electron transport, and hole transport layers of OLEDs [91]. Such polymer/nanotube composites have been successfully exploited for various applications including OPV [92–95], OLEDs [96], and organic field-effect transistors [97, 98]. Among different transport models [99–104], percolation of the nanotube network within the polymer matrix has been suggested to play a primary role behind improved charge mobility of up to two orders of magnitude compared with that in the pristine polymer. This provides a technologically simple pathway to improving the performance of organic electronic and optoelectronic devices, while keeping their fabrication costs as low as possible [90].

The low dielectric constant of conjugated polymers results in large Coulomb interactions between charge carriers, increasing exciton binding energy and photo-response characteristics. The majority of OPV devices operate based on exciton dissociation at the interface formed by two dissimilar materials with a type-II band alignment that favors interfacial charge separation and formation of free polarons. If the rate of bound electron–hole pair separation is low, other, that is, radiative and non-radiative recombinations will prevail, which is a primary reason behind efficiency loss. Internal electric fields at the polymer–metal interfaces (interface dipoles) or dissociation centers, for instance, oxygen impurities that can act as electron traps (monopoles) promote fast exciton dissociation. As the electron affinity remains smaller for conjugated polymers [105], percolated CNTs act as high mobility electron extraction paths or excitonic antennas. Even at low doping levels, highly conductive pathways can be still established due to a large aspect ratio and propensity of SWNT to bundling. While photo-generated electrons will tend to transfer to SWNT, the photo-generated holes are to remain in the polymer matrix that helps to lower the rate of internal recombinations and to mitigate charge-carrier losses [13, 106].

The first solid evidence of the charge transfer between SWCNTs and conjugated polymers (MEH-PPV) was provided by Yang et al. [107] by performing photoinduced absorption spectroscopy. In their study, photoinduced charge transfer was deduced by observing a reduction of the emission from the polymer accompanied by an increase of the polaron peak in the MEH-PPV-SWCNT hybrids. Bindl et al. [108] examined exciton dissociation and charge transfer at s-SWCNT heterojunction formed with archetypical polymeric photovoltaic materials including fullerenes, poly(thiophene), and poly (phenylenevinylene) using an exciton dissociation-sensitive photo-capacitor measurement technique that is advantageously insensitive to optically induced thermal photoconductive effects. It was found that fullerene and polythiophene derivatives induce exciton dissociation resulting in electron and hole transfer away from optically excited s-SWCNTs. Significantly weaker and almost no charge transfer was observed for large bandgap polymers largely due to insufficient energy band offsets.

In another study, Ham et al. [109] fabricated a planar nano-heterojunction comprising well-isolated millimeter-long SNWTs placed underneath a poly(3-hexylthiophene) (P3HT) layer. The resulting junctions displayed photovoltaic efficiencies per nanotube in the range of 3–4%, which exceeded those of polymer/nanotube bulk heterojunctions by almost two orders of magnitude. The increase was attributed to an absence of aggregates in case of the planar device geometry. It was shown that the polymer/nanotube interface itself can be responsible for the exciton dissociation with the best efficiency realized for ~60 nm thick P3HT layer.

Among different classes of nanomaterials, semiconducting CNTs remain the primary candidates to enhance the charge separation when interfaced with conjugated polymers. The difference in the behavior of semiconducting and metallic CNTs in polymer was studied theoretically by Kanai et al. [110] who employed a density functional theory. Case studies involving poly-3-hexylthiophene (P3HT) interfaced with semiconducting and metallic CNTs were carried out. In case of semiconducting nanotubes, the theory predicts a formation of type-II heterojunction, critical to photovoltaic applications. In contrast, in case of the metallic nanotubes, substantial charge redistribution occurred and the built-in-potential was quite small, whereas P3HT became electrostatically more attractive for electrons. These observations confirm that in case of mixed single-walled nanotubes, a majority of interfaces would be made by metallic components to compromise the device performance. Similar conclusions were drawn by Holt et al. [111] in his study of P3HT-polymer/SWNT blends containing varying ratios of metallic to semiconducting SWNTs.

4. Electronic and optoelectronic applications of carbon nanotube/polymer composites

4.1. Organic photovoltaic devices

OPV devices based on π-conjugated polymers have been suggested as low-cost alternatives to silicon-based solar cells [106, 112]. Unlike to energy conversion devices based on semiconductors, in organic solar cell devices, a donor/acceptor (D/A) interface is required to break free photo-generated excitons into free charges carries before they can be collected by the electrodes [113, 114]. The list of the requirements for the materials for application in bulk PV devices includes the following: (1) strong light absorption over the whole solar emission spectrum; (2) sufficient separation between HOMO and LUMO; (3) large electron and hole mobilities within the device active layer; and (4) low device fabrication cost [22, 115]. In addition to a detailed consideration of intrinsic electronic aspects of the constituent components, geometric aspects and chemical stability play equally important role. For example, the dimensions of active layer should not exceed the exciton diffusion length, reportedly on the order of ~10 nm [22, 113].

In CNTs/polymer photovoltaic devices, the dissociation of excitons can be accomplished through the formation of a staggered gap donor/acceptor, type-II heterojunction formed between the s-SWCNTs and the polymer in which the energy offsets at the hetero-interface exceed the exciton binding energy, E_B. Recent experimental and theoretical studies by Schuettfort [116] and Kanai [110], respectively, demonstrate that a type-II band alignment only exists for certain interfaces, such as between small diameter semiconducting SWNTs and P3HT. Even for such blends, energy transfer from the polymer to SWNTs remains one of the fastest de-excitation channels that compete with the charge transfer processes, with the former facilitated by larger surface area and electron affinity of the nanotubes vs. polymers [105, 117].

Kymakis et al. [118] examined both dark and photocurrent–voltage (J–V) characteristics of poly(3-octylthiophene) (P3OT)/SWNT composite photovoltaic cells as a function of SWNT concentration. An open-circuit voltage (V_{OC}) as high as 0.75 V was obtained for 1% doped SWNTs/ P3OT composite which served as a device active layer. An almost 500-fold increase in the photo-response was partly attributed to a 50-fold increase in the hole mobility due to a reduction in the density of the localized states in P3OT matrix, and in part due to enhanced exciton extraction at the polymer/nanotube junctions. Despite the improvement in the rate of the charge separation, the power conversion efficiency was only 0.04% under 100 mW/cm^2 illumination conditions. A poor dispersion of SWNT and the presence of a mixture of metallic and semiconducting tubes were believed the primary factors behind the low efficiency numbers. In 2011, the same group investigated the use of spin-coated SWNTs as a hole transport layer (HTL) in organic bulk heterojunction photovoltaic devices shown schematically in **Figure 2** to raise the conversion efficiency [119]. Varying thickness SWNT films were repetitively spin coated with dichloroethane and next evaluated as the HTL in P3HT and 1-(3-methoxy-carbonyl)-propyl-1-phenyl-(6,6)C61 (PCBM) photovoltaic devices. It was shown that insertion of ~12-nm-thick SWNT layer led to power conversion efficiencies as high as 3.0%, compared with 1.2 and 2.8% for the devices without and with the traditional PEDOT:PSS acting as the HTL. The improved efficiency was attributed to improved hole transport in the polymer matrix due to a higher degree of crystallinity provided with SWNT.

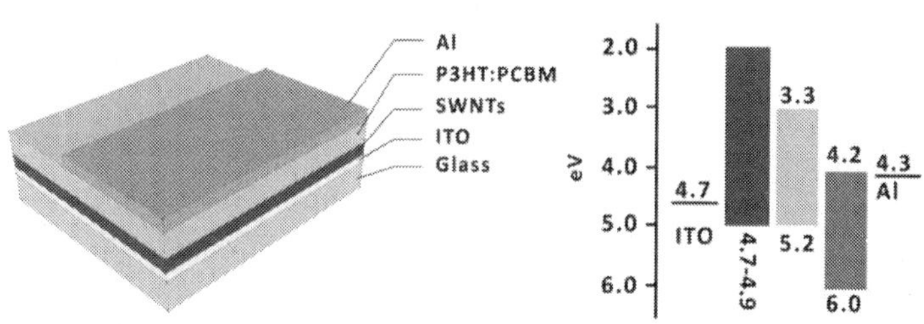

Figure 2. (a) Schematic drawings of the P3HT:PCBM photovoltaic cell with the SWNTs acting as the HTL. (b) Energy level diagrams of photovoltaic device components referenced to the vacuum level.

In another study, June et al. implemented homogeneously dispersed CNTs using alkyl-amide groups to chemically modify nanotubes to improve their dispersion in organic medium [16].

http://dx.doi.org/10.5772/62692

The resultant composites and their based OPV cells exhibited gain in their optical and electrical properties with the device efficiency approaching ~4.4%. The schematic of the fabricated solar cell is shown in **Figure 3**.

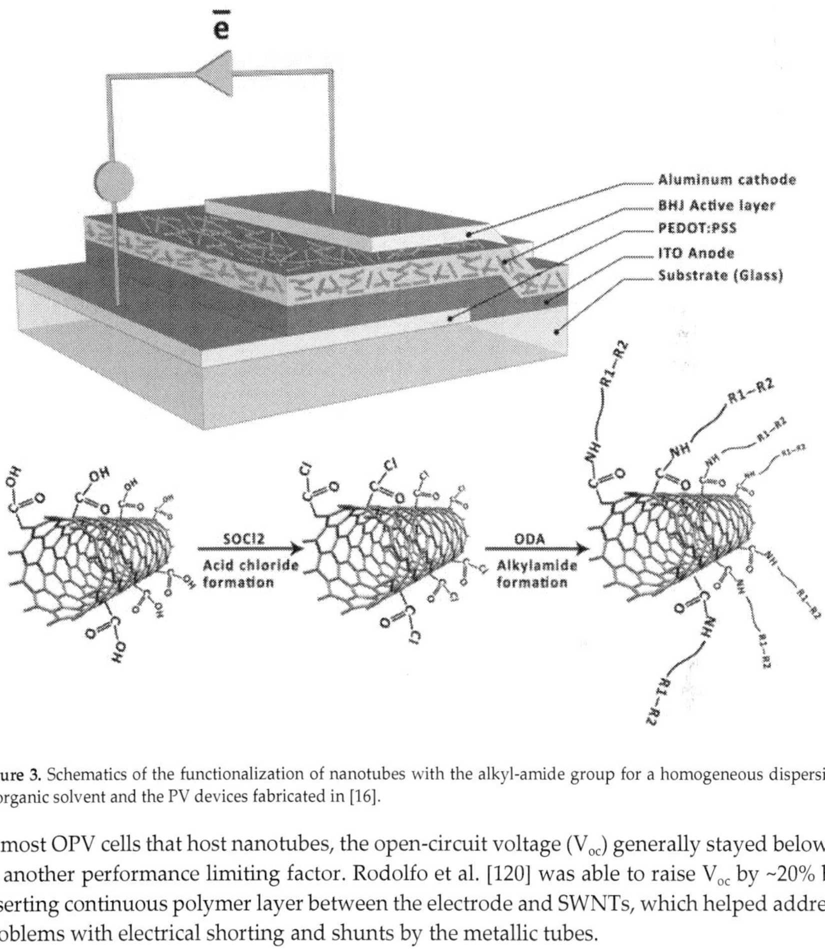

Figure 3. Schematics of the functionalization of nanotubes with the alkyl-amide group for a homogeneous dispersion in organic solvent and the PV devices fabricated in [16].

In most OPV cells that host nanotubes, the open-circuit voltage (V_{oc}) generally stayed below 1 V, another performance limiting factor. Rodolfo et al. [120] was able to raise V_{oc} by ~20% by inserting continuous polymer layer between the electrode and SWNTs, which helped address problems with electrical shorting and shunts by the metallic tubes.

Some prior studies pointed out that uncontrolled interactions at the CNT–polymer interface can not only reduce the ability of the tubes to transport charge but also interfere with the photo-physical processes, which act as a source of recombination centers for excitons (metallic tubes) and energy quenchers (polymer–s-SWCNT), or by electrically shorting the circuit (long tubes). From the standpoint of device engineering practices, a more rational design of the CNTs–polymer interface across different length scales, that is, nano to meso and careful consideration/control of intermolecular level interactions via dispersion will be required [107, 121, 122].

On this front, Arranz-Andres and Blau [122] investigated the influence of the nanotube dimensions (length and diameter) and concentration on the performance of a CNT–polymer device. They found that adding 5% of nanotubes by weight increased the power conversion efficiency (PCE) by three orders of magnitude compared with that of the native polymer. The incorporation of nanotubes into the P3HT matrix favorably affected the energy levels of the P3HT and the morphology of the active layer. They also found that the nanotubes can act as nucleation sites for P3HT chains, improving charge separation and electron transport.

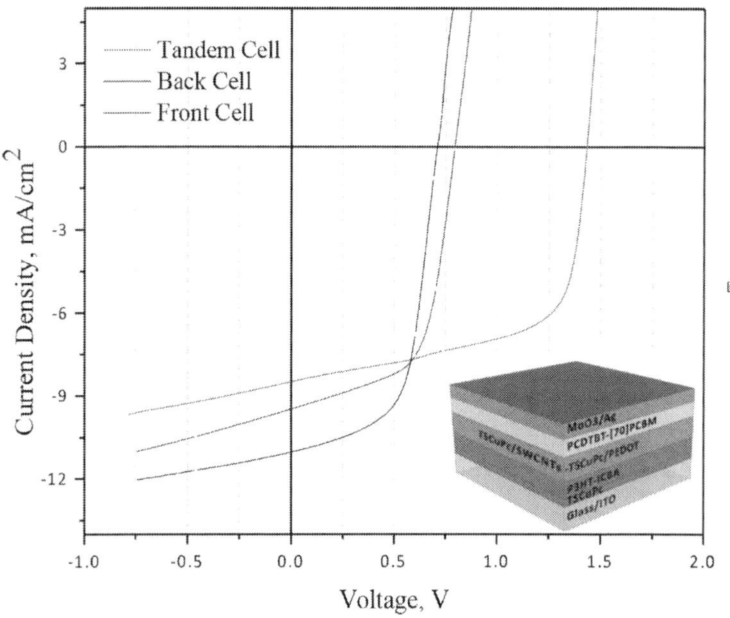

Figure 4. J–V characteristics of series-connected inverted tandem solar cell. Tandem cell curves (red), back cell (blue), front cell (black).

The three-component architectures based on nanotubes-fullerene–conjugated polymer composites were proposed to achieve better photovoltaic efficiencies. Li et al. [125] suggested using C_{60} as an electron acceptor and nanotubes for the photo-generated charge transport. Two types of chemically functionalized nanotubes were tested: carboxylated and octadecylamine functionalized multi-walled nanotubes, in short c-MWNT and o-MWNT. All three photovoltaic parameters, namely short-circuit current density, open-circuit voltage, and fill factor of the P3HT:c-MWNT/C60-based cells showed improvements over those of the P3HT:o-MWNT/C60 cell as a result of a faster electron transfer from C_{60} to the nanotube backbone. Derbal-Habak et al. [94] reported organic PV cells with power conversion efficiency of 3.6% by incorporating functionalized SWNTs within P3HT:PCBM layer that helped improve both the current density J_{sc} and open-circuit voltage, V_{oc} attributed to a partial crystallization of the

RR-P3HT as revealed by XRD studies. Nismy [126] probed the optical and electronic response of the composite devices comprising donor polymer and localized MWNTs also featuring triple heterojunction architecture/scheme. A significant improvement in photoluminescence quenching was observed for the devices with nanotubes embedded into the polymer matrix, with the former facilitating the formation of the trap states. The triple scheme is generally confirmed to yield a lower dark current and hence a significantly improved photovoltaic performance with the PCE approaching ~3.8%.

Relatively high PCEs of ~7.4% were demonstrated by introducing copper-phthalocyanine derivative (TSCuPc)/SWNT layer into the series-connected inverted tandem devices featuring front P3HT–ICBA and back PCBM–PCDTBT active layers, **Figure 4**.

As summarized in **Table 1**, repeated results from studies on CNTs/polymer OPV devices reveal that the performance of nanotubes incorporated OPV cells is dependent on several factors such as the device architecture, treatment or functionalization method of nanotubes, type of CNTs, concentration of nanotubes as well as thickness of the nanotube-incorporated active layer.

Type of CNT	Type of polymer	Preparation method	PCE (%)	FF (%)	References
SWNT	P3HT:PCBM	Spin coating of surfactant-free CNTs as the hole transport layer	3.04	50.7	[119]
SWNT	P3HT:PCBM	Homogenous dispersion by alkyl-amide functionalization of CNT	3.2–4.4	38–52	[16]
SWNT 5 wt%	P3HT	Dispersion in chloroform	52.3	20.3	[122]
SWNT	RR-P3HT: PCBM	Solution mixing	3.66	52.2	[94]
DWNT (double-walled CNT)	P3HT	Dispersion in chloroform	47.2	20.3	[122]
c-MWNT/C_{60}	P3HT	Dispersion of carboxylated nanotubes in o-dichlorobenzene (ODCB) solutions	0.8	44	[125]
MWCNT	P3HT	Dispersion in chloroform	26.8	22	[122]
O-MWCNTs	P3HT, PCBM	Solution mixing	3.8	70	[126]

Table 1. Summary of organic photovoltaic devices reviewed in this chapter

To overcome problems with poor performance of bi-layer devices that stem from short exciton diffusion length in polymers, poor exciton dissociation and absence of a percolated network required for improved photo-generated charge transport, the devices incorporating polymer-fullerene-based donor–acceptor (D–A) material have been reconsidered. Comparative studies on bulk heterojunction devices vs. those with a nanotube-incorporated active layer formed by sequential deposition show that the latter architecture is prone to a higher recombination of carriers due to the introduction of trap states associated with the nanotubes. Photo-generated excitons are also quenched at the D/A material interface due to these additional energy levels and render lower J_{sc} values. On the other hand, the heterojunction scheme yields lower dark currents and better photovoltaic performance confirming a very critical role of the heterojunction in devices with organic/hybrid architectures. For the nanotube/polymer-based OPV cells, the nanotube type is also to influence the performance of such devices. While there is no clear link between the number of walls or the diameter of the nanotubes and the performance of the OPV device, the semiconducting nanotubes were concluded to form a needed, type-II heterojunction. In contrast, in case of metallic nanotubes, a substantial charge redistribution is to take place at the interface. As a result, the built-in-potential is quite small and unlikely to contribute significantly to the subsequent charge separation at this interface, leading to an inefficient PV device. The photovoltaic characteristics of the PV cells are also to depend on the concentration of nanotubes. In particular, the incorporation of low concentrations of nanotubes in the photoactive layer leads to an increase of the current density J_{sc}. The functional groups as well as the preparation methodology are among the other factors that were found to influence the performance of OPV cells.

4.2. Organic light-emitting diodes (OLED)

OLEDs are indispensible to flexible light displays because of their excellent properties: They are lightweight and feature low power consumption, wide angle of view, fast response, low operational voltage, and excellent mechanical flexibility [127, 128]. Light-emitting polymers demonstrate excellent quantum efficiencies and can be solution processed to build electroluminescent devices of very low cost. OLEDs are generally considered as "dual-injected" devices as holes and electrons are injected from the anode and cathode, respectively, into active molecular/macromolecular medium, where they form excitons that recombine radiatively [128, 129].

Recent progress in OLEDs stems not only from the advancement of the polymer science but also from achieving better control over the charge transport in the electroluminescent layers and doping of the emissive materials [22]. A proper layer sequence in OLEDs ensures that the injected charges are properly balanced within the emissive layer to achieve high external efficiency. SWNTs introduced into conducting polymers lower the charge injection barrier formed at the electrode–organic interface and hence favorably affect the device performance [130].

One of the first studies to combine SWNT with conjugated polymer-based OLEDs was attempted by Curran et al. [131]. The observed increase in the quantum yield was attributed to intermolecular π–π stacking interactions that take place between the polymer and nanotubes. A polymer stiffening is another factor that can lead to an increase in the luminescence output. Moreover, when SWNTs are added the strength of the polymer–polymer interaction becomes weaker, which is a source of self-quenching effects. The concentration of SWNTs of 1% (by weight) is considered optimal/sufficient for the polymer strands to experience interaction with the nanotubes. Excess concentrations of SWNTs lead to a drop in the luminescence. Woo et al. [132] prepared double-emitting OLEDs (DE-OLEDs) based on SWNTs-PmPV. A low bias I-Vs obtained on the devices made from the composites were quadratic, while in the devices with pure PmPV, the dependence was significantly more nonlinear: $I \sim V^5$; the result was explained by the presence of structural and chemical defects in the PmPV composite that favors continuous trap-limited charge transport.

In a recent study, Gwinner et al. [134] investigated the influence of small amounts of semiconducting SWNTs on characteristics of ambipolar light-emitting field-effect transistors (LEFETs) comprising polyfluorenes such as poly (9,9-di-n-octylfluorene-alt-benzothiadiazole (F8BT) and poly(9,9-dioctylfluorene) (F8)-conjugated polymers, **Figure 5**. Incorporating SWNTs within a semiconducting layer at the concentrations below the percolation limit significantly augments both hole and electron injections, even for a large bandgap semiconductor such as F8, without invoking a significant luminescence quenching. In general, owning to lower contact resistance and threshold voltage, larger ambipolar currents and in turn higher output/light emissions can be realized.

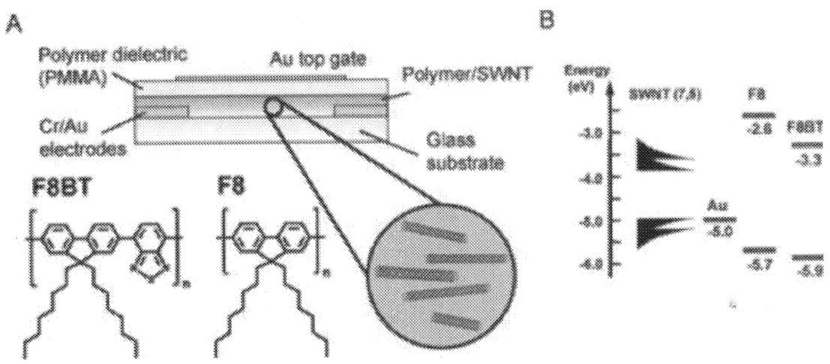

Figure 5. (A) Schematic illustration of bottom contact/top gate polymer field-effect transistor with carbon nanotubes dispersed in the semiconducting polymers F8BT and F8. (B) Energy level diagram of a semiconducting SWNT with (7, 5) chirality, gold (injecting electrode), and HOMO/LUMO levels of both F8 and F8BT. Reprinted from [134] with permission. Copyright © 2011 American Chemical Society.

Divya et al. [134] investigated the use of a diketone ligand, 4,4,5,5,5-pentafluoro-3-hydroxy-1-(phenanthren-3-yl)pent-2-en-1-one (Hpfppd), containing a polyfluorinated alkyl group, by covalently immobilizing it onto the multi-walled CNT host via carboxylic acid functionalization pathway. The resultant nanocomposite displayed intense red emissions with an overall quantum yield of 27% under a wide excitation range from UV to visible (~330–460 nm), making it prime candidate for application in OLEDs.

Indium tin oxide (ITO) features a high transmittance at a low sheet resistance [127] and is ubiquitously employed as an OLED anode but not without drawbacks. ITO is brittle and can suffer from cracks that lead to electrical shorting; it can serve as a source of oxygen that diffuses into emissive layers, while it has insufficiently high work function of ~4.7 eV [129, 135].

On this front, SWNT sheets have been considered as viable alternative and were studied for possible use as anodes in OLEDs, **Figure 6** [136]. Some recent prototypes exhibited brightness of ~2800 cd m^{-2} that was comparable to that of OLED featuring ITO anodes.

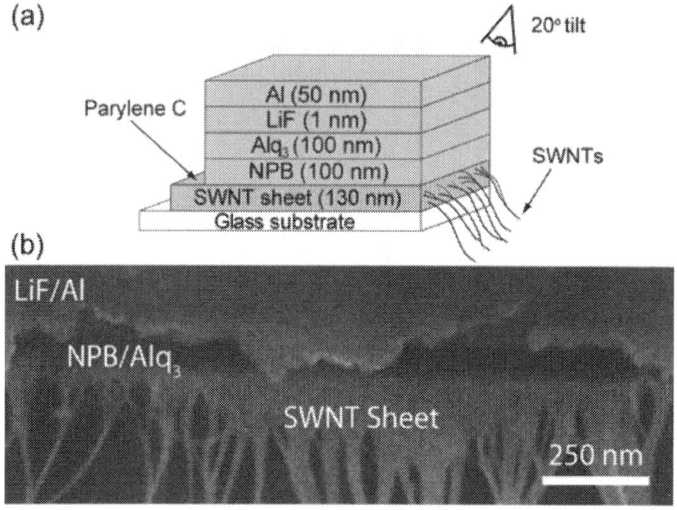

Figure 6. (a) Schematic of the SWNT OLED device and (b) corresponding cross-sectional scanning electron microscopy image at a broken edge taken at a 20° angle from the surface normal. Reprinted from [136] with permission. Copyright © 2006 American Institute of Physics.

Zhang et al. [137] showed arc-discharge nanotubes were overwhelmingly better electrodes than HiPCO-nanotube-based films in all of the critical aspects, including surface roughness, sheet resistance, and transparency. Arc-discharge nanotube films that were PEDOT passivated showed high surface smoothness and featured sheet resistance of ~160 Ω/sq at 87% transparency.

Parekh et al. [138] was able to improve the conductivity of transparent SWNT thin films by treating the samples with nitric acid and thionyl chloride. Geng et al. [139] was able to achieve a fourfold sheet conductance improvement by exposing SWNT films produced by spray technique to a nitric acid with the treated samples demonstrating sheet resistance of ~40 and 70 Ω/sq, at 70 and 80% transmittance, respectively. To break interdependence of the sheet conductance and the transparency, a magnetic field was applied during drop-casting of SWNT–polymer films onto ITO-coated glass and ITO-coated PET substrates [140]. This led to sample de-wetting and enhancement in the electrical conductivity of the films. For a functionalized SWNT–PEDOT:PSS film formed on an ITO-coated PET substrate, a sheet resistance of 90 Ω/sq at 88% transmittance was obtained. SWNT–PEDOT:PSS composite devices formed on PET substrate were proposed as a way to combat the problem, with the films featuring a sheet resistance of 80 Ω/sq, and having a transmittance of 75% at ~550 nm. The ratio of DC to optical conductivity was higher for composites with mass fractions of 5560 wt% than for nanotubes only films. For ~80-nm-thick composite filled with 60 wt% arc discharge nanotubes, this conductivity ratio was maximized at $\sigma_{DC}/\sigma_{0P} = 1$, with the transmittance (at 550 nm) and sheet resistance of 75 and 80 Ω/sq, respectively. These composites also have excellent electromechanical stability, with <1% resistance change over 130 bend cycles.

As outlined above, CNTs/polymer composites could be incorporated into conducting polymers as the buffer layer, or in the form of plain sheets as flexible anode electrode in OLEDs. The characteristics exhibited by the CNTs/polymer composite as the transport layer in OLEDs have been observed to change with the polymer system as influenced by the nature of the polymer–nanotube interactions. Additionally, nanotube sheets can serve as transparent electrodes in OLEDs which make them a viable alternative to the conventional ITO electrodes.

4.3. Infrared sensors

"Infra" from Latin means *"below"*; thus, IR refers to a spectral range beyond the red boundary of the visible electromagnetic spectrum, which corresponds approximately to ~0.8 μm. Since all objects emit IR radiation, the effect is known as a black body radiation, seeing in the dark or through obscured conditions, by detecting the IR energy emitted by objects is possible. IR imaging has therefore become a cornerstone technology for many military and civilian applications including night vision, target acquisition, surveillance, and thermal photovoltaic devices. Biomedical imaging and light-activated therapeutics represent another critical area that particularly benefits from high tissue transparency to IR light. Despite a recent progress in the field of IR sensing and imaging, high cost, requirement for cryogenic cooling, and spectrally limited sensitivity still remain the main disadvantages of this technology today.

Two primary methods of IR detection exist: energy and photon detection. Energy detectors respond to temperature changes generated from incident IR radiation through changes in material properties. Energy detectors, the well-known examples of which are bolometers, pyroelectric, and thermopile detectors, are normally low cost and primarily used in single-detector applications; such applications include fire and motion detection systems as well as automatic light switches and remote thermometers. In contrast to energy detectors, light interacts directly with the semiconductors in photon detectors to generate electrical carriers. More specifically, incident light with energy greater than or equal to the energy gap of the semiconductor drives the semiconductor out of equilibrium by generating excess majority electrical carriers. This translates into a change in the net resistance of the detector. The well-established examples of photon detector materials are lead sulfide (PbS), lead selenide (PbSe). Since these detectors do not function by changing temperature, they respond much faster than energy detectors and in principle can be sensitive to a single photon, if used, for instance, in conjunction with the emerging class of single electron devices. Both, increased sensitivity and reduced response time provided with the use of small bandgap semiconductor materials, have recently led to the development of advanced and very sophisticated IR detection systems, which are of high technological relevance today.

The higher the temperature of an object, the larger the amount of thermal radiation it emits, while its peak intensity also shifts to a shorter wavelength. The demonstrated strong spectral dependence of thermal radiation on the temperature, also known as a Wien's law, necessitates the use of materials with optimized sensitivity at multiple wavelengths for two primary reasons: (1) to increase sensitivity and (2) to enable highly selective military/civilian target identification and acquisition. Until recently, the problem was addressed through simultaneous use of several materials with peak sensitivity corresponding to different wavelengths. As fabrication and processing change dramatically from one material system to another, engineering of wavelength-specific and ultra-sensitive IR detectors currently remains uneconomical.

A recent progress in the field of nanotechnologies, and in particular, in the area of non-lithographic fabrication of multi-functional nanomaterials such as quantum wells, wires, dots and CNTs opens new opportunities for advancing IR sensing technology beyond today's confines. Unlike semiconductor alloys, the effective energy bandgap of nanomaterials and particularly CNTs can be easily tailored by simply changing their size which enables engineering of future IR-devices with expected spectral range of operation: from ~15 to ~0.6 μm (i.e., from ~0.1 to 2eV). Furthermore, as electron scattering is suppressed in materials featuring one-dimensional electronic configurations, nanotube-based IR photo-detectors are expected to demonstrate orders of magnitude improved sensitivity at room temperature as compared with the detectors operating on thin films or quantum wells. This property could potentially mitigate the requirement for cryogenic cooling currently implemented in most IR photon-type sensing devices.

For IR-detection application, aligning of many nanotubes would be highly critical from two points of view: to increase packing density of nanotubes and thus device sensitivity and to realize polarization sensitive IR optical devices. In contrast to conventional semiconductors, conjugated polymers provide dramatic benefits for engineering active optical nano-electronic and photonic devices; this includes reduced processing cost, excellent physical flexibility, and large area coverage. Until now, application of polymers in electronic devices was primarily limited to a visible range of electromagnetic spectrum [142, 143]. While stability of most polymers represents a barrier to their use as UV sensors, extending their use in the IR range becomes possible by implementing CNTs for both light absorption and free carrier generation. The exciton dissociation rate can be increased by introducing heterojunctions or applying external electric fields. The former can be realized by incorporating p-type nanotubes into n-type polymer matrix, such as PPy (pyridine-2,5-diyl) conjugated polymer, which is also known to exhibit relatively high resistance to oxidation.

Composites of CNTs/polymer feature relatively high absorption in a wide spectral range of ~0.2–20 μm and an emissivity coefficient close to unity while. Moreover, such composites are resistive to hard radiation damages and can work in high magnetic fields [144]. Unlike MWNTs and graphene which possesses featureless visible/NIR absorption, semiconducting SWNTs in particular exhibit strong and discrete absorption in the visible/NIR region owing to first optically active interband transition (S_{11}) with its energy scaling inversely proportional to the nanotube diameter. Lu et al. [148] reported a very large photocurrent in the device comprising semiconducting single-walled carbon nanotube (s-SWCNT)/polymer with type-II interface,

Figure 7. The detector featured significantly enhanced NIR detectivity of ~10^8 cm·$Hz^{1/2}$/W, which is comparable to that of the many conventional uncooled IR sensors, **Figure 8**.

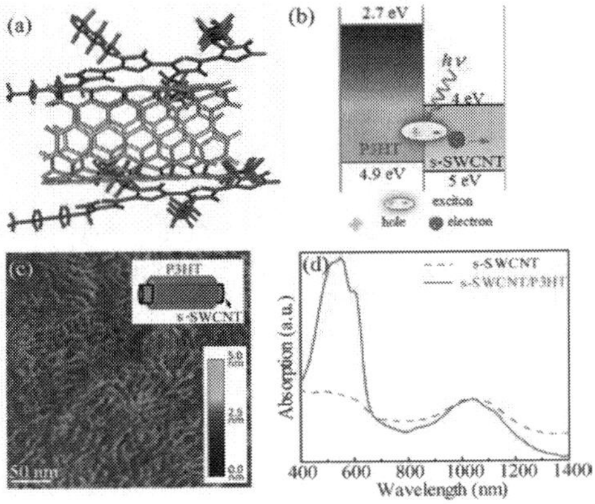

Figure 7. (a) Diagram of s-SWCNT/P3HT nanohybrid. (b) Band structure of the s-SWCNT/P3HT type-II heterojunction. (c) AFM image of s-SWCNT/P3HT. (d) Optical absorbance spectra of sSWCNT and s-SWCNT/P3HT. Reprinted from [148] with permission. Copyright © 2012 American Chemical Society.

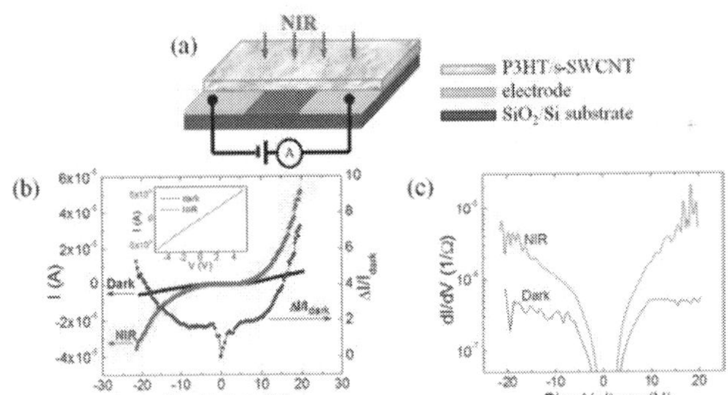

Figure 8. (a) Brief diagram of the electrical setup for IR detections. (b) Representative V–I curves of s-SWCNT/P3HT in dark and under NIR illumination of 2 mW/mm². The inset shows the voltage-biased V–I curves of s-SWCNT control sample. (c) Differential conductance dI/dV of s-SWCNT/P3HT: The data were calculated from (b). Reprinted from [148] with permission. Copyright © 2012 American Chemical Society.

Among other composites, polyaniline–CNTs composite thin film sensors showed an IR photosensitivity enhancement of more than two orders of magnitude under ambient conditions [144]. The attained enhancement in the sensitivity (bolometric effect) is attributed to a higher heat generation by CNTs and large temperature dependence of the resistance of polyaniline. In another study, Aliev [143] built an uncooled bolometric sensor based on SWNTs/polymer composite with voltage responsivity of ~150 V/W. Another, all-printed NIR sensor was engineered by Gohier et al. [146] by depositing multi-walled CNTs on a flexible polyimide substrate; the sensor showed ultra-high responsivity of ~1.2 kV/W. A strong dependence of the device response on the surrounding atmosphere was though noted and attributed to desorption of water molecules that negatively affected the photosensitivity. Glamazda et al. [147] reported on a strong bolometric response in SWNT–polymer composite featuring higher degree of internal alignment. A better alignment dramatically increased the temperature sensitivity of the resistance explained within the framework of fluctuation-induced tunneling theory. A spectrally flat mid-IR responsivity of 500 V W^{-1} was observed and is among the highest reported for nanotube-based bolometers.

Author details

T. Hosseini and N. Kouklin*

*Address all correspondence to: nkouklin@uwm.edu

Departments of Electrical Engineering Computer Science, University of Wisconsin-Milwaukee, Milwaukee, USA

References

[1] J. Xue, "Perspectives on organic photovoltaics," *Polym. Rev.*, 50, 4, 411–419, 2010.

[2] Hosel M., Angmob D., Krebs F. C., Espinosa N., "Solar cells with one-day energy payback for the factories of the future," *Energy Environ. Sci.*, 5, 1, 5117, 2012.

[3] Ranab S., Cho J. W., Lia L., Chan S. H., Sahooa N. G., "Polymer nanocomposites based on functionalized carbon nanotubes," *Prog. Polym. Sci.* , 35, 1, 837–867, 2010.

[4] Emmott C. J. M., Urbina A., Krebs F. C., Mutalea J., Nelsonb J., Azzopardi B., "Economic assessment of solar electricity production from organic-based photovoltaic modules in a domestic environment," *Energy Environ. Sci.* , 4, 10, 3741–3753, 2011.

[5] Sainz R., Martínez M. T., Benito A. M., Maser W. K., "Electroactive polymer–carbon nanotube composites: smart organic materials for optoelectronic applications," *Contrib. Sci.* , 4, 2, 187–192, 2008.

[6] Fincher C. R., Gao F., Blanchet G. B., "Polyaniline nanotube composites: a high-resolution printable conductor," *Appl. Phys. Lett.* , 82, 1290, 2003.

[7] Sik Choi Y., Kang S. Ju, Hwan Cho S., Lee T. W., Park C., Sung J., "AC field-induced polymer electroluminescence with single wall carbon nanotubes," *Nano Lett.*, 11, 966–972, 2011.

[8] Jen A. K. Y., Yipa H. L., "Recent advances in solution-processed interfacial materials for efficient and stable polymer solar cells," *Energy Environ. Sci.*, 5, 5994, 2012.

[9] Z. A. ALOthman, J. G. Shapter, S. M. Wabaidur, H. A. Alturaif, "Use of carbon nanotubes (CNTs) with polymers in solar cells," *Molecules*, 19, 17329–17344, 2014.

[10] Xue J., Cao W., "Recent progress in organic photovoltaics: device architecture and optical design," *Energy Environ. Sci.*, 7, 2123, 2014.

[11] Cherni J. A., Urbina A., Garcia-Valverde R., "Life cycle analysis of organic photovoltaic technologies," *Prog. Photovolt. Res. Appl.*, 18, 535–558, 2010.

[12] Zhu R., Yang Y., Li G., "Polymer solar cells," *Nat. Photonics*, 6, 153–161, 2012.

[13] Weisspfennig C., Parkinson P., Johnston M. B., Herz L. M., Nicholas R. J., Stranks S. D., "Ultrafast charge separation at a polymer-single-walled carbon nanotube molecular junction," *Nano Lett.*, 11, 66–72, 2011.

[14] Prezhdo O. V., Long R., "Asymmetry in the electron and hole transfer at a polymer–carbon nanotube heterojunction," *Nano Lett.*, 14, 3335–3341, 2014.

[15] Chu S., Majumdar A., "Opportunities and challenges for a sustainable energy future," *Nature*, 488, 294–303, 2012.

[16] Jin S. H., Park S. H., Jeon S., Hong S. H., Jun G. H., "Highly dispersed carbon nanotubes in organic media for polymer: fullerene photovoltaic devices," *Carbon*, 50, 40–46, 2012.

[17] P. H. de Oliveira Neto, Gargano R., G. Magela e Silva, W. F. da Cunha, *Int. J. Quantum Chem.*, 108, 2448–2453, 2008.

[18] Issi J. P., Devaux J., Billaud D., Chen X., *Mater J.. Sci.*, 32, 1515–1518, 1997.

[19] Freitag M., Perebeinos V., Avouris P., "Carbon-nanotube photonics and optoelectronics," *Nat. Photonics*, 2, 341–350, 2008.

[20] Meyer W., Baeriswyl D., Harbeke G., Kiess H., *Electron J. . Mater.*, 9, 763–781, 1980.

[21] Gomez N., Schmidt C. E., Guimard N. K., *Prog. Polym. Sci.*, 32, 876–921, 2007.

[22] Seo J. M., Jeon I. Y., Park Y. B., Baek J. B., Hwang S. H., "Conducting Polymer-based Carbon Nanotube Composites: Preparation and Applications," in *Carbon Nanotube–Polymer Composites*, D. Tasis, Ed. Cambridge, UK: The Royal Society of Chemistry, 1–21, 2013.

[23] Zhang L. C., Mylvaganam K., "Fabrication and application of polymer composites comprising carbon nanotubes," *Recent Pat. Nanotechnol.*, 1, 59–65, 2007.

[24] Wang B., Wang H., Zhang Y., Ni W., "Fabrication and properties of carbon nanotube and poly(vinyl alcohol) composites," *J. Macromol. Sci. Part B Phys.* , 45, 659–664, 2006.

[25] Gomes V. G., Altarawneh I. S., Khan M. U., "Synthesizing polystyrene/carbon nanotube composites by emulsion polymerization with non-covalent and covalent functionalization," *Carbon* , 48, 2925–2933, 2010.

[26] Tao Y., Lu J., Zhang Z., Xiao G., "Highly conductive and transparent carbon nanotube composite thin films deposited on polyethylene terephthalate solution dipping," *Thin Solid Films* , 518, 2822–2824, 2010.

[27] Ghislandi M., de With G., Koning C. E., Tkalya E. E., "The use of surfactants for dispersing carbon nanotubes and graphene to make conductive nanocomposites," *Curr. Opin. Colloid Interface Sci.* , 17, 225–232, 2012.

[28] Jung J. W., Lee J. U., Jo W. H., Jo J. W., "Fabrication of highly conductive and transparent thin films from single-walled carbon nanotubes using a new non-ionic surfactant via spin coating," *ACS Nano* , 4, 9, 5382–5388, 2010.

[29] Pan Y., Sahoo N. G., Chong K., Li L., Chan S. H., Zhao J., Cheng H. K. F., "Improvement in properties of multiwalled carbon nanotube/polypropylene nanocomposites through homogeneous dispersion with the aid of surfactants," *J. Appl. Polym. Sci.* , 124, 1117–1127, 2012.

[30] Tawfick S. H., Baughman R. H., Hart A. J., De Volder M. F. L., "Carbon nanotubes: present and future commercial applications," *Science* , 339, 535–539, 2013.

[31] Sahoo S., Ghosh D., Das C. K., Singh R., Moniruzzaman M., "Preparation and characterization of polypyrrole/modified multiwalled carbon nanotube nanocomposites polymerized in situ in the presence of barium titanate," *J. Appl. Polym. Sci.* , 128, 1, 698–705, 2013.

[32] J. Keith, ., "Effect of high-shear mixing on alignment of carbon nanofiber/epoxy nanocomposites," *J. Polym. Compos.* , 2, 1, 9–18, 2014.

[33] Tub J., Wanga X., Yua W., Zhenga W., Zhao Z., Liua Q., "Electrical conductivity of carbon nanotube/poly(vinylidene fluoride) composites prepared by high-speed mechanical mixing," *Carbon* , 50, 1, 339–341, 2012.

[34] Wanga Y., Liua X. Q., Caoa J., Luob Y., Yanga W., Xiea B. H., Yanga M. B., Kea K., "A comparison of melt and solution mixing on the dispersion of carbon nanotubes in a poly(vinylidene fluoride) matrix," *Compos. Part B Eng.* , 43, 3, 1425–1432, 2012.

[35] Thostenson E. T., Pandeya G., "Carbon nanotube-based multifunctional polymer nanocomposites," *Polym. Rev.* , 52, 3, 355–416, 2012.

[36] Lia Y., ., "Interfacial enhancement of carbon fiber composites by poly(amido amine) functionalization," *Compos. Sci. Technol.* , 74, 37–42, 2013.

[37] Shi Z., ., "Functionalization of unzipped carbon nanotube via in situ polymerization for mechanical reinforcement of polymer," *J. Mater. Chem.* , 22, 17663–17670, 2012.

[38] "Enhanced thermal and electrical properties of poly (D,L-lactide)/multi-walled carbon nanotubes composites by in-situ polymerization," *Trans. Nonferrous Metals Soc. China*, 23, 5, 1421–1427, 2013.

[39] Pan L., Fuller G. G., Bao Z., Allen R., "Using in-situ polymerization of conductive polymers to enhance the electrical properties of solution-processed carbon nanotube films and fibers," *ACS Appl. Mater. Interfaces* , 6, 13, 9966–9974, 2014.

[40] Behura S. K., Singh B. P., Bhattacharjee S., Nayak S., "Flexible polymer-multiwall carbon nanotubes composite developed by in situ polymerization technique," *Polym. Compos.* , 1–11, 2015.

[41] Downes R., Liang Z., Li Z., "In situ polymerized pCBT composites with aligned carbon nanotube buckypaper: structure and properties," *Macromol. Chem. Phys.* , 216, 292–300, 2015.

[42] Naumenko V., Kuznetsova A., Yates J. T. Jr, Liu J., Smalley R. E., Mawhinney D. B., "Surface defect site density on singale walled carbon nanotubes by titration," *Chem. Phys. Lett.* , 324, 213–216, 2000.

[43] Bhowmik P., Zhao B., Hamon M. A., Itkis M. E., Haddon R. C., Hu H., "Determination of the acidic sites of purified single-walled carbon nanotubes by acid–base titration," *Chem. Phys. Lett.* , 35, 25–28, 2001.

[44] Ranab S., Whan Cho J., Lia L., Chan S. H., Sahooa N. G., "Polymer nanocomposites based on functionalized carbon nanotubes," *Prog. Polym. Sci.* , 35, 837–867, 2010.

[45] Patole S. P., ., "Self assembled graphene/carbon nanotube/polystyrene hybrid nanocomposite by in situ microemulsion polymerization," *Eur. Polym. J.* , 48, 2, 252–259, 2012.

[46] Patole S. P., ., "Fabrication of polystyrene/multiwalled carbon nanotube composite films synthesized by in situ microemulsion polymerization," *Polym. Eng. Sci.* , 43, 6, 1327–1336, 2013.

[47] Qiu J., ., "Synthesizing multi-walled carbon nanotube-polymethyl methacrylate conductive composites and poly(lactic acid) based composites," *Polym. Compos.* , 37, 2, 503–511, 2016.

[48] Premkumar T., Geckeler K. E., Shin J. Y., "Dispersion of single-walled carbon nanotubes by using surfactants: are the type and concentration important?" *Chem. Eur. J.* , 14, 6044–6048, 2008.

[49] Choi S. M., Kline S. R., Jang H. S., Kim T. H., Doe C., "Charged rod-like nanoparticles assisting single-walled carbon nanotube dispersion in water," *Adv. Funct. Mater* , 18, 2685–2691, 2008.

[50] Park M., Kim J., Lee H., Lee M. S., Park I., "Multiwalled carbon nanotubes functionalized with PS via emulsion polymerization," *Macromol. Res.* , 15, 498–505, 2007.

[51] Lee W., Kim J., Park M., Lee H., Park I., "Selective sequestering of multi-walled carbon nanotubes in self-assembled block copolymer," *Sens. Actuat. B Chem.* , 126, 301–305, 2007.

[52] Such C. H., Hawkett B. S., Nguyen D., "Polymer coating of carboxylic acid functionalized multiwalled carbon nanotubes via reversible addition-fragmentation chain transfer mediated emulsion polymerization," *J. Polym. Sci. Part A Polym. Chem.* , 51, 250–257, 2013.

[53] Song J., ., "Soporous carbon–carbon nanotube–sulfur composite microspheres for high-areal-capacity lithium–sulfur battery cathodes," *ACS Appl. Mater. Interfaces* , 5, 21, 11355–11362, 2013.

[54] Zou W., Du Z., Li H., Zhang C., Yang W., "Enhanced conductive polymer nanocomposite by foam structure and polyelectrolyte encapsulated on carbon nanotubes," *Compos. Sci. Technol.* , 123, 106–114, 2016.

[55] Tagmatarchis N., Karousis N., "Current progress on the chemical modification of carbon nanotubes," *Chem. Rev.* , 110, 5366–5397, 2010.

[56] Strano M. S., Barone P. W., "Reversible control of carbon nanotube aggregation for a glucose affinity sensor," *Angew. Chem. Int. Ed.* , 45, 8138–8141, 2006.

[57] Sugikawa K., Kaneko K., Shinkai S., Numata M., "Creation of hierarchical carbon nanotube assemblies through alternative packing of complementary semi-artificial β-1, 3-glucan/carbon nanotube composites," *Chem. Eur. J.* , 14, 2398–2404, 2008.

[58] Poon Y. F., Chan-Park M. B., Chen Y., Zhang Q., Yan L. Y., "Individually dispersing single-walled carbon nanotubes with novel neutral pH water-soluble chitosan derivatives," *J. Phys. Chem. C* , 112, 7579–7587, 2008.

[59] R. Shvartzman-Cohen, ., "Physical adsorption of block copolymers to SWNT and MWNT: a nonwraping mechanism," *Macromolecules* , 40, 3676–3685, 2007.

[60] Sun J. Z., ., "Wrapping carbon nanotubes in pyrene-containing poly(phenylacetylene) chains: solubility, stability, light emission, and surface photovoltaic properties," *Macromolecules* , 39, 8011–8020, 2006.

[61] Kouklin N., Hosseini T., "Synthesis and spectroscopic and photoconduction characteristics of coaxial poly[2-methoxy-5-(20-ethylhexyloxy)21,4-phenylene vinylene] single-walled carbon nanotube films with ohmic-like transport attributes," *J. Appl. Polym. Sci.*, 131, 6, 40029(1–6), 2014.

[62] Shukla S. K., Saleh T. A., Gupta V. K., Tiwari A., "Covalent and Non-Covalent Functionalization of Carbon Nanotubes," in *Advanced Carbon Materials and Technology*, S. K. Shukla, A. Tiwari, Eds. Hoboken, NJ: John Wiley & Sons, Inc., 317–330, 2013.

http://dx.doi.org/10.5772/62692

[63] Goswami A., ., "Print Share/bookmark Covalent functionalization of monolayered transition metal dichalcogenides by phase engineering," *Nat. Chem.* , 7, 45–49, 2015.

[64] Ménard-Moyon C., ., "Covalent functionalization of multi-walled carbon nanotubes with a gadolinium chelate for efficient t1-weighted magnetic resonance imaging," *Adv. Funct. Mater.* , 24, 7173–7186, 2014.

[65] Lopez-Bezanilla A., "Electronic transport properties of chemically modified double-walled carbon nanotubes," *J. Phys. Chem. C* , 117, 29, 15266–15271, 2013.

[66] Janssen J. L., Bouilly D., "Graft-induced midgap states in functionalized carbon nanotubes," *ACS Nano* , 9, 3, 2626–2634, 2015.

[67] Shayla S., ., "Effect of dsDNA wrapped single-walled carbon nanotubes on the thermal and mechanical properties of polycaprolactone and polyglycolide fiber blend composites," *Polymer* , 56, 476–481, 2015.

[68] Oh C. H., Parka D. S., "Synthesis and characterization of polycaprolactone-grafted carbon nanotubes via click reaction," *Compos. Interfaces* , 22, 3, 193–201, 2015.

[69] Xue L., Ma H., "Carbon nanotubes reinforced poly(L-lactide) scaffolds fabricated by thermally induced phase separation," *Nanotechnology*, 26, 025701(1–10), 2015.

[70] Gattia T., Vicentinia N., "Covalent functionalization enables good dispersion and anisotropic orientation of multi-walled carbon nanotubes in a poly(L-lactic acid) electrospun nanofibrous matrix boosting neuronal differentiation," *Carbon* , 95, 725–730, 2015.

[71] Chao C., ., "Preparation and characterization of thin-film nanocomposite membranes embedded with poly(methyl methacrylate) hydrophobic modified multiwalled carbon nanotubes by interfacial polymerization," *J. Membr. Sci.* , 442, 18–26, 2013.

[72] Teyssandier J., Mammeria F., "Carbon nanotube–poly(methyl methacrylate) hybrid films: Preparation using diazonium salt chemistry and mechanical properties," *J. Colloid Interface Sci.* , 443, 115–122, 2014.

[73] Choi H. J., Zhanga K., "Fabrication and viscoelastic characteristics of amino-functionalized multi-walled carbon nanotube/poly(methyl methacrylate) nanocomposites," *Compos. Struct.* , 125, 170–175, 2015.

[74] Wang X., ., "Surface charge and cellular processing of covalently functionalized multiwall carbon nanotubes determine pulmonary toxicity," *ACS Nano* , 7, 3, 2352–2368, 2013.

[75] Xi L., ., "Covalent functionalization of multi-walled carbon nanotubes with quaternary ammonium groups and its application in ion chromatography," *Carbon* , 62, 127–134, 2013.

[76] Akbarivakilabadi A., Majumder M., Zeimaran E., "Polystyrene Carbon Nanotube Nanocomposites," in *Handbook of Polymer Nanocomposites. Processing, Performance and*

Application: Volume B: Carbon Nanotube Based Polymer Composites, J. K. Pandey, S. Rana, K. K. Kar, Eds. Heidelberg, Germany: Springer Berlin Heidelberg, 213–244, 2015.

[77] Shuai Y., Guo Z., Feng Y., Su X., "Functionalization of multi-walled carbon nanotubes with thermo-responsive azide-terminated poly(N-isopropylacrylamide) via click reactions," *Molecules* , 18, 4, 4599–4612, 2013.

[78] Ren Y., Sun J., Feng L., Wu J., "Carbon nanotube-coated macroporous poly(N-isopropylacrylamide) hydrogel and its electrosensitivity," *ACS Appl. Mater. Interfaces* , 5, 9, 3519–3523, 2013.

[79] Kang I.-K., Gupta K. C., "ATRP graft copolymerization of poly(N-isopropylacrylamide-co-acrylic acid) on multiwalled carbon nanotubes," *Macromol. Res.* , 22, 9, 948–957, 2014.

[80] Liu L., Cox M. A., Grunlan J. C., Etika K. C., "Clay-mediated carbon nanotube dispersion in poly(N-Isopropylacrylamide)," *Colloids Surf. A Physicochem. Eng. Asp.* , 489, 19–26, 2016.

[81] Shin D. G., et. al., "The mechanical and electrical properties of carbon nanotube-grafted polyimide nanocomposites," *J. Polym. Sci. Part B Polym. Phys.* , 52, 960–966, 2014.

[82] Chang Y. H., ., "Conductive-on-demand: Tailorable polyimide/carbon nanotube nanocomposite thin film by dual-material aerosol jet printing," *Carbon* , 98, 397–403, 2016.

[83] Muradov M. B., Malikova E. Y., "Synthesis and characterization of polyvinyl alcohol based multiwalled carbon nanotube nanocomposites," *Phys. E Low Dimens. Syst. Nanostruct.* , 61, 129–134, 2014.

[84] Zhang S., Jewell D., Chen G. Z., Peng C., "Carbon nanotube and conducting polymer composites for supercapacitors," *Prog. Nat. Sci.* , 18, 777–788, 2008.

[85] Zhang W., ., "Energy transfer from photo-excited fluorene polymers to single-walled carbon nanotubes," *J. Phys. Chem. C* , 113, 14946–14952, 2009.

[86] Umeyama T., Kadota N., Tezuka N., Matano Y., Imahori H., "Photoinduced energy transfer in composites of poly[(p-phenylene-1,2-vinylene)-co-(p-phenylene-1,1-vinylidene)] and single-walled carbon nanotubes," *Chem. Phys. Lett.* , 444, 263–267, 2007.

[87] Nish A., Hwang J. Y., Doig J., Nicholas R., "Direct spectroscopic evidence of energy transfer from photo-excited semiconducting polymers to single-walled carbon nanotubes," *J. Nanotechnol.*, 112, 20227(1–6), 2008.

[88] Park S. M., Park J. Y., "DNA hybridization sensors based on electrochemical impedance spectroscopy as a detection tool," *Sensors* , 9, 9513–9532, 2009.

[89] Wéry J., Massuyeau F., "Electronic interaction in composites of a conjugated polymer and carbon nanotubes: first-principles calculation and photophysical approaches," *Beilstei Nanotechnol.* , 6, 1138–1144, 2015.

[90] Lucas B., Marzouk J., "Simple strategy to tune the charge transport properties of conjugated polymer/carbon nanotube composites using an electric field assisted deposition technique," *Polym. Int.* , 63, 1378–1386, 2014.

[91] Ryu K., Zhang D., "Transparent, conductive, and flexible carbon nanotube films and their application in organic light-emitting diodes," *Nano Lett.* , 6, 9, 1880–1886, 2006.

[92] Parbaile E., Chakaroun M., Ratier B., Aldissi M., Moliton A.. R. Radbeh, "Enhanced efficiency of polymeric solar cells via alignment of carbon nanotubes," *Polym. Int.*, 59, 1514–1519, 2010.

[93] R. de Bettignies, Bailly S., Guillerez S., Jousselme B., Berson S., "Elaboration of P3HT/CNT/PCBM composites for organic photovoltaic cells," *Adv. Funct. Mater.* , 17, 3363–3370, 2007.

[94] Bergeret C., Cousseau J., Nunzi J. M., Derbal-Habak H., "Improving the current density Jsc of organic solar cells P3HT:PCBM by structuring the photoactive layer with functionalized SWCNTs," *Sol. Energy Mater. Sol. Cells* , 95, S53–S56, 2011.

[95] Hwang Y. H., Javey A., Pyoa M., Jung Y. S., "PCBM-Grafted MWNT for enhanced electron transport in polymer solar cells," *J. Electrochem. Soc.* , 158, A237–A240, 2011.

[96] Tao X., Wang R., Wang G., "Fabrication and characterization of OLEDs using PEDOT:PSS and MWCNT nanocomposites," *Compos. Sci. Technol.* , 68, 2837–2841, 2008.

[97] Liu J., Zhai L., Khondaker S. I., Sarker B. K., "Fabrication of organic field effect transistor by directly grown poly(3 hexylthiophene) crystalline nanowires on carbon nanotube aligned array electrode," *ACS Appl. Mater. Interfaces* , 3, 1180–1185, 2011.

[98] Nam S., ., "Solution-processed organic field-effect transistors composed of poly(4-styrene sulfonate) wrapped multiwalled carbon nanotube source/drain electrodes," *Org. Electron* , 10, 363–367, 2009.

[99] Li C. M., Zhang J., Gamota D., Gan Y., "Organic thin-film transistors based on conjugated polymer/carbon nanotube composites," *Int. J. Nanotechnol.* , 4, 4, 441–449, 2007.

[100] Singha R. K., Kumara V., Rastogib R. C., Singha R., Kumara J., "Self-assembly of SWCNT in P3HT matrix," *Diam. Relat. Mater.* , 16, 3, 446–453, 2007.

[101] Silvab G. G., ., "Structure and conductivity of multi-walled carbon nanotube/poly(3-hexylthiophene) composite films," *Polymer* , 48, 6, 1667–1678, 2007.

[102] Mata J., Matsuno T., Tsukamoto J., "Organic field effect transistors using composites of semiconductive polymers and single-walled carbon nanotubes," *Jpn. J. Appl. Phys.* , 46, 17, L396–L398, 2007.

[103] Karim M. R., Lee C. J., Kim H. J., "Optical and transport properties of poly(3-hexylthiophene)-single-walled-carbon-nanotube composites," *J. Nanoelectron. Optoelectron.*, 6, 3, 288–292, 2011.

[104] Baghar M., Barnes M. D., "Optical probes of chain packing structure and exciton dynamics in polythiophene films, composites, and nanostructures," *J. Polym. Sci. Part B Polym. Phys.*, 50, 1121–1129, 2012.

[105] Zeng T., Geng J., "Influence of single-walled carbon nanotubes induced crystallinity enhancement and morphology change on polymer photovoltaic devices," *J. Am. Chem. Soc.*, 128, 16827–16833, 2006.

[106] Prezhdo O. V., Long R., "Asymmetry in the electron and hole transfer at a polymer –carbon nanotube heterojunction," *Nano Lett.*, 14, 3335–3341, 2014.

[107] Katza E. A., Yerushalmi Rozen R., Bouniouxa C., "Conjugated polymers–carbon nanotubes-based functional materials for organic photovoltaics: a critical review," *Polym. Adv. Technol.*, 23, 1129–1140, 2012.

[108] Bindl D. J., Safron N. S., Arnold M. S., "Dissociating excitons photogenerated in semiconducting carbon nanotubes at polymeric photovoltaic heterojunction interfaces," *ACS Nano*, 4, 10, 5657–5664, 2010.

[109] Paulus G. L. C., ., "Evidence for high-efficiency exciton dissociation at polymer/single-walled carbon nanotube interfaces in planar nano-heterojunction photovoltaics," *ACS Nano*, 4, 10, 6251–6259, 2010.

[110] Grossman J. C., Kanai Y., "Role of semiconducting and metallic tubes in P3HT/carbon-nanotube photovoltaic heterojunctions: density functional theory calculations," *Nano Lett.*, 8, 3, 908–912, 2008.

[111] Ferguson A. J., ., "Prolonging charge separation in P3HT-SWNT composites using highly enriched semiconducting nanotubes," *Nano Lett.*, 10, 4627–4633, 2010.

[112] Zheng Y., Huang J., Yu J., "Towards high performance organic photovoltaic cells: a review of recent development in organic photovoltaics," *Polymers*, 6, 2473–2509, 2014.

[113] Liu Q., ., "Organic photovoltaic devices based on a novel acceptor material: graphene," *Adv. Mater.*, 20, 3924–3930, 2008.

[114] Domınguez R. M., Kamat P. V., Kongkanand A., "Single wall carbon nanotube scaffolds for photoelectrochemical solar cells. capture and transport of photogenerated electrons," *Nano Lett.*, 7, 3, 676–680, 2007.

[115] Neugebauer H., Sariciftci N. S., Gunes S., "Conjugated polymer-based organic solar cells," *Chem. Rev.*, 107, 1324–1338, 2007.

[116] Nish A., Nicholas R. J., Schuettfort T., "Observation of a type II heterojunction in a highly ordered polymer–carbon nanotube nanohybrid structure," *Nano Lett.*, 9, 11, 3871–3876, 2009.

[117] Srivastava R., ., "Low electrical percolation threshold and PL quenching in solution-blended MWNT–MEH PPV nanocomposites," *J. Exp. Nanosci.* , 5, 5, 412–426, 2010.

[118] Servati P., ., "Effective mobility and photocurrent in carbon nanotube–polymer composite photovoltaic cells," *Nanotechnology* , 18, 435702–435708, 2007.

[119] Stylianakis M. M., ., "Spin coated carbon nanotubes as the hole transport layer in organic photovoltaics," *Sol. Energy Mater. Sol. Cells* , 96, 298–301, 2012.

[120] Patyk B. S. L., ., "Carbon nanotube–polybithiophene photovoltaic devices with high open-circuit voltage," *Phys. Status Solidi Rapid Res. Lett.* , 1, 1, R43–R45, 2007.

[121] Hwang J. Y., ., "Influence of solvent on the dispersion of single-walled carbon nanotubes in polymer matrix and the photovoltaic performance," *J. Phys. Chem. C* , 114, 10932–10936, 2010.

[122] Blau W., Arranz-Andres J., "Enhanced device performance using different carbon nanotube types in polymer photovoltaic devices," *Carbon* , 46, 15, 2067–2075, 2008.

[123] Bhatnagar P. K., ., "Optical and electrical characterization of conducting polymer-single walled carbon nanotube composite films," *Carbon* , 46, 8, 1141–1144, 2008.

[124] Lyer S. S. K., Mazhari B., Mallajosyula A. T., "A comparative study of poly(3-octylthiophene) and poly(3-hexylthiophene) solar cells blended with single walled carbon nanotubes," *Jpn. J. Appl. Phys.*, 48, 1R, 011503 (1–6), 2009.

[125] Y. Chen, Ntim S. A., Mitra S., Li C., "Fullerene-multiwalled carbon nanotube complexes for bulk heterojunction photovoltaic cells," 96, 143303 (1–3), 2010.

[126] Imalka Jayawardena K. D. G., Damitha A. A., Adikaari T., Ravi S., Silva P., Aamina Nismy N., "Nano-engineering of hybrid organic heterojunctions with carbon nanotubes to improve photovoltaic performance," *Org. Electron.* , 22, 35–39, 2010.

[127] Mu X., ., "Modification of carbon nanotube transparent conducting films for electrodes in organic light-emitting diodes," *Nanotechnology*, 24, 435201 (1–8), 2013.

[128] Silva S. R. P., Shi S., "High luminance organic light-emitting diodes with efficient multi-walled carbon nanotube hole injectors," *Carbon* , 50, 4163–4170, 2012.

[129] Hu L., Li J., "Organic light-emitting diodes having carbon nanotube anodes," *Nano Lett.* , 6, 11, 2472–2477, 2006.

[130] Tangonan A., Abdul Baki M. K., "Properties of single-walled carbon nanotube-based poly(phenylene vinylene) electroluminescent nanocomposites," *J. Polym. Sci. Part B Polym. Phys.* , 50, 272–279, 2012.

[131] Daveya A. P., ., "Evolution and evaluation of the polymer/nanotube composite," *Synth. Metals* , 103, 2559–2562, 1999.

[132] Czerw R., ., "Carbon nanotube–polymer composite organic light emitting diodes based on poly(m-phenylene vinylene-co-2,5-dioctoxy-p-phenylene vinylene)," *Appl. Phys. Lett.* , 77, 1393–1395, 2000.

[133] Jakubka F., ., "Enhanced ambipolar charge injection with semiconducting polymer/ carbon nanotube thin films for light-emitting transistors," *ACS Nano* , 6, 1, 539–548, 2012.

[134] Reddy M. L. P., Divya V., "Visible-light excited red emitting luminescent nanocomposites derived from Eu3+-phenathrene-based fluorinated b-diketonate complexes and multi-walled carbon nanotubes," *J. Mater. Chem. C* , 1, 160–170, 2013.

[135] Srivastava R., ., "Carbon nanotube-based organic light emitting diodes," *Nanoscale* , 1, 317–330, 2009.

[136] Auvray S., ., "Carbon nanotube sheets as electrodes in organic light-emitting diodes," *Appl. Phys. Lett.*, 88, 183104 (1–3), 2006.

[137] Ryu K., ., "Transparent, conductive, and flexible carbon nanotube films and their application in organic light-emitting diodes," *Nano Lett.* , 6, 9, 1880–1886, 2006.

[138] Fanchini G., Eda G., Chhowalla M., Parekh B. B., "Improved conductivity of transparent single-wall carbon nanotube thin films via stable postdeposition functionalization," *Appl. Phys. Lett.*, 90, 121913 (1–3), 2007.

[139] Kim K. K., Geng H. Z., "Effect of acid treatment on carbon nanotube-based flexible transparent conducting films," *J. Am. Chem. Soc.* , 129, 7758–7759, 2007.

[140] De S., Lyons P. E., "Transparent, flexible, and highly conductive thin films based on polymer–nanotube composites," *ACS Nano* , 3, 3, 714–720, 2009.

[141] Xu X., Forrest S. R., Renshaw C. K., "A monolithically integrated organic photodetector and thin film transistor," *Org. Electron.* , 11, 1, 175–178, 2010.

[142] Li Y. L., ., "Composites of functional poly(phenylacetylene)s and single-walled carbon nanotubes: preparation, dispersion, and near infrared photoresponsive properties," *Macromolecules* , 46, 8479–8487, 2013.

[143] Aliev A. E., "Bolometric detector on the basis of single-wall carbon nanotube/polymer composite," *Infrared Phys. Technol.* , 51, 541–545, 2008.

[144] Lee J., ., "Effectively enhanced sensitivity of a polyaniline–carbon nanotube composite thin film bolometric near-infrared sensor," *J. Mater. Chem.* , 22, 3215–3219, 2012.

[145] Menamparambath M. M., Nikolaev P., Baik S., Ma R., "Transparent stretchable single-walled carbon nanotube–polymer composite films with near-infrared fluorescence," *Adv. Mater.* , 25, 2548–2553, 2013.

[146] Dhar A., Gohier A., "All-printed infrared sensor based on multiwalled carbon nanotubes," *Appl. Phys. Lett.*, 98, 063103 (1–3), 2011.

[147] Karachevtsev V. A., Euler W. B., Levitsky I. A., Glamazda A. Y., "Achieving high mid-IR bolometric responsivity for anisotropic composite materials from carbon nanotubes and polymers," *Adv. Funct. Mater.*, 22, 2177–2186, 2012.

[148] R. Lu, C. Christianson, A. Kirkeminde, S. Ren, J. Wu, "Extraordinary photocurrent harvesting at type-II heterojunction interfaces: toward high detectivity carbon nanotube infrared detectors," *Nano Lett.*, 12, 6244–6249, 2012.

[149] P. Gómez-Romero, M. Baibarac, "Nanocomposites based on conducting polymers and carbon nanotubes from fancy materials to functional applications," *J. Nanosci. Nanotechnoln*, 6, 1, 1–14, 2006.

Chapter 5

Optical, Mechanical, and Electrical Properties of Polymer Composites Doped by Multiwalled Carbon Nanotubes

Gülşen Akın Evingür and Önder Pekcan

Additional information is available at the end of the chapter

http://dx.doi.org/10.5772/63054

Abstract

Three-dimensional networks can be hydrophilic and/or hygroscopic. Optical, mechanical, and electrical properties of these materials encompass many fields of technology. Composites of carbon nanotubes (CNTs) in polymeric materials have attracted considerable attention in the research and industrial communities due to their unique optical, mechanical, and electrical properties. CNT polymer nanocomposites possess high stiffness, high strength, and good electrical conductivity at relatively low concentrations of CNT filler. Here, in this chapter, we survey the optical, mechanical, and electrical mechanisms for various polymeric systems. Composite gels and films were prepared with various molar percentages of multiwalled carbon nanotubes (MWCNTs). The optical, mechanical, and electrical behaviors of various composite gels doped by MWCNT are also discussed in the each sections of the chapter. The optical behaviors of the composites were performed by the UV-Vis spectroscopy and fluorescence spectroscopy in the first part of the chapter. On the other hand, compressive testing technique and rheological measurements were employed to determine the variations of mechanical properties of the composites in the second part of the chapter. Lastly, we review the electrical properties of the composites improved significantly by addition of MWCNTs researches.

Keywords: multiwalled carbon nanotubes (MWCNTs), drying, swelling, elasticity, conductivity, polyacrylamide (PAAm), polystyrene (PS), poly(vinyl acetate-co-butyl acrylate (P(VAc-co-BuA)), latex, PET(poly(ethylene terephthalate)

1. Introduction

Polymer-carbon nanotube (CNT), discovered by Iijima [1], composites were studied by Ajayan firstly. Carbon nanotubes and their composites have many application areas such as batteries, flat panel screens, sensors, and nanoprobes [2]. When comparing with different materials, nanotubes are stronger and more conductive than others. Therefore, there are many studies on polymer-CNT composites at last decade [3]. Nanotubes can be described as long and slender fullerenes, in which the walls of the tubes are hexagonal carbon (graphite structure). These tubes can either be single-walled carbon nanotube (SWCNT) or multiwalled carbon nanotube (MWCNT).

N-vinylcarbazole (NVC) was polymerized in bulk or in toluene in the presence of MWCNT without any extraneous catalyst [4]. A composite of PNVC with MWCNT isolated from the polymerization system showed high dc conductivity varying from 1.3 to 33 S cm^{-1} depending upon the extent of MWCNT loading in the composite. In situ polymerization and photo-physical properties of poly(p-phenylene benzobisoxazole) (PBO)/MWCNT composites were performed by optical techniques such as UV-Vis absorption, photoluminescence [5]. The investigation of UV-Vis absorption and fluorescence emission spectra exhibited that in situ PBO-MWCNT composite had a stronger absorbance and obvious trend of red-shift compared with blend PBO/MWCNT composites for all compositions. The characterization of poly(3-octylthiophene) (P3OT)/polystyrene (PS)-MWCNT polymer hybrid system performed by Fourier transform infrared spectroscopy, ultraviolet (UV)-visible absorption, and electrical measurements shows significant effects [6]. This study presents that the incorporation of 5 wt % functionalized MWCNT onto P3OT/PS polymer hybrid blend will convert this material from insulator to conductor. Modulation of single-walled carbon nanotubes photoluminescence by hydrogel swelling demonstrated that the shift of nanotubes photoluminescence occurred in a hydrogel matrix. As the hydrogel cross-linking density and hydration state is changed, the nanotubes experience lattice deformations and a shift in photoluminescence emission maxima [7]. CNTs were added into poly(vinyl alcohol) (PVA) hydrogels to modify their mechanical properties. Freezing/thawing method was used for preparing the hydrogels. The tensile modulus, tensile strength, and strain at break of CNTP-0.5 specimen with 0.5% (w/w) CNTs are increased by 78.2, 94.3, and 12.7%, respectively, [8]. Dual stimuli responsive poly(N,N-diethylacrylamide-co-acrylic acid) composite hydrogels functionalized with MWCNTs were obtained. Swelling, deswelling, compression properties of the composites were greatly improved [9]. The relaxation process of the composites was modeled by Fickian diffusion mechanism.

Synthesis, electrical, and mechanical properties of polyethylene oxide (PEO)-MWCNT composites were investigated [10]. The conductivity measurements on the PEO-MWCNT composite films with the highest concentration of MWCNT (50 wt%) showed an increase of eight orders 6.52 S cm^{-1} of magnitude in conductivity from bare PEO film. The elastic modulus and tensile strength of a PEO-MWCNT film were increased by about fivefold and tenfold, respectively, as compared to the corresponding values for a PEO film.

Polyvinyl alcohol (PVA)-vapor growth carbon fiber (VGCF) and PVA-MWCNT were prepared by gelation/crystallization [11]. The percolation threshold of electrical conductivity for the PVA/MWCNT was <1 wt% MWCNT loading which was much lower than that of PVA/VGCFs composites. The mechanical properties of the PVA composite films were significantly by adding VGCFs and MWCNTs. Poly(ethylene terephthalate)(PET)-MWCNT nanocomposites were prepared by coagulation method [12]. The percolation threshold 0.6 wt% for rheological property and 0.9 wt% for electrical conductivity has been found. The less rheological percolation threshold than electrical percolation threshold is mainly attributed to the fact that a denser MWCNT network is required for electrical conductivity, while a less dense MWCNT network sufficiently impedes PET chain mobility related to the rheological percolation threshold.

The electrical and rheological characteristics of poly(vinyl acetate) (PVAc)/multiwall carbon nanotube nanocomposites were investigated [13]. Small amount of MWCNT was observed to remarkably decrease resistivity of the nanocomposites. The G′ and G″ increase with the addition of MWCNT loading compared with that of the PVAc matrix.

The rheological and conductivity threshold of semicrystalline syndiotactic polystyrene (sPS) composites filled with CNTs and carbon nanocapcules (CNCs) were determined dynamic rheological tests of samples in the melt state and from electrical tests in the solid state, respectively [14]. The sPS composites filled with CNT with a higher aspect ratio exhibited a lower threshold than the CNC filled ones in both rheological and conductivity percolation. The electrical and rheological properties of polypropylene (PP)-CNT composites were studied [15]. Thus, the conductivity of the composites was increased with CNT content and the content percolation threshold is between 1 and 2 wt% MWCNTs.

Poly(N-isoproplacrylamide) (PNIPAAm) containing single-walled carbons and single-walled nanohorns showed phase transitions [16]. Preparation and characterization of PAAm/MWCNTs monohybrid hydrogels with microporous structures were presented by mechanical, pH, and temperature sensitive response and swelling kinetics [17]. The addition of nanotubes produced interesting properties, including tailor ability of temperature responsive swelling and mechanical strength of the PNIPAAm-MWCNT composites. Polystyrene (PS)-MWCNT composites have been widely studied and published in the literature. The results of Yu et al. [18] yield an increment in conductivities of such composites which were produced by latex technology. Poly(vinyl acetate) (PVAc) is a non-crystalline, amorphous thermoplastic polymer. The PVAc is generally fabricated via mixing process with certain polymeric materials in order to reinforce the structural properties [19, 20]. PVAc-based composites are widely used in adhesive, paper, emulsifier, paint, and textile industries due to its high-bond reinforced, film-like, nonflammable, and odorless characteristics [20].

In this chapter, after the introduction section, optical properties of various polymer-CNTs composite will be given and then mechanical properties of them which doped by MWCNTs will be provided. In the later section, electrical properties of them will be discussed.

2. Optical properties of different polymer-CNTs composites

Polyacrylamide (PAAm)-MWCNT composite gels were prepared by free radical cross-linking copolymerization [21]. PAAm was doped with various (0.1–15 wt%) of MWCNTs. Before drying was started [22], composites were cut into discs with 10 mm in diameter and 4 mm in thickness from the injector. Disc-shaped gel samples were placed on the wall of a 1 cm path length, square quartz cell filled with air and water for drying and swelling experiments, respectively.

Drying [22] and swelling [23, 24] process were performed by a Model LS-50 spectrometer of PerkinElmer, equipped with temperature controller. All measurements were made at 90° position, and spectral bandwidths were kept at 5 nm. Pyranine as a fluorescence probe in the composite gels was excited at 340 nm during *in situ* fluorescence experiments, and emission intensities of the pyranine were monitored at 427 nm as a function of drying, and swelling time, respectively. It was observed that the fluorescence intensity of pyranine increased as drying time was increased during the drying process. By combining the Stern-Volmer equation with the moving boundary model, water desorption coefficients, D, were determined for the drying gels prepared with various MWCNT content at different temperatures.

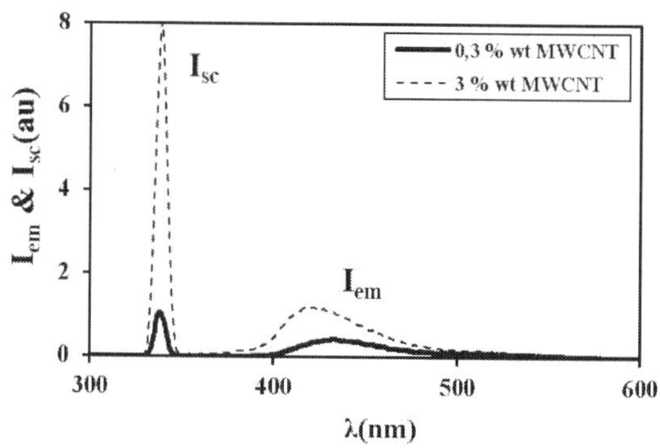

Figure 1. Emission spectra of pyranine from composites during drying in air, for 0.3 and 3 wt% MWCNT content gels at 60°C and in 135 min [22].

Figure 1 shows the emission spectra of pyranine from PAAm-MWCNT composite gel with 0.3 and 3 wt% MWCNT contents during drying in air at 60°C and in 135 min. It can be seen that as the MWCNT content is increased, fluorescence intensity, I_{em}, decreases relative to the scattered light intensity, I_{sc}. Since the decrease in I_{sc} corresponds to the decrease in turbidity of the drying gel [25], the corrected fluorescence intensity, I, was introduced as I_{em}/I_{sc} to eliminate the turbidity effect. As far as the correction of fluorescence emission is concern, totally empirical formula was introduced to produce the meaningful results for the fluorescence quenching mechanisms. In order to quantify these results, a collisional type of quenching mechanism may be proposed for the fluorescence intensity, I, from the gel samples during the drying process using Eq. (1) [26].

$$W = \left(1 - \frac{I}{I_0}\right)\frac{\upsilon}{k_q \tau_0} \tag{1}$$

τ_0 (=5 ns) is known for pyranine [26] so W can be calculated using Eq. (1) and the measured I values, in each drying step.

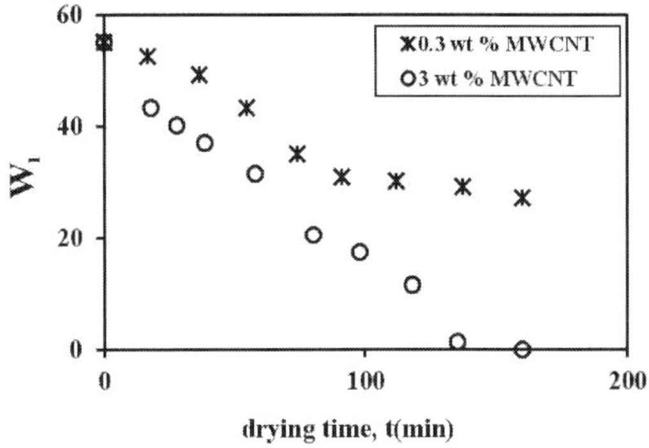

Figure 2. The plots of the water release, *W* versus drying time, *t* for PAAm-MWCNT composite gel dried in air measured by fluorescence technique for 0.3 and 3 wt% MWCNT content samples at 60°C [22].

Figure 2 presents the W_I versus drying time. The plots of W versus $t^{1/2}$ for 0.3 and 3 wt% MWCNT content sample at 60°C are presented in **Figure 2**, where the fit of the data to Eq. (2) produced the desorption coefficients, D_{dI} which are plotted in **Figure 3**.

$$\frac{W_I}{W_{If}} = 2\left[\frac{D_{dI}}{\pi a^2}\right]^{1/2} t^{1/2} \tag{2}$$

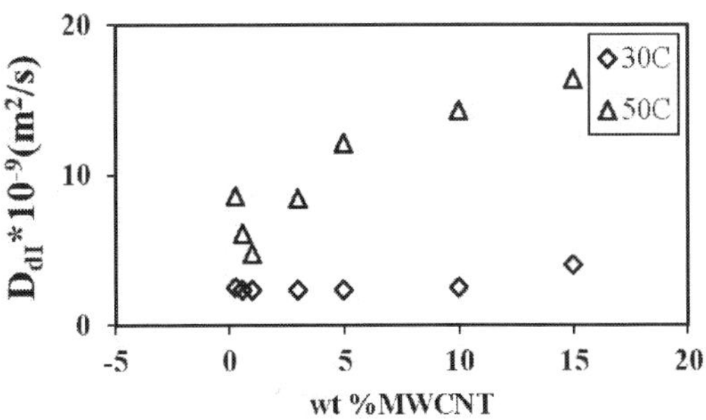

Figure 3. Desorption diffusion coefficients, D_{dI} versus wt% MWCNT content measured by fluorescence technique at 30 and 50°C [23].

Fluorescence technique measures the D_{dI} values at a molecular level. The desorption coefficients, D_{dI}, were obtained from Eq. (2) and measured by fluorescence technique for various

MWCNT content samples, where it was observed that the desorption coefficient decreased as the MWCNT content is increased up to 1 wt% MWCNT and reached to a plateau by presenting the different behaviors below and above the critical MWCNT (1 wt%) content at which the conducting percolation cluster starts to appear [21].

Monitoring of swelling experiments shows that emission light intensity, I_{em}, was decreased which is modeled by the Stern-Volmer equation. The Li-Tanaka equation was used to determine the swelling time constants, τ_S, and cooperative diffusion coefficients, D_{SI}, from fluorescence intensity (**Table 1**).

Figure 4 presents the fluorescence spectra of pyranine from the PAAm-MWCNT composite during the swelling process in pure water at 60°C for 1 and 10 wt% MWCNT at 80 min, respectively. When the water uptake is increased, the emission light intensity, I_{em}, was decreased and the scattered light intensity, I_{sc}, was increased, respectively, because of the quenching of excited pyranines and turbidity.

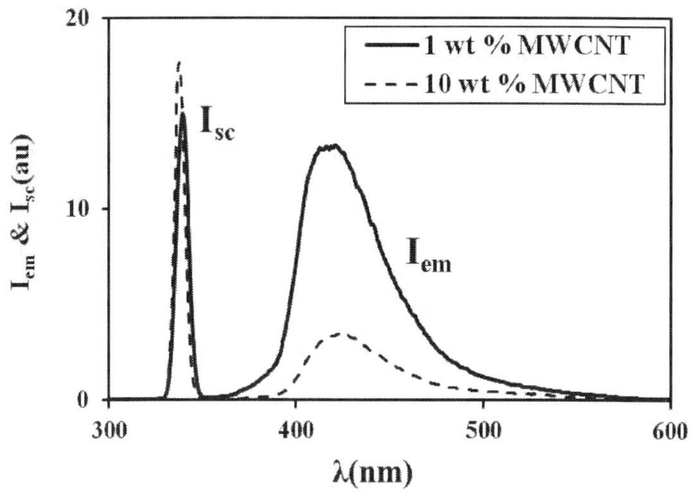

Figure 4. Fluorescence spectra of pyranine from the composite during the swelling process at 60°C for 1 and 10 wt% MWCNT content samples at 80 min [24].

To produce the meaningful results for the fluorescence quenching mechanisms, empirical equation was introduced [27–29]. The corrected fluorescence intensity, I, is produced by dividing emission light intensity, I_{em}, to scattering intensity, I_{sc}, to exclude the effect of turbidity. On the other hand, when quenching of excited pyranines increased, the swelling time, t, increased. This behavior of the composites was modeled by Stern-Volmer Model have been proposed using Eq. (2) [26].

http://dx.doi.org/10.5772/63054

		$D_{dI} \times 10^{-9}$ (m²/s)	τ_S (min)	$D_{SI} \times 10^{-9}$ (m²/s)
0	30	14.10	85	0.95
	40	15	48.5	4
	50	26	37.7	7.1
	60	48.67	25	7.3
0.3	30	2.46	58.82	0.6
	40	4.53	55	1.25
	50	8.56	35.71	1.73
	60	9.15	26.31	1.82
0.6	30	2.30	55	0.65
	40	3.17	50	1.7
	50	6.08	33.33	1.9
	60	7.37	25.64	2.5
1	30	2.29	48	1
	40	2.57	47.61	2.07
	50	4.70	32.25	2.44
	60	5.28	22.72	2.95
3	30	2.30	96	0.4
	40	6.23	95.23	0.44
	50	8.39	90.9	0.68
	60	10.2	83.33	0.76
5	30	2.35	98	0.38
	40	8.65	97	0.39
	50	12.1	95.23	0.59
	60	12.7	89	0.62
10	30	2.46	100	0.36
	40	9.88	98	0.37
	50	14.3	95	0.55
	60	14.2	91	0.6
15	30	4	120	0.32
	40	10.4	111.11	0.35
	50	16.4	105.26	0.44
	60	17.6	103.09	0.46

Table 1. Experimentally measured parameters of PAAm-MWCNT composites for various temperature and wt% MWCNT contents during drying and swelling processes, respectively.

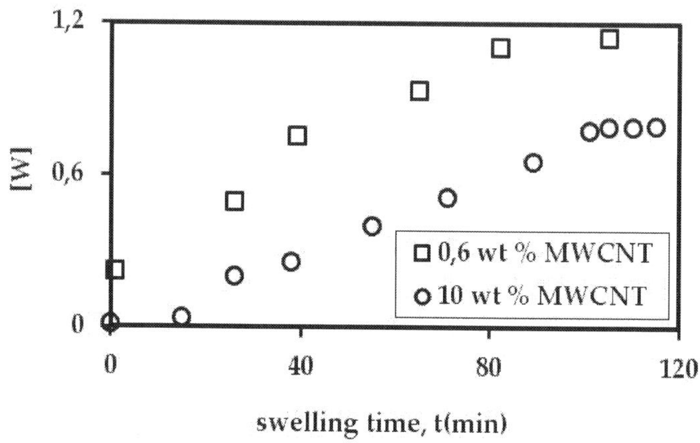

Figure 5. The plots of fluorescence data using Eq. (2) versus swelling time, *t*, for PAAm-MWCNT composite gels swollen in water measured by fluorescence technique for 0.6 and 10 wt% MWCNT content sample at 60°C, respectively.

Plots of water uptake, W, versus swelling time are presented in **Figure 5** for 0.6 and 10 wt% MWCNT content samples at 60°C, respectively. These are typical solvent uptake curves, obeying the Li-Tanaka equation is Eq. (3) [27–29].

$$\frac{W}{W_f} = 1 - B_1 \exp(-t / \tau_S) \tag{3}$$

The logarithmic form of the data was fitted to the following relation produced from Eq. (3)

$$\ln(1 - \frac{W}{W_f}) = \ln B_1 - \frac{t}{\tau_s} \tag{4}$$

where τ_S is the time constant which was calculated from linear regression of the curves in **Figure 5** and Eq. (4) described by Li and Tanaka [28].

$$D_{SI} = \frac{3a_f^2}{\tau_S \alpha_I^2} \tag{5}$$

Using Eq. (5), cooperative diffusion coefficients D_{SI} were determined for these disc-shaped composites and found to be around 10^{-9} m^2/s. All D_{SI} values first increased up to 1 wt% MWCNT content and reached its highest value at this critical point, where the percolation cluster from MWCNT starts to form as given in **Figure 6** and **Table 1**.

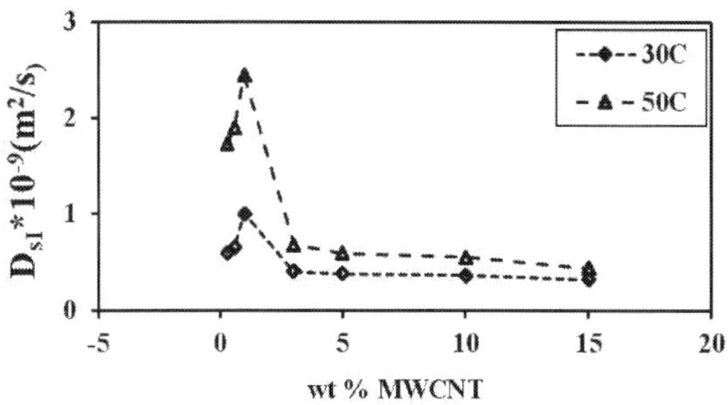

Figure 6. Cooperative diffusion coefficient, D_{SI}, versus MWCNT content measured by fluorescence methods at 30 and 50°C, respectively [23].

The percolation cluster formed from CNTs helps water molecules flow faster in their channels and causes the composite gel swell faster presenting large D_{SI} values for all samples under consideration. However, above the critical point (1 wt% MWCNT), composite gel is quite stiff due to the formation of infinite network from MWCNT. The formation of inelastic composite gel above the critical point then lowers the D_{SI} values to the smaller numbers.

The results were interpreted in terms of the swelling time constants; τ_s, (decreased) and the cooperative diffusion coefficient, D_{SI}, (increased) versus wt% MWCNT content. It was observed that high MWCNT content composites swell much slower producing smaller D_{SI} coefficients for all measurements at a given temperature.

The effect of MWCNT addition in insulating PS matrix was investigated [30]. Composite films were prepared using polystyrene (3.5% w/v) with various (0.13–18 wt%) of MWCNTs stock content on 2.5 × 3.0 cm^2 glass plates at room temperature. Variations in optical properties of very-thin PS-MWCNT composite films were measured using photon transmission techniques. Electrical and optical percolation thresholds were determined. Classical and site percolation theories were used to calculate the critical exponents for optical transmission data.

UV spectrophotometer (Lambda 2 S of PerkinElmer, USA) was used to monitor the variation of optical transparency of the composite films. The transmittances of the composite films were detected at 400 nm wavelength. Measurements were performed at six different positions on the film surface in order to lower the error. All the photon transmission measurements were carried out at room temperature. **Figure 7** shows that the behavior of transmitted (I_{tr}) and scattered (I_{sc}) light intensities versus the ratio of MWCNT contents (M) in composite films. When transmitted light intensity, I_{tr}, sharply decreases, and scattering centers in the film are increased, respectively, due to refractive indices between two medium in the composite system [30]. In **Figure 7**, the scattered light intensity increases rapidly, even though the MWCNT content was 0.13 wt% at the beginning which shows the percolation threshold value is in between 0 and 0.13 wt%. Since $M - M_{op} \rightarrow M$ for the extremely low M_{op} values, the percolation probability for the optical data can be written as follows:

$$I_{sc} = I_0 M^{\beta op} \tag{6}$$

The critical exponent β_{op} was calculated and found to be as 0.32 from the slope of log I_{sc}/I_0 versus log M plot.

The effect of MWCNT addition in insulating PVAc matrix was investigated [31]. Mass fractions of the composites were obtained between 0.5 and 18 wt%. Variations in optical properties of P(VAc-co-BuA)/MWCNT composite films were measured using photon transmission and fluorescence emission measurement techniques. The optical and fluorescence percolation thresholds were determined. The classical and site percolation theories were used to calculate the critical exponents for two data sets from the techniques, respectively.

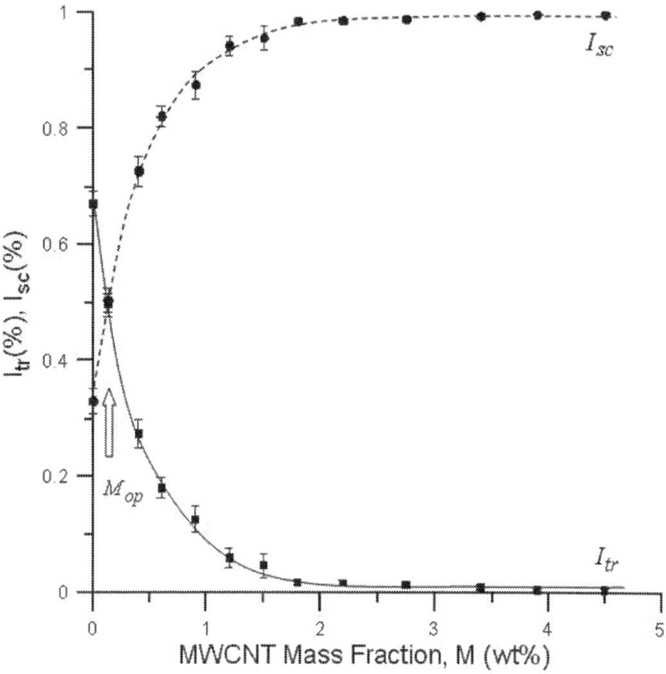

Figure 7. Variation in transmitted (I_{tr}) and scattered (I_{sc}) light intensities on PS-MWCNT composite films versus mass fractions [30].

UV spectrophotometer (Lambda 2S of PerkinElmer, USA) was used to monitor the variation of optical transparency of the composite films. Since the absorbance region of the pyranine molecule was 300–430 nm, the transmittances of the composite films were detected at 470 nm wavelength. Transmission measurements were performed at six different positions on the film surface in order to lower the error. Thus, the average value of transmitted light intensity (I_{tr})

was obtained. Transparency variations of the composite films versus mass fractions (M) of MWCNT were monitored by transmitted light intensity, I_{tr}, from the films. It was observed that I_{tr} sharply decreases as MWCNT content increases in the composite system, and there is almost no light transmission from the films above the mass fraction of 7.0 wt% ($M \geq 7.0$ wt%).

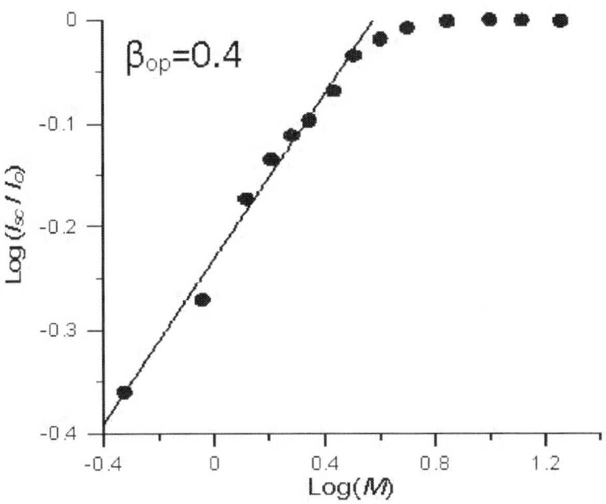

Figure 8. The log-log plot of I_{sc} versus M. The slope of the straight line produces the optical critical exponent, β_{op} as 0.40 [31].

The behavior of I_{tr} versus M predicts that the composite system owns a percolative structure having a percolation threshold at 0.5 wt% MWCNT content. As M is increased, the scattered light intensity, $I_{sc} = I_o - I_{tr}$, increases due to the concentration fluctuations.

The variation of log I_{sc}/I_o versus log M is given in **Figure 8**. The critical exponent, β_{op}, was calculated as 0.40 from the slope of **Figure 8** using Eq. (6). The obtained value of $\beta_{op} = 0.40$ is not far from the theoretical site percolation value of 0.42 [32].

The variation of fluorescence emission of the composite films was measured using Varian Cary Eclipse fluorescence spectrophotometer. Excitation and emission wavelengths were used as 370 and 507 nm, respectively. The emission wavelength at maximum intensity of pyranine is 510 nm, which varies up to 5–10 nm depending on the structure of the polymer molecules. The fluorescence emission measurements were performed at six different positions on the film surface in order to lower the error level, and the average value of the fluorescence emission intensity (I_{fl}) at the maximum was obtained. The maxima of the emission intensity (I_{fl}) versus mass fractions (M) of the samples are shown in **Figure 9**, where it can be seen that the emission intensities of pyranine show a rapid decrease for further addition of MWCNTs in the composite films. As the MWCNT concentration is increased, the number of the scattering centers in the film also increases.

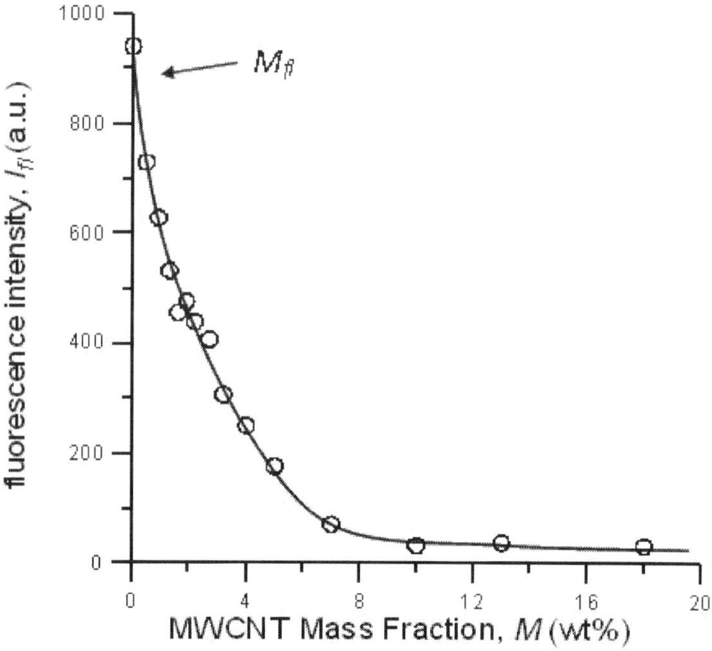

Figure 9. The fluorescence emission intensity, I_{fl}, versus MWCNT mass fraction, M [31].

The fluorescence emission data in **Figure 9** can be treated by the percolation theory. The percolation probability for the data of fluorescence emission could be arranged as follows:

$$\frac{1}{I_{fl}} = (M - M_{fl})^{\beta_{fl}} \tag{7}$$

where M is identical to lattice occupation probability, p, then the percolation threshold value, p_c, is equal to M_{fl}.

Since $(M - M_{fl}) \rightarrow M$ for low M_{fl} then Eq. (7) becomes as follows:

$$\frac{1}{I_{fl}} = M^{\beta_{fl}} \tag{8}$$

By assuming, the percolation probability P_∞ (p) is inversely proportional to the fluorescence emission intensity. The variation of log ($1/I_{fl}$) – log (M) is presented in **Figure 10**, where two different percolation regions can be detected.

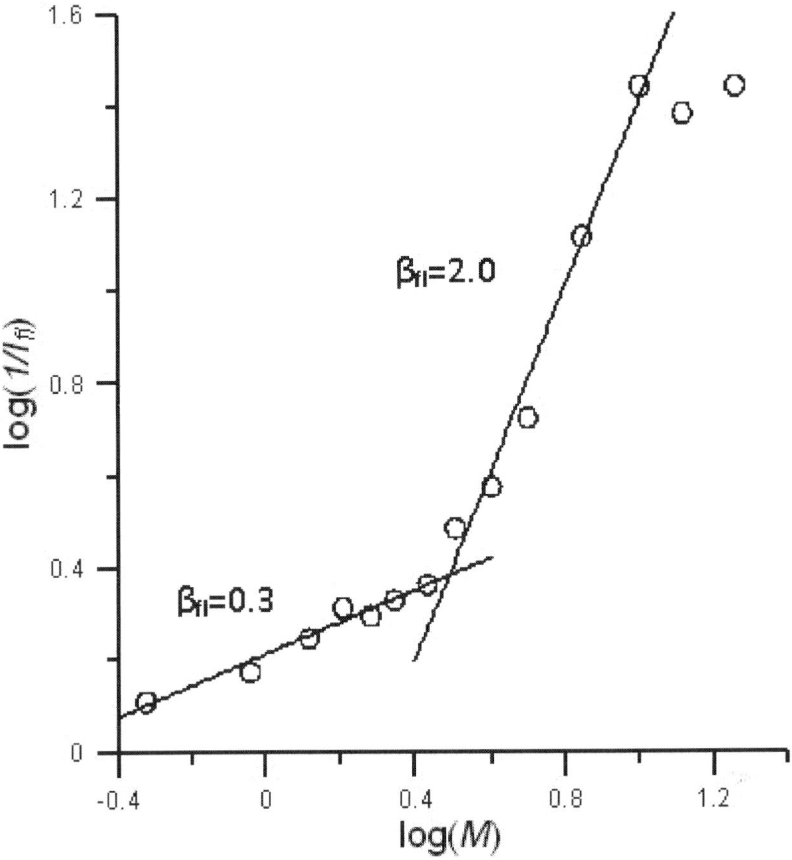

Figure 10. The log-log plot of $1/I_{fl}$ versus M. The slopes of the straight lines produce the fluorescence critical exponents, β_{fl}, as 0.35 and 2.0 for the low and high MWCNT concentration regions, respectively [31].

The critical exponents, β_{fl}, was calculated and found to be as 0.35 and 2.0 from the slopes of the straight lines in **Figure 10**, according to Eq. (8).

At low MWCNT content, pyranine simply probes the scattering sites in the composite film; however, at high MWCNT concentration, pyranine now traces the conducting network [31].

Polystyrene-MWCNT composite was prepared by Ugur et al. showed that healing and inter diffusion process using photon transmission techniques. After annealing step, the transmitted light intensity, I_{tr}, was monitored to observe the film formation process as shown in **Figure 11**. The increase in I_{tr} up to healing temperature, T_h, and above T_h during annealing was explained by void closure and inters diffusion processes, respectively, [33].

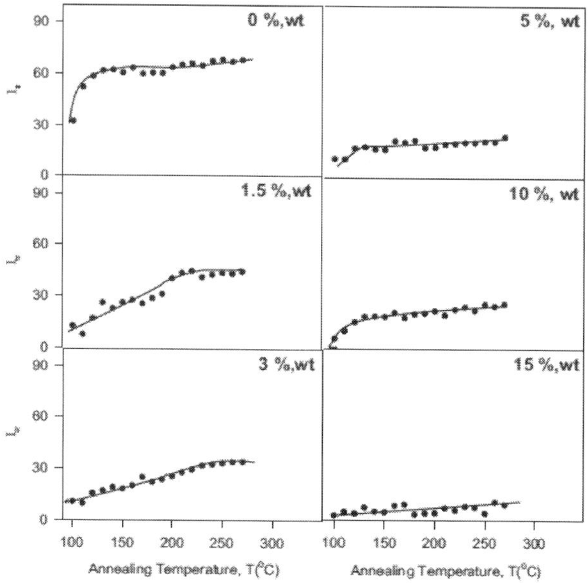

Figure 11. Transmitted photon intensities, I_{tr}, versus annealing temperatures depending on MWCNTs content in the films. Numbers on each figure shows the MWCNTs content [33].

Temperature dependence of oxygen diffusion into polymer-MWCNT composite films was examined by fluorescence spectroscopy [34]. The diffusivity of the composite films was determined by performing oxygen (O_2) diffusion measurements within a temperature range of 24–70°C for each film, and pyrene(P) was used as a fluorescence probe (**Figure 12**).

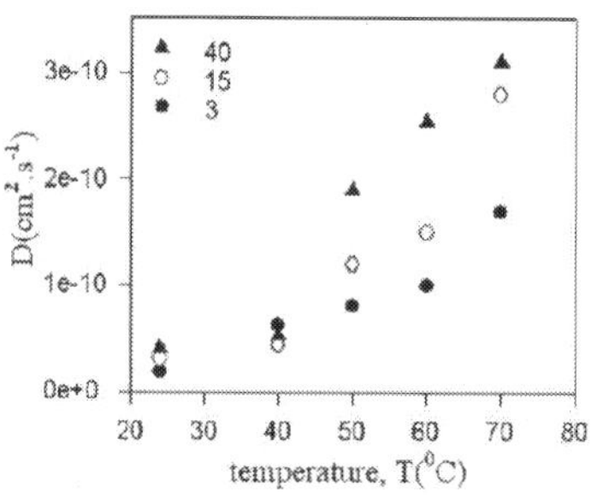

Figure 12. Plot of the diffusion coefficients, D, versus temperatures, T, for the 3, 15, and 40 wt% MWCNT content films [34].

The diffusion coefficients increased drastically with both increases of MWCNT content and also of the temperature, and this increase was explained via the existence of large amounts of pores in composite films which facilitate oxygen penetration into the structure. Therefore, PS/MWCNT nanocomposites have useful properties as fluorescent oxygen sensors, and a simple SSF technique can be used to measure the diffusion coefficient of oxygen molecules into these films quite accurately [34]. PS/MWCNT films were prepared by various contents of MWCNT at room temperature. After annealing at 170°C which is above glass transition (T_g) temperature of PS, fluorescence quenching processes were realized on oxygen diffusion [35]. **Figure 13** illustrates the logarithmic plots of the fluorescence intensity of pyrene behavior within time during oxygen diffusion into the composite films for 15 wt %MWCNT content.

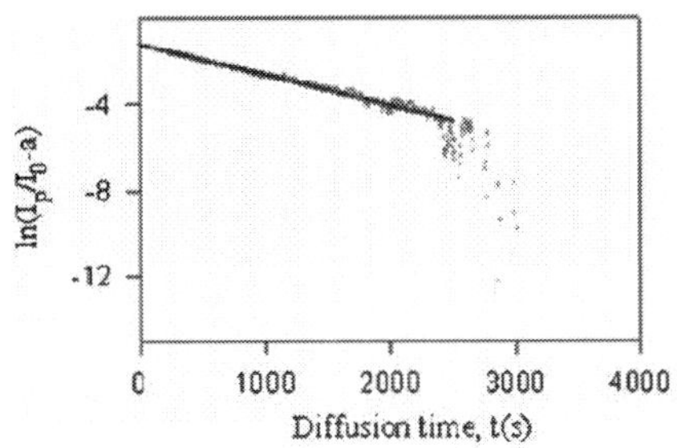

Figure 13. Logarithmic plots of the fluorescence intensity of pyrene behavior within time during oxygen diffusion into the composite films for 15 wt %MWCNT content [35].

The diffusion coefficients increased drastically with the increase of MWCNT content, and this increase was explained via the existence of large amounts of pores in composite films which facilitate oxygen penetration into the structure.

Oxygen permeability of nanocomposite films consisting of MWCNT and PS were determined to investigate the oxygen diffusion depending on MWCNT and temperature [36].

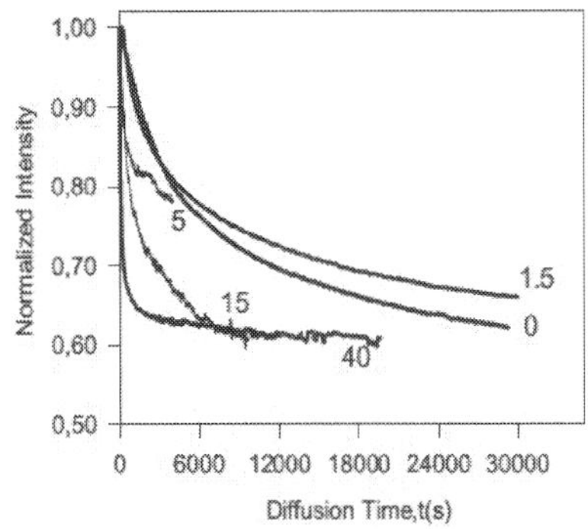

Figure 14. The time behavior of pyrene, fluorescence intensity, I, during oxygen diffusion into the composite films with different MWCNT content. Numbers on each curve indicates the MWCNT content (%) in the film [36].

In **Figure 14** normalized pyrene intensity, I_p curves are presented against diffusion time for films having different MWCNT content exposed to oxygen. It is seen that as oxygen diffused through the planar film, the emission intensity, I_{em}, of the pyrene decreased for each MWCNT content film. After completing oxygen diffusion, I_{em} was saturated. As shown in **Figure 14**, the quenching rate depends on the MWCNT contents in the film. Rapid quenching of excited pyrenes by O_2 molecules is possible for the high MWCNT content composite films [36].

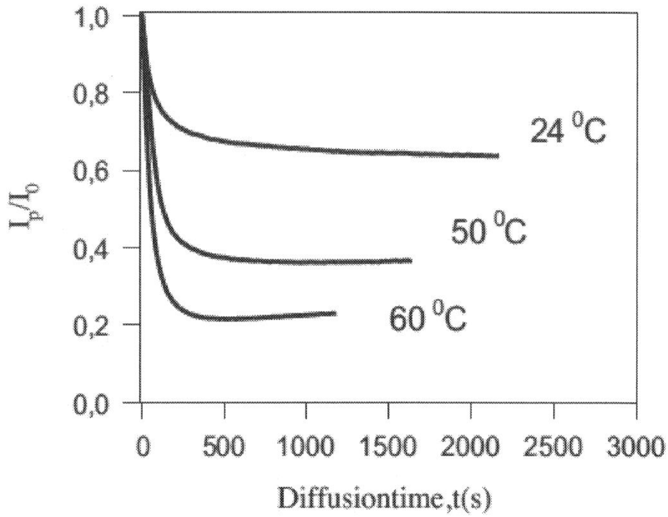

Figure 15. The time behavior of the pyrene, P, fluorescence intensity, I, during oxygen diffusion into the 40 wt% MWCNT content film at various temperatures. Numbers on each curve indicate the temperature [36].

The rate of decrease in intensity is higher at higher temperatures predicting the more rapid quenching of excited pyrene molecules by O_2 molecules diffused into the films. It is worthy to note that in **Figure 15** as expected the D increases with increase in temperature for all composite films. Increase in temperature naturally increases the Brownian motion of oxygen molecules given them more chance to meet the P molecules in the composite film [36].

The results of **Figures 14** and **15** showed that the diffusion of oxygen was accelerated by both increase in MWCNT fraction and temperature.

PS-MWCNT latex composite films were prepared by mixing of pyrene (P)-labeled PS latex with different amounts of MWCNTs varying in the range between 0 and 20 wt% [37]. To monitor the stages of film formation of PS/MWCNT latex composite films, scattered light (I_s) and fluorescence intensities (I_P) from P were performed after each annealing step. Some temperatures such as minimum film formation (T_o), void closure (T_v), and healing, (T_h) temperatures were decided as given in **Figure 16**.

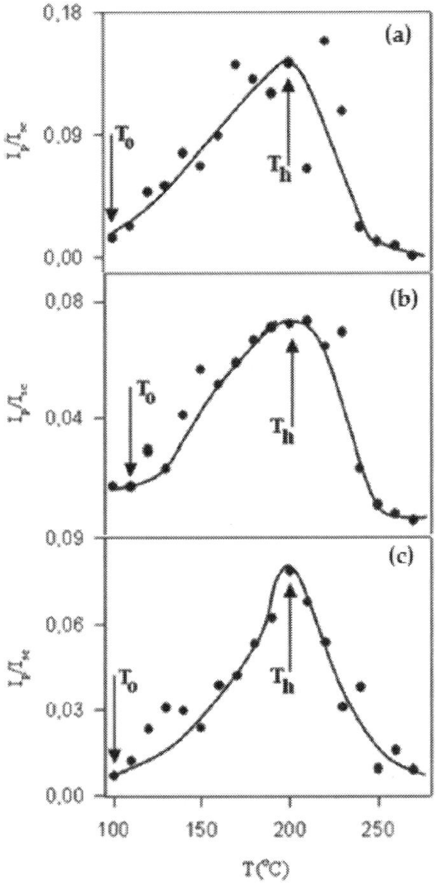

Figure 16. Plot of corrected fluorescence intensity (=fluorescence intensity/scattered fluorescence intensity) of composite films with different MWCNT content versus annealing temperatures. **(a)** 0, **(b)** 1.5, **(c)** 3% MWCNT content, respectively [37]. (T_0, minimum film formation temperature; T_h, healing temperature).

The existence of MWCNT delays the latex film formation process because of the behavior of T_0. However, healing processes are not affected by the presence of MWCNT [37].

3. The mechanical properties of polymer-CNTs composites

The elastic modulus of the swollen PAAm-MWCNT composites was measured to determine the effect of MWCNTs content. Elasticity measurement was performed by compressive testing technique and modeled by the theory of rubber elasticity [38, 39].

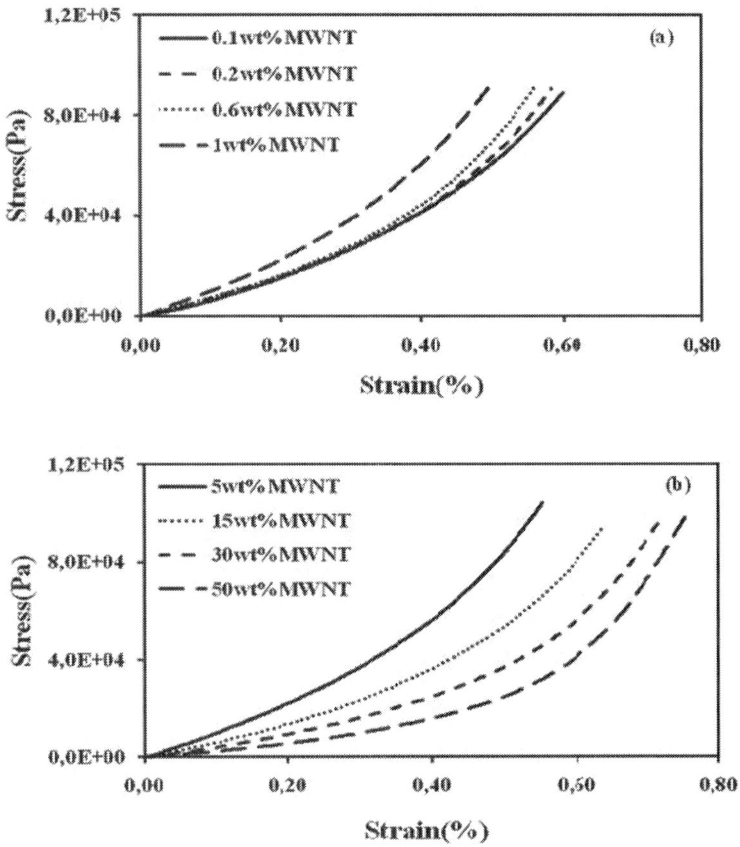

Figure 17. Stress versus strain curves **(a)** lower and **(b)** higher MWCNT contents at 25°C, respectively [39].

Stress (Pa)-strain plots of low and high MWCNTs content gels produced using the data obtained from the linear region, in the plots of F (N) versus compression curves for PAAm-MWCNT composites at 25°C, are presented in **Figure 17**, respectively. The stress versus strain displays a good linear relationship at 25°C, which agrees with Eq. (9).

$$\tau = G\lambda \tag{9}$$

The elastic moduli were obtained by a least square fit to the linear region of **Figure 17**. The addition of MWCNT into PAAm caused an increase in elastic modulus of the composite as expected. In PAAm–1 wt% MWCNT composite, the measured elastic modulus is found to be 0.105 MPa, two times larger than pure PAAm sample. It is seen in **Figure 17a** that PAAm–0.1

wt% MWCNT composite has smaller initial slope than 1 wt% MWCNT content composite. In this case, it appears that the alignment effect of MWCNT plays an important role for getting the different onset behavior [40]. The stress of the PAAm-MWCNT (1–5 wt% MWCNT composites) increases dramatically when the strain exceeds 0.6%, where the alignment is taking place in the composite. On the other hand, at high MWCNT region, the random motion of MWCNT impedes alignment as predicted in **Figure 17b**. Therefore, in a gel with low MWCNT content, alignment of MWCNTs with each other is much easier than in high content MWCNT composite.

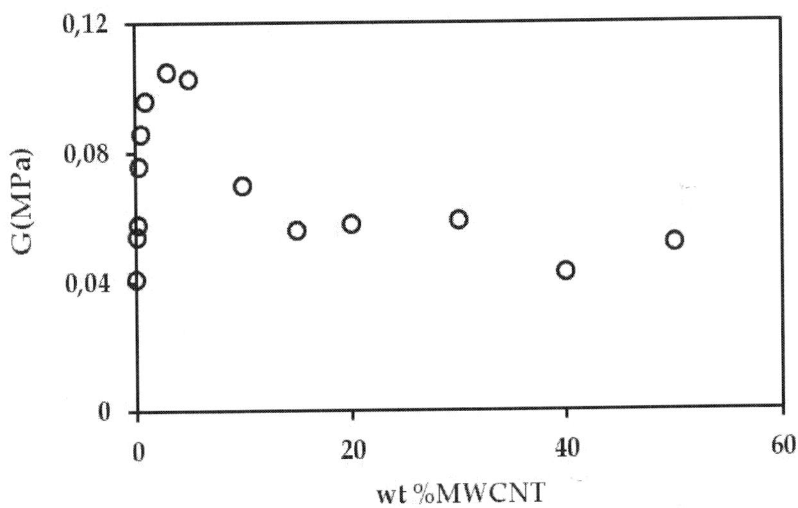

Figure 18. Dependence of elastic modulus on content of wt% MWCNT in the composite [39].

Figure 18 shows the plot of elastic modulus, G versus MWCNTs content in the composite gel. Elastic modulus increases dramatically up to 1 wt% MWCNT with increasing nanotube content and decreases presenting a critical MWCNT value indicating that there is a sudden change in the material elasticity. The sudden change in G predicts that the composites have reached a super-elastic percolation network [41]. At contents above 1 wt% MWCNTs, the elastic modulus is decreased marginally with increasing MWCNTs content. On the other hand, at low MWCNT content (≤1 wt%), the elastic modulus increases up to 0.105 MPa, only exceeding it when the MWCNT content is above 1 wt%, and then decreases further as the MWCNTs content is raised. At the percolation threshold, 1 wt% the nanotubes form an interconnecting structure, call percolation cluster exhibiting a high degree of nanotube interactions and/or entanglement [42]. The increasing MWCNT content produces infinite network in reducing the swelling and decreasing elastic modulus as was expected for the composites at high MWCNT content. The decrease in elastic modulus, G, can be explained by the formation of a carbon nanotube network which significantly improves the stiffness of composite gel. That is due to the high stiffness the mechanical properties should be influenced substantially [43]. On the other hand, when the MWCNT content is below 1 wt%, the elastic modulus is presenting lower values and increases as MWCNT is added. Because of molecular tube-tube and tube-polymer interactions, the mechanical properties of MWCNT-polymer composites were affected. On the other hand, to determine mechanical properties of MWCNT-polymer composites, load transfer and interfacial bonding should be known. Therefore, nanotube dispersion in the polymer plays a critical role for this phenomenon [40].

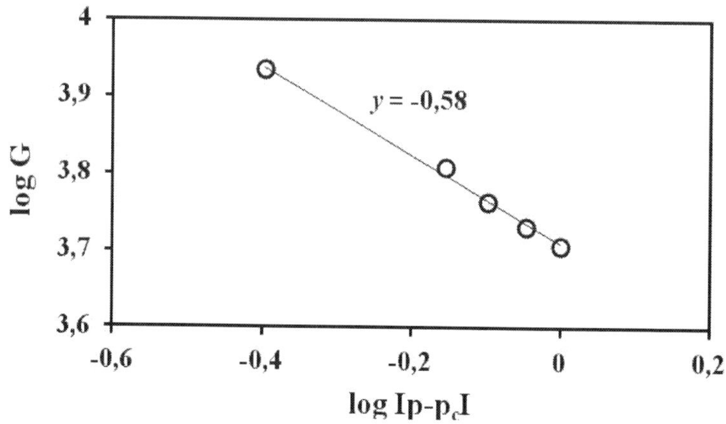

Figure 19. Logarithmic plot of the elastic modulus versus MWCNTs contents curves for $p < p_c$. The exponent, y, was determined from the slope of the straight line [39].

In **Figure 19**, it is understood that 1 wt% is the critical percolation threshold, p_c, at which gel system owns a percolation cluster formed from MWCNTs. Here, the composite gel passes the highest elasticity presenting the highest G value.

$$G(p) \approx (p_c - p)^{-y} \tag{10}$$

Equation (10) now can be used to fit the G versus wt% MWCNT curve below the critical point (below 1 wt%). The value of the fitting exponent y in Eq. (10) was estimated from the slope of the linear relation between log G and log Ip_pcI, as shown in **Figure 19**. Elastic percolation occurs below 1 wt% MWCNTs (**Figure 18**) with a critical exponent around y = 0.58, which is close to the theoretical prediction of this value in the 3D percolated system known as a SEPN. The critical exponent, y, agrees with the literature values [44]. Here, x is the critical exponent which is taken from literature by the upper limit x = 2.1. Moreover, Δ was found to be 0.74, very close to the upper limit of Δ, as given in range between 0.61 and 0.75. The mechanical properties of PVAc-MWCNT nanocomposites [13] were investigated as shown in **Figure 20**.

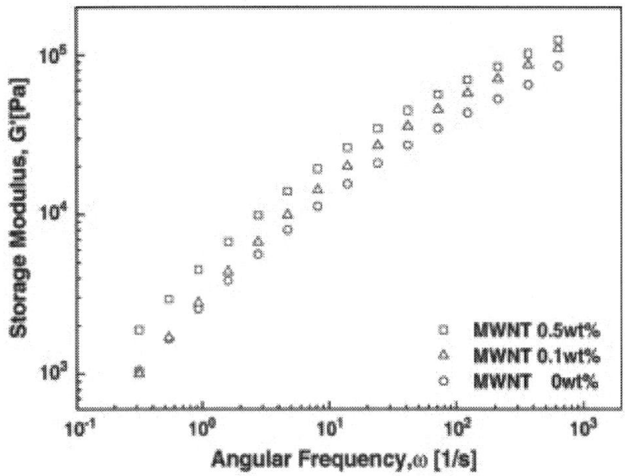

Figure 20. Storage modulus as a function of angular frequency at 145°C for PVAc-MWCNTs nanocomposites [13].

As shown in **Figure 20**, the storage modulus G′ increases with an addition of MWCNT loading compared with that of the PVAc matrix.

The rheological properties of PET-MWCNT nanocomposites were determined using low strain values with frequency 0.1–500 rad/s [46]. The storage modulus versus frequency is given in **Figure 21**.

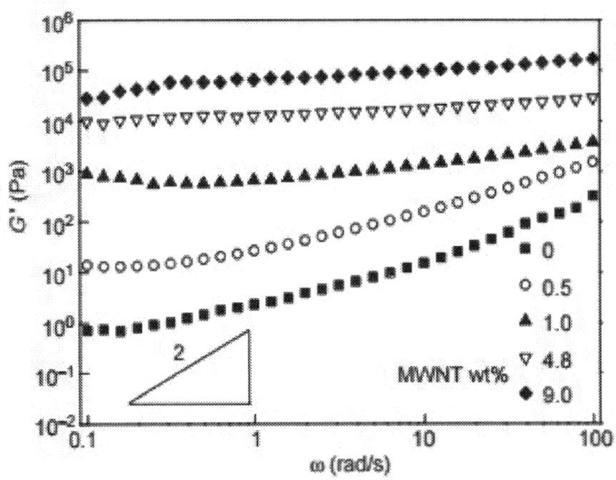

Figure 21. Storage modulus (G') of PET-MWCNT nanocomposites versus frequency (w) at 265°C [12].

The rheological behavior of the nanocomposites depends on the MWCNT content with respect to the frequencies because of the restriction of PET chain relaxation and the short-range dynamics or local motion of the PET chains in the nanocomposites [12].

4. The electrical properties of polymer-CNTs composites

The electrical resistivity of the PVAc/MWCNT composite films was performed by a Keithley Model 6517A electrometer and Keithley Model 8009 resistivity test fixture. 10–100 V DC potential for every 15 s periods was applied to measure their surface resistivity, R_s, (Ohm/square or Ohm) which was four different orientations and repeated measurements were repeated. The surface conductivity values were calculated from the following equation.

$$\sigma = 1/R_s \tag{11}$$

Then, the obtained results from Eq. (11) are presented in **Figure 22**. It can be seen from **Figure 10** that the conductivity, σ, dramatically increases above M_σ = 1.8 wt%. This behavior can be explained by the existence of vertical conductive paths of MWCNTs in the composite film. After M_σ = 1.8 wt%, the insulating system starts to transform to the conductive system. Therefore, M_σ = 1.8 wt% is the percolation threshold of conductivity. For a dilute composite structure, classical percolation theory was given in Eq. (12) can be used [25].

$$\sigma = \sigma_0 \left(M - M_\sigma\right)^{\beta}{}_{\sigma} \tag{12}$$

Here, σ is the conductivity (Siemens), σ_0 is the conductivity of pure MWCNT film, M is the volume or mass fraction of MWCNT, M_σ is the percolation threshold value, and β_σ is the critical exponent for the conductivity, respectively, which was calculated from Eq. (12) and determined from the slope of the log σ and log $(M - M_\sigma)$ plot. The $\beta_\sigma = 2.25$ value is well agreed with the theoretical and the experimental data in the literature [45].

The average values of the surface resistivity, R_s, of PVAc/MWCNT composite films versus mass fractions of MWCNT, M, were measured and observed that the surface resistivity (R_s) of the composite films do not change much below 1.0 wt% ($M \leq 1.0$ wt%)[45]. However, R_s values of the composite films dramatically decrease from 10^{11} to 10^4 Ohm/square in the bandgap of M = 1.0–4.0 wt%.

This behavior indicates that the electrical percolation occurs at low levels of M. The surface conductivity values were calculated from Eq. (11). The percolation threshold of the surface conductivity (M_σ) is 1.0 wt%. β_σ was calculated from the slope of the curve in **Figure 23**, which was drawn from the logarithms of the surface conductivity data treated with Eq. (12), and found to be as 2.1 which is well agreed with the theoretical and the experimental results in the literature [45].

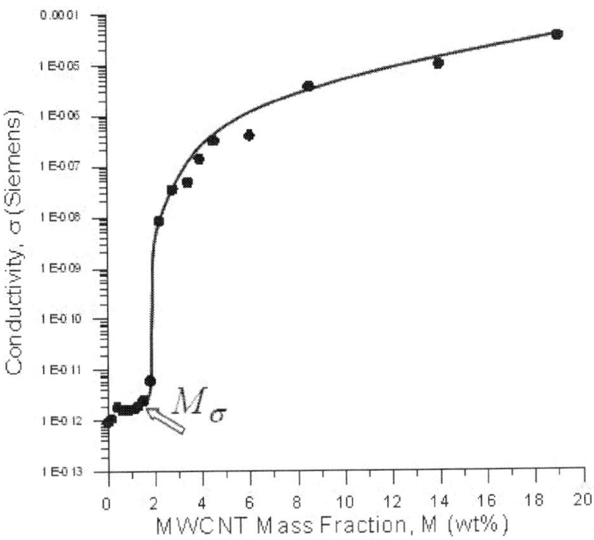

Figure 22. Variations in conductivities, σ, which were calculated by Eq. (11) versus mass fractions [30].

The surface conductivity properties of PS-MWCNT composite films were measured at room temperature using a two probe technique [30]. **Figure 24** shows the electrical conductivity (σ) of PS-MWCNT composite films as a function of MWCNT ratio, R. While low MWCNT content composites (R < 0.04) show similar conductivity between 10^{-13} and 10^{-12}, the conductivity of high MWCNT content films (R > 0.04) increases dramatically to 10^{-7}–10^{-6} S.

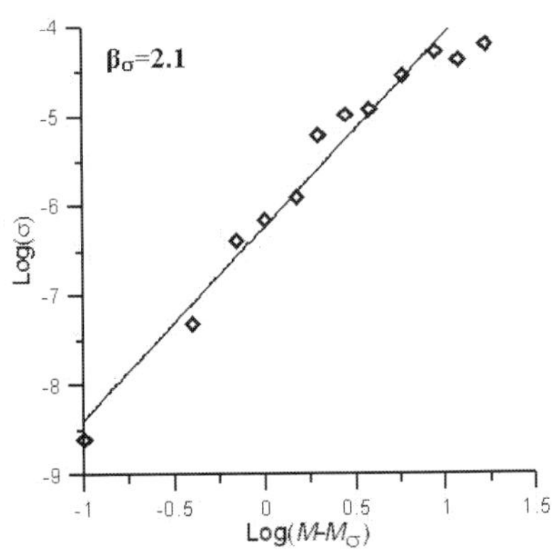

Figure 23. The log-log plot of σ versus $M \rightarrow M_{\sigma}$. The slope of the straight line produces the electrical critical exponent, β_{σ} as 2.1 [31].

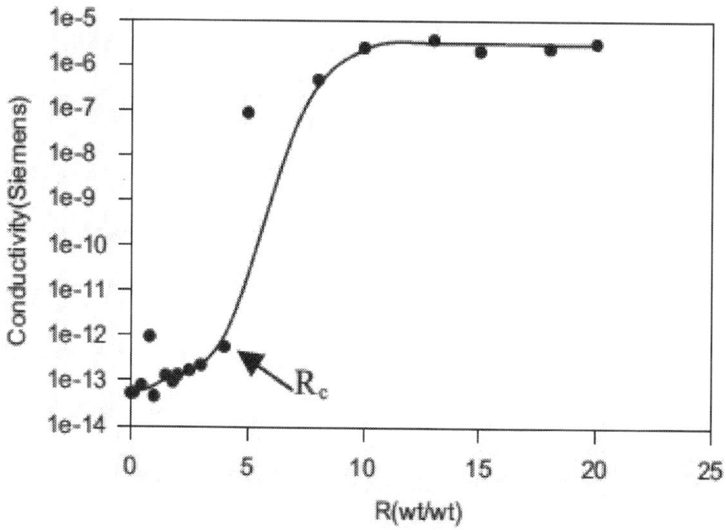

Figure 24. Conductivity versus MWCNT content, R (w/w) [30].

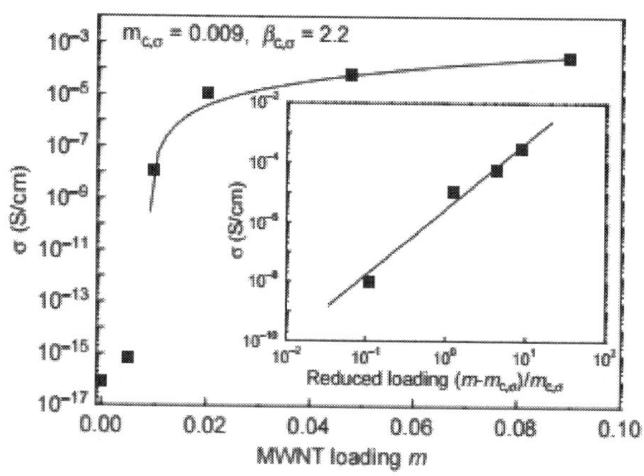

Figure 25. Electrical conductivity (σ) of the PET-MWCNT nanocomposites as a function of MWCNT loading. Inset: a log-log plot of electrical conductivity versus reduced MWCNT loading. The solid lines are fits to a power law dependence of electrical conductivity on the reduced MWCNT loading [12].

Figure 25 shows the electrical conductivity (σ) of the PET-MWCNT [12]. The electrical conductivity of pure PET is 8.6×10^{-17} S cm^{-1}. By loading MWCNT from 0.5 to 1 wt%, the conductivity of PET-MWCNT nanocomposites increases in 8 orders of magnitude, because percolative path was formed in the nanocomposite [12].

5. Conclusion

In this chapter, we review the optical, electrical, and mechanical behaviors of polymers doped by multiwalled carbon nanotubes (MWCNTs). This chapter covered the works produced from different types of polymers. We try to give the experimental results based on the mentioned theoretical models. In the first section of the chapter, some examples were given about optical behavior of polymer-MWCNT composites which were performed by fluorescence and UV-Vis spectroscopy. Therefore, the behavior of them at a molecular level were discussed, and measured some physical quantities were reported. The second section of the chapter has demonstrated that the mechanical and rheological measurements can be used to determine the variations of the elastic and storage modulus of the composites. The third part of the chapter has shown that the insulator-conductor transition takes place by the addition of a small amount of MWCNT in the polymer composite system. The insulator system starts to transform to a more conductive state by consisting of conductive paths of MWCNTs between the electrodes. The size of MWCNTs and electron hopping and/or tunneling effects play important roles in the early percolation behavior of the films.

In conclusion, we try to give the experimental evidences based on the mentioned theoretical models. As far as the engineering applications are concerned, optical, mechanical, and electrical properties of polymer composites prepared by MWCNT contents are very important in coating, food, electronic, and pharmaceutical industries. This chapter introduces basic parameters for the given processes in the polymer composites which can find important applications in the mentioned fields.

Author details

Gülşen Akın Evingür[1*] and Önder Pekcan[2]

*Address all correspondence to: gulsen.evingur@pirireis.edu.tr

1 Piri Reis University, Tuzla, İstanbul, Turkey

2 Kadir Has University, Cibali, İstanbul, Turkey

References

[1] Lijima, S. Helical microtubules of graphitic carbon. Nature. 1991; 354: 56–58.

[2] Ajayan PM, Zhou OZ. Applications of carbon nanotubes. Carbon Nanotubes. 2001; 80: 391–425.

[3] Olek M. Carbon Nanotube Composites Mechanical, Electrical and Optical Properties. Mathematics and Nature Department, Bonn University, Bonn, Germany. 2006.

[4] Maity A, Biswas M. Polymerization of N-vinylcarbazole by multiwalled carbon nanotube. Journal of Applied Polymer Science. 2007; 104: 4121–4126.

[5] Zhou C, Qiu X, Zhuang Q, Han Z, Wu Q. In situ polymerization and photophysical properties of poly(p-phenylene benzobisoxazole)/multiwalled carbon nanotubes composites. Journal of Applied Polymer Science. 2012; 124: 4740–4746.

[6] Ibrahim SS, Ayesh AS. Electrical and optical properties of functionalized multiwalled carbon nanotubes/poly(3-octylthiophene)/polystyrene composites. Journal of Thermoplastic Composite Materials. 2015; 28(2): 225–240.

[7] Barone PW, Yoon H, Garcia RO, Zhang J, Ahn JH, Kim JH, Strano MS. Modulation of single walled carbon nanotube photoluminescence by hydrogel swelling. ACS Nano. 2009; 3(12): 3869–3877.

[8] Tong X, Zheng J, Lu Y, Zhang Z, Cheng H. Swelling and mechanical behaviors of carbon nanotube/poly(vinyl alcohol) hybrid hydrogels. Materials Letters. 2007; 61: 1704–1706.

[9] Liu H, Liu M, Zhang L, Ma L, Chen J, Wang Y. Dual stimuli sensitive composites based on multi walled carbon nanotubes and poly(N,N-diethylacrylamide-co-acrylic acid) hydrogels. Reactive & Functional Polymers. 2010; 70: 294–300.

[10] Awasthi K, Awasthi S, Srivastava A, Kamalakaran R, Talapatra S, Ajayan PM, Srivastava ON. Synthesis and characterization of carbon nanotube-polythylene oxide composites. Nanotechnology. 2006; 17: 5417–5422.

[11] Bin Y, Mine M, Koganemaru A, Jiang X, Matsuo M. Morphology and mechanical and electrical properties of oriented PVA–VGCF and PVA–MWNT composites. Polymer. 2006; 47: 1308–1317.

[12] Hu G, Zhao C, Zhang S, Yang M, Wang Z. Low percolation thresholds of electrical and rheology in poly(ethylene terephthalate through the networks of multi walled carbon nanotubes). Polymer. 2006; 47: 480–488.

[13] Choi CS, Park BJ, Choi HJ. Electrical and rheological characteristics of poly(vinyl acetate)/multiwalled carbon nanotube nanocomposites. Diamond Related Materials. 2007; 16: 1170–1173.

[14] Huang CL, Wang Chi. Rheological and conductive percolation laws for syndiotactic polystyrene composites filled with carbon nanocapsules and carbon nanotubes. Carbon. 2011; 49: 2334–2344.

[15] Seo MK, Park SJ. Electrical resistivity and reological behaviors carbon nanotubes filled polypropylene composites. Chemical Physics Letters. 2005; 395: 44–48.

[16] Miyako E, Nagata H, Hirano K, Hirotsu T. Photodynamic thermoresponsive nanocarbon–polymer hybrids. Small (Weinheim an der Bergstrasse, Germany). 2008; 4(10): 1711–1715.

[17] Luo YL, Zhang CH, Chen YS, Yang W. Preparation and characterization of polyacrylamide/MWNTs nanohybrid hydrogels with microporous structures. Materials Research Innovations. 2009; 13(1): 18–27.

[18] Yu J, Lu K, Sourty E, Grossiord N, Koning CE, Loos J. Characterization of conductive multiwall carbon nanotube/polystyrene composites prepared by latex technology. Carbon. 2007; 45: 2897–2903.

[19] Choi CS, Park BJ, Choi HJ. Electrical and rheological characteristics of poly(vinyl acetate)/multi-walled carbon nanotube nanocomposites. Diamond Related Materials. 2007; 16(4–7):1170–1173.

[20] Verma SK, Bisarya SC. Improvement in properties of poly(vinyl acetate)-emulsion with dibasic acids. Journal of Applied Polymer Science. 1986; 31(8): 2675–2684.

[21] Pekcan Ö, Evingür GA. Conductivity Percolation of Carbon Nanotubes in Polyacrylamide Gels. In: Siva Yellampalli, editor. Carbon Nanotubes–Polymer Nanocomposites. Crotia: InTech; 2011. 197 p. ISBN 978-953-307-498-6.

[22] Evingür GA, Pekcan Ö. Drying of polyacrylamide–multiwalled carbon nanotube (MWNT) composites with various MWNTs contents: a fluorescence study. Journal of Polymer Engineering. 2013; 33(1): 33–39.

[23] Evingür GA, Pekcan Ö. Monitoring of dynamical processes in PAAm–MWNTs composites by fluorescence method. Advanced Composite Materials. 2012; 21(2): 193–208.

[24] Evingür GA, Pekcan Ö. Effect of multiwalled carbon nanotube (MWNT) on the behavior of swelling of polyacrylamide–MWNT composites. Journal of Reinforced Plastics and Composites. 2014; 33(13): 1196–1206.

[25] Kara S, Pekcan Ö. Photon transmission technique for monitoring drying processes in acrylamide gels formed with various cross linker contents. Journal of Applied Polymer Science. 2001; 80(11): 1898–1906.

[26] Birks JB. Photophysics of aromatic molecules. New York, NY: Wiley; 1971.

[27] Tanaka T, Fillmore DJ. Kinetics of swelling of gels. Journal Chemical Physics. 1979; 70(3): 1214–1218.

[28] Li Y, Tanaka T. Kinetics of swelling and shrinking of gels. Journal Chemical Physics. 1990; 92(2): 1365–1371.

[29] Zrinyi M, Rosta J, Horkay F. Studies on the swelling and shrinking kinetics of chemically crosslinked disk-shaped poly(vinyl acetate) gels. Macromolecules. 1993; 26: 3097–3102.

[30] Kara S, Arda E, Dolaştır F, Pekcan Ö. Electrical and optical percolations of polystyrene latex–multiwalled carbon nanotube composites. Journal of Colloid and Interface Science. 2010; 344: 395–401.

[31] Arda E, Kara S, Pekcan Ö. Electrical, optical, and fluorescence percolations in P(VAc-co-BuA)/MWCNT composite films. Phase Transitions. 2013; 86(10): 1017–1032.

[32] Stauffer D, Aharony A. Introduction to Percolation Theory. London: Taylor & Francis. 1994.

[33] Ugur S, Yargı Ö, Pekcan Ö. Percolation and film formation behaviors of MWNT/PS nanocomposites. Procedia Engineering. 2011; 10: 1709–1717.

[34] Yargı Ö, Ugur S, Pekcan Ö. Temperature dependence of oxygen diffusion into polymer/carbon nanotube. Polymer Engineering & Science. 2012; 52: 172–179.

[35] Yargı Ö, Ugur S, Pekcan Ö. Oxygen diffusion into multiwalled carbon nanotube doped polystyrene latex films using fluorescence technique. Journal of Fluorescence. 2013; 23: 357–366.

[36] Yargı Ö, Ugur S, Pekcan Ö. Fluorescence quenching method for monitoring oxygen diffusion into PS/CNT composite films. Progress in Organic Coatings. 2013; 76(12), 1805–1809.

[37] Yargı Ö, Ugur S, Pekcan Ö. Polymer/carbon nanotube composite film formation: a fluorescence study. Polymer Composites. 2014; 35: 817–826.

[38] Evingür GA, Pekcan Ö. Temperature effect on elasticity of swollen composite formed from polyacrylamide (PAAm)–multiwalled carbon nanotubes (MWNTs). Engineering. 2012; 4: 619–624.

[39] Evingür GA, Pekcan Ö. Elastic percolation of polyacrylamide (PAAm)–multiwall carbon nanotubes (MWNTs) composites. Phase Transition. 2012; 85: 553–564.

[40] Awasthi K, Awasthi S, Srivastava A, Kamalakaran R, Talapatra S, Ajayan PM, Srivastava ON. Synthesis and characterization of carbon nanotube–polyethylene oxide composites. Nanotechnology. 2006; 17: 5417–5422.

[41] Du F, Scogna RC, Zhou W, Brand S, Fischer JE, Winey K. Nanotube networks in polymer nanocomposites: rheology and electrical conductivity. Macromolecules. 2004; 37: 9048–9055.

[42] Pötschke P, Fornes TD, Paul DR. Rheological behavior of multiwalled carbonnanotube/polycarbonate composites. Polymer. 2002; 43: 3247–3255.

[43] Meincke O, Kaempfer D, Weickmann H, Friedrich C, Vathauer M, Warth H. Mechanical properties and electrical conductivity of carbon–nanotube filled polyamide-6 and is blends with acrylonitrile/butadiene/styrene. Polymer. 2004; 45: 739–748

[44] Sahimi M. Application of Percolation Theory. London: Taylor and Francis. 1994.

[45] Weber M, Kamal MR. Estimation of the volume resistivity of electrically conductive composites. Polymer Composites. 1997; 18: 711–725.

[46] Grunlan JC, Mehrabi AR, Bannon MV, Bahr JL. Water-based single-walled-nanotube-filled polymer composite with an exceptionally low percolation threshold. Advanced Material. 2004; 16: 150–152.

Chapter 6

Mechanical Properties of Carbon Nanotubes-Polymer Composites

Lixing Dai and Jun Sun

Additional information is available at the end of the chapter

http://dx.doi.org/10.5772/62635

Abstract

Carbon nanotubes (CNTs), as a one-dimensional material, have outstanding mechanical properties such as extreme tensile strength and Young's modulus. At present, to prepare pure CNTs materials is quite difficult and the mechanical properties of the materials are limited in a low level. Because of their extraordinary mechanical properties and high aspect ratio, CNTs are considered to be ideal candidates for polymer reinforcement. In addition, CNTs/polymer composite materials are much easier to prepare than pure CNTs materials, so they have been paid much attention by researchers recently. However, challenges must be faced to prepare the CNTs/polymer composite with ultimate mechanical properties. So in this chapter, the main concerns are how to disperse CNTs in polymer matrix to prepare homogenous composite dispersions, how to prepare homogenous CNTs/polymer composite using possible fabricating processes based on the homogenous dispersions, how to increase the fiber mechanical properties especially through the enhancing interaction between polymer and CNTs, controlling the amount of CNTs and enhancing their orientation in the matrix.

Keywords: mechanical property, carbon nanotube, polymer, composite, dispersibility

1. Introduction

In recent years, because stronger, lighter, or less expensive compared to traditional materials, high-performance composite materials have increasingly become an essential in a wide range of structural applications [1, 2]. As early as in ancient times, people have realized that the combination of two or more materials could form new materials which had better properties than either of the materials compositions. For example, ancient people in China began to use straw to reinforce mud to form bricks as building materials, and to use hemp or silk to reinforce paint and so on [3, 4]. Nowadays, composite materials are used extensively in various areas, not only in the aerospace industry but also in a large and increasing number of commercial mechanical engineering applications.

Composites are made up of reinforcements and matrix materials, in which the latter surrounds and supports the former by maintaining their relative positions, and conversely, the reinforcements impart their special mechanical and physical properties to enhance the matrix properties [1]. With their rapid development and because of their advantages in good processibility and high mechanical properties, polymers have gradually replaced many of the conventional materials as matrix of composites in various applications. In order to obtain high-strength, high-modulus, and light-weight composites used in harsh loading conditions such as aerospace components, anti-bullet, high temperature resistance, etc., strong and light reinforcing fibers such as Kevlar, carbon fiber, ultrahigh-molecular-weight polyethylene fiber and glass fiber are always selected to suit the requirements [5]. Provided the fibers are mechanically well attached to the matrix, they can greatly improve the overall properties of composites. Recently, fiber-reinforced composite materials have increasingly gained popularity in high-performance materials.

Carbon nanotubes (CNTs) as one of the strongest and stiffest materials in terms of tensile strength and elastic modulus in nature have received much attention since their discovery by Iijima in 1991 [6]. Their excellent mechanical properties result from the covalent sp2 bonds formed between the individual carbon atoms. It has been shown that CNTs are very strong in the axial direction, Young's modulus on the order of 270–950 GPa and tensile strength 11–63 GPa [7]. Furthermore, CNTs as one-dimensional materials possess some basic characteristic and advantages of fiber, which have inspired interest in using CNTs as a filler in polymer-based composites to obtain ultra–high performance structural materials with enhanced mechanical properties [8, 9]. A lot of efforts have been made to explore CNTs/polymer composite materials in both academy and industry. The structure, morphology, mechanical properties, and possible applications of these composite materials have been extensively investigated with some very promising results. They have been proven to be very effective fillers in polymers.

However, CNTs have large specific surface area, bending fiber-like shape, and strong van der Waals interactions which are easy to make the CNTs agglomerate and entangled, and difficult to be dissolved in water or organic solvents, or disperse in polymer [10]. The agglomeration of CNTs will lead to difficulties of stress transfer from matrix to the reinforced materials CNTs efficiently, and adversely lower the mechanical properties of the CNTs/polymer composite. So, how to enhance the dispersibility of CNTs in polymer matrix has been one of the most concerns in the field of CNTs reinforced polymer-based composite materials. Therefore, functionalization of CNTs is extremely important for their dispersion, stress-transfer, and potential applications in polymer composites [11]. In this chapter, we will focus on the

http://dx.doi.org/10.5772/62635

dispersibility of CNTs in polymer matrix through enhancing the interaction between CNTs and polymer, the processing of the composites, and the mechanical properties of the composites.

2. Dispersion of CNTs in polymers

Dispersity of CNTs in polymer and the stress transfer from the matrix to CNTs have significant effects on reaching optimum mechanical properties for CNTs/polymer composites. Only if CNTs disperse homogeneously in polymer matrix, effective stress transfer in the whole composite materials will be guaranteed; conversely, poor dispersity of CNTs in polymer matrix will lead to poor stress transfer, which not only cannot play their role in increasing mechanical properties but also can possibly weaken the original properties of the matrix. Therefore, the reinforcement of CNTs to polymer matrix is to great extent dependant on the dispersity of CNTs in the matrix.

To improve the dispersity of CNTs has two approaches: macroscopical mechanical mixing and functionalizing CNTs to build interaction between the interfaces of CNTs and polymer matrix. The interaction through bonding has another significant effect, that is, strengthening stress transfer effect. Obtaining a good dispersion of CNTs in the polymer matrix is very challenging. Ultrasonication, physical or chemical functionalization are common to achieve a good dispersity of CNTs in polymer matrix with enhanced mechanical properties.

2.1. Ultrasonic dispersion

Ultrasonication generates alternating low-pressure and high-pressure waves in liquids, causing high-speed impinging liquid jets and strong hydrodynamic shear-forces. So, ultrasonic device can be used as a high-speed mixer and agitator [12]. Ultrasonication is a common tool used to break up CNTs agglomerates in solution, but pure ultrasonication treatment of CNTs is not common, for the dispersion effect is not so satisfactory: some of CNTs re-aggregate or some of them deposit in a short period of time once sonication stops. **Figure 1** shows the photos of MWCNTs aqueous solutions in different standing time after sonication. The solutions sonicated 20 min, 1 hr and 2 hr all deposited after standing 30 min.

Generally, ultrasonication of CNTs always combines with application of surfactants to obtain better dispersion effect: ultrasonication is used to disperse CNTs, and surfactants are used as stabilizers. Sodium dodecyl sulfate (SDS) [13, 14], dodecyl benzene sulfonic acid (DBSA) [15, 16], and sodium dodecyl benzene sulfonate (SDBS) [17] are commonly used as surfactants. The process is simply that CNTs mix with surfactants and the mixture are then sonicated. During sonication, CNTs are gradually exfoliated and disentangled from aggregates and bundles and

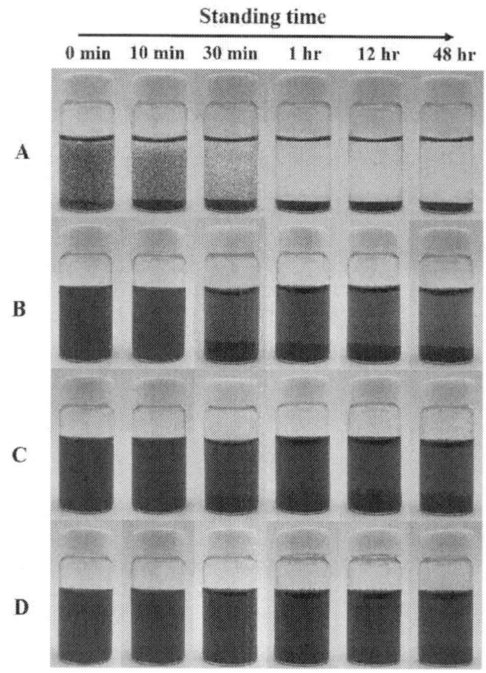

Figure 1. Digital images of MWNTs aqueous suspensions after ultrasonication and standing for different time. Ultrasonication time: A. 0 min; B. 20 min; C. 1 hr; D. 2 hr.

stabilized by surfactants. **Figure 2** shows the photos of MWCNTs aqueous solutions added SDBS in different standing time after sonication. Clearly, the addition of SDBS distinctly improves the dispersity and stability of MWCNTs in water.

CNTs dispersions often foam heavily during ultrasonication, and these foams consequently reduce the dispersion efficiency. This bubble problem can often be solved by adding antifoam agents such as oligomers of polyether and polysiloxane to eliminate air layers or to reduce foams [18].

The destruction of CNTs in suspensions by use of ultrasonication was described as early as 1996. Ultrasonic treatment of CNTs causes a considerable amount of defects including buckling, bending, and dislocations in the carbon structures [19]. So suitable ultrasonication time and energy output have to be considered before CNTs are treated.

2.2. Covalent functionalization

Covalent functionalization of CNTs can greatly improve the stress transfer from the matrix to CNTs besides dispersity, but it usually introduces structural defects to the surface of CNTs. The functionalization of CNTs can be achieved by either direct introduction of oxygen

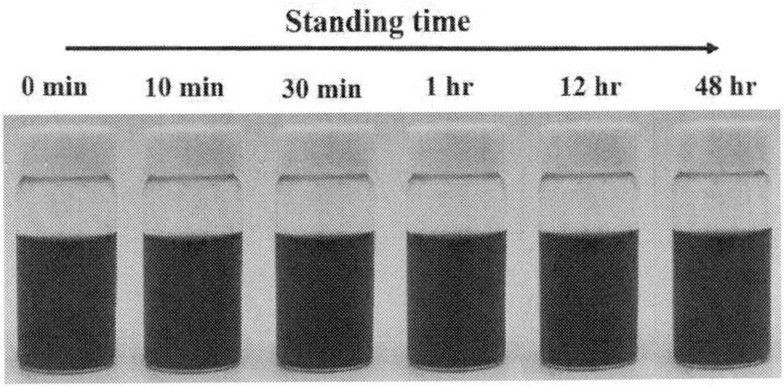

Figure 2. Digital images of MWNTs aqueous suspensions changed with the standing time after ultrasonication for 10 min in the presence of SDBS.

containing groups to the sidewalls of CNTs by acid treatment, or on this basis further introduction of polymer chains through its chain ends directly reacting with the oxygen containing groups on CNTs (ex situ), or commonly through the monomers reacting with the groups on CNTs to introduce polymer chains, that is, in situ polymerization.

2.2.1. Acid treatment

The most common procedure used for covalent attachment of oxygen containing reactive groups to CNTs is the treatment with inorganic acids. Usually, the nanotubes are treated through being refluxed with a mixture of concentrated nitric and sulfuric acid with the ratio of volume 1 to 3 although their concentrations may be slightly different [20–23]. However, accidently, the ratio of the acids changes, for example, CNTs were oxidized with the mixture of 83 mL of sulfuric acid and 133 mL of nitric acid (65%) [24]. Concentrated nitric acid [25] and the mixture of hydrogen peroxide and sulfuric acid (vol = 1:4) [26] are also selections for oxidizing CNTs. These oxidative treatments usually result in shortening of the CNTs' length and formation of surface reactive groups, such as hydroxyl, carbonyl, and mainly Carboxyl. Furthermore, the inorganic impurities of the Al-Fe catalyst used in the preparation of CNTs are solubilized by the acids and the concentration of the impurities is gradually decreased with treatment time [20, 23, 26, 27]. It has been shown that acid functionalized CNTs has good dispersibility in water and organic solvents, and the carboxylic functional groups give a stronger nanotube-polymer interaction, leading to enhanced values in mechanical properties of the CNTs/polymer composite [27, 28]. Acid-treated MWCNTs show homogeneous dispersion and scarcely any precipitates are observed. Because the sidewalls of the MWCNTs carry more dissociated carboxyl groups after oxidization, the nanotubes can stabilize via an electrostatic stabilization mechanism [29]. Saito et al. consider that this shortened MWCNTs after acid treatment could be dispersed in the polar solvents such as ethanol, DMSO, and DMF more easily than the crude MWCNTs. Moreover, these dispersion liquids of the MWCNTs are stable without aggregation for more than one month. This phenomenon can be ascribed to the reduction of their length and the effect of the solvation brought by the introduced hydrophilic group, i.e., the carboxyl group [20].

At moderate temperature, only little sidewall damage happens with the increase of the acid treatment time, whereas at high temperature, much shorter nanotubes are resulted [26].

Being different from reaction of CNTs with concentrated acids, carboxylic functional groups can be introduced on the CNTs through a photosensitized dry oxidation induced by ultraviolet radiation [30, 31]. Organic acid could also be used as an oxidative, for example, trifluoroacetic anhydride/H_2O_2 (3:1, 60% H_2O_2) were used to functionalize single-walled carbon nanotubes (SWCNTs), in addition to oxygen-based functional groups, trifluoroacetic groups were covalently attached to the SWCNTs. Moreover, these modified SWCNTs were shortened into ca. 300 nm in length during functionalization, resulting in exfoliation of nanotube ropes to yield small bundles and individual nanotubes. The resultant SWCNTs were easily dispersed in polar solvents such as dimethylformamide, water, and ethanol [32].

2.2.2. *Grafting macromolecular chains*

Polymer-functionalized CNTs, compared to the unmodified ones, usually show relatively good dispersibility in organic solvents and high compatibility with polymer matrix in CNTs/polymer composites. The functionalized CNTs containing active groups (predominantly carboxyl) can further corporate with polymers through esterification or amidation reactions between the groups on the CNTs and those on the polymer chains, [21, 33, 34] and conversely, treated or functionalized macromolecular chains can also directly attach on pristine CNTs [29, 35, 36].

Most of the approaches require the pre-modification of CNTs to introduce functional groups to CNTs surfaces, which can react to the groups on polymer chains, leading to the chains grafted onto CNTs. For example, after CNTs are acidic functionalized, polyvinyl alcohol (PVA) grafts to the tubes through esterification as shown in **Figure 3**. PVA-functionalized CNTs are soluble in the same solvent of neat PVA, thus allowing the intimate mixing of the nanotubes with the matrix polymer, so the functionalization of CNTs by polymer is an effective way for the homogeneous nanotube dispersion to obtain high-quality polymeric carbon nanocomposite materials [37]. Grafting different kind of polymer onto CNTs from the matrix polymer which is however compatible with the former polymer is another example. Massoumi et al first prepared carboxylated MWCNTs by acid treatment, then poly(ethylene glycol)-grafted MWCNTs (PEG-MWCNTs), whereas finally obtained PVA/MWCNTs-PEG nanocomposite and PVA/starch/MWCNTs-PEG nanocomposite [21].

Sometimes, polymer chains are not necessary to graft onto CNTs, but only some groups which are well compatible with polymer are needed to connect with CNTs. For instance, amino groups grafted MWCNTs possess good compatibility with the matrix PEEK, so simply by mixing the MWCNTs with PEEK, the composite with significantly increased storage modulus, glass transition temperature, and friction coefficient was prepared [27].

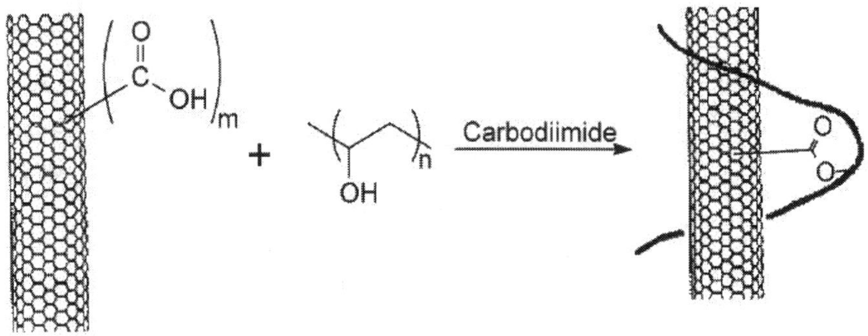

Figure 3. Schematic of functionalized CNTs with PVA in carbodiimide-activated esterification reactions [37].

On the other hand, the use of pristine or unmodified CNTs in the preparation of polymer functionalized CNTs has shown several advantages such as preventing the CNTs from damage. It is reported that polymer grafted CNTs have better reinforcement than pristine CNTs. This kind of polymer functionalized CNTs can be realized through free-radical polymerization, ozonization of polymer, Friedel-Crafts reaction and etc. [29, 35, 36, 38]. Through a radical coupling reaction involving polymer-centered radicals, polystyrene and poly[(tert-butyl acrylate)-b-styrene] are used to functionalize shortened SWCNTs [36]. Ozonization of polymers produces alkylperoxide and hydroperoxide groups in polymer chains, which can decompose into radicals by heating. As the radicals are reactive toward the sp2 hybrid carbons of CNTs, the ozonized polymer chains can be grafted onto the CNTs through the reaction between CNTs sidewalls and the radicals of polymer chains as shown in **Figure 4** [35].

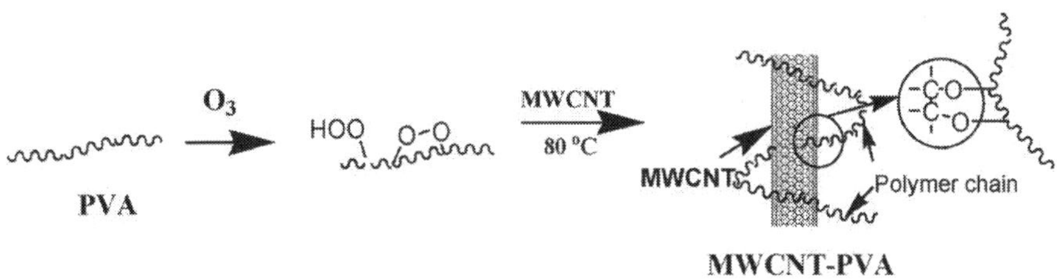

Figure 4. Reaction mechanism of ozonized PVA grafting onto MWCNT [35].

After Friedel–Crafts reaction has been successfully applied for the chemical functionalization of CNTs through grafting small molecules onto CNTs [39, 40], researchers are increasingly interested in application of Friedel-Crafts for functionalization of CNTs using polymer. The approach is more feasible than the traditional carboxylation because the surface treatment of

CNTs before the grafting of polymer chains to the CNTs is not needed. Wu et al. [41] successfully functionalized MWCNTs with high percentage of grafting via Friedel-Crafts alkylation. The PVC-grafted MWCNTs could be dispersed well in organic solvent and the dispersion was quite stable. Moreover, it has been shown that Friedel-Crafts alkylation is a less destructive or nondestructive reaction for efficient dispersion and functionalization of CNTs.

In our previous work [29], PVA was grafted onto MWCNTs by Friedel-Crafts alkylation reaction with anhydrous aluminum chloride ($AlCl_3$) as a catalyst. MWCNTs, PVA, and DMSO were first mixed and dissolved, then the catalyst $AlCl_3$ was added, and the mixture reacted and terminated. The reaction product was centrifuged and washed with distilled water until no $AlCl_3$ could be detected, and finally, the functionalized MWCNTs (f-MWCNTs) were obtained. In the experimental procedures, $AlCl_3$ reacted with the hydroxyl on PVA and formed carbocations, which then attacked the benzene rings on MWCNTs surface due to their inherent eletrophilicity. Subsequently, the PVA chains were grafted onto MWCNTs as depicted in **Figure 5**.

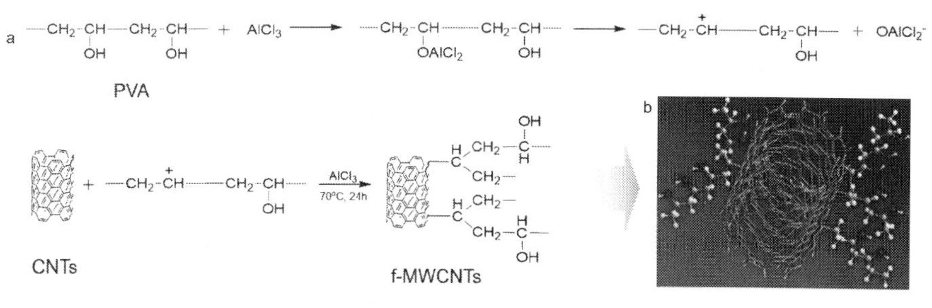

Figure 5. (a) Functionalization of MWCNTs with PVA by Friedel-Crafts alkylation reaction and (b) cartoon depicting of f-MWCNTs structure [29].

Figure 6 shows TEM images of pristine MWCNTs (p-MWCNTs), f-MWCNTs and carboxyl group functionalized MWCNTs (c-MWCNTs) treated by a mixture of nitric and sulfuric acid in DMSO/H_2O. p-MWCNTs are aggregate obviously as shown in **Figure 5a**, whereas f-MWCNTs are homogenously dispersed without distinct destructive structural damages (**Figure 5b**). In comparison, c-MWCNTs present serious fragmentation (**Figure 5c**). It is clear that functionalization of MWCNTs by Friedel-Crafts alkylation reaction is relatively much gentler than acid treatment. Images of single p-MWCNT and f-MWCNT tubes clearly show that the surface of the p-MWCNT is clean and smooth, while the surface of f-MWCNT is surely wrapped with PVA chains (**Figure 5d** and **5e**).

UV detector can be used to measure the transmittance of MWCNTs in the supernatant, which is used to characterize the dispersibility of the MWCNTs, that is, the higher the transmittance, the worse the dispersibility. The UV-vis spectra of the samples after storage for different periods of time are recorded as shown in **Figure 7**. For p-MWCNTs, the transmittance is almost

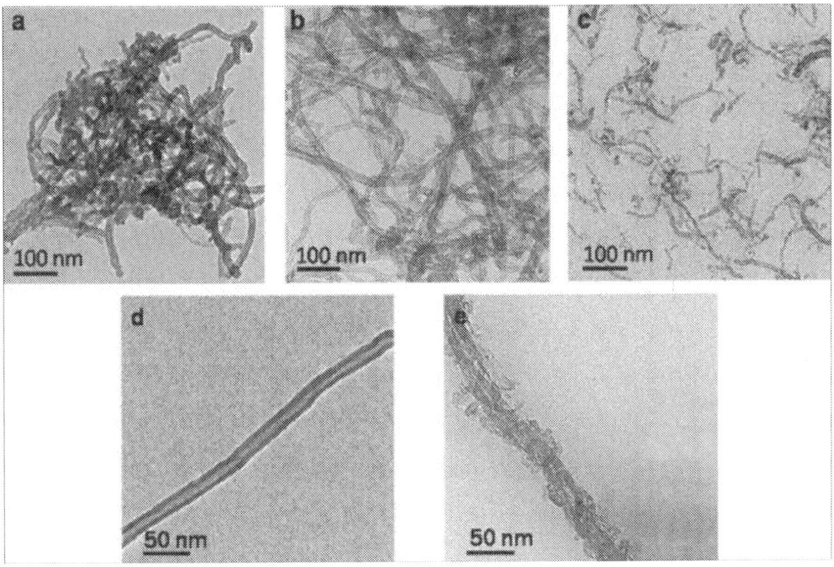

Figure 6. TEM images of (a) p-MWCNTs, (b) f-MWCNTs, (c) c-MWCNTs, (d) single p-MWCNTs, and (e) single f-MWCNTs in DMSO/H_2O [29].

100% after the first three days, indicating that the MWCNTs are relatively rapidly precipitated. However, the transmittance of modified MWCNTs, whatever f-MWCNTs or c-MWCNTs, increases only about 20% in the first three days and keeps stable in the following days, proving their good dispersibility which is apparently related with the interaction between PVA and MWCNTs.

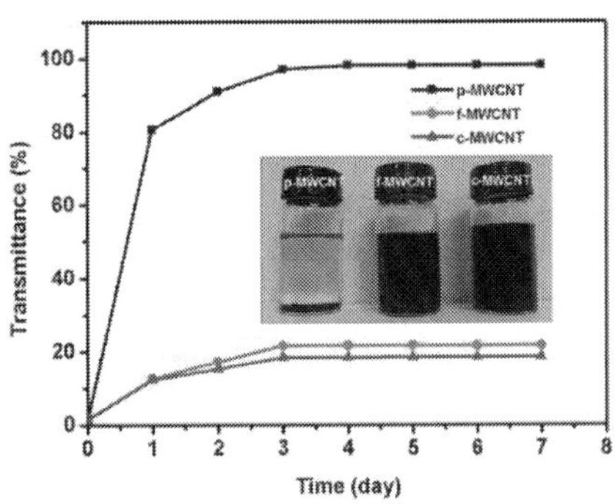

Figure 7. UV–vis transmittances of p-MWCNTs, f-MWCNTs and c-MWCNTs. The content of MWCNT is 0.05 wt%. Inset: image of the dispersed samples after 7 days [29].

2.2.3. *In situ polymerization*

The main feature of in situ polymerization is that it enables polymer macromolecules from the starting material of monomers to graft onto the convex walls of CNTs, which can provide a better nanotube dispersion and formation of a strong interfacial interaction between the nanotube and the polymer matrix. In addition, in situ polymerization is a very convenient technique that allows the preparation of composites with high nanotube loading and provides very good miscibility with almost any polymer type [8, 38]. It is believed that this simple in situ polymerization method is in general scalable for large-scale production of CNTs/polymers composites [42]. Step polymerization is often applied to prepare CNTs reinforced polymer nanocomposites by in situ polymerization of monomers [43, 44]. In the in situ grafting reaction, CNTs first have to graft reactive groups on the side wall, which then react with reactive groups of monomers to form covalent bond. For example, synthesis of MWCNTs-reinforced polyimide (PI) nanocomposites by in situ polymerization of monomers is as follows: MWCNTs associated with acyl groups were formed and then participated in the reaction with the monomers through the formation of amide bonds. The mechanical properties of the resultant MWCNTs-PI nanocomposites were significantly enhanced at a very low loading (0.5 wt. %) as shown in **Figure 8** [45].

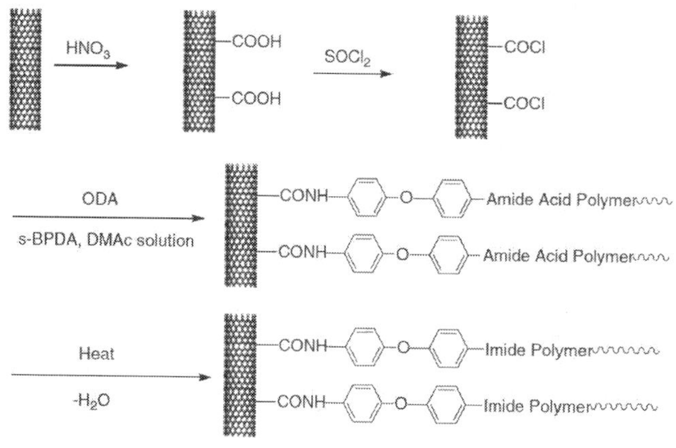

Figure 8. Outline of the preparation of MWNT-polyimide nanocomposite films [45].

MWCNTs modified by functional amine groups via ozone oxidation followed by silanization were incorporated into a vegetable oil-based polyurethane (PU) network via in situ polymerization to prepare CNTs/PU nanocomposites. Storage modulus, glass transition temperature, Young's modulus, and tensile strength of the nanocomposites increased with increasing the MWCNTs loading up to 0.8%. However, increasing the MWCNTs content to 1.2 wt % resulted in a decrease in thermomechanical properties of the nanocomposites [38]. A SWCNTs/nylon 6 graft copolymer was prepared by in situ polymerization of caprolactam with SWCNTs possessing carboxylic acid and amide functionalities. The results show that SWCNTs with a higher concentration of carboxylic acid groups can form a stronger SWCNTs-polymer interfacial interaction, which consequently improves the mechanical properties [43]. Diamine and dianhydride were reacted with carboxylated MWCNTs in situ to give a homogeneous MWCNTs/poly(amic acid) mixture and then to give a series of MWCNTs/PI composites after imidization. Tensile tests show the elastic modulus and the yield strength of the composites increase, and the failure strain decreases [44].

Besides, in situ polymerization can also be realized through radical polymerization [46]. The composites of f-MWCNTs/polyacrylonitrile (PAN) copolymer were prepared by in situ solution polymerization. Acrylonitrile (AN) and acrylamide (AM) dissolved in DMSO/H_2O solvent and f-MWCNTs mixture suspension was initiated by AIBN, reacted and finally PAN grafted MWCNTs were obtained. Compared with p-MWCNTs as shown in **Figure 9a**, the f-MWCNTs (**Figure 9b**) are not destructed after the functionalization and have a more compact morphology. As shown in **Figure 9c** and **9d**, p-MWCNTs settle obviously, whereas f-MWCNTs are disentangled and dispersed uniformly.

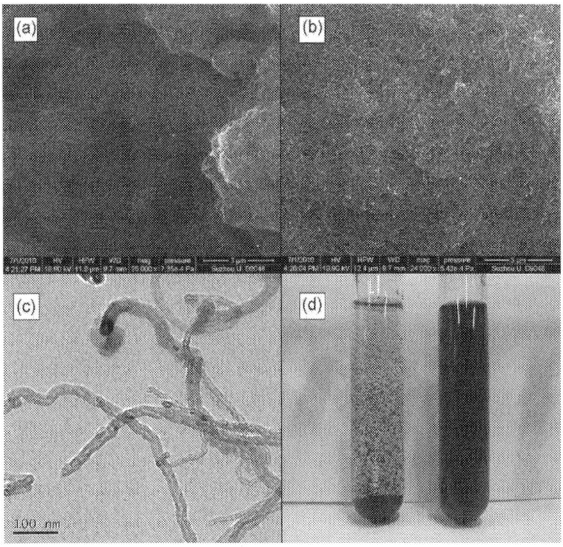

Figure 9. SEM images of (a) p-MWCNTs (b) f-MWCNTs, TEM images of (c) f-MWCNTs, and (d) photograph of p-MWCNTs (left) and f-MWCNTs (right) [46].

2.3. Noncovalent functionalization

Functionalization of CNTs by chemical processes not only causes damage to different extent which will restrict the reinforcement of the CNTs but also is difficult to be commercially available due to the complicated operations. Noncovalent modification is now adding its appeal for many researchers since it is more feasible than covalent modification [47, 48]. The dispersion of CNTs noncovalent modified in polymer is generally achieved by interaction of secondary van der Waals bonding or hydrogen bonding among polymer, dispersant, and CNTs surface [49]. Non-covalent treatment has the possibility of adsorbing various groups on CNTs surface without disturbing the π system of the graphene sheets of CNTs. So the non-covalent surface treatment by surfactants or polymers has been widely used in the preparation of both aqueous and organic solutions containing high weight fraction of well dispersed CNTs. Generally, ionic surfactants are suitable for CNTs/watersoluble while nonionic surfactants are used in the case of organic solvents [50].

2.3.1. Synthetic surfactant

2.3.1.1. Ionic surfactants

Several synthetic ionic surfactants are reported to efficiently disperse bundled nanotubes into suspensions of individual nanotubes, particularly for SWCNTs [13, 51, 52]. For example, SWCNTs could be suspended in aqueous media as individuals surrounded by SDS adsorbed phase [53]. Diouri et al. prepared CNTs/SDS aqueous solution with the help of ultrasonication, then the solution mixed with PVA aqueous solution to achieve the CNTs/PVA composites with required CNTs loadings [14]. An extensive all-atom molecular dynamics study on the morphology of SDS surfactant aggregates adsorbed on SWCNTs has been carried out. The calculations reveal that the nanotube diameter is the primary factor that determines the morphology of the aggregates [54]. SDS makes CNTs have satisfied dispersibility, but compared with oxidized CNTs, CNTs stabilized by surfactant SDS exhibit weaker interactions with the matrix PVA [55].

SDBS could be used to solubilize high weight fraction SWCNTs in water by the nonspecific physical adsorption. A series of anionic, cationic, and nonionic surfactants and polymers have been tested for their ability to suspend individual SWCNTs by Moore et al [49]. For the ionic surfactants, SDBS gives the most well resolved spectral features, while for the nonionic systems, surfactants with higher molecular weight suspend more SWCNTs and have more pronounced spectral features. The dispersing power of a range of surfactants has been explored by Islam et al. [13]. The results show that NaDDBS (i.e. SDBS)-CNTs dispersions are by far the most stable. The adsorption and the self-assembly of SDBS on SWCNTs are investigated via all-atom molecular dynamics simulations [56]. The results show that the self-assembly of SDBS depends on the surface coverage but to a small extent on the SWCNTs diameter, and controlling the morphology of the surfactant aggregates will lead to the selective stabilization of aqueous CNTs dispersions. The aggregation kinetics of SWCNTs and MWCNTs, initially dispersed by SDBS, are evaluated. The results show that the CNTs could be effectively suspended in aqueous solution using the surfactant (SDBS), and that increased electrolyte concentrations will induce aggregation [57].

Regarding cationic surfactant, Regarding cationic surfactant, dodecyl trimethylammonium bromide (DTAB) could form exceptionally stable SWCNTs dispersions [58]. Cetyltrimethylammonium bromide (CTAB) modified CNTs in H_2SO_4 aqueous solution assisted by sonication led to uniform dispersion of the embedded CNTs in polydiphenylamine (PDPA) and a reinforced PDPA nanocomposite. [59].

The ratio of surfactant to CNTs is sure to have effect on the dispersibility of CNTs, but the concentration of the surfactant is found to be a more important factor on the resulting dispersibility than the ratio [51].

2.3.1.2. Non-ionic surfactant

Another fraction of surfactants is nonionic surfactants such as polyoxyethylene 8 lauryl (CH_3 $(CH_2)_{11}(OCH_2$-$CH_2)_7OCH_2CH_3$), nonylphenol ethoxylate (Tergitol NP-7), polyoxyethylene octyl phenyl ether (Triton X-100), Tween-60, polysorbate-80 and so on [50, 52, 60–62].

It is clearly seen that after the surface treatment by Triton X-100, large agglomerates and closely packed CNTs are significantly loosened without breakage or shortening of CNTs. The surfactant-treated CNTs/epoxy nanocomposites exhibit much better performances including storage modulus, flexural strength, and flexural modulus than those without treatment for CNTs. The above observations are attributed to the 'bridging' effects between the CNTs and epoxy introduced by the hydrophobic and hydrophilic segments of the nonionic surfactant [60]. With the surfactant polyoxyethylene 8 laurylor as an additive, the nanotubes are dispersed better. Through the introduction of only 1 wt. % CNTs in epoxy, the glass transition temperature increases from 63 to 88°C and the elastic modulus increases by more than 30% [63]. Some polymers such as PEO-PPO-PEO triblock can also enhance the suspendability of CNTs. This is easily explained with steric stabilization by the adsorbed surfactant or polymer layer [49].

A molecule with a π-conjugated backbone built from aromatic thiophene and dialkoxyphenylene units and substituted imidazolium groups (TPO) is recently designed to obtain ultra-stable SWCNTs dispersion in aqueous medium. TPO provides synergistic π-π stacking, charge transfer and cation-π interactions. The dispersions even prepared using very low concentration of TPO (0.5 mg/mL) or SWCNTs (0.02 mg/mL) show long-time stability for twelve months [64]. Blanch et al. [51] systematically studied into the ability of some well-known surfactants and polymers to disperse SWCNTs. The smaller ionic surfactants are generally more effective dispersants than larger polymer and surfactant molecules. However, an effective dispersant for the CNTs by a certain technique may not necessarily perform well by a different method. Optimal concentrations for dispersion of the CNTs are determined for the anionic surfactants SDBS and DOC as well as some nonionic such as Triton X-405, for example, the optimal surfactant concentrations for CNTs dispersion are approximately 1.6% (DOC), 0.5% (SDBS), and 3% (Triton X-405). Exceeding the optimum concentrations is detrimental as it leads to agglomeration of the CNTs. As stated above, SDBS is excellent in dispersing CNTs, whereas SDS has turned out to be the most promising surfactant out of the following four sets of surfactants: SDS, CTAB, SDS + CTAB and Tween, for the effective dispersion of hydrophobic MWCNTs in rubber latex [62].

2.3.2. Non-synthetic surfactants

Recently, natural molecules as surfactants have been used in functionalization of CNTs by physical interaction. Liu et al. [65] described a non-destroyable surface decoration of CNTs with biopolymer chitsoan via a controlled surface-deposition and cross-linking process. Nakamura et al. [66] discovered that a green tea solution could dissolve SWCNTs without aggregation in an aqueous medium. In our previous study [67, 68], the strength and modulus of PVA fibers are extremely increased by introduction of MWCNTs functionalized with natural surfactants. π-π stacking between the benzene rings of surfactants and CNTs is the major interaction between CNTs and the surfactants.

2.3.2.1. Tea polyphenols

As non-covalent modification, using green tea to disperse SWCNTs in water was first reported by Nakamura et al. [66] and now has drawn a great deal of attentions. Tea polyphenols (TP) is the extract powder from tea, which is biocompatible and biodegradable. TP-functionalized MWCNTs and the composites with PVA were prepared in our previous work [67]. TP dissolved in DMSO/H_2O (vol = 3:1) was mixed with MWCNTs (MWCNTs:TP wt. = 1:3). After the mixture was sonicated and homogenized, the prepared PVA solution was poured into the dispersion of MWCNTs/TP and stirred to form MWCNTs/TP/PVA dispersion. **Figure 10** illustrates the interaction between PVA and MWCNTs: π-π interaction between MWCNT and TP, and hydrogen bond between TP and PVA.

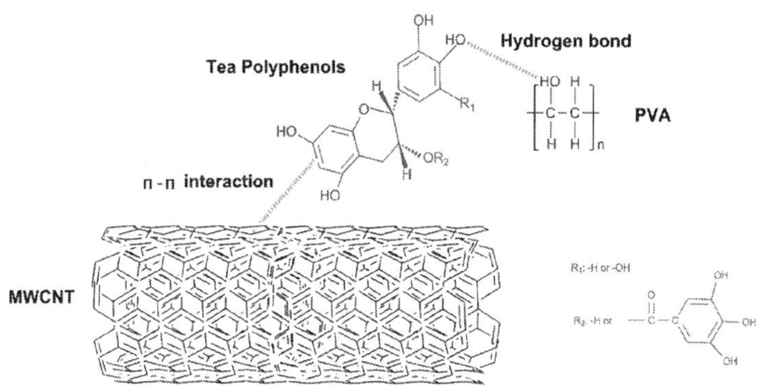

Figure 10. Structure of PVA/MWCNTs composites with TP as a surfactant [67].

As shown in **Figure 11a**, TEM images of pristine MWCNTs in DMSO/H_2O display apparent aggregation and entanglement, while MWCNTs modified by TP in DMSO/H_2O are separated and homogeneously distributed (**Figure 11b**). After adding PVA, the MWCNTs are still well distributed as shown in **Figure 11c**. Because the non-covalent interactions exist between MWCNTs and PVA through the 'bridge' effect of TP, TP-functionalized MWCNTs can be well dispersed in PVA, leading to a stable dispersion with the help of sonication and mechanical homogenization. As further evidences, the dispersions in the bottles in **Figure 11** show the similar results to the TEM images.

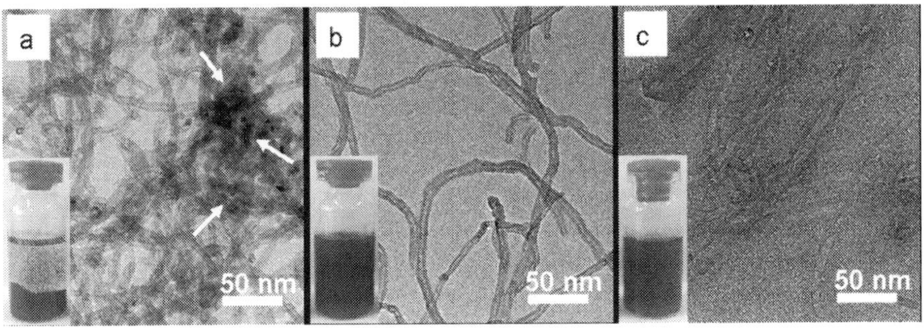

Figure 11. TEM images of (a) MWCNTs/DMSO/H_2O, (b) MWCNTs/TP/DMSO/H_2O, (c) MWCNTs/PVA/TP/DMSO/H_2O. MWCNTs:TP (wt) is 1:3, and the concentrations of MWCNTs are 0.2 wt.% (insets are MWCNTs dispersions in the corresponding media after staying 2 hr) [67].

2.3.2.2. *Rosemary acid*

Rosemary acid (RosA) is an extract from rosemary, an evergreen shrub with aromatic linear leaves, which is green, non-toxic and environment-friendly. Like TP, it possesses the conditions: π-π stacking between RosA and CNTs and hydrogen bonds between RosA and polymer matrix. Besides, in comparison with TP, its more hydroxyl groups and added carboxyl group provide CNTs more chances to interact with polymer. **Figure 12** shows the illustration of the interaction between PVA and RosA-modified MWCNTs (m-MWCNTs). π-π stacking exists between MWCNTs and RosA on m-MWCNTs, and hydrogen bonds are supposed to be formed between hydroxyl groups of RosA and PVA, so that PVA and MWCNTs can be connected.

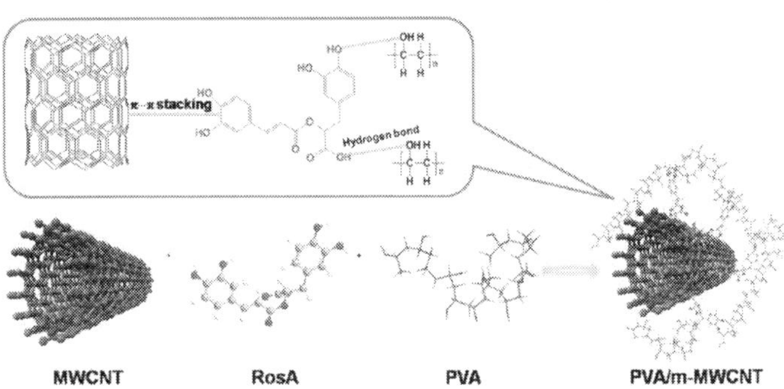

Figure 12. Schematic illustration of the structure of PVA/m-MWCNTs composites [68].

As shown in **Figure 13a**, pristine MWCNTs (p-MWCNTs) exhibit almost 100% UV transmittances after staying one day, indicating poor dispersibility in the solution, while as the introduction of RosA, the transmittance quickly decreases, proving obviously improved dispersibility of m-MWCNTs. Apparent aggregation of p-MWCNTs can be seen in **Figure 13b**, whereas m-MWCNTs are clearly separated and homogeneously distributed as shown in **Figure 13c**. The inset picture provides much intuitive evidences for the dispersibility of p-MWCNTs and m-MWCNTs.

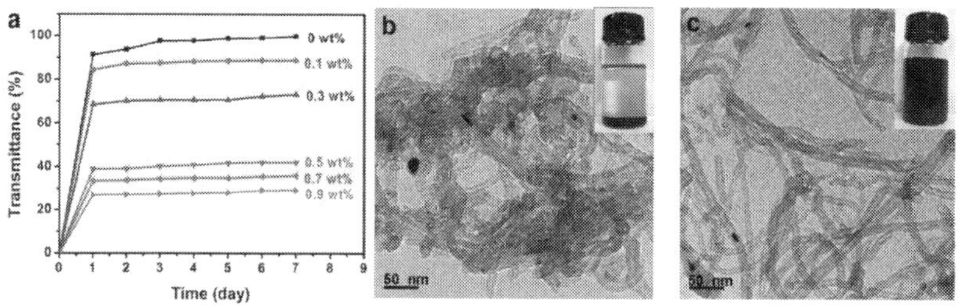

Figure 13. UV–vis transmittances of the MWCNTs dispersions in DMSO/H_2O (vol ratio = 3/1) with different concentration of RosA marked on the ends of the curves (a) and TEM images of p-MWCNTs, (b) m-MWCNTs (c) in DMSO/H_2O. The inserts are the corresponding MWCNTs dispersions after staying for 7 days [68].

3. Processing of CNTs/polymer composites

Polymeric materials mainly include plastics, fiber, and rubber (elastomer), so they can be the representatives as matrices to composite with CNTs. This section will briefly describe the fabrication of the CNTs/polymer composites based on three kinds of typical synthetic materials: fiber, film, and elastomer.

3.1. Composite fiber

Fiber processing has its own advantages, one of which is that it can be drawn in large draw ratio. That means it has higher possibilities to enhance mechanical properties of the composites through the orientation of related structural units. Ideal CNTs fibers, comprising axially aligned and highly packed CNTs, could have much higher specific modulus and specific strength than those of commercial carbon and polymeric fibers, but so far the modulus and strength values are still low [69]. To date, no breakthrough has been reported in the specific strength and specific stiffness of CNTs fibers. The outstanding mechanical and physical properties of individual CNTs have provided the impetus for researchers in developing high-performance fiber polymeric materials based on CNTs because they can be handled much more conveniently than the individual CNTs [70].

3.1.1. Solution spinning

Manufacturing CNTs-based fibers commonly uses solution spinning, which can be divided into two processes: dry spinning during which the solvent in fiber forced from the spinning hole is removed by evaporation by heating, and wet spinning during which the solvent in fiber is removed by coagulation in another fluid. In general, mechanical properties of neat CNTs fibers prepared by dry spinning are far higher than those of the fibers prepared by wet spinning [71]. However, dry spinning has disadvantages such as difficulties in process scalability and mass production, which thus affects the structural characteristics and physical properties of CNTs-based fibers, while the wet spinning is a relatively simple technique with advantages of commercially ready availability [72, 73]. So a number of researches have been conducted to make CNTs/polymer composite fibers by wet spinning. Vigolo et al. firstly reported the wet spinning of SWCNTs/PVA composite fibers via a coagulation method [74] and the composite fibers prepared by wet spinning were measured to have 230 MPa of tensile strength and 40 GPa of Young's modulus. Dalton et al. [75] manufactured mechanically strong SWCNTs/PVA gel fibers by using a modified coagulation-based wet spinning method. The composite fibers had 1.8 GPa of tensile strength and 80 GPa of Young's modulus. Fibers with high loading, as high as 23 wt% of SWCNTs, were successfully produced through wet spinning [76].

Besides, a dry-jet wet spinning is generally used to prepare high strength and high modulus fibers, and it is expected to use this spinning process to produce high-performance polymer/CNTs composite fibers. In our previous work [29, 67], f-MWCNTs/PVA composite dispersion was forced through a spinning hole into an air gap of 5 mm and then into cold methanol coagulating bath (**Figure 14a**). Postdrawing was carried out in a tube oven between two 6.5-cm-diameter rollers at a temperature about 200°C and draw ratio 8–12. Postdrawing could lead to increased orientation and crystallization of the fibers and thus to improvement in both strength and modulus. The spinning dope was extremely uniform due to good dispersibility of MWCNTs in PVA, the fiber formation was quite easy to be controlled, and the drawn fibers

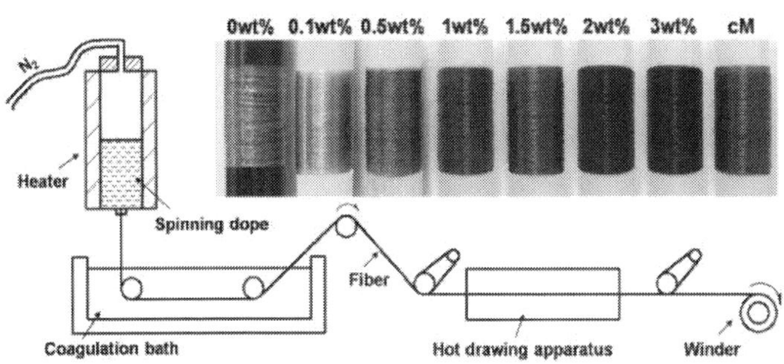

Figure 14. (a) Spinning apparatus for preparing PVA/f-MWCNTs composite fibers. (b) Photograph of the composite fibers with different f-MWCNTs loadings (marked on the tops) and cM represents the composite fiber with 0.5 wt% MWCNTs containing carboxyl groups [29].

were quite smooth and uniform at certain MWCNTs loadings (**Figure 14b**), and thus the fibers obtained expected mechanical properties.

A method for manufacturing sheath-core structured composite fibers was developed using wet spinning techniques as shown in **Figure 15**. The core portion of a fiber was prepared using CNTs solution while the sheath used a fiber-forming polymer such as PVA. The CNTs solution was injected into the inner nozzle and the PVA polymer solution to the outer nozzle. The inner and outer diameters of the spinneret were 0.3 and 0.5 mm. The interfacial contact between the core and sheath was found to be good and did not hinder the flexibility of the resulting fibers so they could be easily woven into fabric. The spinning technology used here is scalable and compatible with mass production methods in industry [77].

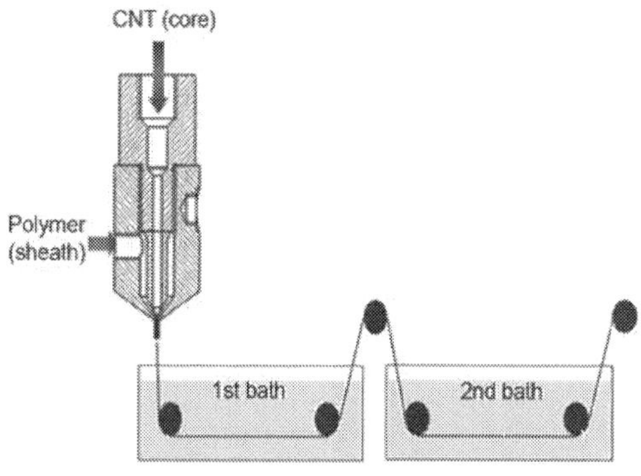

Figure 15. The wet spinning process used to produce the sheathcore structured CNT/PVA fiber [77].

Composite fibers can also be differently prepared through wet spinning. MWNTs solution was forced through a spinneret into a rotating bath of acidic PVA solution, the pH of which was 2 adjusted by adding concentrated hydrochloric acid. The MWNTs/PVA composite fibers were then collected in a water bath and drawn upward for drying. The toughness of the composite fibers is approx. 563 J. g^{-1} which exceeds by far the toughness of materials such as aramid or spider silk [24].

3.1.2. Melt spinning

Melt spinning is the simplest method for fiber manufacturing. If a polymer can be melted under reasonable conditions, its conversion to a fiber by melt spinning is preferred over the solution-spinning process, mainly as the former does not involve the use of solvents and the problems associated with their use, namely, their removal, recovery, the associated environmental concerns and the low spinning speeds [78]. Melt spinning is also a common approach used for preparation of CNTs/polymer composite fiber, but it is not well suited to make composite fibers containing a large fraction of CNTs. Indeed, the presence of CNTs results in a strong increase in the viscosity of the polymer, which makes the polymer containing CNTs extrusion and the fiber spinning particularly difficult. In contrast, wet spinning of CNTs/polymer solutions allows inclusion of greater fractions of CNTs, provided that the nanotubes are homogeneously dispersed in the polymer solutions [79]. Moreover, the macromolecular chains and CNTs in the melt are seriously entangled, so to achieve high orientation of the chains and CNTs is extremely difficult, which hinders enhancement of the mechanical properties of the composite fiber.

The melt spinning of CNTs/polymer can generally be completed in following three steps. The melt-compounding of polymer and CNTs is conducted in a twin-screw extruder at a high CNTs concentration to firstly produce a masterbatch, then the masterbatch is diluted to the desired CNTs concentration by mixing with the desired amount of neat polymer, and finally, the mixture is melted and spun in suitable apparatus and the obtained as-spun fiber is drawn continuously or separately. The processing of CNTs/polymer composite is commonly needed to transfer to a screw extruder, and then the molten CNTs/polymer composite is forced through a spinneret with a constant-speed drive, which could be adjusted to modify the mass flow rate of the polymer. The spinning hole on the spinneret is a cylindrical capillary, which is generally less than 1 mm in diameter. The extruded thread is wound-up on a rotating package at a speed of several hundred or thousands meters per minute [80]. Sometimes, if melt spinning is in small scale and in low melt viscosity, a melting apparatus, piston-barrel system, such as capillary rheometer, can also be used (**Figure 16**) [81].

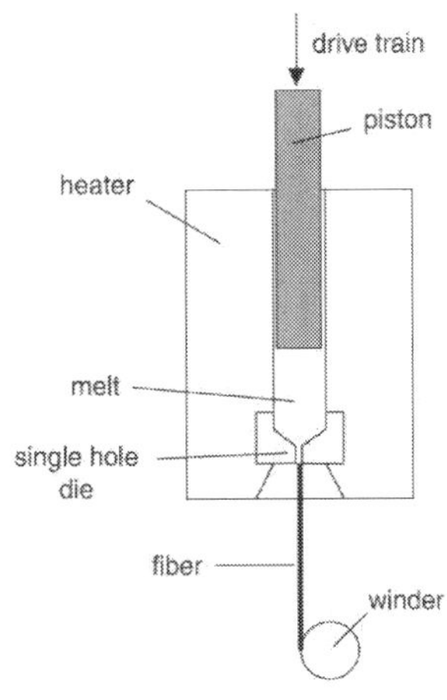

Figure 16. Schema of the piston type spinning device [81].

3.1.3. *Electrospinning*

Electrospinning is an electrostatically induced self-assembly process wherein ultra-fine fibers are produced. Recently, the electrospinning technique has also been used for the alignment of CNTs in a polymer matrix, leading to high-strength, high-modulus, and even high electrical conductivity. It has been established that electrospinning a polymer solution containing well-dispersed CNTs leads to nanocomposite fibers with the embedded CNTs oriented parallel to the nanofiber axis due to the large shear forces in a fast fiber-drawing process [82].

Fiber collection methods for electrospinning can be divided into two classes: continuous filament wound on roller and felt piled by short fibers collected on nets or sheets. Commonly, the nanofibers are collected in felt due to ease of processing. For example, when MWCNTs/PAN dispersion in DMF was electrospun, a flat metal net covered in aluminum foil was served as a grounded counter electrode to collect the short fibers to form a felt after several hours' spinning. Because the MWCNTs have good dispersibility in PAN solution, even if the loadings of MWCNTs increase, the fibers are formed in uniform diameter and smooth surface as shown in **Figure 17a–Figure 17e**. On the other hand, with the increase of the loadings of MWCNTs, that is, the conductivity of the spinning solution rises, the improved Coulomb force and static electricity facilitate the formation of small-diameter fibers (**Figure 17f**).

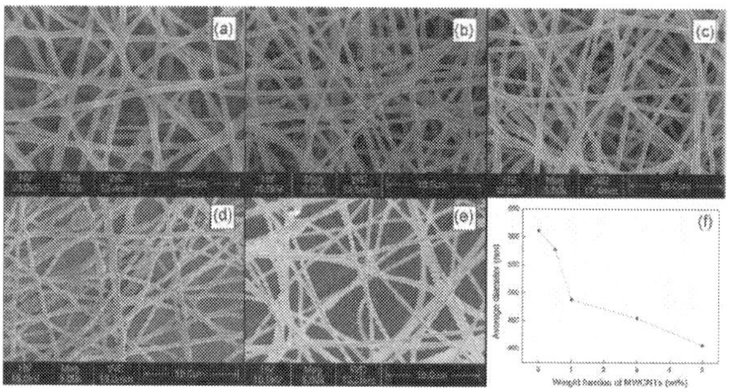

Figure 17. SEM images of PAN/f-MWCNTs composite nanofibers with different f-MWCNTs concentrations (a) 0 wt%, (b) 0.5 wt%, (c) 1 wt%, (d) 3 wt%, (e) 5 wt%, and (f) diameter distribution of nanofibers [46].

Particularly, electrospinning set-up with water as collector (**Figure 18**) was used to obtain aligned fiber samples and then the fiber was drawn out using a mechanical roller to obtain continuous fiber bundles [83]. As evident from **Figure 18**, individual fibers within the fiber bundles are tightly packed and aligned together. The stiffness and strength are seen to increase by 32% and 28%, respectively, when 3 wt.% CNTs is added to neat PVDF fibers, which can be attributed to the nano-reinforcement effect of the CNTs.

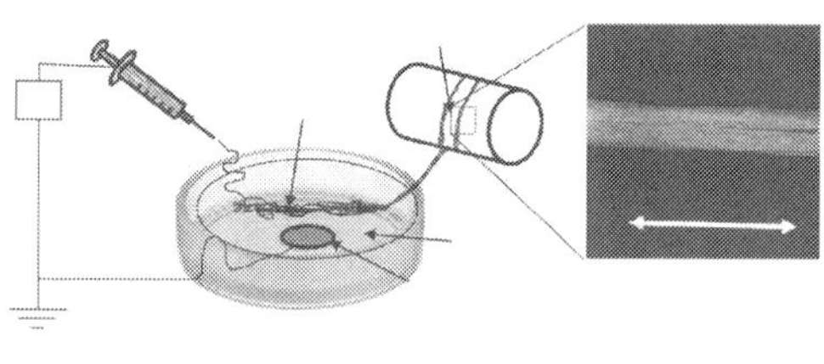

Figure 18. Schematic of the electrospinning setup used to obtain continuous self-assembled yarn structure [83].

3.2. Composite film

Film or membrane is one of the major products of plastics family. The composite film consisting of CNTs and polymer has extensive application because of its excellent mechanical and electrical properties. Casting is a common process to prepare the composite film, and a micrometer order uniform thickness film can be obtained after being dried naturally or at heating condition. For example, uniform MWCNTs/PVA, MWCNTs/cellulose, and MWCNTs/PS composite films can be prepared by casting [35, 84, 85]. The storage modulus for a 6 wt. % MWCNTs/PS composite cast film at 80°C is up to 122% of neat PS and the glass transition temperature increased significantly with an increase in MWCNTs concentration [85]. In order to manufacture highly stretchable, twistable, transparent, and conductive polymer and CNTs bilayer films, an efficient spin-coating and curing method is applied. In general, the spin-coating technique is known as one of the most promising processes to provide desired film uniformity, easy control of thickness, short operation time, and high reproducibility without environment limit [86]. For instance, defect-free MWCNTs were dispersed in deionized water to obtain a stable aqueous solution with the aid of SDBS. After spin-coating the aqueous MWCNTs solution on glass plates, polydimethyl siloxane (PDMS) was applied on the surface of the MWCNTs layer and then was cured. Finally, the MWCNTs/PDMS bilayer films were peeled off easily from the glass plates (**Figure 19**) [87].

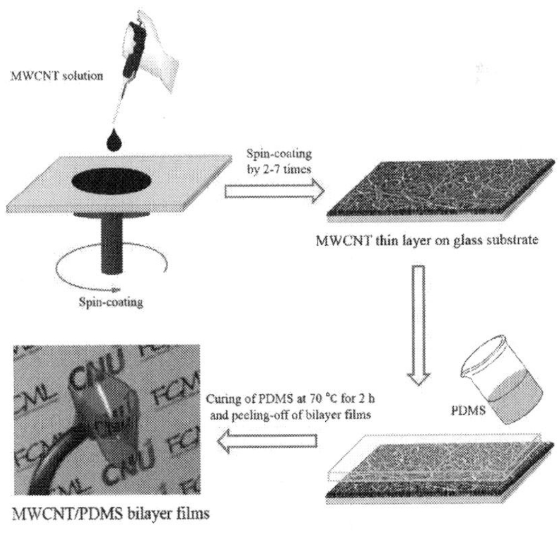

Figure 19. Schematic procedure to manufacture MWCNTs/PDMS bilayer films by spin-coating of MWCNTs solution and following curing of PDMS [87].

A novel in situ bulk polymerization method to prepare vertically aligned carbon nanotubes (VACNTs)/polymer composite films was developed to prevent CNTs condensation that could disturb CNTs orientation during liquid phase processing. A VACNTs array was infiltrated with styrene monomer with a certain amount of polystyrene-polybutadiene (PS-PB) copolymer that acted as a plasticizer, which confirms that the addition of PS-PB into the matrix can improve the elongation at break of the CNTs/PS composite film. These CNTs/polymer composite membranes show high gas and water permeability comparable to the other VACNTs composite membranes, potentially enabling applications that may require membranes with high flux, flexibility, and durability [88].

3.3. Elastomers

The unique properties of elastomers especially their high and reversible deformability are of great industrial importance. However, on account of their low elastic modulus, they are generally compounded with a reinforcing filler. The reinforcement of elastomers is probably one of the most important processes in rubber industry, especially in modern tire manufacture industry. Generally, reinforced elastomers show an increase in modulus, hardness, tensile strength, abrasion, and tear resistance as well as resistance to fatigue and cracking [89]. The conventional reinforcing fillers such as carbon blacks (CBs) and silicas have been widely used, while due to the advantages of CNTs as stated above, they as reinforcing fillers of elastomers have attracted much researchers' attention. Potential applications of CNT-filled rubber composites range from industrial applications such as rubber hoses, tire components, and sensing devices, to electrical shielding and electrical heating devices [90].

3.3.1. Reinforced natural rubber

Toluene is often used as a dispersant for CNTs in natural rubber (NR). For example, NR containing all the formulation ingredients could be dissolved in the suspension of CNTs and toluene. After NR/CNTs dispersion was mixed, the toluene was carefully removed, and the mixture was cured and formed sheets under certain pressure and temperature. The CNTs bring significant improvements in the mechanical properties with regard to the pure polymer. It is demonstrated that the intrinsic potential of CNTs is excellent reinforcing filler in elastomeric materials and small filler loadings substantially improve the mechanical behavior of the soft matrix [89]. Similarly, Fakhru'l-Razi reported that the preparation of MWCNTs/NR nanocomposites was carried out by a solvent-casting method using toluene as a solvent. Using this technique, CNTs can be dispersed homogeneously in the NR matrix in an attempt to increase the mechanical properties of these nanocomposites. There is an increase in the initial modulus for up to 12 times over pure NR. Application of the CNTs can result in rubber products having improved mechanical, physical, and chemical properties, compared with existing rubber products reinforced with CB or silicone [91].

Silane functionalization of MWCNTs could offer NR better performance in terms of property improvement when loaded in elastomeric composites. The MWCNTs are initially subjected to aminopropyltriethoxysilane (APS) treatment to bind amine functional groups ($-NH_2$) on the nanotube surface. Successful grafting of APS on the MWCNTs surface through Si-O-C linkages is confirmed. Loading of silane-functionalized MWCNTs in the ENR (epoxidized NR) matrix results in a significant improvement in the mechanical, electrical, and thermal degradation properties of the composite materials, when compared to gum or pristine MWCNTs-loaded materials. As expected, the modulus of the composites at various strains increases significantly with an increase in nanotube loading. For instance, the 100% modulus increases to 106% when pristine CNTs are loaded, whereas it increases to about 120% when aminosilane-functionalized CNTs are loaded [92].

3.3.2. Reinforced synthetic elastomer

The process of reinforcing silicone rubber by MWCNTs needs following several steps. Silicone rubber resin was dissolved in petroleum ether to obtain homogeneous solution; fumed silica, KH550, and water were added to the above solution and agitated; Fe_2O_3 was added to the above mixture and agitated; MWCNTs or their suspension was added into the above mixtures and dispersed by ultrasonic treatment and agitation; the above mixture was desolventized in vacuum oven; curing agent and catalyst were added into the mixtures and stirred followed by moving into a mould for curing to obtain the composites. When 5 phr of MWCNTs was added, the room temperature vulcanized (RTV) silicone rubber composite reinforced by MWCNTs achieved good comprehensive performance. Its tensile strength, tear strength, elongation, and onset decomposition temperature reach 2.0 MPa, 11.7 kN/m, 238%, and 510°C, respectively, while the above said values for the composite with untreated MWCNTs are only 1.1 MPa, 7.0 kN/m, 83%, and 484°C, respectively [93].

Reinforcement of thermoplastic polyurethanes (TPUs) by MWCNTs was also carried out by solution method. MWCNTs were ultrasonicated by a dipping tip sonicator at room temperature in tetrahydrofuran-THF (0.1% volumetric solution of nanotube). TPU pellets were then added to the ultrasonicated solution and mixed with a magnetic stirrer. The solution was then poured in a Petri dish in order to allow THF evaporation at room temperature and to obtain reinforced TPU films. Such films were further dried in a vacuum oven. Typical mechanical properties (Young's modulus and yield strength) are improved and the strain energy dissipation is also increased [94].

Although solution blending is common to incorporate MWCNTs into an elastomer, this approach is not suitable for industrial practice due to its higher environmental and economic costs. Melt blending is the most convenient and efficient technique for the preparation of CNTs reinforced elastomers although it also has some disadvantages, so melt blending is a good selection to prepare CNT-reinforced elastomers. For example, appropriate amounts of CNTs were compounded into the elastomers styrene-butadiene rubber (SBR) and nitrile-butadiene rubber (NBR) by melt-mixing using a two-roll open mill. The elastomer compounds were vulcanized with the aid of sulphur at 150°C and at 3000 psi using a compression molding to obtain thin sheets (0.8 mm thick). The composites of CNTs/rubber show the improved properties such as resistance to solvent swelling, enhanced glass transition temperature, and improved storage and loss moduli [95].

4. Mechanical properties of the composites

4.1. The amount of CNTs

As it can be understood, mechanical properties of CNTs/polymer composites are improved due to effective CNTs reinforcement in the composites. In comparison with pure polymer, addition of functionalized CNTs causes an increasing in elastic modulus and yield strength, which is due to good dispersibility of CNTs and stress-transfer effect from the matrix to the CNTs caused by good interfacial interaction between the CNTs and the matrix. However, the optimum added amount of CNTs to polymer has to be carefully considered to obtain ultimate mechanical properties of the composite. For instance, Young's modulus, tensile strength, and storage modulus of the CNTs/PP nanocomposites can be increased with increasing CNTs content, but CNTs tend to aggregate into bundles in the matrix and hence the mechanical properties of the composite are reduced when CNTs over a certain amount, usually 2–3 wt. % [96].

Liu et al. [61] reported that mechanical properties of CNTs/polymer nanocomposite materials are greatly dependent on the content of CNTs. Within a limited weight fraction, mechanical properties of the nanocomposites increase significantly with addition of CNTs, and they reach the optimum when the content of CNTs is around 2 wt. %. The breaking strength of MWCNTs/poly(p-phenylene terephthalamide) (PPTA) nanocomposites containing 2 wt.% MWCNTs increases from 212 MPa up to 327 MPa, with a 54.2% increase compared to the pure aramid membranes. Being different from strength, with low volume fraction of CNTs, the modulus of the nanocomposite material exhibits a linear relationship with the content of the CNTs, and when the content of CNTs is further increased, the enhancement effects are weakened. The enhancement mechanism of MWCNTs/PPTA nanocomposites can be simply described as follows. The thread morphology of CNTs which can overlap with each other and strong π-π bond effects and van der Waals forces among adjacent nanotubes can result in the interweaving between nanotubes, leading to network-like structure. The structure has excellent stress transfer effect due to the strong interfacial forces between CNTs and the aramid matrix, and the CNT network can inter-pass throughout the layered structures of MWCNTs/PPTA nanocomposite membranes, which can greatly enhance the interaction between adjacent layers, hindering the slippage under tensile stress, that is to say, these layered structures are mechanically locked, improving the stiffness and tensile strength of nanocomposites. However, when the loading of CNTs exceeds a certain value, cluster structures inducing stress concentration will be formed which will become the sources of break, weakening the reinforcement of CNTs.

Similarly, the ultimate tensile strength of MWCNTs/ABS composites is enhanced with the increase in MWCNTs loading up to 70 MPa at a load of 3 wt.% MWCNTs and the values slightly decrease with further increase as shown in **Figure 20a**. When the MWCNTs loading is 10 wt. %, the strength significantly declines to 57 MPa, but this value is still higher than that of neat ABS. The variation of Young's modulus with the different loadings of MWCNTs loading is different from the strength (**Figure 20b**) and the modulus of the composites shows a linear increase till the largest loadings in the experiment. **Figure 20c** shows that percentage elongation decreases with an increase in MWCNT loading [2].

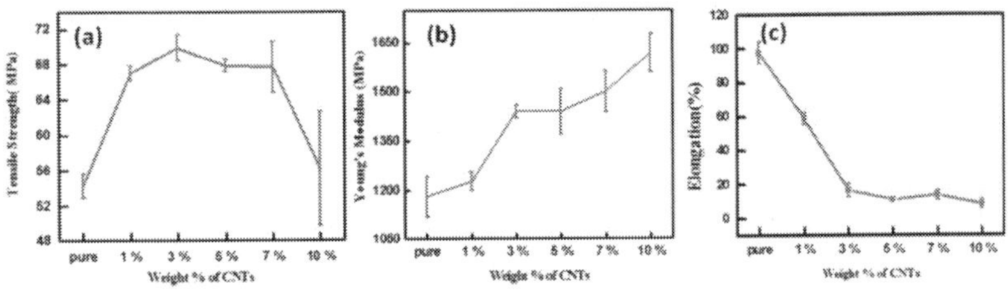

Figure 20. (a) Tensile strength, (b) Young's modulus, and (c) % elongation of MWCNTs/ABS composites with respect to the weight percentage of MWCNTs [2].

Similar trends are shown in our previous study as shown in **Figure 21** [29]. The addition of small amount of f-MWCNTs except acid-treated MWCNTs (c-MWCNTs) leads to a significant increase in the tensile strength of f-MWCNTs/PVA composite fibers, but the strength decreases when f-$MWCNT_S$ further increase from ultimate value 926MPa at 2 wt.% f-MWCNTs to about 805 MPa at 3 wt.% f-MWCNTs. However, Young's modulus keeps increasing with the increase of the MWCNTs loadings, and the break strain shows gradually decrease. The increase of Young's modulus and the decrease of the break strain reflect that at the same time of reinforcement the ductility decreases with the increasing MWCNTs loadings.

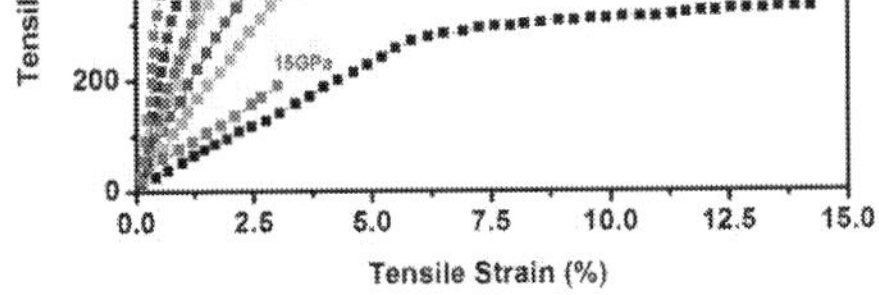

Figure 21. Stress–strain curves of PVA/f-MWCNTs composite fibers, the values on the tips of curves are Young's modulus of the corresponding fibers [29].

In order to explain this, a cross-sections of pure PVA- and TP-treated MWCNTs and PVA composite fibers (**Figure 22a** and **22b**) were prepared by breaking the fibers in liquid nitrogen to give an intact surface fracture (**Figure 22a′** and **22b′**). The pure PVA fiber shows evidently ductile fracture (**Figure 22a′**), whereas the fiber containing MWCNTs exhibits a typical feature of stiff and rigid fracture behavior with clear-cut fracture cross section [97]. On the contrary, for brittle polymer, such as cellulose, with a CNT loading of 5 wt %, the tensile toughness reaches 5.8 MJ/m^3, about 346% higher than that of neat cellulose film. The explanation is that the interfacial hydrogen bonding facilitates the stress transfer and simultaneously reduces the density of hydrogen bonding network of cellulose so as to obviously enhance the plastic deformation of the nanocomposite [98].

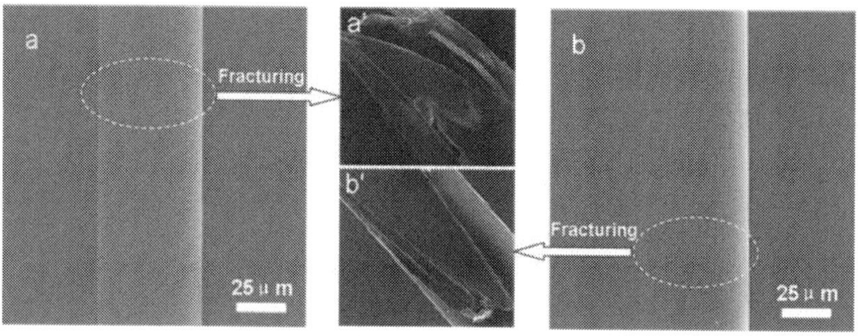

Figure 22. SEM images of (a) pure PVA and (b) MWCNTs/PVA composite drawn fibers prepared with 0.2 wt% MWCNTs loading; (a′) the fractured cross-section of PVA fiber (a) and (b′) is the fractured cross-section of MWCNTs/PVA composite fiber [67].

Different from the statement above, the incorporation of large amount CNTs in polymer matrix can still significantly enhance the nanocomposite tensile strength. The CNTs (1 wt. %)/PP nanocomposite shows 18% improvement in the tensile strength, while the 20% CNTs (20 wt. %)/PP nanocomposite exhibits an extremely high tensile strength, 1.8 times that of the unfilled polymer. This finding proves that high concentrations of CNTs can be incorporated in PP without degrading the tensile strength [97].

Sample		Specific tensile strength (kgf•cm^{-2})	Specific modulus (kgf•cm^{-2})	Elongation at break (%)
	PAN	188.2	88.5	45.2
In situ composite nanofibers	0.5% f-MWCNTs/PAN	213.4	309.2	42.8
	1% f-MWCNTs/PAN	310.3	553.9	40.1
	3% f-MWCNTs/PAN	246.8	371.9	39.6
	5% f-MWCNTs/PAN	208.6	287.5	35.7
Ex situ composite nanofibers	0.5% f-MWCNTs/PAN	194.7	280.3	40.7
	1% f-MWCNTs/PAN	297.4	455.1	38.9
	3% f-MWCNTs/PAN	219.6	292.8	36.0
	5% f-MWCNTs/PAN	201.7	217.4	33.4

Table 1. Mechanical properties of PAN, ex situ composite and in situ composite nanofibers.

At the same CNTs loadings, the mechanical properties of covalent treated CNTs-based composite are obviously higher than those of non-covalent treated CNTs based composite. For instance, the strength of TP-treated MWCNTs and PVA composite fiber are 643 MPa when the MWCNTs loadings are 0.6 wt.%, whereas the strength of PVA grafted MWCNTs and PVA composite fiber is 695 MPa at a load of 0.5 wt.% MWCNTs, indicating that the latter has stronger stress transfer effect due to covalent interaction between PVA and MWCNTs than the former [16, 46]. On the other hand, at the same CNTs loadings, the mechanical properties of the polymer composite containing CNTs functionalized by in situ polymerization are higher than those by ex situ polymerization as shown in **Table 1**. [46].

4.2. Orientation of CNTs

As we know, the mechanical properties such as tensile strength and Young's modulus of polymeric materials depend strongly on the orientation of the polymer chains. Similarly, as far as conventional fiber reinforcement is concerned, unidirectional composites show the highest improvements in strength and modulus. Therefore, it is expected that alignment of CNTs is important for improving mechanical properties [81]. So, the approach to improving the mechanical properties of CNTs/polymer nanocomposites is through alignment of CNTs by taking advantage of their exceptional anisotropic properties of CNTs arising from the one-dimensional structure of CNTs with extremely high aspect ratios. CNTs alignment can be achieved using various routes, including mechanical force, magnetic field, electric field, shear flows, and electrospinning [83, 99].

In the route of mechanical force, the composite materials are stretched usually by drawing, making CNTs orient along the drawing direction. Hot stretching of the MWCNTs/epoxy prepregs can markedly improve the mechanical properties of the composites. The improved mechanical properties of stretched composites derive from the increased MWCNTs volume fraction and the reduced MWCNT waviness caused by stretching. With a 3% stretch ratio, the MWCNTs/epoxy composites achieve the best tensile strength and elastic modulus [100]. Fiber drawing allows both the nanotubes and the polymer chains to be aligned, leading to enhance-

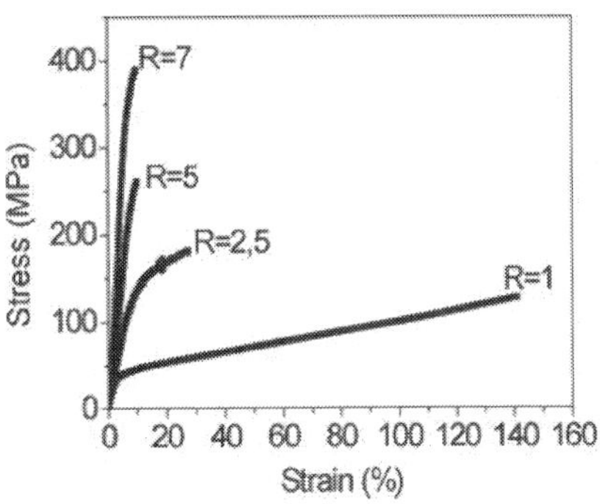

Figure 23. Stress-strain curves of MWCNTs/PA12 (7 wt. %) fibers with different draw ratio (R) values [101].

ment of the mechanical properties of the composites. As shown in **Figure 23**, the stress of MWCNTs (7 wt. %)/PA12 composite fibers increases with the draw ratio obviously [101].

The alignment of MWCNTs in bulk epoxy matrices can be realized by application of external electric field. The alignment gives rise to much improved electrical conductivity, elastic modulus, and quasi-static fracture toughness compared to those with CNTs of random orientation. The storage modulus of MWCNTs/epoxy composite can increase by 50% [102]. The Young's moduli of the 0.3 wt.% CNTs/epoxy nanocomposites with and without CNTs orientation are about 40% and 15% higher than that of the neat epoxy, respectively. The additional 25% enhancement in modulus clearly demonstrates the effectiveness of the alignment [99].

In addition, the shear forces during electrospinning present in the liquid jet result in automatic alignment of well-dispersed nanotubes [103]. **Figure 24** clearly demonstrates the significant increases in Young's modulus and tensile strength of the functionalized MWCNTs reinforced PMMA nanofibers obtained by electrospinning. The significant improvements in modulus and strength are likely related to the good dispersion and orientation of the CNTs within the polymer and to the strong interfacial adhesion due to the nanotube surface modification [104].

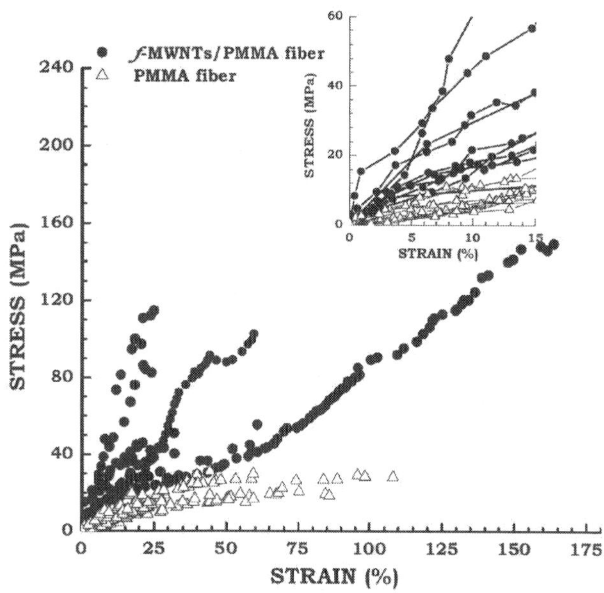

Figure 24. Stress-strain curves of electrospun PMMA nanofibers and CNT/PMMA nanofibers. The inset is a magnified view of the initial range (up to 15 %) of strain from which Young's modulus was calculated by linear regression [104].

http://dx.doi.org/10.5772/62635

There is an obvious difference in the intensity of the G band on polarized Raman spectra for different polarization directions after hot-stretching, which can be correlated to the alignment of CNTs in the composites. Normally, the degree of CNT alignment can be evaluated by the depolarization factor R, the ratio of the peak intensities of the G band in the two polarization directions, i.e., parallel (VV configuration) and perpendicular (VH configuration) to the fiber axis. The marked increase in R can be ascribed to the better alignment of CNTs after hot-stretching. For instance, the *R* values of MWCNTs/PAN composite are 1.03 and 2.95 for samples before and after hot-stretching, respectively, indicating the big difference between both R values, that is, much better alignment of CNTs after hot-stretching [105]. Besides, the orientation of CNTs can directly be characterized by TEM image as shown in **Figure 25**, in which the oriented CNTs can clearly be seen.

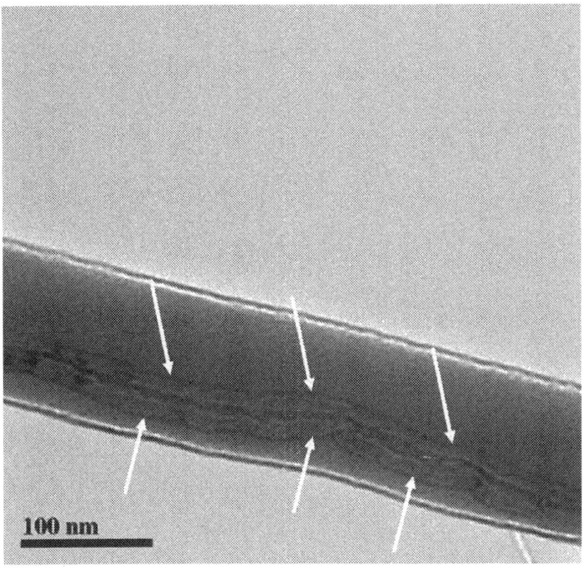

Figure 25. Representative TEM image of PVDF fibers filled with 2 wt.% CNTs [83].

In our previous paper [67], we presented a structure model of MWCNTs/PVA composite fiber as shown in **Figure 26**. TP-functionalized MWCNTs connects with PVA by non-covalent interactions and disperses uniformly in PVA matrix. It is considered that there are 'constrained regions' in the composite, which are composed of MWCNTs, constrained PVA molecular chains, and the surfactant TP molecules combined by H-bonding and π-π stacking among them. After the fiber is drawn, all the elements in the 'constrained regions' as a whole align along the fiber axis, which greatly increases orientation effect of all the elements in the composite, further strengthens the interactions between MWCNTs and PVA chains, and consequently forms a structure possessing high mechanical properties.

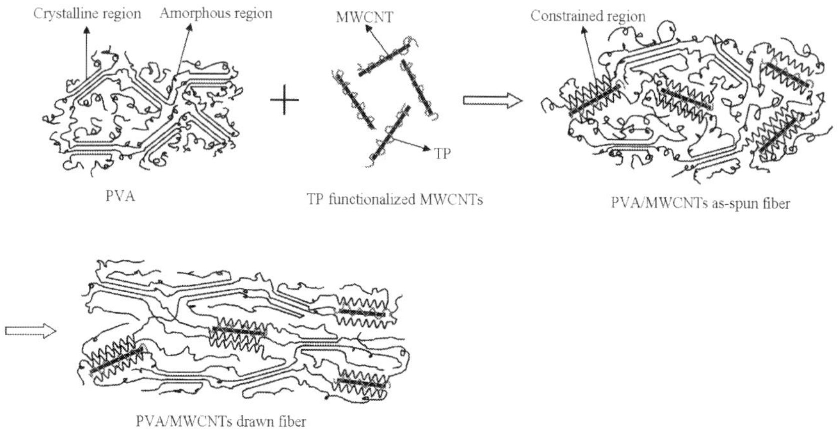

Figure 26. Proposed structural model of PVA/MWCNTs composite fiber [67].

5. Concluding remarks

Recently, CNTs have been widely used to incorporate into polymers to develop high-performance composite materials. Dispersibility of CNTs in polymer and the stress transfer from the matrix to CNTs have significant effects on reaching ultimate mechanical properties for CNTs/polymer composites. Mechanical processes to disperse CNTs such as ultrasonication have limited dispersing effect, and they generally are applied together with other dispersing processes so as to enhance the disperse effect of CNTs. Both covalent and non-covalent functionalization of CNTs can well disperse CNTs in polymer matrix, but at the same time have their disadvantages, such as, CNTs structure damage for covalent functionalization or weak interaction between matrix and CNTs for non-covalent functionalization. Due to better stress transfer, the mechanical properties of covalent functionalized CNTs/polymer composite are higher than non-covalent functionalized composite. As a one-dimensional material, CNTs are quite suited to reinforce polymeric fiber, which moreover can be easily drawn in a high draw ratio, leading to high orientation of CNTs and polymer chains. In addition to dispersibility of CNTs in the matrix, the amount and orientation are two important factors which cannot be ignored. Optimum amount and high orientation of CNTs should be considered in order to achieve excellent mechanical properties of CNTs/polymer composite. CNTs reinforcement makes the mechanical properties of polymer increase several times higher than the neat polymer in some of literatures, which conforms that CNTs are very effective fillers for polymers. However, the challenges the researcher should still face are how to optimize various factors in preparing CNTs/polymer composite in order to reach its ultimate mechanical properties, and how to scale up from laboratory and thus realize mass production.

Acknowledgements

This chapter was performed with the support of a project funded by the Priority Academic Program Development of Jiangsu Higher Education Institutions and State and Local Joint Engineering Laboratory for Novel Functional Polymeric Materials. The authors also thank Mr. Yajun Li, Mr. Yao Cheng and Miss Juan Li for their kind assistances.

Author details

Lixing Dai* and Jun Sun

*Address all correspondence to: dailixing@suda.edu.cn

College of Chemistry, Chemical Engineering & Materials Science, Soochow University, Suzhou, Jiangsu, P.R. China

References

[1] Wikipedia. Composite material [Internet]. 2012. Available from: https://en.wikipedia.org/wiki/Composite_material

[2] Jyoti J., Basu S., Singh B. P., Dhakate S. R. Superior mechanical and electrical properties of multiwall carbon nanotube reinforced acrylonitrile butadiene styrene high performance composites. Composites Part B: Engineering. 2015;83:58–65. DOI: 10.1016/j.compositesb.2015.08.055

[3] Jiao G. Q., Jia P. R. Mechanics of Composite Materials. Xi'an, China: Northwestern Polytechnical University Press; 2008. 166 p. ISBN: 9787561223321 (in Chinese).

[4] Guo Y. C. The history of composites. Fiber Composites. 1989;1:55–57. ISSN: 1003-6423-CN:23-1267/TQ (in Chinese).

[5] Gillespie J. W., Devault J. B., Edie D. D., Gabara V., Haulik T. J., Kardos J. L., Schadler L. S. High-performance structural fibers for advanced polymer matrix composites. Washington, DC: National Academies Press; 2005. DOI: 10.17226/11268

[6] Iijima S. Helical microtubules of graphitic carbon. Nature. 1991;354:56–58. DOI: 10.1038/354056a0

[7] Yu M. F., Lourie O., Dyer M. J., Moloni K., Kelly T. F., Ruoff R. S. Strength and breaking mechanism of multiwalled carbon nanotubes under tensile load. Science. 2000;287(5453):637–640. DOI: 10.1126/science.287.5453.637

[8] Coleman J. N., Khan U., Gun'ko Y. K. Mechanical reinforcement of polymers using carbon nanotubes. Advanced Materials. 2006;18:689–706. DOI: 10.1002/adma.200501851

[9] Wu C. C., Su C. C., Yang C. F. Preparation, structure and properties of carbon nanotube reinforced polymer nanocomposites. Synthetic Metals. 2015;205:98–105. DOI: 10.1016/j.synthmet.2015.03.021

[10] Grossiord N., Loos J., Regev O., Koning C. E. Toolbox for dispersing carbon nanotubes into polymers to get conductive nanocomposites. Chemistry of Materials. 2006;18(5): 1089–1099. DOI: 10.1021/cm051881h

[11] Byrne M. T., Gun'ko Y. K. Recent advances in research on carbon nanotube–polymer composites. Advanced Materials. 2010;22:1672–1688. DOI: 10.1002/adma.200901545

[12] Wikipedia. Sonication [Internet]. Available from: https://en.wikipedia.org/wiki/Sonication

[13] Islam M. F., Rojas E., Bergey D. M., Johnson A. T., Yodh A. G. High weight fraction surfactant solubilization of single-wall carbon nanotubes in water. Nano Letters. 2003;3(2):269–273. DOI: 10.1021/nl025924u

[14] Diouri N., Baitoul M. Effect of carbon nanotubes dispersion on morphology, internal structure and thermal stability of electrospun poly(vinyl alcohol)/carbon nanotubes nanofibers. Optical and Quantum Electronics. 2014;46:259–269. DOI: 10.1007/s11082-013-9787-9

[15] Zelikman E., Narkis M., Siegmann A., Valentini L., Kenny J. M. Polyaniline/multiwalled carbon nanotube systems: dispersion of CNT and CNT/PANI interaction. Polymer Engineering & Science. 2008;48(10):1872–1877. DOI: 10.1002/pen.21208

[16] Mechrez G., Suckeveriene R. Y., Tchoudakov R., Kigly A., Segal E., Narkis M. Structure and properties of multi-walled carbon nanotube porous sheets with enhanced elongation. Journal of Materials Science. 2012;47:6131–6140. DOI: 10.1007/s10853-012-6533-9

[17] Krause B., Mende M., Potschke P., Petzold G. Dispersability and particle size distribution of CNTs in an aqueous surfactant dispersion as a function of ultrasonic treatment time. Carbon. 2010;48:2746–2754. DOI: 10.1016/j.carbon.2010.04.002

[18] Sato H., Sano M. Characteristics of ultrasonic dispersion of carbon nanotubes aided by antifoam. Colloids and Surfaces A: Physicochemical and Engineering Aspects. 2008;322:103–107. DOI: 10.1016/j.colsurfa.2008.02.03

[19] Lu K. L., Lago R. M., Chen Y. K., Green M. L., Harris P. J., Tsang S. C. Mechanical damage of carbon nanotubes by ultrasound. Carbon. 1996;34(6):814–816. DOI: 10.1016/0008-6223(96)89470-X

[20] Saito T., Matsushige K., Tanaka K. Chemical treatment and modification of multi-walled carbon nanotubes. Physica B: Condensed Matter. 2002;323(1–4):280–283. DOI: 10.1016/S0921-4526(02)00999-7

[21] Massoumi B., Jafarpour P., Jaymandb M., Entezami A. A. Functionalize multiwalled carbon nanotubes as reinforcing agents for poly(vinyl alcohol) and poly(vinyl alcohol)/starch nanocomposites: synthesis, characterization and properties. Polymer International. 2015;64:689–695. DOI: 10.1002/pi.4867

[22] Zhang W. W., Li W. L., Wang J. J., Qin C. X., Dai L. X. Composites of polyvinyl alcohol and carbon nanotubes decorated with silver nanoparticles. Fibers and Polymers. 2010;11(8):1132–1136. DOI: 10.1007/s12221-010-1132-3

[23] Weng B. C., Xu F. H., Salinas A. L., Lozano K. Mass production of carbon nanotube reinforced poly(methyl methacrylate) nonwoven nanofiber mats. Carbon. 2014;75:217–226. DOI: 10.1016/j.carbon.2014.03.056

[24] Ne´ri W., Maugey M., Miaudet P., Derre´ A., Zakri C., Poulin P. Surfactant-free spinning of composite carbon nanotube fibers. Macromolecular Rapid Communications. 2006;27:1035–1038. DOI: 10.1002/marc.200600150

[25] Amr I. T., Al-Amer A., Selvin T. P., Al-Harthi M., Girei S. A., Rachid S. Effect of acid treated carbon nanotubes on mechanical, rheological and thermal properties of polystyrene nanocomposites. Composites Part B: Engineering. 2011;42:1554–1561. DOI: 10.1016/j.compositesb.2011.04.013

[26] Ziegler K. J., Gu Z., Peng H., Flor E. L., Hauge R. H., Smalley R. E. Controlled oxidative cutting of single-walled carbon nanotubes. Journal of the American Chemical Society. 2005;127:1541–1547. DOI: 10.1021/ja044537e

[27] Cao Z. S., Qiu L., Yang Y. Z., Chen Y. K., Liu X. G. The surface modifications of multi-walled carbon nanotubes for multi-walled carbon nanotube/poly(ether ether ketone) composites. Applied Surface Science. 2015;353:873–881. DOI: 10.1016/j.apsusc.2015.07.025

[28] Gao J. B., Bin Z., Itkis M. E., Elena B., Hui H., Verina K., et. al. Chemical engineering of the single-walled carbon nanotube-nylon 6 interface. Journal of the American Chemical Society. 2006;128(23):7492–7496. DOI: 10.1021/ja057484p

[29] Zhang P., Qiu D. L., Sun J., Wang J. J., Qin C. X., Dai L. X., et al. Preparation of MWCNTs grafted with polyvinyl alcohol through Friedel-Crafts alkylation and their composite fibers with enhanced mechanical properties. Journal of Materials Chemistry A. 2015;3(4):1442–1449. DOI: 10.1039/c4ta03979c

[30] Cai L. T., Bahr J. L., Yao Y. X., Tour J. M. Ozonation of single-walled carbon nanotubes and their assemblies on rigid self-assembled monolayers. Chemistry of Materials. 2002;14(10):4235–4241. DOI: 10.1021/cm020273o

[31] Chen C. Y., Zepp R. G. Probing photosensitization by functionalized carbon nanotubes. Environmental Science & Technology. 2015;49:13835–13843. DOI: 10.1021/acs.est.5b01041

[32] Liu M., Yang Y., Zhu T., Liu Z. Chemical modification of single-walled carbon nanotubes with peroxytrifluoroacetic acid. Carbon. 2005;43(7):1470–1478. DOI: 10.1016/j.carbon.2005.01.023

[33] Karousis N., Tagmatarchis N. A., Tasis D. Current progress on the chemical modification of carbon nanotubes. Chemical Reviews. 2010;110:5366–5397. DOI: 10.1021/cr100018g

[34] Mallakpour S., Abdolmaleki A., Rostami M. Morphology and thermal properties of environmental friendly nanocomposites using biodegradable poly(amide–imide) based on N-trimellitylimido-S-valine matrix reinforced by fructose-functionalized multi-walled carbon nanotubes. Colloid & Polymer Science. 2015;293:545–553. DOI: 10.1007/s00396-014-3442-2

[35] Pan W. H., Lue S. J., Chang C. M., Liu Y. L. Alkali doped polyvinyl alcohol/multi-walled carbon nano-tube electrolyte for direct methanol alkaline fuel cell. Journal of Membrane Science. 2011;376(1–2):225–232. DOI: 10.1016/j.memsci.2011.04.026

[36] Liu Y., Yao Z., Adronov A. Functionalization of single-walled carbon nanotubes with well-defined polymers by radical coupling. Macromolecules. 2005;38(4):1172–1179. DOI: 10.1021/ma048273s

[37] Lin Y., Zhou B., Fernando K. A., Liu P., Allard L. F., Sun Y. P. Polymeric carbon nanocomposites from carbon nanotubes functionalized with matrix polymer. Macromolecules. 2003;36(19):7199–7204. DOI: 10.1021/ma0348876

[38] Zhang C. Q., Vennerberg D., Kessler M. R. In-situ synthesis of biopolyurethane nanocomposites reinforced with modified multiwalled carbon nanotubes. Journal of Applied Polymer Science. 2015;132(36):42515–42523. DOI: 10.1002/app.42515

[39] Balaban T. S., Balaban M. C., Malik S., Hennrich F., Fischer R., Rösner H., et al. Polyacylation of single-walled nanotubes under Friedel–Crafts conditions: an efficient method for functionalizing, purifying, decorating, and linking carbon allotropes. Advanced Materials. 2006;18(20):2763–2767. DOI: 10.1002/adma.200600138

[40] Nanjundan A. K., Jeon I. Y., Sohn G. J., Jain R., Kumar S., Baek J. B. Highly conducting and flexible few-walled carbon nanotube thin film. ACS Nano. 2011;5(3):2324–2331. DOI: 10.1021/nn103630y

[41] Wu X. L., Liu P. Poly(vinyl chloride)-grafted multi-walled carbon nanotubes via Friedel-Crafts alkylation. Express Polymer Letters. 2010;4(11):723–728. DOI: 10.3144/expresspolymlett.2010.87

[42] Huang F., Vanhaecke E., Chen D. In-situ polymerization and characterizations of polyaniline on MWCNT powders and aligned MWCNT films. Catalysis Today. 2010;150(1–2):71–76. DOI: 10.1016/j.cattod.2009.05.017

[43] Gao J. B., Bin Z., Itkis M. E., Elena B., Hui H., Verina K., et al. Chemical engineering of the single-walled carbon nanotube-nylon 6 interface. Journal of the American Chemical Society. 2006;128(23):7492–7496. DOI: 10.1021/ja057484p

[44] Takassi M. A., Zadehnazari A., Farhadi A., Mallakpour S. Highly stable polyimide composite films based on 1,2,4-triazole ring reinforced with multi-walled carbon nanotubes: study on thermal, mechanical, and morphological properties. Progress in Organic Coatings. 2015;80:142–149. DOI: 10.1016/j.porgcoat.2014.12.001

[45] Hu N. T., Zhou H. W., Dang G. D., Rao X. H., Chen C. H., Zhang W. J. Efficient dispersion of multi-walled carbon nanotubes by in-situ polymerization. Polymer International. 2007;56(5):655–659. DOI: 10.1002/pi.2187

[46] Wang K. T., Gu M. B., Wang J. J., Qin C. X., Dai L. X. Functionalized carbon nanotube/polyacrylonitrile composite nanofibers: fabrication and properties. Polymers for Advanced Technologies. 2012;23(2):262–271. DOI: 10.1002/pat.1866

[47] Vaisman L., Wagner H. D., Marom G. The role of surfactants in dispersion of carbon nanotubes. Advances in Colloid & Interface Science. 2006;128:37–46. DOI: 10.1016/j.cis.2006.11.007

[48] Fujigaya T., Nakashima N., Fujigaya T., Nakashima N. Methodology for homogeneous dispersion of single-walled carbon nanotubes by physical modification. Polymer Journal. 2008;40(7):577–589. DOI: 10.1295/polymj.PJ2008039

[49] Moore V. C., Strano M. S., Haroz E. H., Hauge R. H., Smalley R. E., Schmidt J., et al. Individually suspended single-walled carbon nanotubes in various surfactants. Nano Letters. 2003;3(10):1379–1382. DOI: 10.1021/nl034524j

[50] Deng Z., Yu H. J., Wang L., Zhai X. T. A novel ferrocene-containing polymer based dispersant for noncovalent dispersion of multi-walled carbon nanotubes in chloroform. Journal of Organometallic Chemistry. 2015;791:274–278. DOI: 10.1016/j.jorganchem.2015.06.012

[51] Blanch A. J., Lenehan C. E., Quinton J. S. Optimizing surfactant concentrations for dispersion of single-walled carbon nanotubes in aqueous solution. Journal of Physical Chemistry B. 2010;114(30):9805–9811. DOI: 10.1021/jp104113d

[52] Lin S. C., Hilmer A. J., Mendenhall J. D., Strano M. S., Blankschtein D. Molecular perspective on diazonium adsorption for controllable functionalization of single-walled carbon nanotubes in aqueous surfactant solutions. Journal of the American Chemical Society. 2012;134:8194–8204. DOI: 10.1021/ja301635e

[53] O'Connell M. J., Bachilo S. M., Huffman C. B., Moore V. C., Strano M. S., Haroz E. H., et al. Band gap fluorescence from individual single-walled carbon nanotubes. Science. 2002;297(5581):593–596. DOI: 10.1126/science.1072631

[54] Tummala N. R., Striolo A. SDS surfactants on carbon nanotubes: aggregate morphology. ACS Nano. 2009;3(3):595–602. DOI: 10.1021/nn8007756

[55] Bartholome C., Miaudet P., Derré A., Maugey M., Roubeau O., Zakri C., Poulin P. Influence of surface functionalization on the thermal and electrical properties of nanotube–PVA composites. Composites Science and Technology. 2008;68:2568–2573. DOI: 10.1016/j.compscitech.2008.05.021

[56] Suttipong M., Tummala N. R., Kitiyanan B., Striolo A. Role of surfactant molecular structure on self-Assembly: aqueous SDBS on carbon nanotubes. Journal of Physical Chemistry C. 2011;115:17286–17296. DOI: 10.1021/jp203247r

[57] Ju L., Zhang W., Wang X., Hu J., Zhang Y. Aggregation kinetics of SDBS-dispersed carbon nanotubes in different aqueous suspensions. Colloids and Surfaces A: Physicochemical and Engineering Aspects. 2012;409:159–166. DOI: 10.1016/j.colsurfa.2012.06.015

[58] Whitsitt E. A., Barron A. R. Silica coated single walled carbon nanotubes. Nano Letters. 2003;3(6):775–778. DOI: 10.1021/nl034186m

[59] Bagheri H., Ayazi Z., Es'haghi A., Aghakhani A. Reinforced polydiphenylamine nanocomposite for microextraction in packed syringe of various pesticides. Journal of Chromatography A. 2012; 1222: 13–21. DOI: 10.1016/j.chroma.2011.11.063

[60] Geng Y., Liu M. Y., Li J., Shi X. M., Kim J. K. Effects of surfactant treatment on mechanical and electrical properties of CNT/epoxy nanocomposites. Composites Part A: Applied Science & Manufacturing. 2008;39(12):1876–1883. DOI: 10.1016/j.compositesa.2008.09.009

[61] Liu S. D., Luo G. H., Wei F. Poly(p-phenylene terephthalamide)/carbon nanotube composite membrane: preparation via polyanion solution method and mechanical property enhancement. Composites Science and Technology. 2015;118:135–140. DOI: 10.1016/j.compscitech.2015.09.002

[62] Ponnamma D., Sung S. H., Hong J. S., Ahn K. H., Varughese K. T., Thomas S. Influence of non-covalent functionalization of carbon nanotubes on the rheological behavior of natural rubber latex nanocomposites. European Polymer Journal. 2014;53:147–159. DOI: 10.1016/j.eurpolymj.2014.01.025

[63] Gong X. Y., Liu J., Baskaran S., Voise R. D., Young J. S. Surfactant-assisted processing of carbon nanotube/polymer composites. Chemistry of Materials. 2000;12(4):1049–1052. DOI: 10.1021/cm9906396

[64] Nogueira S. L., Sahoo S. K., Jarrosson T., Serein-Spirau F., Lère-Porte J. P., Moujaes E. A. A new designed π conjugated molecule for stable single walled carbon nanotube

dispersion in aqueous medium. Journal of Colloid and Interface Science. 2016;464:117–125. DOI: 10.1016/j.jcis.2015.11.016

[65] Liu Y., Tang J., Chen X., Xin J. H. Decoration of carbon nanotubes with chitosan. Carbon. 2005;43(15):3178–3180. DOI: 10.1016/j.carbon.2005.06.020

[66] Nakamura G., Narimatsu K., Niidome Y., Nakashima N. Green tea solution individually solubilizes single-walled carbon nanotubes. Chemistry Letters. 2007;36(9):1140–1141. DOI: 10.1246/cl.2007.1140

[67] Lu L. L., Hou W. J., Sun J., Wang J. J., Qin C. X., Dai L. X. Preparation of poly (vinyl alcohol) fibers strengthened using multiwalled carbon nanotubes functionalized with tea polyphenols. Journal of Materials Science. 2014;49(49):3322–3330. DOI: 10.1007/s10853-014-8039-0

[68] Zhang P., Zhou T. F., Sun J., Wang J. J., Qin C. X., Dai L. X., et al. Dispersion of multiwalled carbon nanotubes modified by rosemary acid into poly(vinyl alcohol) and preparation of their composite fibers. RSC Advances. 2015;5(68):55492–55498. DOI: 10.1039/c5ra06804e

[69] Lu W. B., Zu M., Byun J. H., Kim B. S., Chou T. W. State of the art of carbon nanotube fibers: opportunities and challenges. Advanced Materials. 2012;24(14):1805–1833. DOI: 10.1002/adma.201104672

[70] Ryu S., Chou J. B., Lee K., Lee D., Hong S. H., Zhao R., et al. Direct insulation-to-conduction transformation of adhesive catecholamine for simultaneous increases of electrical conductivity and mechanical strength of CNT fibers. Advanced Materials. 2015;27(21):3250–3255. DOI: 10.1002/adma.201500914

[71] Behabtu N., Green M. J., Pasquali M. Carbon nanotube-based neat fibers. Nano Today. 2008;3:24–34. DOI:10.1016/S1748-0132(08)70062-8

[72] Min H. J., Jin U. C., Park S. H., Jeong Y. G., Baik D. H. Influences of tensile drawing on structures, mechanical, and electrical properties of wet-spun multi-walled carbon nanotube composite fiber. Macromolecular Research. 2012;20(6):650–657. DOI: 10.1007/s13233-012-0094-2

[73] Qi H., Schulz B., Vad T., Liu J., Mäder E., Seide G., et al. Novel carbon nanotube/cellulose composite fibers as multifunctional materials. ACS Applied Materials & Interfaces. 2015;7(40):22404–22412. DOI: 10.1021/acsami.5b06229

[74] Vigolo B., Pénicaud A., Coulon C., Sauder C., Pailler R., Journet C., et al. Macroscopic fibers and ribbons of oriented carbon nanotubes. Science. 2000;290(5495):1331–1334. DOI: 10.1126/science.290.5495.1331

[75] Dalton A. B., Collins S., Munoz E., Razal J. M., Ebron V. H., Ferraris J. P., et al. Supertough carbon-nanotube fibres – these extraordinary composite fibres can be woven into electronic textiles. Nature. 2003;423(6941):703–703. DOI: 10.1038/423703a

[76] Sa V., Kornev K. G. A method for wet spinning of alginate fibers with a high concentration of single-walled carbon nanotubes. Carbon. 2011;49(6):1859–1868. DOI: 10.1016/j.carbon.2011.01.008

[77] Park G., Jung Y., Lee G. W., Hinestroza J. P., Jeong Y. Carbon nanotube/poly(vinyl alcohol) fibers with a sheath-core structure prepared by wet spinning. Fibers & Polymers. 2012;13(7):874–879. DOI: 10.1007/s12221-012-0874-5

[78] Gupta V. B. Solution-spinning processes. In: Gupta V. B., Kothari V. K., editors. Manufactured fibre technology. 1st ed. Netherlands: Springer; 1997. pp. 124–138. DOI: 10.1007/978-94-011-5854-1

[79] Mercader C., Denis-Lutard V., Jestin S., Maugey M., Derre' A., Zakri C., et al. Scalable process for the spinning of PVA–carbon nanotube composite fibers. Journal of Applied Polymer Science. 2012;125:E191–E196. DOI: 10.1002/app.36308

[80] Yang J., Chen Q., Chen F., Zhang Q., Wang K. Fu Q. Realizing the full nanofiller enhancement in melt-spun fibers of poly(vinylidene fluoride)/carbon nanotube composites. Nanotechnology. 2011;22:355707–355717. DOI: 10.1088/0957-4484/22/35/355707

[81] Potschke P., Brunig H., Janke A., Fischer D., Jehnichen D. Orientation of multiwalled carbon nanotubes in composites with polycarbonate by melt spinning. Polymer. 2005;46:10355–10363. DOI: 10.1016/j.polymer.2005.07.106

[82] Naebe M., Lin T., Feng L. F., Dai L. M., Abramson A., Prakash V., Wang X. G. Conducting polymer and polymer/CNT composite nanofibers by electrospinning. Chapter 4. In: Nagarajan R., Zukas W., Hatton T. A., Lee S., editors. Nanoscience and nanotechnology for chemical and biological defense. Washington, DC: American Chemical Society; 2009. pp. 39–58. DOI: 10.1021/bk-2009-1016.ch004

[83] Baji A., Mai Y. W., Abtahi M., Wong S. C., Liu Y., Li Q. Microstructure development in electrospun carbon nanotube reinforced polyvinylidene fluoride fibers and its influence on tensile strength and dielectric permittivity. Composites Science & Technology. 2013;88(10):1–8. DOI: 10.1016/j.compscitech.2013.08.021

[84] Lee T. W., Jeong Y. G. Regenerated cellulose/multiwalled carbon nanotube composite films with efficient electric heating performance. Carbohydrate Polymers. 2015;133:456–463. DOI: 10.1016/j.carbpol.2015.06.053

[85] Srivastava R. K., Vemuru V. S. M., Zeng Y., Vajtai R., Nagarajaiah S., Ajayan P. M., et al. The strain sensing and thermal–mechanical behavior of flexible multi-walled carbon nanotube/polystyrene composite films. Carbon. 2011;49:3928–3936. DOI: 10.1016/j.carbon.2011.05.031

[86] Jo J. W., Jung J. W., Lee J. U., Jo W. H. Fabrication of highly conductive and transparent thin films from single-walled carbon nanotubes using a new non-ionic surfactant via spin coating. ACS Nano. 2010;4(9):5382–5388. DOI: 10.1021/nn1009837

[87] Yan J., Jeong Y. G. Highly elastic and transparent multiwalled carbon nanotube/polydimethylsiloxane bilayer films as electric heating materials. Materials & Design. 2015;86:72–79. DOI: 10.1016/j.matdes.2015.07.089

[88] Kim S., Fornasiero F., Park H. G., In J. B., Meshot E., Giraldo G., et al. Fabrication of flexible, aligned carbon nanotube/polymer composite membranes by in-situ polymerization. Journal of Membrane Science. 2014;460(12):91–98. DOI: 10.1016/j.memsci.2014.02.016

[89] Bokobza L. Mechanical, electrical and spectroscopic investigations of carbon nanotube-reinforced elastomers. Vibrational Spectroscopy. 2009;51(1):52–59. DOI: 10.1016/j.vibspec.2008.10.001

[90] Fletcher A., Gupta M. C., Dudley K. L., Vedeler E. Elastomer foam nanocomposites for electromagnetic dissipation and shielding applications. Composites Science and Technology. 2010;70:953–958. DOI: 10.1016/j.compscitech.2010.02.011

[91] Fakhru'l-Razi A., Atieh M. A., Girun N., Chuah T. G., El-Sadig M., Biak D. A. Effect of multi-wall carbon nanotubes on the mechanical properties of natural rubber. Composite Structures. 2006;75(s 1–4):496–500. DOI: 10.1016/j.compstruct.2006.04.035

[92] Shanmugharaj A. M., Ryu S. H. Influence of aminosilane-functionalized carbon nanotubes on the rheometric, mechanical, electrical and thermal degradation properties of epoxidized natural rubber nanocomposites. Polymer International. 2013;62(10):1433–1441. DOI: 10.1002/pi.4437

[93] Liu Y., Chi W., Duan H., Zou H., Yue D., Zhang L. Property improvement of room temperature vulcanized silicone elastomer by surface-modified multi-walled carbon nanotube inclusion. Journal of Alloys & Compounds. 2016;657:472–477. DOI: 10.1016/j.jallcom.2015.10.129

[94] Aurilia M., Sorrentino L., Iannace S. Interface dissipative mechanisms in an elastomeric matrix reinforced with MWCNTs. Macromolecular Theory & Simulations. 2013;22(3):198–206. DOI: 10.1002/mats.201200070

[95] Perez L. D., Zuluaga M. A., Kyu T., Mark J. E., Lopez B. L. Preparation, characterization, and physical properties of multiwall carbon nanotube/elastomer composites. Polymer Engineering & Science. 2009;49(5):866–874. DOI: 10.1002/pen.21247

[96] Bikiaris D. Microstructure and properties of polypropylene/carbon nanotube nanocomposites. Materials. 2010;3(4):2884–2946. DOI: 10.3390/ma3042884

[97] Al-Saleh M. H. Electrically conductive carbon nanotube/polypropylene nanocomposite with improved mechanical properties. Materials & Design. 2015;85:76–81. DOI: 10.1016/j.matdes.2015.06.162

[98] Huang H. D., Liu C. Y., Zhang L. Q., Zhong G. J., Li Z. M. Simultaneous reinforcement and toughening of carbon nanotube/cellulose conductive nanocomposite films by

interfacial hydrogen bonding. ACS Sustainable Chemistry & Engineering. 2015;3(2): 317–324. DOI: 10.1021/sc500681v

[99] Khan S. U., Pothnis J. R., Kim J. K. Effects of carbon nanotube alignment on electrical and mechanical properties of epoxy nanocomposites. Composites Part A: Applied Science & Manufacturing. 2013;49(3):26–34. DOI: 10.1016/j.compositesa.2013.01.015

[100] Nam T. H., Goto K., Nakayama H., Oshima K., Premalal V., Shimamura Y., et al. Effects of stretching on mechanical properties of aligned multi-walled carbon nanotube/epoxy composites. Composites Part A: Applied Science & Manufacturing. 2014;64(21):194–202. DOI: 10.1016/j.compositesa.2014.05.013

[101] Perrot C., Piccione P. M., Zakri C., Gaillard P., Poulin P. Influence of the spinning conditions on the structure and properties of polyamide 12/carbon nanotube composite fibers. Journal of Applied Polymer Science. 2009;114(6):3515–3523. DOI: 10.1002/app.30875

[102] Zhu Y. F., Ma C., Zhang W., Zhang R. P., Koratkar N., Liang J. Alignment of multiwalled carbon nanotubes in bulk epoxy composites via electric field. Journal of Applied Physics. 2009;105:54319–54325. DOI: 10.1063/1.3080243

[103] Dror Y., Salalha W., Khalfin R. L., Cohen Y., Yarin A. L., Zussman E.. Carbon nanotubes embedded in oriented polymer nanofibers by electrospinning. Langmuir. 2003;19(17): 7012–7020. DOI: 10.1021/la034234i

[104] Liu L. Q., Tasis D., Prato M., Wagner H. D. Tensile mechanics of electrospun multi-walled nanotube/poly(methyl methacrylate) nanofibers. Advanced Materials. 2007;19(9):1228–1233. DOI: 10.1002/adma.200602226

[105] Ji J. Y., Sui G., Yu Y. H., Liu Y. X., Lin Y. H., Du Z. J., et al. Significant improvement of mechanical properties observed in highly aligned carbon-nanotube-reinforced nanofibers. Journal of Physical Chemistry C. 2009;113(12):4779–4785. DOI: 10.1021/jp8077198

Chapter 7

Protective Coatings Based on PMMA–Silica Nanocomposites Reinforced with Carbon Nanotubes

Samarah V. Harb, Fábio C. dos Santos,
Sandra H. Pulcinelli, Celso V. Santilli,
Kevin M. Knowles and Peter Hammer

Additional information is available at the end of the chapter

http://dx.doi.org/10.5772/62808

Abstract

Polymethylmethacrylate–silica hybrids have been prepared using the sol–gel route by the radical polymerization of methyl methacrylate(MMA) using benzoyl peroxide (BPO) as a thermal initiator and 3-(trimethoxysilyl)propyl methacrylate(MPTS) as a coupling agent, followed by acid-catalyzed hydrolytic condensation of tetraethoxysilane (TEOS). Carbon nanotubes (CNTs) were first dispersed either by surfactant addition or by functionalization with carboxyl groups and then added at a carbon (CNT) to silicon (TEOS and MPTS) molar ratio (C_{CNT}/Si_{Hybrid}) of 0.05% to two different hybrid matrices prepared at BPO/MMA molar ratios of 0.01 and 0.05. Films of 2–7 μm thickness deposited onto carbon steel by dip-coating were characterized in terms of their microstructure and their mechanical, thermal and anticorrosive behavior. Atomic force microscopy and optical microscopy confirmed that there was a homogeneous dispersion of CNTs in the nanocomposites and that the surfaces of the films were very smooth. X-ray photoelectron spectroscopy (XPS) confirmed the nominal composition of the films while nuclear magnetic resonance showed that the connectivity of the silica network was unaffected by CNT loading. Thermogravimetric analysis and mechanical measurements confirmed an increase of thermal stability, hardness, adhesion and scratch resistance of CNT-loaded coatings relative to those without CNTs. Electrochemical impedance spectroscopy measurements in 3.5% NaCl solution interpreted in terms of equivalent circuits showed that the reinforced hybrid coatings, prepared at the higher BPO/MMA molar ratio used in this work, act as a very efficient anticorrosive barrier, with an impedance modulus up to $10^9\ \Omega\ cm^2$.

Keywords: organic–inorganic hybrids, carbon nanotubes, mechanical reinforcement, structural properties, anticorrosive coating

1. Introduction

Organic–inorganic hybrids are a class of nanocomposite materials, which combine different components on the molecular or nanometric scale, resulting in hybrid systems that not only reflect the sum of the properties of the individual components but also are new materials with unique features. The blend of organic and inorganic phases interacting on the molecular scale combines properties such as processability, flexibility and hydrophobicity of the polymeric organic phases with thermal, chemical and mechanical stability of inorganic ceramic compounds. The nature, size and compatibility of the organic and inorganic phases are of critical importance, because they determine the transparency, homogeneity and stability of the hybrid material. The nature of the bonding at the interface between the phases is of particular significance for this class of nanocomposites. This can be used to classify these hybrid materials: the presence of relatively weak bonding such as van der Waals, dipole–dipole, hydrogen or ionic bonding is characteristic of a class I hybrid material, while strong covalent or ionic–covalent chemical bonding are both characteristic of class II hybrid materials [1].

The sol–gel process is possibly the most suitable method for the synthesis of hybrid materials because of the relatively mild synthesis conditions, the environmental compatibility, and, in particular, the possibility of combining a large number of precursors in different proportions. The simultaneous hydrolytic condensation of the inorganic precursor and polymerization of the organic species produces homogeneous nanocomposites with tunable properties. The multifunctionality of hybrid materials enables them to be used in a variety of applications such as drug delivery systems, optical and electrical devices, catalysts, photochromic devices and protective coatings [1, 2].

Among the large number of reported organic–inorganic nanocomposite systems, in which polymers such as epoxy, polyimide, acrylic and polyethylenimine phases are combined with inorganic oxides such as silica, alumina, zirconia, titania and ceria, one important hybrid class is the polymethylmethacrylate–silica (PMMA–silica) system. PMMA–silica nanocomposites have recently received considerable attention because of their ability to protect a wide variety of metal surfaces such as steels, stainless steels, aluminum alloys and magnesium alloys in an efficiently and environmentally compliant manner [3–5]. These alloys are particularly important for key industries such as the aerospace, automotive and offshore companies. However, most of these alloys suffer severe corrosion in maritime environments and even humid environments and therefore need appropriate surface passivation to survive for long periods in aggressive environments.

Corrosion is a spontaneous and irreversible reaction between a metal surface and its environment, resulting in significant economical losses, the failure of critical components and environmental problems. The prevention of corrosion, or at least its mitigation, is therefore one of the main challenges industrially worldwide. The application of protective coatings such as paints or resins, or those based on ceramic materials, is the most common way to improve the durability of metallic alloys significantly. However, organic coatings are relatively thick and can suffer poor thermal and mechanical stability and also a lack of adhesion, while coatings based solely on ceramic materials are likely to be porous and suffer from intrinsic stress-induced cracking, leading to thickness limitations [3, 6].

Therefore, organic–inorganic hybrids have been developed to overcome the limitations of traditional coatings, forming an efficient and durable corrosion protection system for metallic surfaces. In the case of PMMA–silica hybrid nanocomposites, this corrosion protection is a consequence of the covalent bonding between PMMA and silica nodes through the coupling agent 3-(trimethoxysilyl)propyl methacrylate (MPTS), formed by three methoxy-silane groups linked by a nonhydrolysable Si–C bond to a methacrylate tail. This bonding mechanism produces a class II hybrid with a nanostructure of dense silica cross-link nodes bridged by short polymeric chains. As a consequence, the closely packed nanostructure acts as an efficient corrosion barrier against the uptake of aggressive agents [3, 4].

One drawback of most organic-inorganic hybrids is their relatively high polymer content of 60–80%; this leads to a reduced mechanical and thermal stability of these materials relative to ceramic systems. To overcome this limitation, carbon nanotubes (CNTs), known for their exceptional mechanical and thermal properties, are regarded as being the most suitable nanostructures to reinforce polymeric and hybrid materials. Thus, for example, in a recent study, Nafion® modified functionalized multiwall CNTs were dispersed in a PMMA–silica nanocomposite at carbon-to-silicon molar ratios of 0.1%, 1.0% and 5.0% [7]. The results of this study showed that the CNTs could be dispersed efficiently within the nanocomposite and that their presence did not affect the connectivity of the hybrid network. In addition, the coatings were able to maintain their high corrosion resistance, with an impedance modulus of about 10^7 Ω cm^2 in 3.5% NaCl solution [7]. However, no mechanical tests were performed in this study.

Other studies also report on hybrids and, in particular, on polymers modified by CNTs [8–10]. The development of nanocomposites with improved electrical conductivity, thermal stability and mechanical strength by incorporation of CNTs are the most cited objectives in these studies. For polyethylene–CNT composites containing CNTs in the range of 1–2.5 wt.%, an increase in electrical conductivity up to six orders of magnitude has been observed [8]. A uniform dispersion of CNTs in a polypropylene (PP) matrix has been shown to produce a substantial increase in thermal stability at extremely low loading levels of CNTs, attributed to the relatively large interfacial area common to the PP chains and the free radial scavenging CNTs [9]. For epoxy–CNT nanocomposites, in which the epoxy resin matrix was modified with 0.1 wt.% of amino-functionalized CNTs, an improvement in strain to fracture and an increase in Young's modulus from 3.29 GPa for the neat resin to 3.50 GPa for the nanocomposite have both been reported [10].

Protective hybrid coatings, modified with CNTs, have also been subject of a number of recent studies. Thus, for example, Fe_3O_4 nanoparticles attached to CNTs have been incorporated successfully at a concentration level of 3 wt.% into epoxy resin coatings deposited on carbon steel [11]. Experimental results showed a significant increase of coating adhesion and corrosion protection efficiency relative to coatings without both the Fe_3O_4 nanoparticles and the CNTs [11]. Epoxy–CNT composite coatings deposited on aluminum alloy 2024-T3 substrates at CNT levels of 0.1 wt% or 0.5 wt% showed a similar result with an improvement in adhesion strength, wear resistance and rate of corrosion with CNT loading, the latter explained by a CNT-induced decrease of the level of open porosity within the coating [12]. Polypyrrole (PPy) coatings containing low levels of CNTs and chitosan deposited on St-12 steel have shown a significant improvement in corrosion protection relative to PPy coatings with an increase in the corrosion resistance in 3.5% NaCl solution from 176 Ω cm^2 for pure PPy to 861 Ω cm^2 for PPy–CNT–chitosan coatings [13]. This was attributed to the improved density and more compact structure of the PPy–CNT–chitosan composite relative to the pure PPy coatings [13]. A study of the corrosion protection performance of poly(N-methylpyrrole)-dodecylsulfate/CNT composite coatings on SAE 304 stainless steel was performed in 0.5 mol L^{-1} H_2SO_4 solution [14]. In these coatings, the CNTs were added as a second layer on top of the poly(N-methylpyrrole)-dodecylsulfate base layer, either by electrodeposition or by dispersing the CNTs in a Nafion® solution. The results also confirmed a significantly improved corrosion protection of the base layer coated with the Nafion®-dispersed CNTs. This was attributed by these authors to the electrostatic repulsion of corrosive anionic species by the negatively charged CNTs and Nafion® containing surface layers [14]. Using a similar strategy, a conductive coating based on PPy has been modified with 0.25–1 at.% of functionalized and nonfunctionalized CNTs and coated on 60/40 α/β brass [15]. The observed improvement of corrosion protection efficiency of the brass in 3.5% NaCl solution relative to coatings without CNTs was explained by the authors in terms of an increase in electrical conductivity of the CNT-loaded coatings to help form anodically protecting passive oxide films on the metal and also to the increase in tortuosity of the paths corrosive ions have taken through the coating to reach the passive film in order to attack it chemically [15]. The ability for deliberately undercured coatings with 20 wt.% of CNTs and microcapsules containing electrically conductive epoxy resin with self-healing property has also been demonstrated [16]. In this work, Bailey et al. used a novel electrotensile test. Upon cracking of the undercured coating during tensile testing, microcapsules in the crack path release the healing solvent ethyl phenyl acetate (EPA), enabling the subsequent reaction with residual hardener in the vicinity of the crack to make the matrix swell locally and cause cracks to be closed [16].

It is important to note that most as-synthesized CNTs consist of large aggregates or bundles insoluble both in water and in common organic solvents because of their enhanced polarizability induced by their cylindrical shape and hence the strong van der Waals' interactions between individual nanotubes [17]. Efficient dispersion of CNTs in a polymer matrix requires the initial disentanglement of these large aggregates and chemical compatibility between the CNTs and the polymer matrix to maintain a homogeneous and stable composite structure. This chemical compatibility can be induced either by preselecting a matrix which interacts electrostatically with CNTs or by modifying the interaction potential between the CNTs and the polymer by functionalization. A suitable functionalization of CNTs is able to increase their electrostatic potential, thereby reducing their tendency to agglomerate [9]. However, it is evident that if harsh treatment conditions are used, such as prolonged sonication or excessive chemical treatment, a high level of damage to the hexagonal nanotube structure can occur, leading to a significant loss of mechanical and electrical performance of the CNTs.

http://dx.doi.org/10.5772/62808

Alternative approaches have been developed to disperse individual CNTs by noncovalent functionalization employing a wrapping agent, typically a surfactant or an organic polymer. A successful separation of CNTs leading to a stable suspension in aqueous solutions of sodium dodecyl sulfate (SDS) surfactant with coaddition of saturated fatty acids was recently demonstrated [17]. Following on from previous work on PMMA–silica coatings containing CNTs [7], we have successfully prepared CNT-reinforced protective hybrid coatings on carbon steel in this study. The uniform dispersion of CNTs in the PMMA–silica matrix was accomplished using two different pretreatments: the functionalization by carboxylic groups and surfactant assistance using SDS for subsequent introduction in the PMMA–silica hybrids. Particular attention was paid to the extent to which CNTs could be incorporated successfully into the PMMA–silica matrix without compromising the excellent anticorrosive characteristics of the hybrids. The effects of the inclusion of CNTs on the morphological, structural, thermal, mechanical and electrochemical properties of the hybrid matrix were evaluated by optical and atomic force microscopy (AFM), nuclear magnetic resonance, X-ray photoelectron spectroscopy (XPS), mechanical testing and electrochemical impedance spectroscopy (EIS).

2. Experimental

2.1. Synthesis

All reagents were purchased from Sigma-Aldrich and used as received, apart from the methyl methacrylate monomer, which had been distilled before use to remove the ≤30 ppm amount of 4-methoxyphenol added as a polymerization inhibitor. The PMMA–silica hybrid synthesis consisted of radical polymerization of methyl methacrylate (MMA) and 3-(trimethoxysilyl)propyl methacrylate (MPTS) using the thermal initiator benzoyl peroxide (BPO) and tetrahydrofuran (THF) as a solvent, followed by hydrolysis and polycondensation of tetraethoxysilane (TEOS) and MPTS silane sites, catalyzed by nitric acid (pH 1). The following molar ratios were kept constant: MMA/MPTS = 8, TEOS/MPTS = 2, H_2O/Si = 3.5 and ethanol/H_2O = 0.5. The BPO/MMA molar ratio was fixed at a value of 0.01 and 0.05 to study the influence of CNTs in two different matrices designated BPO0.01 and BPO0.05.

The TEOS, MPTS and MMA molecular structures are shown in **Figure 1**. The siloxane bridges (C–Si–O) between the organic and the inorganic phase were derived from MPTS, a modified silicon alkoxide with a methacrylate group which acts as a coupling agent between the organic component, PMMA (polymerized MMA), and the inorganic component, silica. In the presence of acidified water, TEOS and MPTS form a silica network through sol–gel hydrolysis and condensation reactions, a process that converts a colloidal suspension (the sol) into a three-dimensional network (the gel). First, the alkoxy groups (O–CH_2–CH_3 and O–CH_3) are hydrolysed, forming silanol groups (Si–OH) and eliminating alcohol molecules (HO–CH_2CH_3 and HO–CH_3), and then silanol groups can react with one another or the initial reagent to yield Si–O–Si bonds [18].

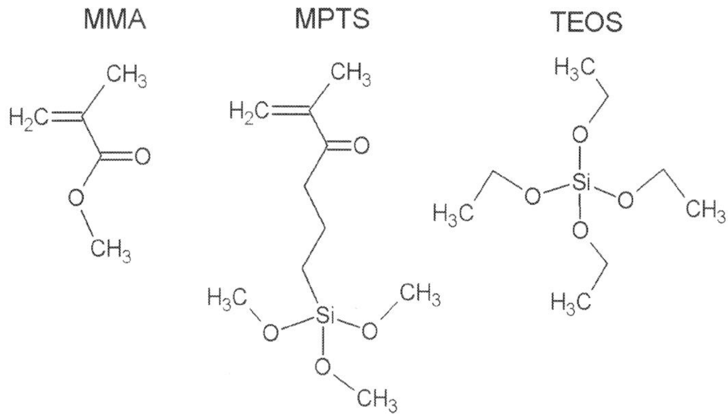

Figure 1. Molecular structures of the synthesis reagents.

Single-wall CNTs were purchased from Dropsens for incorporation into the two hybrids BPO0.01 and BPO0.05. In one pretreatment prior to their incorporation, the CNTs were dispersed using the method described by Alves da Cunha et al. [17], in which aqueous solutions of SDS surfactant (Sigma-Aldrich) and hexadecanoic acid (palmitic acid, Sigma-Aldrich) are used. The dispersion procedure, schematized in **Figure 2a**, starts from raw CNTs and is based on the nonpolar groups of SDS and palmitic acid promoting physical interaction with CNTs, while polar groups of these two chemicals interact with water [17]. After dispersion in SDS and palmitic acid, the CNTs were added at the end of the PMMA–silica hybrid synthesis, at a C_{CNT}/Si_{Hybrid} molar ratio of 0.05%, to the two matrices BPO0.01 and BPO0.05. The two nanocomposites produced in this manner were designated BPO0.01_CNT_SDS and BPO0.05_CNT_SDS, respectively.

In addition to the SDS method, dispersion through functionalization with carboxyl groups was also studied. In the second method, 0.1 g of CNTs was first put in a flask containing 75 ml of concentrated sulfuric acid (H_2SO_4, Sigma-Aldrich) and 25 ml of concentrated nitric acid (HNO_3, Sigma-Aldrich). The CNT-containing solution was then heated and stirred under reflux at 70°C for 2 h followed by 30°C for 4 h. Then, the functionalized CNTs were filtered through an ANOPORE 0.02 μm pore size membrane and washed with distilled water until the pH was 6. After this, drying was carried out at 70°C for 4 h under vacuum and at 200°C for 4 h in air (**Figure 2b**). The oxidation procedure with nitric acid and sulfuric acid adds carboxyl groups at the walls of the CNTs and enhances their solubility in the PMMA–silica hybrid. These functionalized CNTs (CNTCOOH) were dispersed using Nafion® and incorporated into BPO0.05 matrix in the inorganic phase at a C_{CNT}/Si_{Hybrid} molar ratio of 0.05%. The nanocomposite produced in this manner was designated BPO0.05_CNTCOOH.

After synthesis, the five homogeneous and transparent hybrid sols were used to deposit films onto 2.5 cm x 2.5 cm x 0.4 cm A1020 carbon steel substrates by dip-coating (3 immersions, each of 1 min, at a withdrawal rate of 14 cm min^{-1}, with air-drying intervals of 10 min between dips), with the remainder of the solutions placed in Teflon holders to obtain unsupported films, and then heat-treated initially at 60°C for 24 h, followed by 160°C for 3 h. Prior to being dipped, the carbon steel substrates had all been sanded with 100, 300, 600 and 1500 grit sandpaper, washed with isopropanol for 10 min in an ultrasound bath and dried under nitrogen.

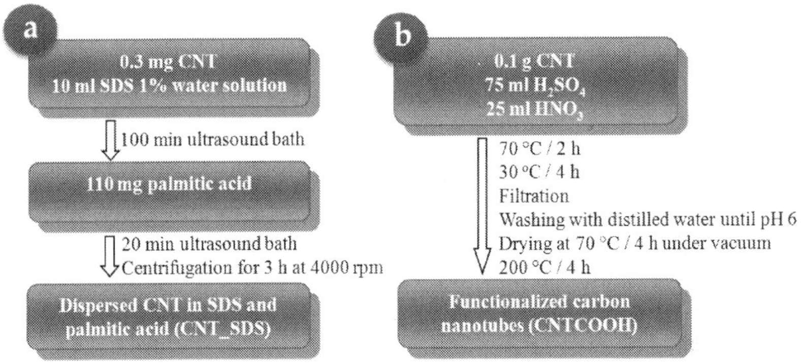

Figure 2. Experimental procedure for the dispersion of carbon nanotubes by (a) interaction with SDS and palmitic acid and (b) functionalization with carboxyl groups.

2.2. Characterization techniques

A JEOL 71500F field-emission gun scanning electron microscope (FEG-SEM) was used to acquire micrographs of raw and dispersed CNTs.

XPS was used for characterization of the CNTs and the hybrids by extracting the elemental compositions and monitoring the local bonding structure of carbon (C 1s), oxygen (O 1s) and silicon (Si 2p) Measurements were performed in a UNI-SPECS UHV surface analysis system, using Mg Kα radiation (hν = 1253.6 eV) and a pass energy of 10 eV for high-resolution spectra. The inelastic background of the C 1s, O 1s and Si 2p photoemission peaks was subtracted using the Shirley baseline. The displacement due to charge accumulation was corrected by fixing the C–H component of the C 1s spectrum at 285.0 eV. The surface composition was determined from peak intensities corrected by the orbital sensitivity factors of the corresponding elements. The CasaXPS processing software was used for deconvolution of the spectra using combinations of Gaussian and Lorentzian functions (Voigt profiles) for analysis of the chemical bonding structure.

^{29}Si nuclear magnetic resonance spectroscopy (^{29}Si-NMR) measurements in the solid state were performed in a Varian Inova spectrometer operating at 300 MHz and 7.05 T, using a Larmor frequency of 59.59 Hz and tetramethyl silane (TMS) as an external standard. The spectra were obtained from Fourier transforms following a single excitation pulse of $\pi/2$ with a relaxation time of 2 s. The CasaXPS processing software was used for deconvolution of the spectra using Voigt profiles.

The thickness of each coating was determined using a Filmetrics F3-CS optical interference system. AFM was used to evaluate the surface morphology of the coatings and to determine their roughness. Agilent Technologies Model 5500 and NX10 Park System atomic force microscopes were used in tapping mode with a silicon cantilever. The results were analyzed using Gwyddion software. RMS (root mean square) roughness values were obtained from 1 μm × 1 μm topography images of the hybrid coatings deposited on the A1020 carbon steel.

Thermogravimetric analysis (TGA) of the five unloaded and CNT-loaded hybrids, each in the form of unsupported films, were carried out in a TA Instruments STD Q600 analyzer. The samples were heated at a rate of 5°C min^{-1} from 25°C to 800°C, under 100 mL min^{-1} of nitrogen flow.

Nanoindentation measurements were carried out in a Nano Indenter® XP system, MTS, equipped with TestWorks 4 Professional level software. A diamond tip with Berkovich geometry was used. For each sample, nine measurements were performed, with 100 μm spacing between each indentation. The input parameters were Poisson ratio (0.35), depth limit (140 nm), allowable drift rate (0.8 nm/s), frequency target (45 Hz) and percent unload in stiffness calculation (50%). Use of the continuous stiffness measurement (CSM) method allowed the continuous determination of the contact stiffness during loading, providing more accurate results. This was achieved by superimposing a small oscillation on the primary loading signal and by analyzing the resulting response of the system using a lock-in amplifier. The hardness and elastic modulus were obtained as a continuous function of depth from a comparison of samples indented in the range of 60–120 nm. To avoid effects on the nanomechanical properties of the films from the underlying steel substrates, the maximum penetration depth for the indentation experiments was set at less than 10% of the coating thickness [19].

Microscratch measurements were performed using homemade equipment at the National Physical Laboratory (Teddington, London, U.K.) to evaluate the scratch resistance and the adherence of the coatings to the A1020 carbon steel substrates. For each sample, 3 parallel tracks of 6 mm length with 1 mm spacing between the tracks were made using a linearly increasing load (from 2 mN to 100 mN), with a diamond tip with spherical conical geometry and 10 μm radius. For CNT-loaded hybrids, further microscratch experiments were undertaken where the load was increased up to a maximum of 240 mN. The measurements also provided the coefficient of friction as a function of the track distance. The tracks were analyzed using a Nikon Measuring Microscope MM-60, coupled with a Nikon SC-213 Digital Counter, which enabled the critical load at which delamination started to be determined.

The anticorrosion efficiency of the hybrid coatings deposited on the A1020 carbon steel was analyzed by EIS using a Gamry Potentiostat Reference 600. The impedance data were collected once a week, until failure, over a frequency range from 10^{-2} Hz to 10^{4} Hz with 10 points per decade and signal amplitude of 10 mV (rms) in an electrochemical cell containing 80 ml of 3.5% NaCl solution at 25°C. The electrochemical cell consisted of a Ag|AgCl|KCl_{sat} reference electrode, a platinum mesh counter electrode, a platinum electrode connected to the reference electrode through a 0.1 μF capacitor and the working electrode of either coated or uncoated carbon steel. The experimental data were fitted with equivalent electrical circuits using Zview software to analyze the EIS response.

3. Results and discussion

3.1. CNT characterization

Van der Waals' forces between CNTs cause their agglomeration in the form of dense bundles. Commercial CNT powder consists of dense particles (**Figure 3a**), which comprises the bundles of CNTs (**Figure 3b**). It is evident from **Figure 3c** and **Figure 3d** that both procedures used for dispersing the CNTs were successful.

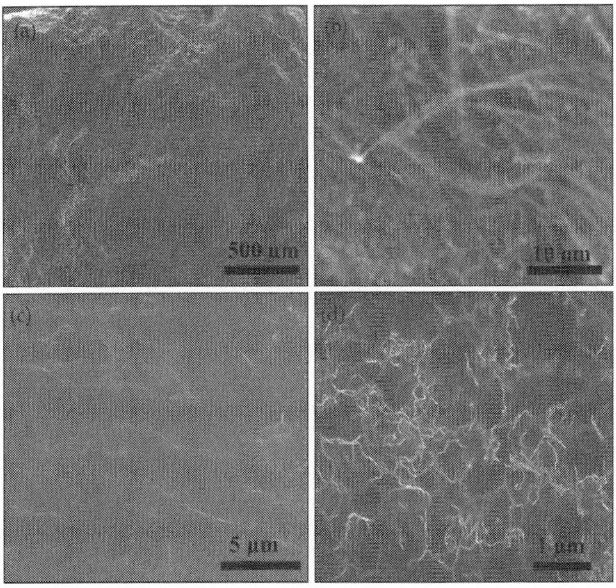

Figure 3. FEG-SEM micrographs of carbon nanotubes (a, b) commercial powder, (c) dispersed in SDS and palmitic acid and (d) functionalized and dispersed in the precursor solution.

CNTs have a peculiar XPS C 1s spectrum with the presence of a predominant aromatic C-C-sp^2 component and characteristic π plasmon transitions, the intensities of which scale with the degree of order of the hexagonal carbon structure. XPS C 1s spectra of pure and functionalized CNTs are presented in **Figure 4**. Quantitative XPS analysis can detect all elements except hydrogen and helium. Discounting the presence of hydrogen, the raw CNTs are composed of 93.2 at.% of carbon and 6.8 at.% of oxygen (**Figure 4a**), partially related to surface contamination by oxygenated hydrocarbon groups of adventitious carbon. The following characteristics indicate a highly aromatic structure: the presence of plasmon peaks (collective $\pi \rightarrow \pi^*$ transitions at ~291 eV and ~394 eV) and the narrow and intense component related to aromatic C-C-sp^2bonds (284.4 eV) with FWHM (full width at half maximum) of about 0.9 eV and a peak area of 65.5% [20]. The high energy components related to C–O, C=O and O–C=O bonds, which can also be observed in the O 1s XPS spectrum (**Figure 4b**), show the presence of ether/alcohol, carbonyl and carboxyl groups on the surface of the nanotubes, associated mainly with the presence of adventitious carbon responsible for the C–H component at ~285 eV. After functionalization, the O–C=O component increases significantly (**Figure 4c** and **4d**) due to the linking of these groups to the walls of the nanotubes, aiding the dispersion of CNTs in the hybrid matrix through their polarity. The degree of functionalization of nanotube walls defined as the intensity ratio I(O–C=O)/I(C-C-sp^2) was 0.3.

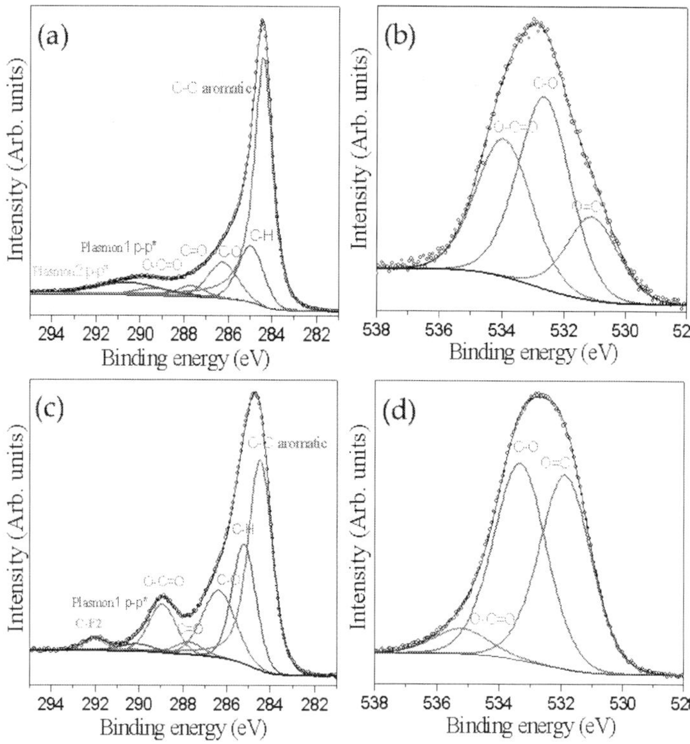

Figure 4. (a) C 1s and (b) O 1s XPS spectra of the as-received CNTs and (c) C 1s and (d) O 1s XPS spectra of functionalized CNTs.

3.2. PMMA–silica hybrid characterization

3.2.1. Surface morphology

All the PMMA–silica hybrid coatings deposited on the A1020 carbon steel were transparent with a homogeneous, colorless appearance. A representative image of one of the 2.5 cm x 2.5

cm x 0.4 cm coated samples is shown in **Figure 5a**. Inspection by optical microscopy performed on free standing hybrids in transmission mode confirmed the uniformity of these coatings (e.g., **Figure 5b**) and indicate a very good dispersion of CNTs in the nanocomposites.

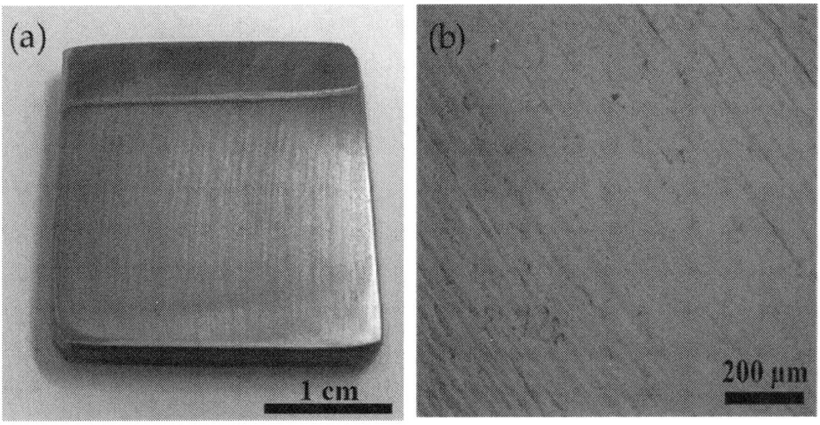

Figure 5. (a) Representative image of BPO0.01_CNT_SDS coating deposited on A1020 carbon steel and (b) optical microscopy image showing a detail of the BPO0.01_CNT_SDS transparent film. Parallel lines in (b) are related to the steel substrate morphology.

One effect caused by increasing the BPO/MMA molar ratio from 0.01 to 0.05 was the reduced gel time of the hybrid sol. This occurred because of the enhanced polymerization rate induced by the increase in the number of radicals of the BPO thermal initiator, leading to a higher viscosity of the solution. In addition to this effect, the inclusion of CNTs also increased the viscosity of the solution prior to dip coating. Together, these two effects account for the trend in the observed hybrid coating thicknesses shown in **Table 1**.

Sample name	BPO/MMA molar ratio	Thickness (μm)	Surface RMS roughness (nm)
BPO0.01	0.01	2.8	0.4
BPO0.01_CNT_SDS	0.01	5.7	0.4
BPO0.05	0.05	3.5	0.4
BPO0.05_CNT_SDS	0.05	6.6	0.5
BPO0.05_CNTCOOH	0.05	4.9	0.2

Table 1. Thickness and surface roughness of pure and CNT-containing PMMA–silica hybrids prepared at different BPO/MMA molar ratios.

AFM topography images obtained for pure PMMA–silica films and for those containing well-dispersed CNTs showed that each of the five hybrids presented a very smooth and uniform surface morphology (**Figure 6**). No defects, pores, cracks or inhomogeneities were observed on the coated samples. The RMS roughness (R_{RMS}) extracted from AFM measurements (Table 1) showed very low values of <0.5 nm for all coatings, confirming the homogeneity of the films and the efficient dispersion of CNTs within the hybrid matrices. The BPO0.01_CNT_SDS surface morphology is shown in the high-resolution AFM image in **Figure 7**. The local smoothness of the surface is confirmed in this image, while there is also indication for a possible presence of a single CNT on the surface.

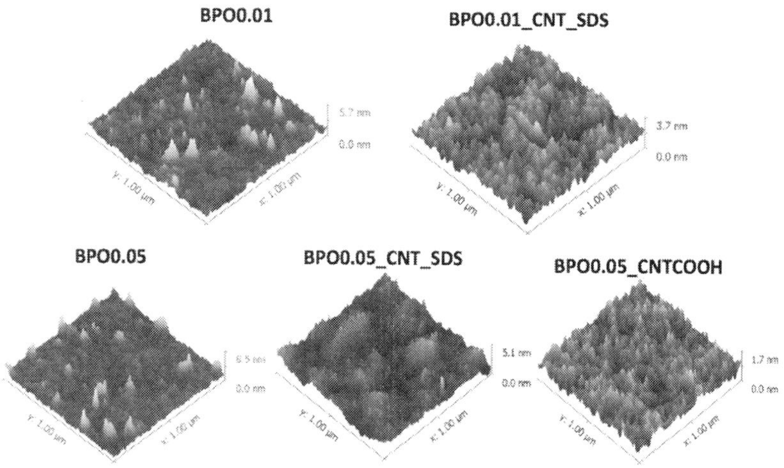

Figure 6. AFM images of all hybrid coatings deposited on carbon steel.

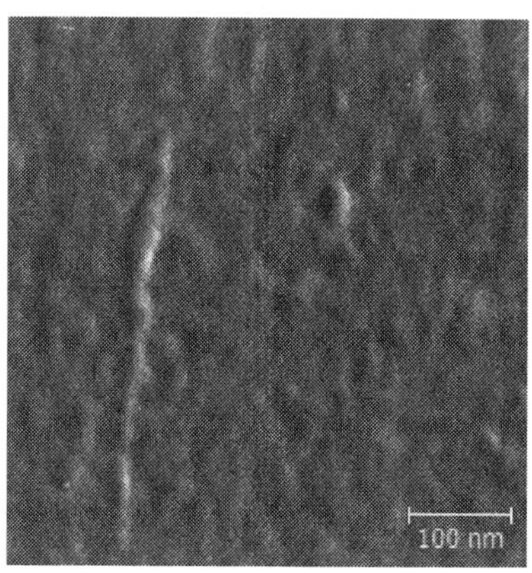

Figure 7. High-resolution AFM image of the BPO0.01_CNT_SDS sample.

3.2.2. Bonding structure

XPS analysis showed that the composition of all hybrids was very similar, with values close to the nominal atomic percentages of 62 at.% of carbon, 32 at.% of oxygen and 6 at.% of silicon, with an experimental error of ±5%. Representative spectra of carbon, oxygen and silicon, deconvoluted into their structural components, are presented in **Figure 8**. The C 1s spectrum, shown in **Figure 8a**, has four components related to C–H, C̲–C–O, C–O and O–C=O bonds present in the PMMA and MPTS molecules (**Figure 8d**) [21]. The carbon underlined corresponds to the atom that was analyzed. The O 1s spectrum (**Figure 8b**) was fitted with three components, associated with O̲–C=O and O=C bonds of PMMA and MPTS, and O–Si bonds of the inorganic network, observed also in the Si 2p spectrum (**Figure 8c**). For completely condensed SiO_2 phase the well-known binding energy of the Si 2p peak is located at 103.5±0.2 eV [21]. As the condensation reaction was incomplete, it is possible that the observed binding energy shift to a lower value of 103.1 eV is caused by some remaining silanol groups (Si–OH) from TEOS and MPTS hydrolysis and condensation. The addition of CNTs to the PMMA–silica hybrids had no effect on the XPS spectra because the additional carbon concentration level of 500 ppm from the nanotubes was below the 3000 ppm detection limit of carbon.

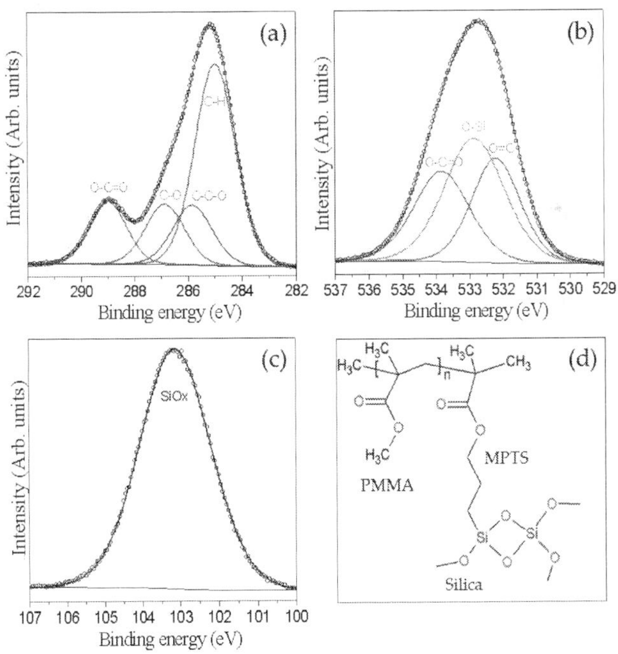

Figure 8. Representative (a) C 1s, (b) O1s and (c) Si 2p XPS spectra of a PMMA–silica hybrid and (d) a schematic of the PMMA–silica hybrid structure.

^{29}Si-NMR analysis also allows the identification of the local chemical bonding structure and the quantitative evaluation of the connectivity of the inorganic phase. To determine the influence of the CNTs on the inorganic silica network, NMR was used to compare the pure and CNT-containing hybrids. Representative spectra are shown in **Figure 9** for a CNT-containing sample and a pure sample. Both spectra have two groups of peaks corresponding to T^i (i = 1, 2, 3) and Q^j (j = 2,3,4) structures shown in **Figure 10**. These two forms of local structures arise as a consequence of the MPTS and TEOS precursor species, respectively. The peaks at chemical shifts of −45, −55 and −63 ppm correspond to T^1, T^2 and T^3 environments, respectively, while the peaks at chemical shifts at −90, −100 and-110 ppm are associated with Q^2, Q^3 and Q^4 environments, respectively (**Figure 10**) [22]. The connectivity of the inorganic phase, defined as degree of polycondensation (C_d), was determined from the fitted Voigt profiles using the following equation:

$$C_d = \left(\frac{T^1 + 2T^2 + 3T^3}{3} + \frac{Q^1 + 2Q^2 + 3Q^3 + 4Q^4}{4} \right) \times 100 \quad (1)$$

The degree of polycondensation determined for BPO0.01 was 78±1%, meaning that about 80% of the silicon atoms are bonded to other silicon atoms through Si–O–Si oxygen bridges. Similar C_d values were obtained for the CNT-containing hybrid BPO0.01_CNT_SDS, indicating that CNT loading did not affect the connectivity of the silica phase.

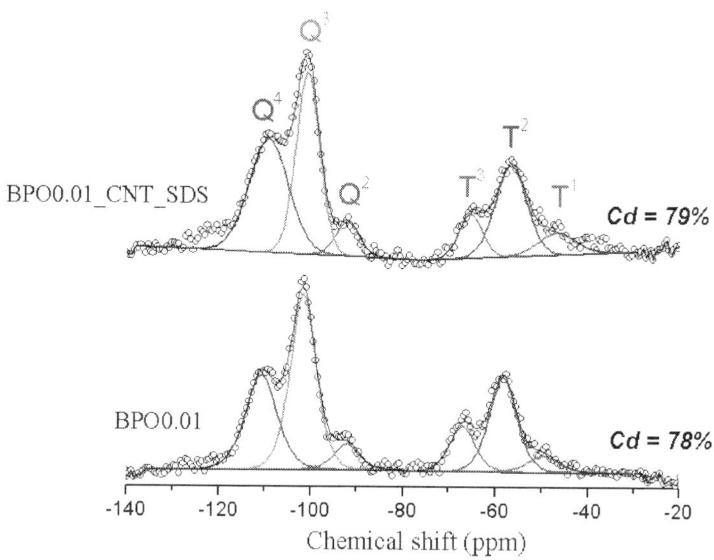

Figure 9. ^{29}Si-NMR spectrum from the BPO0.01 and BPO0.01_CNT_SDS samples.

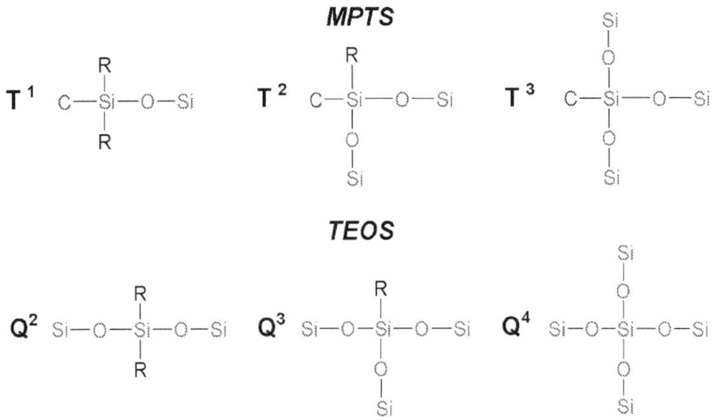

Figure 10. Schematic representation of T^i and Q^j structures. 'R' indicates OH or OCH_3 groups in MPTS or OCH_2CH_3 in TEOS.

3.2.3. *Thermal properties*

Thermogravimetry examines the overall connectivity of the hybrid network in terms of the thermal stability of the hybrid materials in different atmospheres. Under nitrogen, PMMA degrades in 3 events as the temperature increases: scission of head-to-head linkages at about 200°C (T_1), scission of vinylidene chain-ends at about 300°C (T_2) and finally random scissions of the polymer chains due to the rupture of head–tail segments at about 400°C (T_3) [23, 24]. The T_4 event at higher temperatures around 500°C is due to the dehydration of the remaining silanol groups of the silica network, detected in the XPS Si 2p spectra of **Figure 8** [23]. Thermogravimetric (TG) curves and their derivatives (differential thermogravimetry curves – DTG curves) for all the hybrid samples are shown in **Figure 11**. The onset temperature, T_0, which is a measure of the thermal stability of each material, is defined as the temperature at which a 5 % weight loss occurs. The temperatures of all events and the percentages of the silica and graphitic residues at 800°C are listed for the five hybrid samples in **Table 2**.

Comparing the DTG results obtained for the BPO0.01 and BPO0.05 matrices (**Figure 11b**), it is apparent that the higher quantity of BPO leads to an increase in the degree of polymerization and thus to a higher thermal stability, so that the low-temperature events seen in BPO0.01 are suppressed in BPO0.05. It can be seen from the data in Table 2 that all five hybrids are stable up to 200°C, with sample BPO0.01_CNT_SDS having the highest thermal stability (220°C). Furthermore, considering that the T_4 peak related to the dehydration of silanol is almost constant and that all loaded samples have the same CNT concentration, the residue formed mainly of pure silica (SiO_2) and some remaining graphitic phase can be used to estimate the fraction of the organic phase. The observation that hybrids prepared at the lower BPO to MMA ratio of BPO0.01 have about 4% higher amount of residue than those synthesized at the higher ratio of 0.05 is consistent with an increase in polymerization and a higher fraction of the polymeric phase in the BPO0.05-based samples.

It is also interesting to note that the addition of the CNTs to BPO0.01 enhanced its thermal stability. In addition to the 15°C increase in T_0, the first two depolymerization events (T_1 and T_2) shifted by about 50°C to higher temperatures (Table 2). The T_3 disintegration event increased by 20°C. This result is similar to that found by Jin et al. [25], in which T_3 was shifted upwards by 30°C for a PMMA matrix containing 26 wt.% of CNTs, a concentration significantly higher than reported in this work. The retardation effect was attributed by Jin et al. to interactions between the carbon nanostructure and macroradicals generated during the depolymerization, as suggested by Troitskii et al. [26]. In contrast, the BPO0.05 matrix shows values of the thermal degradation events almost unchanged by the presence of CNTs. Compared to the BPO0.01 matrix, this behavior can be understood in terms of a more stable structure of the BPO0.05 matrix induced by the higher degree of polymerization.

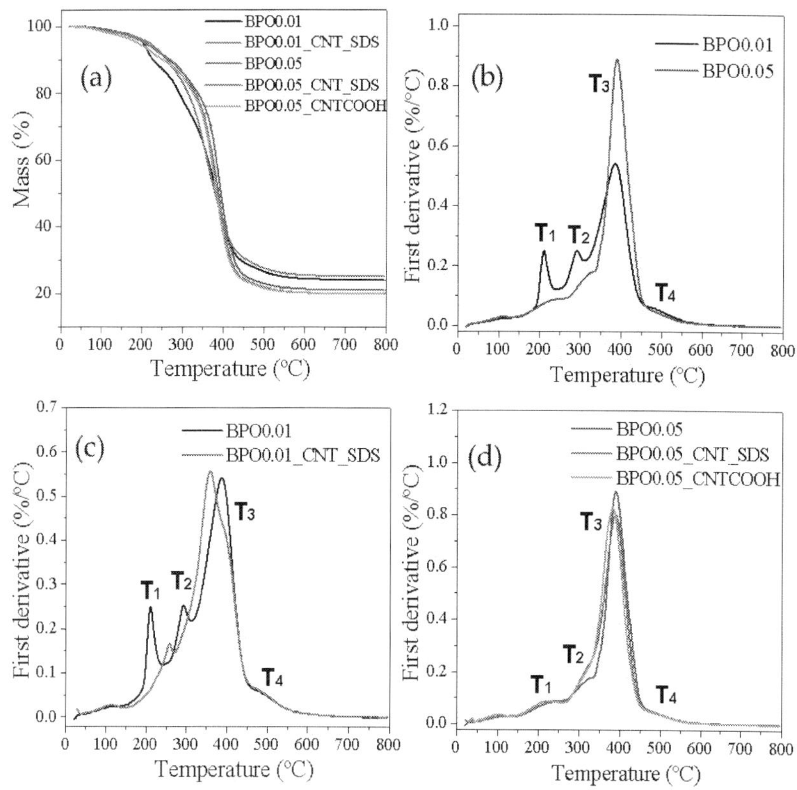

Figure 11. (a) TG curves and (b–d) DTG curves of the five hybrid samples.

Hybrid	T_0 (°C)	T_1 (°C)	T_2 (°C)	T_3 (°C)	Residue (%)
BPO0.01	205	210	290	385	24.0
BPO0.01_CNT_SDS	220	255	355	390	25.1
BPO0.05	208	220	310	390	21.1
BPO0.05_CNT_SDS	209	220	310	390	19.8
BPO0.05_CNTCOOH	192	220	310	385	20.1

T_0: Temperature of 5% weight loss, and temperatures T_1 of the first, T_2 of the second and T_3 of the third degradation event.

Table 2. The characteristic temperatures of the degradation events of PMMA–silica hybrid and the residue percentage obtained by thermal analysis in nitrogen atmosphere.

Our work and a number of other studies investigating the reinforcement effects by CNTs in diverse organic and hybrid matrices all come to the same conclusion: the modification improves the thermal stability of the composite. Thus, for example, in a recent study, Sabet et al. [27] used 5 wt.% of multiwall CNTs (MWCNTs) functionalized with carboxylic groups to reinforce an organic–inorganic hybrid matrix based on polyhedral oligomeric silsesquioxane (POSS). The approach involved a covalent conjugation between the CNTs and the POSS molecules through amide bonds. These authors observed that the decomposition under nitrogen of the neat POSS started at 265°C, and a complete weight loss was observed at 500°C. The POSS–MWCNT composite exhibited a fairly stable thermal behavior from room temperature to 200°C, and only 40% of weight loss by 1000°C measured by TGA.

Zhang et al. [28] coated functionalized MWCNTs with silica nanospheres, and subsequently introduced these into PMMA at a loading level of 4.28 wt.% to make it more flame resistant. TGA of the resultant PMMA/silica/MWCNT nanocomposites heated at 20°C min^{-1} indicated that the MWCNT/silica combination not only increased the temperature indicating 5% weight loss from 300°C of PMMA to 343°C, but also the temperature of the maximum rate of degradation increased from 338°C for PMMA to 387°C for the nanocomposites. These results, supported by cone calorimeter tests and scanning electron microscopy, showed that the MWCNT/silica combination introduced into the PMMA noticeably improved the thermal stability and flame retardancy of PMMA by in effect forming a surface thermal barrier layer during burning which helped to protect the underlying bulk from exposure to the external heat source.

Fraser et al. performed an in situ polymerization of PMMA in the presence of a low 0.1 wt.% concentration of either raw single-wall CNTs (SWCNTs) or acid-treated SWCNTs [29]. Although these transparent nanocomposites had slightly lower temperatures at which 10% weight loss had occurred in TGA when heated at 10°C min^{-1} in comparison with commercial PMMA, the temperature corresponding to the maximum rate of degradation increased by 8°C for the composites with the raw SWCNTs and by 18°C for the composites with acid-treated SWCNTs. Interestingly, these authors also showed using Raman spectroscopy that acid treating of the SWCNTs enabled them to bind covalently with the PMMA, rather than merely be in contact with it, as was the case with the raw SWCNTs.

The thermal analysis results obtained by Xiong et al. for polyurethane (PU) covalently linked with 2 wt.% of amino-functionalized MWCNTs indicate that the temperature at which the maximum rate of degradation occurs increased from 408°C for PU to 419°C for the composite in TGA experiments with heating rates of 20°C min^{-1}, once again indicating an improvement of thermal stability of a polymer matrix with the addition of CNTs [30]. Another TGA study, this time with a heating rate of 10°C min^{-1} on PU nanocomposite coatings modified with 5 wt. % MWCNTs, showed that the temperature at which there was complete decomposition of the matrix PU increased by 21°C with the introduction of the MWCNTs [31]. This was explained by the authors in terms of the relatively inert MWCNTs retarding the free movement of the PU chains.

Overall, it can be concluded that the improvement of the thermal stability of CNT-modified polymers and hybrid nanocomposites, reported by a number of laboratories, can be attributed to a variety of factors, all of which are related to the intrinsic thermal stability of CNTs, the effects of radical scavenging and by forming a physical barrier making it difficult for volatile products in the matrix to escape from the bulk.

3.2.4. *Mechanical properties*

Nanoindentation curves provide continuous values of Young's modulus and hardness during loading as a function of displacement (e.g., **Figure 12**). The average Young's modulus and hardness values and the corresponding standard deviations and coefficients of variation for the five nanocomposites, determined from displacements between 60 and 120 nm, are summarized in **Table 3**.

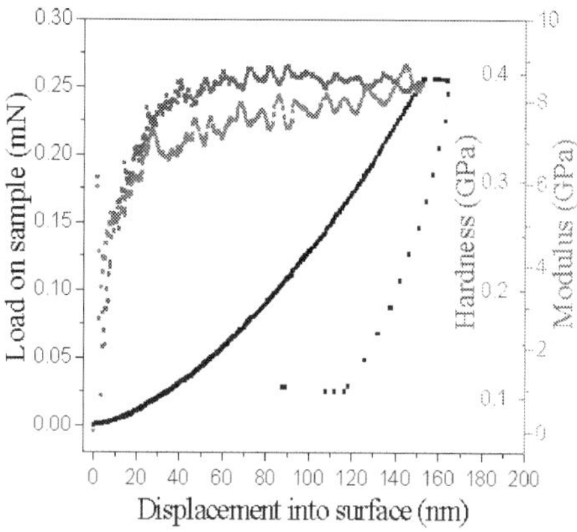

Figure 12. Representative loading and unloading nanoindentation curves for sample BPO0.01 from which hardness and Young´s modulus values were obtained.

Nanoindentation results show a coating hardness between 0.38 ± 0.4 GPa and 0.49 ± 0.6 GPa for all samples, values about twice as high as those for PMMA (0.22–0.26 GPa), but, not surprisingly, significantly lower than amorphous SiO_2 (7–9 GPa) [32]. Young's modulus values were in the range between 6.6 ± 0.3 GPa and 7.7 ± 0.1 GPa, about three times higher than pure PMMA (2.24–3.24 GPa), but about one order of magnitude lower than the elastic modulus of silicon oxide (73 GPa). These values represent a significant improvement of hardness and stiffness compared to pure acrylic, despite the presence of more than 70% polymethacrylate groups in the hybrid [33]. The inclusion of CNTs only produced a significant increase in hardness for the BPO0.05_CNT_SDS coating, with an increase of some 20% in comparison with the hardness of the BPO0.05 reference sample.

Sample	Young's modulus (GPa)			Hardness (GPa)		
	Mean	Std. dev.	% COV	Mean	Std. dev.	% COV
BPO0.01	7.849	0.398	5.07	0.400	0.042	10.41
BPO0.01_CNT_SDS	6.597	0.346	5.25	0.379	0.036	9.58
BPO0.05	7.749	0.137	1.77	0.413	0.012	2.83
BPO0.05_CNT_SDS	7.718	0.454	5.88	0.491	0.059	12.07
BPO0.05_CNTCOOH	7.501	0.154	2.06	0.431	0.014	3.23

Table 3. Young's modulus and hardness values, determined from indentations between 60 and 120 nm deep.

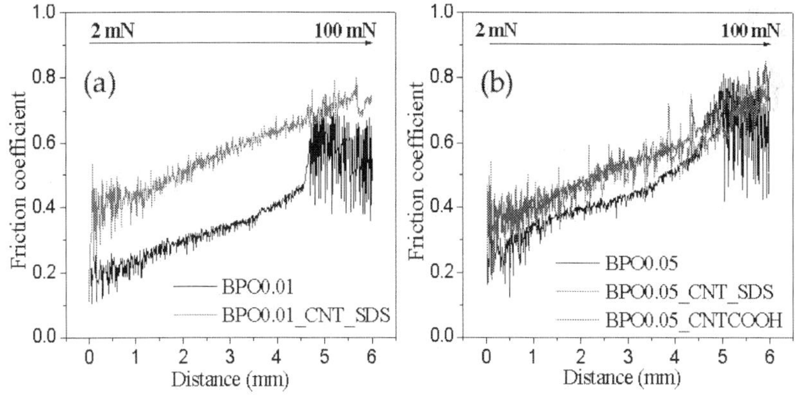

Figure 13. Microscratch curves (a) for BPO0.01 matrix coatings and (b) for BPO0.05 matrix coatings.

Scratch testing is a widely used, fast and effective method to provide information on the level of adhesion, resistance to scratching, the mechanism of fracture, the coefficient of friction and the wear characteristics of coatings. In a typical experiment, a coating is scratched with increasing normal force using a diamond stylus. The track is then analyzed by optical or electron microscopy to determine the mechanism of mechanical failure, such as coating detachment (loss of adhesion to the substrate), cracking and plastic deformation. The scratch test provides the coefficient of friction, defined as the ratio of the applied load and the normal load, directly.

Curves of the coefficient of friction as a function of scratch distance (microscratch curves) for all hybrid coatings are shown in **Figure 13**. The increase of the friction coefficient is associated with an increase in scratch resistance (friction force), while the critical load for delamination is related to the difficulty in breaking the adhesive interaction between the coating and the metal substrates. The scratch tracks, shown in **Figure 14**, were analyzed by optical microscopy to determine the failure mechanism and the critical load for film cracking and delamination.

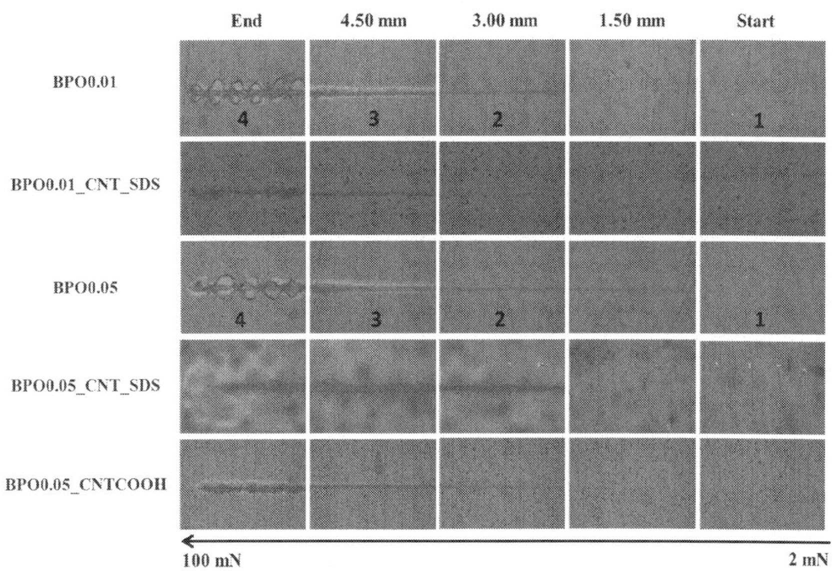

Figure 14. Optical microscopy of the five hybrid coatings deposited on A1020 carbon steel after scratch testing.

As is evident from **Figure 14**, the PMMA–silica reference samples (BPO0.01 and BPO0.05) were the softest coatings, showing four deformation stages with increasing force: (1) elastic deformation, (2) plastic deformation, (3) cracks and (4) delamination. The critical loads for delamination were 78 mN for BPO0.01 and 84 mN for BPO0.05, marked by the start of strong noise on the microscratch curves shown in **Figure 13**. The hybrids containing CNTs showed a higher scratch resistance and better adhesion to the A1020 steel substrate. Most interestingly, the BPO0.01_CNT_SDS coating showed an extreme reinforcement effect, with a higher coefficient of friction than carbon steel (0.5) [34] and no delamination up to a load of 240 mN (**Figure 15a**), the maximum load capacity of the equipment. BPO0.05_CNT_SDS and BPO0.05_CNTCOOH had critical loads for delamination of 133 mN (**Figure 15b**) and 122 mN (**Figure 15c**), respectively, both are higher than those obtained for the BPO0.05 matrix. These results confirm that the intrinsic mechanical properties of CNTs, i.e., the high elastic modulus (~1.4 TPa) and the high strength (50–500 GPa), contribute to a significant reinforcement of the hybrid [35]. The strong adhesion of the hybrid film to the carbon steel is a consequence of the covalent interaction between the hydroxyl groups of the substrate and the silanol groups of the inorganic part of the hybrid. The increased mechanical strength of the hybrids, induced by the incorporation of CNTs, increases the critical force for delamination, thus extending the functionality of protective PMMA–silica coatings to conditions where abrasive forces act in an aggressive environment, such as in reactors for the acidic processing of sugar cane, for example.

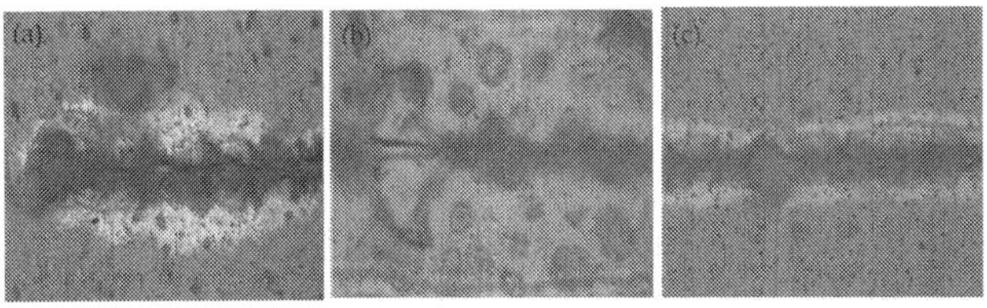

Figure 15. Optical microscopy of (a) the BPO0.01_CNT_SDS coating after scratch testing to a load of 240 mN, (b) the BPO0.05_CNT_SDS coating after scratch testing to a load of 133 mN and (c) the BPO0.05_CNTCOOH coating after scratch testing to a load of 122 mN.

Other studies have also examined the effect of CNT incorporation on the mechanical properties of organic and hybrid coatings. At the low SWCNT loading studied by Fraser et al. in their PMMA–SWCNT composites, no clear benefit was seen in the tensile properties, although there was some evidence to suggest that composites with acid-treated SWCNTs had improved impact strength in comparison with pure PMMA [29].

In their recent study on epoxy–CNT composite coatings deposited on 2024-T3 aluminum alloy substrates, Khun et al. were able to conclude that 0.5 wt.% CNT loading clearly produced composites with improved adhesion to the substrates and improved wear resistance relative to the unloaded epoxy coatings [12]. This improvement was explained in terms of a relaxation of the residual stress within the epoxy coating caused by the incorporation of the CNTs.

Kumar and Gasem have been able to demonstrate the beneficial effects of incorporating 2 wt.% of functionalized MWCNTs into polyaniline (PANI) coatings deposited on mild steel by dip coating [36]. The PANI–MWCNT coatings showed a Vickers micro hardness of 385 HV, compared with 266 HV for pure PANI coatings. Furthermore, PANI–MWCNT coatings showed significantly improved resistance to scratching in comparison with pure PANI coatings.

Despite the results obtained by Fraser et al. [29], the clear trend in the results of the work reported here and elsewhere is that controlled incorporation of CNTs into hybrid and polymer matrices at suitable levels is likely to be beneficial to the matrices in terms of improved mechanical performance. Thus, it is reasonable to expect significant improvements in terms of scratch and wear resistance, adhesion strength and also hardness and Young's modulus of the matrices when incorporating CNTs. For the most part, this can be attributed to the excellent mechanical properties of the CNTs. However, more work is clearly required to understand fully the mechanism responsible for the beneficial effect of incorporating CNTs on the adhesion of these matrices to metallic substrates.

3.2.5. *Anticorrosive properties*

The corrosion protection efficiency of the pure and the modified hybrid coatings was determined by EIS, performed in an electrochemical cell containing aqueous 3.5% NaCl solution at 25°C. The principle of EIS is to impose a small sinusoidal potential with varying frequency and, by measuring the alternating current response, to obtain the impedance of the electrochemical system. The impedance $Z(\omega)$ is composed of a real and an imaginary part, involving the ohmic and capacitive contributions, and can be represented as a vector of length $|Z|$, where $|Z| = (Z_{real}^2 + Z_{imag}^2)^{1/2}$. The angle between the Z vector and the Z_{real} axis is the phase angle ϕ [37]. For each measurement, three graphs were obtained: a Nyquist plot (Z_{real} vs. Z_{imag}) and two Bode graphs of the impedance modulus and phase angle as a function of frequency.

EIS measurements were performed for all hybrid-coated samples after one day of immersion and then at 1 week intervals until a significant drop of the impedance modulus occurred due to pitting. The time interval until the onset of pitting defined the lifetime of the coating. The Nyquist and Bode plots are shown in **Figure 16**, while the equivalent electrical circuits of the electrolyte–coating–substrate system used to fit the EIS data are shown in **Figure 17**. For comparison, EIS characteristics of bare A1020 carbon steel substrate were also recorded, as shown in **Figure 16**. The impedance modulus at low frequency and the phase angle behavior is an indicator of the anticorrosion performance. Coatings with modulus higher than $10^8\ \Omega$ cm^2 typically provide excellent protection, while those below $10^6\ \Omega\ cm^2$ have poor protection efficiency [38].

The BPO0.01 matrix had an initial impedance modulus of $10^8\ \Omega\ cm^2$ which remained unchanged during 56 days it survived testing, while the BPO0.05 matrix had a one order of magnitude higher impedance modulus of $10^9\ \Omega\ cm^2$, remaining stable for its lifetime of 21 days. This finding might be related to the higher overall connectivity of the BPO0.05 hybrid, which, on the basis of the results of NMR, TG/DTG and mechanical testing, had highly polymerized organic moieties densely interconnected with reticulated silica nodes. This excellent performance of the PMMA–silica hybrid coating can be compared with the performance of the bare carbon steel. The coated samples showed up to 5 orders of magnitude higher corrosion resistance, a consequence of the dense structure acting as an efficient diffusion barrier against aggressive agents. In this context, it should be noted that the anticorrosive performance reported for most hybrid coatings, in terms of initial impedance modules, is usually of the order of $10^7\ \Omega\ cm^2$ [5, 39, 40].

The addition of CNTs dispersed in SDS in the BPO0.01 matrix did not change neither the impedance modulus (10^8 Ω cm^2) significantly, nor the lifetime of the coating (43 days). The addition of functionalized and dispersed CNTs to the BPO0.05 matrix increased the impedance modulus, but decreased the lifetime of the coating to 7 days for BPO0.05_CNT_SDS and 10 days for BPO0.05_CNTCOOH. As has been suggested in recent studies [12, 13], CNTs act in the PMMA–silica nanocomposite as structural reinforcements and densifier agents, providing improved thermal and mechanical properties without degrading the chemical barrier characteristics. The extraordinary electrochemical performance of these micron thick films, approaching that of thick paints, is related to their dense hybrid structure, thus providing an efficient passivation of metallic surfaces [4]. Although the CNT-containing coatings were thicker than pure hybrid films, they actually showed shorter lifetimes in the NaCl solution. This can be explained in terms of electrolyte uptake via diffusion paths along the outer nanotube walls and also through the cavities of the nanotubes. Therefore, the degradation of the coatings after long-term exposure is associated with the penetration of the electrolyte, involving Cl^- ions, oxygen and water and subsequent chemical reaction (corrosion) at the coating/metal interface [38], causing a sudden drop of the electrochemical performance.

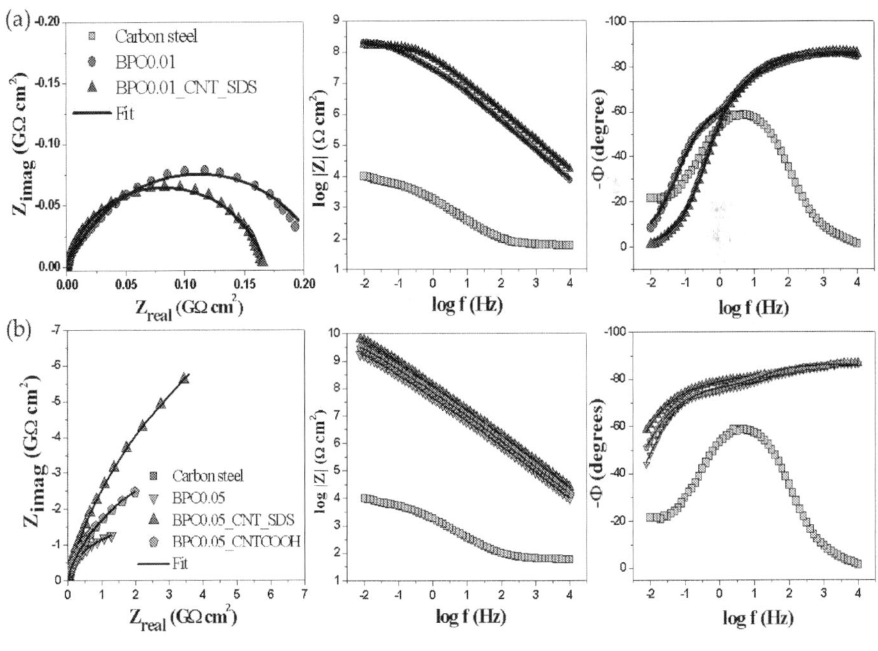

Figure 16. Nyquist and Bode plots for uncoated carbon steel (a) BPO0.01 matrix and (b) BPO0.05 matrix after 1 day of immersion in 3.5% NaCl standard saline solution.

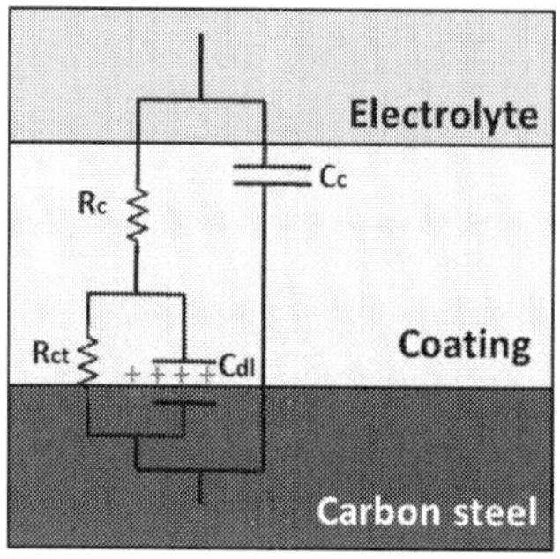

Figure 17. The electrical equivalent circuit used to fit all the EIS experimental data.

To obtain a quantitative model of the electrochemical system, the equivalent circuit of **Figure 17** was used to fit the impedance data of Figure 16. The circuit consists of two time constants, R_c/C_c at high frequency (~10^4 Hz) and R_{ct}/C_{dl} at low frequency (~1 Hz), where R_c is the coating resistance, C_c is the coating capacitance, R_{ct} is the charge transfer resistance and C_{dl} is the capacitance of the electric double layer of the coating/carbon steel interface [38]. As frequently applied for electrochemical systems, the capacitors were replaced by a constant phase element (CPE) to take into account the nonideality of the capacitor representing the coating, expressed by the n_c and n_{dl} exponents. All circuit parameters obtained by fitting the EIS data are shown in **Table 4**. All chi-square (χ^2) values were smaller than 10^{-3}, ensuring a high fit quality. Coatings showing a combination of high corrosion resistance and low phase angle values close to −90°, indicative of ideal capacitive behavior, are very efficient in blocking the electrolyte uptake. All hybrid films showed elevated volume and interface resistances and high-frequency phase angle values below −80° over four decades, all of which are characteristics expected for efficient protective coatings.

	BPO0.01		BPO0.01_CNT_SDS		BPO00.5		BPO005_CNT_SDS		BPO005_CNTCOOH	
χ^2	1.4×10^{-3}		3.0×10^{-4}		3.0×10^{-4}		2.3×10^{-4}		4.4×10^{-4}	
R_C (Ω cm^2)	2.8×10^7	(3.80)	1.3×10^7	(8.77)	1.5×10^7	(5.52)	1.0×10^7	(10.7)	1.5×10^7	(5.52)
CPE_C (Ω^{-1}cm^{-2}s^n)	3.8×10^{-9}	(1.00)	1.5×10^{-9}	(0.96)	2.7×10^{-9}	(0.77)	9.5×10^{-10}	(1.25)	2.7×10^{-9}	(0.77)
n_C	0.95	(0.13)	0.96	(0.10)	0.95	(0.09)	0.96	(0.13)	0.95	(0.09)
R_{ct} (Ω cm^2)	1.9×10^8	(1.32)	1.6×10^8	(1.09)	3.8×10^9	(1.23)	2.6×10^{10}	(1.74)	3.8×10^9	(1.23)
CPE_{dl} (Ω^{-1}cm^{-2}s^n)	7.3×10^{-9}	(1.52)	1.7×10^{-9}	(1.29)	2.2×10^{-9}	(0.91)	6.6×10^{-10}	(1.76)	2.2×10^{-9}	(0.91)
n_{dl}	0.74	(1.27)	0.65	(0.93)	0.70	(0.33)	0.70	(0.43)	0.70	(0.33)

*Values in brackets represent the errors in percentage.

Table 4. Equivalent circuit parameters for all samples after 1 day in 3.5% NaCl solution.

There are a number of related studies which have examined the electrochemical performance of CNT–hybrid and CNT–polymer nanocomposite coatings in contact with saline environments. Liu et al. modified bis-[triethoxysilylpropyl] tetrasulfide (BTESPT) silane films by MWCNTs functionalized with carboxylic groups [41]. EIS results, obtained for coated 304 stainless steel samples in 3.5 % NaCl solution, showed that the addition of different amounts of CNTs (0.5, 2.5 and 5 mg) improved the corrosion resistance relative to the pure BTESPT coatings. The impedance modulus of the BTESPT/MWCNT hybrid film reached about 10^7 Ω cm², a value about three orders of magnitude higher than the bare substrate and one order of magnitude higher than the pure BTESPT film. The authors suggested that carboxylated MWCNTs react with silanol groups from the silane precursor to form covalent bonds between BTESPT and the MWCNTs which strengthen the adhesion strength of the film, increasing its density and thus inhibiting the penetration of the corrosive ions. In addition, it was suggested by these authors that the long chain structure of the CNTs may be able to fill pits and cracks in the film, thus helping to reduce the number of corrosion-induced defects.

Jeon et al. reported EIS results combined with hygrothermal cyclic testing of their CNT-loaded epoxy coatings deposited on carbon steel [42]. In the hygrothermal testing, the temperature was ramped up from 25°C to 85°C and back to 25°C over a period of 12 hours while the coating was in contact with the electrolyte. This procedure was designed to accelerate the cumulative effect of the electrolyte on the coating/substrate interface. They found that after 30 hygrothermal test cycles the impedance modulus for pure epoxy coatings at 0.01 Hz decreased from 10^{10} Ω cm² to 10^6 Ω cm². For the two loadings of 0.25 and 0.5 wt.% that they examined for the CNT-containing samples, there was a lower initial resistance at 0.01 Hz because of the conductive nature of the CNTs, but after 30 cycles the impedance modulus at this frequency was higher than for the pure epoxy coatings. These results were explained by the authors in terms of the decrease in the water uptake of the CNT-loaded coatings, rather than any decrease in chemical attack by the 0.5 wt% NaCl solution, and the increase in adhesion strength of the coatings with the addition of the CNTs.

The anticorrosive performance of epoxy–CNT composite coatings was also reported for aluminum alloy 2024-T3 substrates, tested by EIS in a 0.5 M NaCl solution by Kuhn et al. [12]. Epoxy coatings and epoxy coatings containing 0.1 wt.% of MWCNTs had a impedance modulus of 3×10^4 Ω cm², slightly higher than bare Al, which had an impedance modulus of 10^4 Ω cm². On comparison, epoxy coatings containing 0.5 wt.% of MWCNTs had a higher impedance modulus of 6×10^4 Ω cm², explained by the authors in terms of the 3D dispersion of the MWCNTs in the epoxy matrix at this level contributing to a reduction in the porosity of the coating, and thus reducing its electrolyte uptake.

Deyab [43] studied the effect on the corrosion protection efficiency of coated carbon steel of different CNT concentrations from 0.2 to 0.5 wt.% in alkyd resin films used as the protective coatings. He found that all CNT-loaded films had an improved corrosion resistance in 3.5 wt. % NaCl saline solution relative to the pure alkyd resin films. These findings were attributed to the ability of the functionalized CNTs to absorb resin on their surfaces, thereby enhancing the density of the coatings, eliminating microflaws in the coating and making it more difficult for corrosive species to be transported through the coating to the underlying steel.

Finally, the corrosion protection performance of the PANI–MWCNT coatings studied by Kumar and Gasem demonstrated improved corrosion performance of the coating containing MWCNT in comparison with the pure PANI coatings [36]. This observation was also explained by the authors in terms of a reduced level of porosity in the PANI coating as a consequence of bonding between the PANI matrix and the functionalized CNTs, leading in turn to reduced permeability of coating to corrosive agents.

In comparison with protective coatings based on organic–CNT composites reported to date in the literature, the PMMA–silica–CNT films on carbon steel, discussed in the work, are by far the most effective corrosion protection barrier, with a much higher impedance modulus and lifetime in 3.5% NaCl solution.

Overall, the results obtained in this study have shown that a homogeneous dispersion of single-wall CNTs in PMMA–silica nanocomposites represents a novel and very promising coating system that is able to combine high anticorrosive performance with elevated thermal and mechanical stability, extending the application of these coatings to abrasive environments.

4. Conclusions

Single-wall CNTs have been dispersed through surfactant assistance and by functionalization with carboxylic groups. Both dispersion procedures have proved to be very effective in the modification of PMMA–silica hybrids, producing homogeneous and defect-free nanocomposite coatings with very smooth surfaces (R_{RMS} <0.5 nm) and thicknesses of 2–7 μm. ^{29}Si-NMR results showed that the addition of 500 ppm of CNTs into PMMA–silica hybrid matrices did not affect the high connectivity of the inorganic phase (~80%), while XPS results confirmed the nominal composition and the proportion of bonding environments forming the hybrid network. CNTs were effective in improving the thermal stability of the hybrids, increasing their onset temperature of degradation and shifting all depolymerization events to higher temperatures. Mechanical reinforcement of the hybrid coatings was achieved for both CNT dispersion methods, resulting in a significantly higher scratch resistance and improved adhesion of the coating to A1020 carbon steel substrate relative to the hybrid coatings without the CNTs. EIS results showed that the electrochemical performance of the CNT-loaded coatings is superior to most organic–CNT coating systems reported to date in the literature, being able to act for several weeks as efficient corrosion barriers in aggressive saline environments, maintaining an impedance modulus of up to 10^9 Ω cm^2. These results suggest that CNT-reinforced PMMA–silica nanocomposites have a great potential to extend the applicability of these environmentally compliant, high efficiency anticorrosive coatings to abrasive environments.

Acknowledgements

We would like to thank Dr John Nunn from the National Physical Laboratory, Teddington, U.K. for access to the microscratch equipment used in this work, and the Laboratory for Surface

Science (LCS) at the National Nanotechnology Laboratory (LNNano) in Campinas, Brazil for AFM access. This work was supported by CNPq, CAPES and FAPESP.

Author details

Samarah V. Harb[1], Fábio C. dos Santos[1], Sandra H. Pulcinelli[1], Celso V. Santilli[1], Kevin M. Knowles[2] and Peter Hammer[1*]

*Address all correspondence to: peter@iq.unesp.br

1 Instituto de Química, UNESP-UniversidadeEstadualPaulista, Araraquara, SP, Brazil

2 University of Cambridge, Department of Materials Science and Metallurgy, Cambridge, England

References

[1] Sanchez C., Belleville P., Popall M., Nicole L. Applications of advanced hybrid organic–inorganic nanomaterials: from laboratory to market. Chem Soc Rev. 2011;40(2):696–753. DOI: 10.1039/C0CS00136H.

[2] Molina E.F., Marçal L., Carvalho H.W.P.De, Nassar E.J., Ciuffi K.J. Tri-ureasil gel as a multifunctional organic–inorganic hybrid matrix. Polym Chem. 2013;4(5):1575–1582. DOI: 10.1039/c2py21049e.

[3] Hammer P., Santos F.C.Dos, Cerrutti B.M., Pulcinelli S.H., Santilli C.V. Highly corrosion resistant siloxane-polymethyl methacrylate hybrid coatings. J Sol–gel Sci Technol. 2012;63(2):266–274. DOI: 10.1007/s10971-011-2672-8

[4] Santos F.C.Dos, Harb S.V., Menu M., Turq V., Pulcinelli S.H., Santilli C.V., Hammer P. On the structure of high performance anticorrosive PMMA-siloxane-silica hybrid coatings. RSC Adv. 2015;5(129):106754–106763. DOI: 10.1039/c5ra20885h.

[5] Sarmento V.H.V., Schiavetto M.G., Hammer P., Benedetti A.V., Fugivara C.S., Suegama P.H., Pulcinelli S.H., Santilli C.V. Corrosion protection of stainless steel by polysiloxane hybrid coatings prepared using the sol–gel process. Surf Coat Technol. 2010;204(16–17):2689–2701. DOI: 10.1016/j.surfcoat.2010.02.022.

[6] Sanctis O. De, Gomez L., Pellegri N., Parodi C., Marajofsky A., Duran A. Protective glass coatings on metallic substrates. J Non Cryst Solids. 1990;121(1–3):338–343. DOI: 10.1016/0022-3093(90)90155-F.

[7] Hammer P., Santos F.C.Dos, Cerrutti B.M., Pulcinelli S.H., Santilli C.V. Carbon nanotube-reinforced siloxane-PMMA hybrid coatings with high corrosion resistance. Prog Org Coat. 2013;76(4):601–608. DOI: 10.1016/j.porgcoat.2012.11.015.

[8] Valentino O., Sarno M., Rainone N.G., Nobile M.R., Ciambelli P., Neitzert H.C., Simon G.P. Influence of the polymer structure and nanotube concentration on the conductivity and rheological properties of polyethylene/CNT composites. Phys E. 2008;40(7):2440–2445. DOI: 10.1016/j.physe.2008.02.001.

[9] Chu C.-C, White K.L., Liu P., Zhang X., Sue H.-J. Electrical conductivity and thermal stability of polypropylene containing well-dispersed multi-walled carbon nanotubes disentangled with exfoliated nanoplatelets. Carbon. 2012;50(12):4711–4721. DOI: 10.1016/j.carbon.2012.05.063.

[10] Gojny F.H., Wichmann M.H.G., Köpke U., Fiedler B., Schulte K. Carbon nanotube-reinforced epoxy-composites: enhanced stiffness and fracture toughness at low nanotube content. Compos Sci Technol. 2004;64(15):2363–2371. DOI: 10.1016/j.compscitech.2004.04.002.

[11] Pham G.V., Trinh A.T., To T.X.H., Nguyen T.D., Nguyen T.T., Nguyen X.H. Incorporation of Fe3O4/CNTs nanocomposite in an epoxy coating for corrosion protection of carbon steel. Adv Nat Sci: Nanosci Nanotechnol. 2014;5(3):035016–035022. DOI: 10.1088/2043-6262/5/3/035016.

[12] Khun N.W., Troconis B.C.R., Frankel G.S. Effects of carbon nanotube content on adhesion strength and wear and corrosion resistance of epoxy composite coatings on AA2024-T3. Prog Org Coat. 2014;77(1):72–80. DOI: 10.1016/j.porgcoat.2013.08.003.

[13] Ashassi-Sorkhabi H., Bagheri R., Rezaei-moghadam B. Sonoelectro-chemical synthesis of PPy-MWCNTs-Chitosan nanocomposite coatings: characterization and corrosion behavior. J Mater Eng Perform. 2015;24(1):385–392. DOI: 10.1007/s11665-014-1297-9.

[14] Zeybek B., Aksun E., Üğe A.. Investigation of corrosion protection performance of poly(N-methylpyrrole)-dodecylsulfate/multi-walled carbon nanotubes composite coatings on the stainless steel. Mater Chem Phys 2015;163:11–23. DOI: 10.1016/j.matchemphys.2015.06.022.

[15] Davoodi A., Honarbakhsh S., Farzi G.A. Evaluation of corrosion resistance of polypyrrole/functionalized multi-walled carbon nanotubes composite coatings on 60Cu–40Znbrass alloy. Prog Org Coat 2015;88:106–115. DOI: 10.1016/j.porgcoat.2015.06.018.

[16] Bailey B.M., Leterrier Y., Garcia S.J., Zwaag S.V.D., Michaud V. Electrically conductive self-healing polymer composite coatings. Prog Org Coat 2015;85:189–198. DOI: 10.1016/j.porgcoat.2015.04.013.

[17] Alves da Cunha J.R., Fantini C., Andrade N.F., Alcantara P., Saraiva G.D., Filho A.G.S., Terrones M., Santos M.C.Dos. Enhanced solubilization of carbon nanotubes in aqueous

suspensions of anionic–nonionic surfactant mixtures. J Phys Chem C. 2013;117(47): 25138–25145. DOI: 10.1021/jp4084663.

[18] Brinker C.J., Scherer G.W., editors. Sol–gel science: the physics and chemistry of sol–gel processing. San Diego, CA, Academic Press; 1990. 908 p. DOI: 10.1016/B978-0-08-057103-4.50002-7

[19] Tiwari A., Hihara L.H. Effect of inorganic constituent on nanomechanical and tribological properties of polymer, quasi-ceramic and hybrid coatings. Surf Coat Technol. 2012;206(22):4606–4618. DOI: 10.1016/j.surfcoat.2012.05.020.

[20] Briggs D., Seah M.P., editors. Practical surface analysis: auger and x-ray photoelectron spectroscopy. 2nd ed. Chichester: Wiley; 1990. 674 p.

[21] Naumkin A.V., Kraut-Vass A., Gaarenstroom S.W., Powell C.J. NIST x-ray photoelectron spectroscopy database: NIST standard reference database 20 [Internet]. [Updated: 2012. Available from: http://srdata.nist.gov/XPS/ [accessed: 01/2016].

[22] Sassi Z., Bureau J.C., Bakkali A. Spectroscopic study of TMOS-TMSM-MMA gels: previously identification of the networks inside the hybrid material. Vib Spectrosc. 2002;28(2):299–318. DOI: 10.1016/S0924-2031(02)00009-7.

[23] Wang Y.T., Chang T.C., Hong Y.S., Chen H.B. Effect of the interfacial structure on the thermal stability of poly(methyl methacrylate)-silica hybrids. Thermochim Acta. 2003;397(1–2):219–226. DOI: 10.1016/S0040-6031(02)00327-1.

[24] Kashiwagi T., Inaba A., Brown J.E., Hatada K., Kitayama T., Masuda E. Effects of weak linkages on the thermal and oxidative degradation of poly(methyl methacrylates). Macromolecules. 1986;19(8):2160–2168. DOI: 10.1021/ma00162a010.

[25] Jin Z., Pramoda K.P., Xu G., Goh S.H. Dynamic mechanical behavior of melt-processed multi-walled carbon nanotube/poly(methyl methacrylate) composites. Chem Phys Lett. 2001;337(1–3):43–47. DOI: 10.1016/S0009-2614(01)00186-5.

[26] Troitskii B.B., Troitskaya L.S., Yakhnov A.S., Lopatin M.O., Novikova M.A. Retardation of thermal degradation of PMMA and PVC by C60. Eur Polym J. 1997;33(10–12):1587–1590. DOI: 10.1016/S0014-3057(97)00048-7.

[27] Sabet S.M., Mahfuz H., Terentis A.C., Hashemi J., Boesl B. A facile approach to the synthesis of multi-walled carbon nanotube-polyhedral oligomeric silsesquioxane (POSS) nanohybrids. Mater Lett 2016;168:9–12. DOI: 10.1016/j.matlet.2015.12.149.

[28] Zhang T., Du Z., Zou W., Li H., Zhang C. The flame retardancy of blob-like multi-walled carbon nanotubes/silica nanospheres hybrids in poly (methyl methacrylate). Polym Degrad Stab. 2012;97(9):1716–1723. DOI: 10.1016/j.polymdegradstab.2012.06.014.

[29] Fraser R.A., Stoeffler K., Ashrafi B., Zhang Y., Simard B. Large-scale production of PMMA/SWCNT composites based on SWCNT modified with PMMA. ACS Appl Mater Interfaces. 2012;4(4):1990–1997. DOI: 10.1021/am201824k.

[30] Xiong J., Zheng Z., Qin X., Li M., Li H., Wang X. The thermal and mechanical properties of a polyurethane/multi-walled carbon nanotube composite. Carbon. 2006;44(13):2701–2707. DOI: 10.1016/j.carbon.2006.04.005.

[31] Wei H., Ding D., Wei S., Guo Z. Anticorrosive conductive polyurethane multiwalled carbon nanotube nanocomposites. J Mater Chem A. 2013;1(36):10805–10813. DOI: 10.1039/C3TA11966A.

[32] Mammeri F., Bourhis E.Le, Rozes L., Sanchez C. Elaboration and mechanical characterization of nanocomposites thin films: Part I: Determination of the mechanical properties of thin films prepared by in situ polymerisation of tetraethoxysilane in poly(methyl methacrylate). J Eur Ceram Soc. 2006;26(3):259–266. DOI: 10.1016/j.jeurceramsoc.2004.11.013.

[33] Callister W.D., Rethwisch D.G. Appendix B/Properties of Selected Engineering Materials. In: Callister W.D., Rethwisch D.G., editors. Fundamentals of materials science and engineering: an integrated approach. 4th ed. Hoboken: Wiley; 2012. pp. 835–836.

[34] Harb S.V., Cerrutti B.M., Pulcinelli S.H., Santilli C.V., Hammer P. Siloxane-PMMA hybrid anti-corrosion coatings reinforced by lignin. Surf Coat Technol 2015;275:9–16. DOI: 10.1016/j.surfcoat.2015.05.002.

[35] Eder D. Carbon nanotube-inorganic hybrids. Chem Rev. 2010;110(3):1348–1385. DOI: 10.1021/cr800433k.

[36] Kumar A.M., Gasem Z.M. Effect of functionalization of carbon nanotubes on mechanical and electrochemical behavior of polyaniline nanocomposite coatings. Surf Coat Technol 2015;276:416–423. DOI: 10.1016/j.surfcoat.2015.06.036.

[37] Macdonald J.R., Johnson W.B. Fundamentals of impedance spectroscopy. In: Barsoukov E., Macdonald J.R., editors. Impedance Spectroscopy: Theory, Experiment, and Applications. 2nd ed. Hoboken, NJ, USA: John Wiley & Sons; 2005. 1–26. DOI: 10.1002/0471716243.ch1.

[38] McIntyre J.M., Pham H.Q. Electrochemical impedance spectroscopy; a tool for organic coatings optimizations. Prog Org Coat. 1996;27(1–4):201–207. DOI: 10.1016/0300-9440(95)00532-3.

[39] Sakai R.T., Di F.M., Cruz L.Da, Melo H.G.De, Benedetti A.V., Santilli C.V., Suegama P.H. Electrochemical study of TEOS, TEOS/MPTS, MPTS/MMA and TEOS/MPTS/MMA films on tin coated steel in 3.5% NaCl solution. Prog Org Coat. 2012;74(2):288–301. DOI: 10.1016/j.porgcoat.2012.01.001.

[40] Asadi N., Naderi R., Saremi M., Arman S.Y., Fedel M., Deflorian F. Study of corrosion protection of mild steel by eco-friendly silane sol–gel coating. J Sol–gel Sci Technol. 2014;70(3):329–338. DOI: 10.1007/s10971-014-3286-8.

[41] Liu Y., Cao H., Yu Y., Chen S. Corrosion protection of silane coatings modified by carbon nanotubes on stainless steel. Int J Electrochem Sci 2015;10:3497–3509.

[42] Jeon H., Park J., Shon M. Corrosion protection by epoxy coating containing multi-walled carbon nanotubes. J Ind Eng Chem. 2013;19(3):849–853. DOI: 10.1016/j.jiec.2012.10.030.

[43] Deyab M.A. Effect of carbon nano-tubes on the corrosion resistance of alkyd coating immersed in sodium chloride solution. Prog Org Coat 2015;85:146–150. DOI: 10.1016/j.porgcoat.2015.04.003.

Chapter 8

Carbon Nanotubes and Graphene as Additives in 3D Printing

Steve F. A. Acquah, Branden E. Leonhardt,
Mesopotamia S. Nowotarski, James M. Magi,
Kaelynn A. Chambliss, Thaís E. S. Venzel,
Sagar D. Delekar and Lara A. Al-Hariri

Additional information is available at the end of the chapter

http://dx.doi.org/10.5772/63419

Abstract

3D printing is a revolutionary technology for the consumer and industrial markets. As the technology for 3D printing has expanded, the need for multi-materials that support fused deposition modeling and other forms of additive manufacturing is increasing. 3D printing filaments infused with carbon nanotubes and graphene are now commercially available, with the promise of producing conductive composites. This chapter explores some of the research, products, and challenges involved in bringing the next generation of functional printing materials to the consumer market.

Keywords: 3D printing, additive manufacturing, graphene, carbon nanotubes, carbon black

1. Introduction

Johannes Gutenberg catalyzed one of the first revolutions in education, with the development of the printing press, allowing the dissemination of information to take a more structured pathway. In the modern world, computers, smartphones, and the internet have all contributed to the advancements in science, technology, and engineering. 3D printing is another revolutionary technology for the consumer and industrial markets. Although the consumer market is still relatively young, the industrial sector has matured with its inception in the mid-1980s. The first 3D printer was created by Charles W. Hull of 3D Systems Corp. He published patents detailing many of the concepts involved in 3D Printing, some of which are used today. Developments in the various forms of printing have resulted in the use of the umbrella term of additive manufacturing. At its core, 3D printing simply uses additives to create 3D structures produced from successive layers of additives, such as polymers, and is deposited onto a platform. There are many companies that develop affordable consumer 3D printers such as Makerbot and Flashforge (**Figure 1**).

Figure 1. A Flashforge Creator 3D printer. This is a dual-extrusion printer with a build volume of 300 cubic inches.

The most recognizable form of 3D printing is thermoplastic extrusion, commonly known as fused deposition modelling (FDM) [1], where a filament (typically 1.75 mm) is unwound from a spool and fed into an extrusion assembly. The use of a motor-driven gear helps feed the filament into a temperature controlled melting chamber before the melted plastic is extruded through the tip of the nozzle. A fan is also used to help maintain the temperature as shown in **Figure 2**. The design of extruders continues to evolve, making printing more efficient by improvements in the regulating temperature across the extrusion assembly, and design modifications to reduce wear and blockages.

Polylactic acid (PLA), a sustainable biodegradable polymer, and acrylonitrile butadiene styrene (ABS), a strong terpolymer used in a range of manufactured products, are two of the most popular polymers used for FDM, and have formed the basis of the next generation of mixed material filaments for researchers. Other polymers used include polyvinyl alcohol (PVA), which is water soluble and biodegradable, and are support structures for 3D printing. Polyamides (nylons) are strong, abrasion resistant, and typically used for mechanical parts. All of these polymers have the potential for the inclusion of carbon nanostructured additives to extend their utility in multi-functional 3D printed composites.

There are many crowdfunded initiatives that are helping to further reduce the cost of entry level 3D printers and their accessories. Kickstarter, a world leading crowdfunding platform has seen a large interest in 3D printers providing a unique opportunity for both early adopters and industry specialists to evaluate cutting edge technology. 'Form 1: An affordable, professional 3D printer' is an example of a successful Kickstarter initiative with 2068 backers who pledged a total of $2,945,885 to bring the project to life. The printer retails at $2799 from Formlabs. Another example of a successful Kickstarter initiative was Printrbot which raised $830,827 with 1808 backers, and sold units for $199.

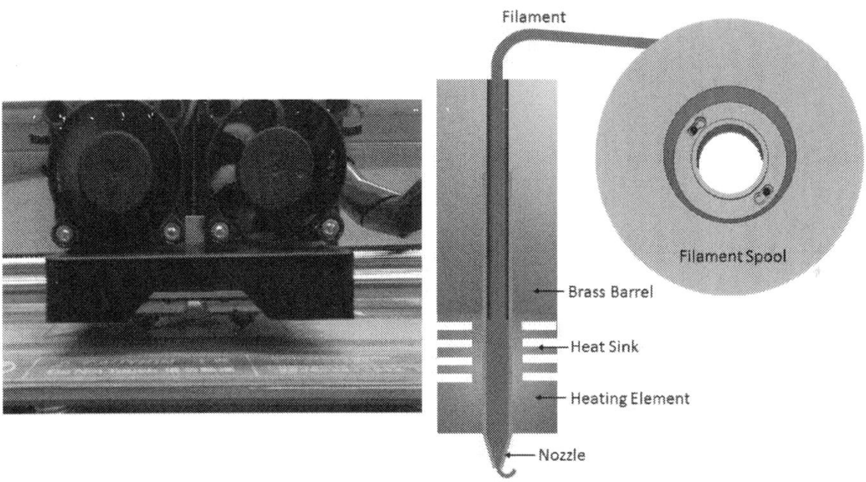

Figure 2. A picture of a dual extruder assembly from a FDM printer and a schematic of the extruder.

Selective laser sintering (SLS) is used to produce 3D components with a laser and powdered thermoplastics. During this process, a laser pulses onto a platform, creating a cross-section of the desired object on the powder. The laser is set to heat the powder just below its melting point causing it to fuse together. Fusing the powder in specific locations defined by the software creates the layers. An advantage of this method is the ability to process combinations of polymers and metals, and unused powder can be easily recycled. However, the particle size of the powder determines the precision of the print. This process has the potential for the incorporation of carbon nanotubes, but more research needs to be applied toward heat transfer processes [2].

Digital light processing (DLP) is another way of producing high resolution 3D objects. The process uses a vat of photopolymer with controlled exposure to the light from a DLP projector or an instrument using a digital micromirror device (DMD) [3]. The stage/build plate moves as the exposed liquid polymer hardens (**Figure 3**). Essentially, a 3D model on a computer is sliced into 2D cross-sectional layers to produce images that are sequentially sent to the DLP. The process continues until the 3D object is complete. Any remaining liquid photopolymer is then purged from the vat and the completed print removed.

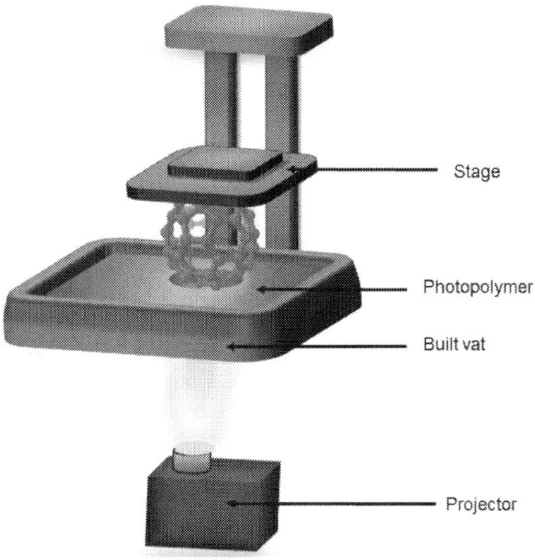

Figure 3. Diagram of a digital light processing printer.

Stereolithography is similar to DLP printing with the use of photopolymers, but differing with the laser based light source. Objects are produced by printed layers, one at a time, by the use of a laser beam on the surface of a vat of the liquid photopolymer. The polymer hardens over the areas where the laser strikes and a stage is moved down into the vat by one layer. Depending on the model of the printer, a recoater blade would move across the surface of the vat, helping to deposit the next layer of liquid polymer. This process is repeated until the print is completed and the printed object is raised from the vat and the liquid polymer drained. The printed object is then cleaned and cured in ultraviolet light to finish the processing. Stereolithography has been used with a range of materials including graphene oxide [4]. There are many potential applications that directly relate to the inclusion of carbon nanotubes and graphene as additives for 3D printing [5, 6] in a sector that is currently undergoing rapid growth. The 3D printing sector took over 20 years to reach $1 billion by 2009 and over the following three years it reached $2 billion. Analysis of the market suggests it may reach $10 billion by 2019 [7].

While there has been much praise for the 3D printing industry, there have also been some negative consequences on the environment and security. The process of thermal extrusion consumes a lot of electricity and the estimates of emission rates of ultra-fine particles were large for printing with PLA feedstock and even higher for printing with ABS thermoplastic feedstock [8]. Security has become an important issue after the fabrication of a 3D printed gun.

The US State Department is trying to implement a ban on the distribution of files for the 3D printing of guns; however, the nature of file distribution on the internet may make this almost

http://dx.doi.org/10.5772/63419

impossible to police. Initially only files for replacement gun parts were available, but a video soon demonstrated the successful operation of a printed gun.

2. Polymers in 3D printing

Additive manufacturing is developing to meet the demands for durable replacement parts, which need to conform to specific mechanical and design requirements. With thermal extrusion, new printing materials must have suitable rheological and thermal properties to be able to be extruded and solidify while maintaining the accuracy of successive layers. The use of polymers in additive manufacturing are governed by the glass transition temperature, melting point, heat capacity, melt viscosity at elevated temperatures, and the shear stress of the material. Understanding the relationship between the structural properties will aid in the development of advanced functional printing filaments and materials.

The glass transition temperature (Tg) is the temperature below which the polymer becomes brittle and hard like glass. The glass transition temperature is not the melting point, that is defined by the temperature at which crystalline or semi-crystalline polymers change from its crystal structure to its liquid form. Amorphous polymers have no sharp melting point due to their non-crystalline structure but they have a Tg temperature. Semi-crystalline polymers have both a Tg and melting temperature. Polymers that have a Tg below room temperature are elastic, while those with a Tg above room temperature tend to be rigid and brittle. The Tg is higher for polymers with stiff backbones and pendant groups that interact with nearby structures. Both of these features result in polymer chains requiring elevated temperatures to move. One example of a rigid backbone is a benzene ring as found in poly(p-phenylene) while a flexible backbone can be found with the CH_2-CH_2 bond in poly(ethylene) (**Figure 4**). A bulky branch may act as a plasticizer and lower the Tg by decreasing the packing of polymer chains. This can be seen with poly(methyl methacrylate) with a Tg = 100-120°C as compared to a Tg of 20°C for poly(butyl methacrylate), a difference of three additional carbons in the branch.

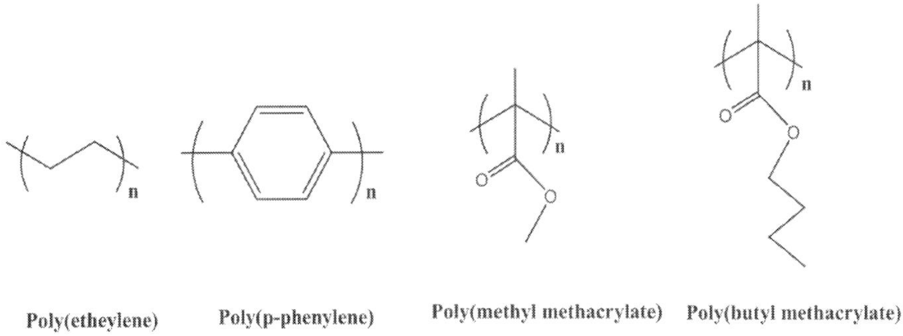

Figure 4. Structure of poly(ethylene), poly(p-phenylene), poly(methyl methacrylate) and poly(butyl methacrylate).

Amorphous polymers decrease in strength above the Tg and start to soften gradually while the viscosity changes.

The lower the Tg of the polymer, the faster it reaches the optimal viscosity for extrusion. Heat capacity also plays a role in homogeneity during the heating process. The Tg should be lower than the degradation temperature of the polymer, or in a temperature range where there are no chemical changes. The heat capacity of semi-crystalline polymers has a greater change compared to amorphous polymers. As a polymer is heated above its Tg, the viscosity will change based on the temperature and shear rate as it reaches the printing nozzle. The shear rate is higher at the printing nozzle (100–200 s^{-1}) compared to the other parts of the extrusion assembly.

The effect of viscosity is even more pronounced as the polymer is extruded from the nozzle as there is a possibility of radial flow. This phenomenon, typically known as swelling at the nozzle, can be quantified by calculating the ratio of the maximum diameter of the extruded material to the diameter of the nozzle. The addition of inelastic (ceramics) and carbon fillers to the polymer reduce swelling at the nozzle and results in an increase in the resolution of printing [9]. A sphere of polymer melt is then deposited on the substrate. The cross-sectional area of that sphere is proportional to the flow rate (q) of the polymer melt and inversely proportional to the velocity of the printing head, as shown in Eq. (**1**). The effect of the flow rate is shown in Eq. (**2**), where k is a constant depending on dimensions of the liquefier/melting chamber, ΔP is the change in pressure, and η is the viscosity of the polymer melt. The higher the viscosity, the lower the flow rate and cross-sectional area of the deposited sphere. A small cross-sectional area results in a higher resolution of the object or part printed.

$$A = \frac{q}{v_{printing\ head}} \tag{1}$$

$$q = \frac{\pi\ k\ \Delta P}{8\eta} \tag{2}$$

The interaction between the printed melt and the surface depends on the surface tension and roughness. The polymer melt should adhere to the substrate allowing it to be removed once the polymer solidifies. The next interaction is between the first deposited polymer layer and the next layer. This interaction will have a critical role in mechanical properties, structure, and success or failure of the object or parts. For bonding to occur at that level, the deposited polymer needs to reach or surpass its Tg. This can be achieved by the transfer of heat from the melt polymer to the deposited polymer layer.

2.1. Thermoplastic resurgence

Thermoplastic polymers like polyamide, polyolefin, polystyrene, polyester, and their co-polymers are favored in many applications from fabrics to packaging due to their excellent mechanical properties, durability, relative ease of processing, and the possibility of recycling. A thermoplastic is a polymer that softens upon heating above a temperature range and then solidifies upon cooling; this process can be repeated several times. At the molecular level, the polymer chains are associated via intermolecular Van der Waals forces that are easily broken at elevated temperatures. In contrast, thermoset polymers harden upon heating and are no longer moldable, decomposing at high temperatures. For 3D printing, thermoplastic polymers are the best candidates since they melt and mold during the extrusion process, and this has caused a resurgence in the production of these polymers with a focus on sustainability for the 3D printing industry [10]. Thermoset polymers are only used with their corresponding monomers, with an initiator added, as the printed materials are cured by ultraviolet light or heat during post processing.

The most common materials used in 3D printing (FDM in specific) are amorphous thermoplastic ter-acrylonitrile-butadiene-styrene (ABS), PLA (**Figure 5**), PVA, polycaprolactone (PCL), and nylon.

Figure 5. Commercially available PLA and ABS pellets.

ABS is a terpolymer that is made from acrylonitrile, 1-3-butadiene, and styrene (**Figure 6**). The ratio of each monomer in ABS can be selectively modified, based on the synthesis method used, to yield different grades of ABS with the required mechanical, thermal, and processing properties. The percentage of styrene can vary from 65–76% and acrylonitrile by 24–35%. ABS is a light-weight, rubber-toughened thermoplastic, with low temperature toughness, and is stronger than polystyrene. It adopts the rubbery properties from polybutadiene, the toughness of acrylonitrile, while maintaining the reflective property of polystyrene, which can be enhanced using acetone. ABS is chemically resistant to water, aqueous acids, and alkali solutions but reacts/dissolves in carbon tetrachloride, concentrated nitric acid, concentrated sulfuric acid, esters, and acetone. For 3D printing, ABS can be produced in a variety of colors by adding pigments as raw ABS is translucent. Due to the reactive double bond in the polybutadiene region of the terpolymer, it may be oxidized in ultraviolet light, so indoor applications are preferred. Commercially available ABS filaments sold for 3D printing are extruded at 220–235°C, below the degradation temperature (400°C) at which it decomposes into its carcinogenic monomers. Due to its amorphous nature, ABS has no specific melting point and glass transition temperature, but it does have a range of 80 to 125°C. ABS, extrusion grade, has a tensile strength of 30–55 MPa and an elastic modulus of 897–2898 MPa.

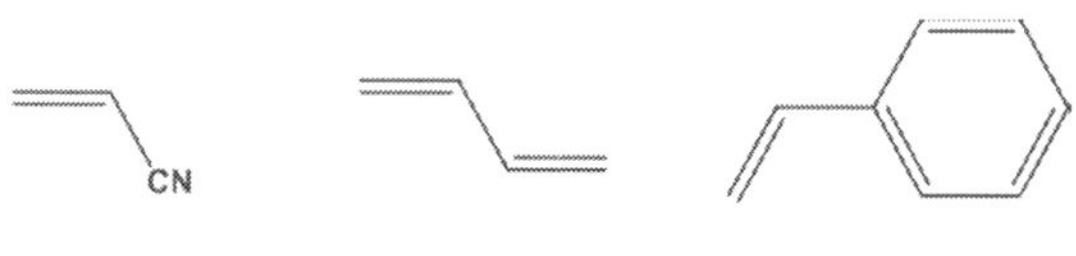

Figure 6. Monomers of ABS.

PLA (**Figure 7**) is a biodegradable and low toxicity polyester thermoplastic made from lactide or lactic acid monomers. Both can be derived from the fermentation of carbohydrates which is a renewable resource, so PLA is considered an ecofriendly thermoplastic. PLA can be produced in amorphous and crystalline form. The most common way to produce high molecular weight PLA is by the ring opening of lactide catalyzed by a metal. PLA produced by this method is a racemic mixture of both L and D PLA (stereocenter is labeled in **Figure 7**–red star). PLA is a hygroscopic thermoplastic; it can undergo degradation with elevated humidity and temperature. PLA has a glass transition temperature of 40–70°C and a melting point of 130–180°C. Various degrees of degradation at 210 and 240°C have been reported [11], and the rate of degradation was high at 240°C. Moisture has a strong effect on degradation at 210°C. PLA undergoes thermal degradation, scission of bonds, and results in weight loss which has an effect on mechanical and rheological properties [12, 13]. It is worth noting that the extent of degradation also depends on other factors ranging from molecular weight to particle size. Commercially available PLA filaments are printed at 180–220°C at which it can react with water, if present, and cause discoloration and degradation, which might affect the end product mechanically. Amorphous PLA is soluble in most organic solvents, while crystalline PLA is soluble at elevated temperatures. PLA has a high tensile strength of 48–52 MPa, a tensile modulus of 3500 MPa, and good heat salability but tends to be brittle.

PVA is a hydrophilic, water soluble, biodegradable thermoplastic polymer (**Figure 7**). PVA is synthesized by the hydrolysis or alcoholysis of poly(vinyl acetate) (PVAc) due to the instability of its vinyl alcohol monomer. It can be produced in two types: partially hydrolyzed or fully hydrolyzed and can reach up to 99% hydrolysis. It has been used in the medical and industrial sectors, as PVA is biocompatible due to its low toxicity and minimal cell adhesion to its surface. The molecular weight of PVA depends on the molecular weight of the PVAc and can vary from 20,000–400,000 g/mol which affects the properties of PVA. When PVA has a molecular weight of 75,000–81,000 g/mol and is 98–99% hydrolyzed, it has a tensile strength of 17 MPa, and a tensile modulus of 1470 MPa [14]. PVA with a molecular weight of 146,000–186,000 g/mol when 98–99% hydrolyzed has a tensile strength of 51 MPa and tensile modulus of 443 MPa. PVA being soluble in water makes it suitable as support material for printing complex objects by forming a support structure. Initially, PVA rafts can be printed or dual extrusion printing techniques can be used to fill voids or intricate fine details during the print. The finished object can then be immersed in water to dissolve the PVA and leave the other extruded polymer in place. Rubber-elastomer filaments made of PVA with another polymer, sold under name of "Lay-felt", are commercially available to produce micro-porous objects as end products by immersing printed objects in water.

PCL (**Figure 7**) is polymerized from a caprolactone (a five member ring cyclic ester) monomer by ring opening catalyzed by stannous octoate. PCL is hydrophobic and soluble in chloroform, carbon tetrachloride, cyclohexanone, benzene, and toluene. PCL biodegrades in the presence of microorganisms. It has a low glass transition temperature of -60°C [15], which makes it less brittle than other polymers with glass transition temperatures above room temperature. It also has a low melting point of 59–64°C with the advantage of low processing temperatures especially when mixed with other materials to form composites.

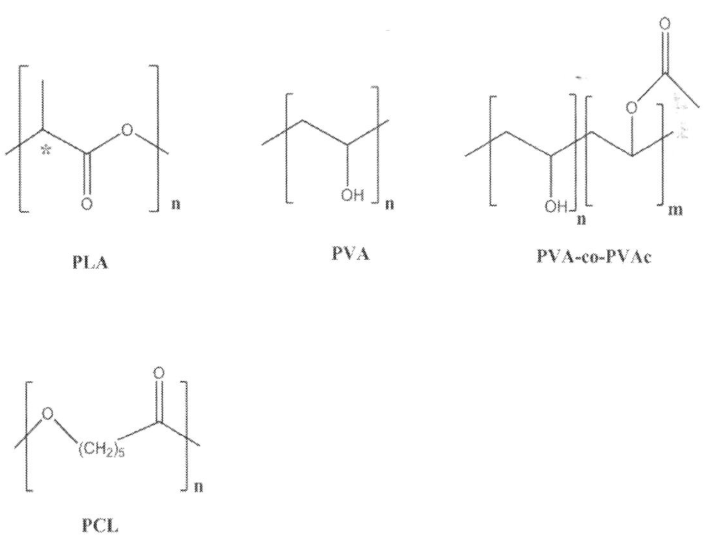

Figure 7. Structures of PLA, PVA, PVA-co-PVAc, and PCL.

3. Carbon nanotubes

Carbon nanotubes are known for their mechanical, electrical, and thermal properties, which initially makes them a suitable candidate to integrate into 3D printing polymers. MWCNTs have structural defects which provide suitable nucleation sites that allow for strong interactions with polymers [16] and for cross-linking and functionalization [17–19]. A homogeneous dispersion of carbon nanotubes in polymeric solutions are essential if they are to be used for enhanced CNT based filaments. The familiar problem of the aggregation of CNTs can be detrimental to FDM, possibly causing blockages at the nozzle and flux instability while printing, so research has focused on determining the concentration of CNTs that would surpass the percolation threshold (the transition between an insulating and conductive polymer) while maintaining the parameters for 3D printing.

The theoretical concentration of carbon nanotubes required to reach the electrical percolation threshold for a CNT/Polymer composite can be obtained, as a first step, by the use of the power law:

$$\sigma \propto (\varphi - \varphi_c)^{\alpha}$$

where σ is the electrical conductivity, φ is the MWCNT volume concentration in the nanocomposite, φ_c is the critical MWCNT volume concentration at electrical percolation, and α is a critical exponent [6]. A comprehensive table of percolation thresholds for CNTs in polymer matrices was assembled by Bauhofer et al. [20]. It was also noted by Bauhofer that there were conflicting results concerning the dependence of the percolation threshold on the aspect ratio. The excluded volume analysis conducted by Celzard et al. showed that the percolation threshold of a fiber suspension would decrease when the aspect ratio increased [21]. Research by Bai et al. demonstrated a decreasing set of values for the percolation threshold when the CNT length increased [22]; however, Martin et al. [23] found an increasing percolation threshold when the CNT length increased. This problem can be reconciled by considering the type of percolation thresholds. Bai et al. possibly obtained statistical thresholds while Martin et al. acquired kinetic percolation. This is important to know as theoretical analyses reported tend to ignore the movement of filler particles and only seem to predict the dependence of the statistical percolation threshold on the filler aspect ratio.

Ultrahigh molecular weight polyethylene (UHMWPE) has been processed with MWCNTs and extruded to produce filaments [24]. However, it is not feasible to produce UHMWPE filaments by conventional extrusion due to the high Weissenberg effects [25] that would likely affect the flux during extrusion both for filament production and 3D printing. An alternative would be to use the gel-spinning technique [26] in which the polymer is dissolved and spun. The addition of MWCNTs can be problematic as they tend to aggregate during the solvent evaporation stage, but the problems were addressed by using a range of techniques including sonication, melt mixing, and extrusion [24].

The compatibilization of a polymer is simply the addition of a material to immiscible blends of polymers resulting in an increase in stability. Compatibilized polyolefin rubber is an example of a composite that can be made conductive by the addition of carbon nanotubes [27] or graphene. The percolation threshold has been widely studied with carbon nanotube systems [28, 29]. The study of PC, ABS, and MWCNT composites confirmed that localization of MWCNTs changes from the ABS to the PC phase when the rubber content was reduced from 60 to 5% [30]. At low concentrations of rubber, MWCNTs localized in the PC phase resulting in an increase in conductivity and a low percolation threshold of around 0.5–1 wt% [28].

Ternary systems could be an important framework for developing advanced 3D printing materials. With the ternary mixture of PCL, PLA, and MWCNT, it has been shown that the localization of carboxylic functionalized nanotubes can be identified at the PCL phase and at the phase interface. When the MWCNTs are not functionalized, they can only be located at the PCL phase resulting in an elevated percolation threshold. Ternary composites exhibit conductivities that are 3–4 orders higher than binary composites when the MWCNT content reaches 1 wt% [31]. Pötschke et al. developed a method to mix MWCNTs into thermoplastic matrices of PC and polyamide-6 (PA6) by melt blending the PE based masterbatch with high MWCNT loading. This improved the CNT dispersion in PC and PA6 and also reduced the percolation threshold [32]. In general, there are many strategies that can be applied to the mixing of CNTs with polymers.

Postiglione et al. reported the assembly of conductive 3D structures using a PLA/MWCNT nanocomposite using liquid deposition modeling (LDM) with dichloromethane (DCM) [6]. They reported a percolation threshold concentration of 0.67% with a conductivity of 10 S/m, and the highest conductivity was obtained with 5 wt% MWCNT with 100 S/m. Postiglione found that at a composition of 35% PLA in DCM with 1 wt% MWCNT, the rheological effects prevent the extrusion through the 3D printer as a higher pressure from the extrusion assembly would be required. Nanocomposites made from 25 wt% of PLA in DCM with 1 wt% MWCNTs can be printed out at a lower shear stress with a higher shear rate (1–50 s^{-1}) when compared to 30 wt% PLA in DCM with 1 wt% MWCNTs (shear rates 5–11 s^{-1}). Although the 25 wt% PLA/CNT mixture prints at a higher speed, a better resolution product was obtained using a 30 wt% PLA/CNT mixture at $10s^{-1}$ and a low speed of 0.1 mm s^{-1}.

Grafting of single wall carbon nanotubes SWCNTs to poly(L-lactic acid) has been investigated [33] but so far MWCNTs are dominant, possibly due to their increased metallic character. However, Vatani et al. reported the fabrication of a highly stretchable sensor by dispersing 1 wt% SWCNT (average diameter 1.5 nm, length 1–5 micrometer) in a matrix of a blend of two photocurable monomers (cyclic trimethylolpropane formal acrylate and acrylate ester) [34]. The monomers/SWCNT composite was printed using direct writing into a polyurethane substrate on which the monomers were photo cured. The wires sustained strain up to 90% elongation and resistivity change increased proportionally with the strain.

3.1. Delivering on the hype

It is well known that the advent of carbon nanotubes caused a flurry of interest with researchers over how these nanoscale tubes could change the way we live. While it is easy to lead research by conjecture, once a plateau is reached, the refinement of the field begins. What we are seeing now is the development of stronger and highly efficient carbon nanotube based composites that, without the fanfare, are making their way into products.

An article in Chemical and Engineering News [35] highlighted the difficulty in working on the field of carbon nanotubes. Phaedon Avouris, a member of the American Academy of Arts and Sciences and an expert in the field of carbon nanotubes and graphene said when scientists work on new materials, there is a rush to characterize it, publish papers in prominent journals, and then move on to a different material, "We're left with a lot of unfinished work and unproven claims". Although a fundamental understanding of the research topic may be achieved, few people are determined to solve the problems that facilitate the transition to applications.

Production capacity of multi-walled tubes has peaked with Timesnano and Showa Denko, each producing over 2500 metric tons every year, but demand is still growing with conductive adhesives and fire retardant plastics leading the requests for applications [35]. In 2015, over 400 metric tons of multi-walled carbon nanotubes were used for conductive polymer composites with estimates reaching 1700 metric tons by the year 2020. It is possible that this may increase further as carbon nanotubes continue to transition into additive manufacturing.

Mihail Roco, the senior adviser for nanotechnology at the National Science Foundation (NSF) said that carbon nanotubes form a component of nanotechnology. "Components are essential in advancing nanoscience, but they're not the end goal in applications." The NSF intends to advance toward the goal of building nanotech systems that provide answers to problems in industry and other areas. Carbon nanotubes have steered researchers toward the development of prototypes, and will continue to promote innovation over the next decade. Roco also said, "Carbon nanotubes focused attention on understanding matter at the nanoscale, on making new tools, on performance, on how to create groups that could cross disciplines". This shows that beyond the hype, the legacy of carbon nanotubes for nanotechnology may prove to be an important milestone.

3.2. Commercial carbon nanotube materials for 3D printing

3DXTech is a company that provides filaments containing carbon nanotubes. Their 3DXNano™ ESD ABS filaments (containing MWCNTs) are available with diameters of 1.75 mm and 2.85 mm. The filament is tailored toward applications that require electrostatic discharge (ESD). The filament is produced using MG-94 Premium ABS and mixed with MWCNTs, and process/dispersion modifiers. 3DXTech state that the Tensile strength is 42 MPa in comparison to 41 MPa for the unfilled ABS. The surface resistance for '3DXNano™ ESD ABS' is 10^{7}–10^{9} Ω. An extrusion temperature of 220–240°C with a FDM platform temperature of 100–110°C is also suggested.

Nanocyl are one of the worldwide leading experts in CNT based materials, producing research and industry grade carbon nanotubes. One of their product lines, PLASTICYL™, is a collection of carbon nanotubes thermoplastic concentrates for applications requiring electrical conductivity with good mechanical properties. The concentrates contain 10–20% of carbon nanotubes

http://dx.doi.org/10.5772/63419

and are available in a diverse range of thermoplastic resins, including PC, PP, PA, PET, HDPE, and others. Although these enhanced thermoplastics were not specifically aimed at the FDM sector, they have a formulation that makes them applicable, subject to the temperature range of the extruder. PLASTICYL™ can be used in many applications and a surface resistivity range of 1–10^{12} Ω, and the typical loading for static dissipative applications are around 2–3% of CNTs in the final compound. The conductivity can be tailored for a given loading of CNTs, depending on the compounding conditions, and the viscosity of the basic resin. Nanocyl optimizes the dispersion of CNTs in a wide range of thermoplastics with the PLASTICYL™ range (www.nanocyl.com).

F-Electric is a PLA-based filament produced by 'Functionalize' that incorporates carbon nanotubes. F-Electric is one of the best conductive 3D filaments available on the market with a 0.75 Ω•cm in volume resistivity. **Table 1** shows some of the commercially available carbon nanotube based materials.

Company	Product
3DXTech	3DXNano™ ESD ABS + Carbon Nanotube Filament
	3DXNano™ ESD PETG + Carbon Nanotube Filament
Functionalize F-Electric	Functionalize F-Electric filament (PLA & Carbon Nanotube)
Filabot	MWCNT1 Multi Walled Carbon Nanotube Pellets
Cheap Tubes Inc.	Carbon Nanotube Masterbatches CNT-ABS-10
Nanocyl	PLASTICYL™ ABS 1501*
	PLASTICYL™ HIPS 1001
	PLASTICYL™ PC 1501
	PLASTICYL™ PP 2001

Table 1. Companies that provide filaments and pellets containing carbon nanotubes.

3.3. Filament production

Consumer 3D printers are perfect for creating replacement parts and tools, and supporting learning from K-12 to higher education, but the process can produce a lot of waste from failed prints, supports, and rafts. Recycling these materials can be highly effective in reducing the costs of printing. Crowdfunded initiatives have been the main driving force in producing filament reclaimers with products such as the Filastruder (filastruder.com) and the Filabot (filabot.com) range of extruders and accessories. The process of reusing filaments is further complicated by the need to break down the waste so that it can be successfully channeled through a feed screw through to the melting chamber (surrounded by a heater) as shown in **Figure 8**. Extruded filaments can be collected on automated systems such as the Filabot Spooler, allowing the spool to be detached and connected to a 3D printer when completed.

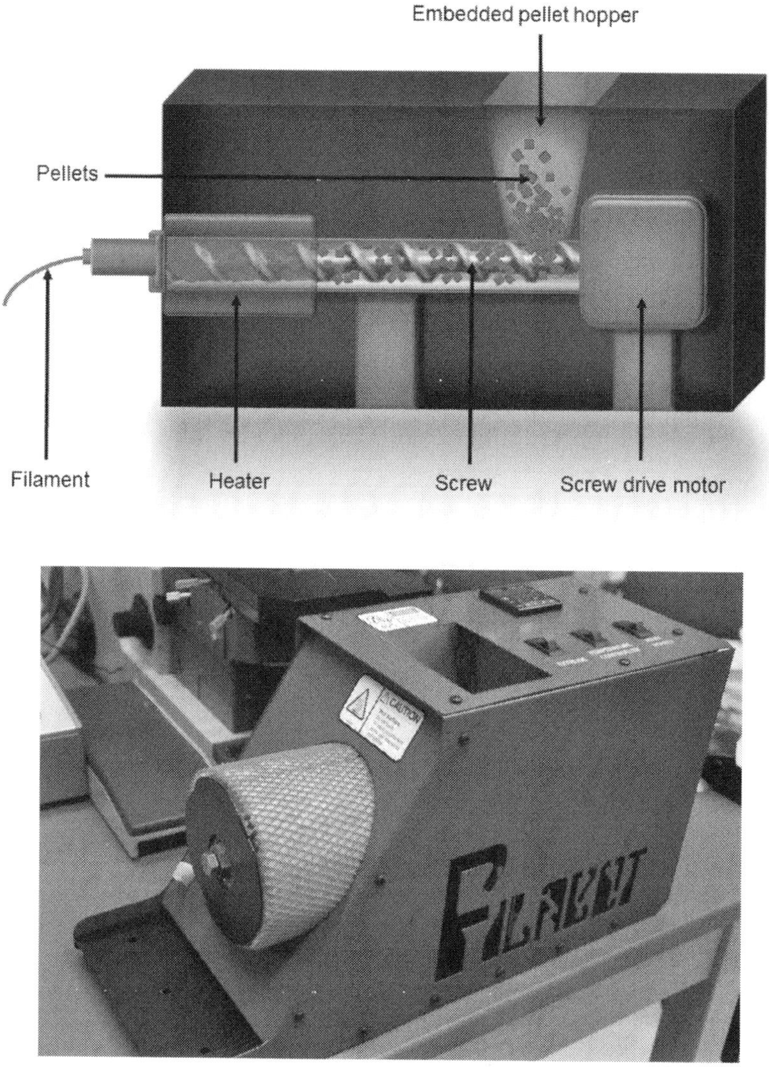

Figure 8. Diagram of a filament maker and a picture of the 'Filabot – Original' used for producing filaments by FDM.

Graphene exhibits a range of exceptional qualities including flexibility and conductivity. 3D printing filaments augmented with graphene have the potential to enhance the manufacturing process of strong conductive composites. There are many applications of these carbon

http://dx.doi.org/10.5772/63419

nanostructured additives in 3D printer filaments including sensors, trackpads, electromagnetic, and RF shielding.

4. Graphene

Graphene has been rising in popularity as a material that would revolutionize electronics; fortunately, graphene has safely passed the peak of overestimated expectations and is now settling on some novel applications. Graphene has many interesting properties such as low resistivity, excellent thermal conductivity, optical transparency, and high electron mobility.

There are only a select few companies that have produced graphene enhanced 3D printing materials, including Angstron Materials and Graphene 3D Labs (**Table 2**).

Company	Product
Black Magic 3D	Conductive Graphene Filament
	Conductive Thermoplastic Graphene/PLA Pellets
Filabot	Graphite Infused Filament—ABS Based
Angstron Materials	Graphene Enhanced Nanocomposites—PP, PC, ABS
Graphene 3D Lab	GRAPH-PLA (Graphene/PLA Pellets)

Table 2. Commercial graphene or graphite infused 3D printing materials.

4.1. Polymer integration

The use of graphene in 3D printing started with the Canadian company Grafoid, resulting in the product MesoGraf, which is produced from raw, unprocessed graphite ore in a one-step process. It is envisaged that the use of graphene in 3D printing will aid in conductivity and strength. Grafoid worked with Altamat to construct a facility to produce MesoGraf graphene-based powders and filaments for 3D printing, and Grafoid intends to supply a diverse catalog of MesoGraf-based powders and filaments to aid companies with additive manufacturing processes to produce their end-product prototypes and end-user products.

There are some limitations with current 3D printing technology, especially when attempting to produce filaments with advance functional materials such as metals, graphene and carbon nanotubes. Problems may be associated with the size of particles and temperature variations of the constituents within a polymer matrix.

Research is still ongoing into making FDM more applicable to producing advanced functional 3D printed materials in Dr. Acquah's research group at Florida State University. The research group is using a combination of graphene-enhanced and carbon nanotube-enhanced 3D printing materials to tailor the properties. Research is also focusing on modifying polymers with graphene [36]. Although graphene has also been referred to as a remarkable material, homogeneous mixtures of graphene and polymers are essential to exploit the unique properties [37]. Wei et al. [38] stated that the major problem confronted by graphene composites was that of phase separation between graphene sheets. They addressed the problem with the use of graphene oxide (GO) to substitute graphene as the additive. GO contains oxygenated functional groups on its basal planes, which may assist graphene's dispersion in polymer phases [38].

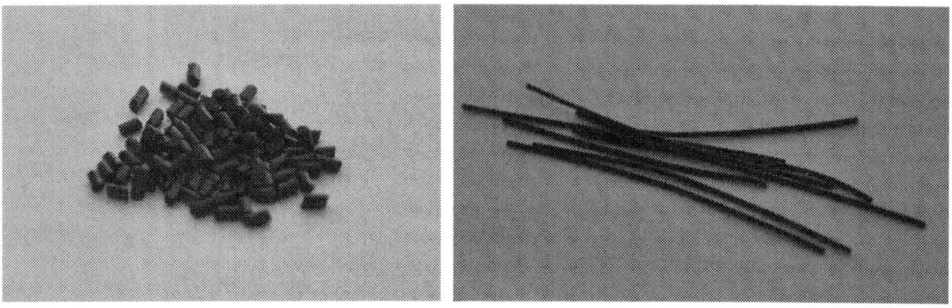

Figure 9. Graphene infused PLA pellets and 1.75 mm filaments.

Graphene 3D Labs is working on 3D printable batteries incorporating graphene, with the potential to surpass current commercially available batteries. They demonstrated a prototype battery in 2014 and the following year launched conductive graphene filaments for 3D printing. The filaments and pellets (**Figure 9**) contain highly conductive proprietary nanocarbon materials with PLA. Both the filaments and pellets extruded through a filament-producer are compatible with commercially available FDM printers. The volume resistivity is listed as 1 Ω•cm which provides an excellent starting point for 3D printed circuitry and capacitive touch sensors.

Research by Seung Kwon Seol of the Korea Electrotechnology Research Institute has demonstrated a process that is capable of 3D printing pure graphene nanostructures [39]. This achievement marked the first time graphene has been printed by itself without being used as the additive. The research available in the journal 'Advanced Materials' shows potential for expansion once the challenges such as reducing the size of the extruded material and increasing the yield are addressed.

5. Carbon structured additives

Carbon Nanotubes and Graphene are some of the popular choices as additives for 3D printing, but carbon black (CB) and carbon fibres are unique carbon structured additives that have an extensive history in manufacturing, tailoring the properties of composites for electronic applications, and structural reinforcement respectively.

5.1. Carbon black

CB is soot-like in its appearance but differs from soot at the molecular level. CB is produced from the incomplete combustion of heavy petroleum products such as coal tar. As such, it is readily available and inexpensive. It is considered one of the most popular conductive additives because of its low cost and chemical stability [40].

A conductive thermoplastic composite called 'carbomorph' which can be extruded through a consumer 3D printer has been produced [15]. In this work, Leigh et al. reported on the 3D printing of a piezo-resistive sensor from a composite of PCL with 15 wt% of CB as filler. Leigh stated that the transition from insulating to non-insulating behavior for composites with conductive filler is generally observed when the volume concentration of filler reaches a threshold of around 25% [41]. However, their decision to use 15 wt% CB was based on optimization, considering the thermal and rheological parameters required for successful printing. The thermoplastic polymer selected for the composite was the polymorph PCL. Bending the sensor resulted in a change of resistivity of 4%. The conductivity of the printed filament made from the PCL/CB composite was 11.1 S/m, which falls within the range of semi-conductors.

Resistance was tested using 5 mm cubes of carbomorph by two-probe measurements with the two opposite cube faces (painted with silver conductive paint). The resistivity of the composite, in-plane with the layers, was 0.09±0.01 ohm m^{-1} and perpendicular to the layers, the resistivity was 0.12±0.01 ohm m^{-1}. This is a significant observation as the reduction in the resistivity of 25% from the perpendicular to parallel orientation is a feature that needs to be considered for the next generation of functional composites. The plane of the layers of the printed filaments provides an unperturbed conductive pathway between the electrodes, while perpendicular to the layers, the conductive pathway depends upon the connection between successive layers.

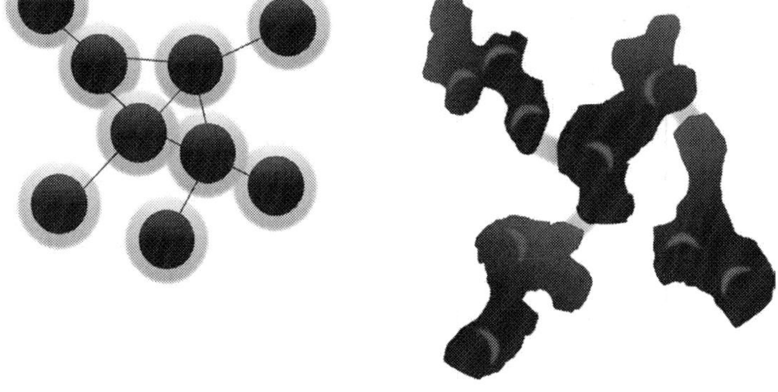

Figure 10. Low-structure and high-structure carbon black. Figure adapted from Balberg [42].

Work by Balberg into the electrical phenomena in CB-polymer composites looked at the difficulties in establishing the percolation threshold for the incorporation of CB into polymers [42]. While he noted that previous studies in his reviews had explained the electrical data within the confines of inter-particle tunneling conduction [43] and/or that of classical percolation theory, the observations were far more convoluted. He noted that for different types of CB, the same volume percent of the CB phase in the composite produces different values for the resistivity [44, 45]. By investigating the characteristics, Balberg stated that the values of the (apparent) percolation threshold and the (apparent) percolation critical resistivity-exponent depended strongly on the particular type of CB. The particles could be treated in terms of how spherical-like they look, with more spherical CB particles termed 'low-structure' CB, in comparison to 'high-structure' CB. **Figure 10** shows the closely packed network of spheres representing the low structure in a polymer composite with nearest-neighbor tunneling. The black spherical structures are the CB particles, and the blue shells represent the tunneling distance. The nearest-neighbor connections (black lines) indicate the dominant conducting elements that result in percolation-like behavior. **Figure 10** also shows a high-structure CB polymer composite. The distances between the nearest inter-particle surfaces have a narrow, non-diverging distribution of the tunneling-resistor values in the network.

PLA filaments with CB as an additive are commercially available (**Figure 11**). Proto-pasta offers filaments that have a volume resistivity of 15 Ω•cm, resulting in 3D prints that are 30 Ω•cm perpendicular to the layers, and 115 Ω•cm through the layers.

Figure 11. Conductive PLA 1.75 mm filament containing carbon black.

http://dx.doi.org/10.5772/63419

5.2. Carbon fiber

As a stalwart of modern engineering, carbon fibers are an important part of reinforcement in composite materials. Properties such as low weight, high tensile strength, and low thermal expansion are some of the advantages of incorporating these fibers in composites. One of the known limitations of FDM is the low transfer of strength in the printed composites, but the use of carbon fibers in 3D printing is an important step toward creating high strength composites. Ning et al. reported printing out parts from an ABS/carbon fiber composite with different percentages by weight. The highest tensile strength reported for a 5 wt% composite of carbon fibers in ABS, printed using FDM was 42 MPa [46]. The tensile strength dropped as the percentage of carbon fiber increased to 10 wt%. The ductilities of all composites were less than the pure ABS. The 7.5 wt% carbon fiber composite had the largest value for the Young's modulus. The reduction in the tensile strength and ductility of composites exceeding 10 wt% was due to the higher porosity of those composites.

A new start-up, MarkForged Inc., has been working on improving the integration of carbon fibers. The company has developed a new printer which uses two separate print-heads. The first head dispenses a polymer such as nylon or PLA, while the other dispenses a carbon fiber tow which is a coated thermoplastic. As the carbon fiber can be introduced during any part of the print process, composites can be produced without reinforced sections. MarkForged reported that tests demonstrated that the parts produced were stronger than 6061-T6 aluminum.

Carbon fiber filaments are also offered by Proto-pasta with fibers mixed with PLA. Proto-pasta states that the abrasive nature of the filament may cause the extrusion nozzle to fail prematurely.

6. Discussion

Additive manufacturing is a multi-disciplinary field that encompasses many aspects of chemistry, physics, and engineering in a manner not so dissimilar from its nanoscale counterpart, in the field of nanotechnology. The two areas of development that steer advances are the composition of the next generation of filaments, pellets, and resins for printing; and the equipment used for filament production. One of the most important examples of the level of advancement in both areas is the 3D printing of bio-inspired materials. Kang et al. recently produced life-sized body parts and tissues with living cells acting as the printing materials. The parts were stable enough to be used as viable replacements that could be tailored to individual needs rather than generic replacements [47].

One of the limitations of consumer thermoplastic extrusion is the temperature range of the extruder assembly. While entry level printers were designed to work with ABS and PLA filaments, this meant that the nozzle temperatures were only required to reach 230–240°C. The latest generation of extruders such as the E3D V6 hotend are capable of high temperature extrusion up to 300°C and 400°C when the thermistor is exchanged for a thermocouple. These temperatures are essential for multi-material filaments that may require elevated temperatures, and polycarbonate and nylon based filaments. The high temperatures would also create additional problems with mixed materials. Polymers may degrade at higher temperatures and the carbon nanostructured additives may cause unfavourable results during extrusion. Stratasys offer a proprietary product called Digital Materials which are a range of several hundred combinations of PolyJet base resins. The Objet260 Connex 3D printing platforms use the PolyJet resin and are capable of depositing three materials with a layer thickness of 16 μm. The build resolution is 600 dpi on both the X-axis and Y-axis, and 1600 dpi on the Z-axis.

The difficulty in producing mixed material filaments with carbon nanostructures, and the additional problems associated with the rheology and flux during FDM extrusion may be addressed by a move toward other additive manufacturing techniques such as the PolyJet system. Carbon nanotubes, graphene, and CB have been explored as additives for SLS [48–50]. The main issue again is creating a homogeneous distribution of the carbon nanostructures, but this can be achieved by simple mixing techniques. Paggi et al. reported on the process for the optimization of PA12/MWCNT nanocomposites by SLS [48]. They described a procedure for dispersing the MWCNT powder initially in chloroform using ultrasonic techniques for one hour. They added the polyamide powder to the suspension and continued mixing on a magnetic stir plate for 50 min to homogenize the solution. After a filtration process with a cellulose filter, the mixture was washed with acetone and placed in an oven at 80°C for four hours. Rotary blades were then used to homogenize the final powder. A CO_2 laser (10 watts) with a beam diameter of 250 μm was used to fuse the powder. Pulsed mode was used to operate the laser at 5 kHz and the average layer thickness produced was 200 μm.

The worldwide market for 3D printing, including the services sector, will likely increase at a compound annual growth rate (CAGR) of 32.2% from 2014 to 2019 [51]. Some additive manufacturing processes show good potential for growth from 2014 to 2019. The 3D stereolithography sector is estimated to rise to $1.8 billion with a CAGR of 27.5%. PolyJet technology will rise to a modest $334 million, but with a CAGR of 40.7%. FDM will almost reach $1 billion with a CAGR of 30%, and SLS is expected to reach $504 million with a CAGR of 32.3%, so development into additives for these industries will continue to be strong. The industry segment focusing on services including design and development, manufacturing, and equipment repair will contribute to 53% of revenues. BCC Research also estimates that 'services' will reach $7.8 billion in 2019 with a CAGR of 31.4% from 2014 to 2019 [51].

The major advantage in this market is the high level of diversification in the technology and applications. Advances in the design and function of 3D printers have catalysed the development of multi-material filaments and printing techniques that overcome some of the limitations of the durability of printed composites. 3D printing may also have a pronounced effect on more traditional manufacturing processes, affecting many aspects from the design and development of materials to the cost savings through the rapid production of custom parts. For consumers, it is the potential to be able to print replacement parts for household appliances.

7. Summary

Polymer composites with conductive fillers have the potential to be used in many areas from engineering to consumer product development, with tunable properties that include elasticity, durability, water wettability (hydrophobic/hydrophilic), and conductance.

The use of metal fillers in polymer composites have been reported [52], and although the ability to print with conductive filler such as a metal is a clear advantage, the difficulty resides in the processability of the composite which may affect the quality and resolution of the 3D object. To reach a workable conductivity, the percentage weight of filler may need to increase, which causes an increase in the density and viscosity of the composite. The addition of the filler affects the rheological properties of the polymer. Metallic fillers are usually susceptible to oxidation, and most of the conductive fillers used are in a 5–120 μm diameter range, which can be a challenge by itself causing blockages at the nozzle-opening of most types of FDM printers, which typically have a nozzle diameter of 400 μm.

Carbon nanotubes as an additive for 3D printing are relatively new but an essential step toward the production of mixed-materials 3D printing. Much of the work in this area has focused on producing homogenous mixtures with PLA and ABS, with the goal of reducing the percolation threshold for conductivity. Polyhydroxylated fullerenes may also be an important bridge toward creating homogenous mixtures in polymers [53].

Graphene has been incorporated into filaments for FDM by a small but growing number of commercial entities. Its incorporation into filaments and pellets has been marketed primarily for their electron transport properties, as an improvement over CB based filaments.

Author details

Steve F. A. Acquah[1*], Branden E. Leonhardt[1], Mesopotamia S. Nowotarski[1], James M. Magi[1], Kaelynn A. Chambliss[1], Thaís E. S. Venzel[1], Sagar D. Delekar[1] and Lara A. Al-Hariri[2]

*Address all correspondence to: sacquah@chem.fsu.edu

1 Department of Chemistry & Biochemistry, Florida State University, Tallahassee, FL, USA

2 Department of Chemistry, University of Massachusetts Amherst, Amherst, MA, USA

References

[1] Gross BC, Erkal JL, Lockwood SY, Chen CP, Spence DM. Evaluation of 3D printing and its potential impact on biotechnology and the chemical sciences. Anal Chem. 2014;86(7): 3240–53.

[2] Bai JM, Goodridge RD, Yuan SQ, Zhou K, Chua CK, Wei J. Thermal influence of CNT on the polyamide 12 nanocomposite for selective laser sintering. Molecules. 2015;20(10):19041–50.

[3] Shallan AI, Smejkal P, Corban M, Guijt RM, Breadmore MC. Cost-effective three-dimensional printing of visibly transparent microchips within minutes. Anal Chem. 2014;86(6):3124–30.

[4] Lin D, Jin S, Zhang F, Wang C, Wang Y, Zhou C, et al. 3D stereolithography printing of graphene oxide reinforced complex architectures. Nanotechnology. 2015;26(43): 434003.

[5] Lawes S, Riese A, Sun Q, Cheng NC, Sun XL. Printing nanostructured carbon for energy storage and conversion applications. Carbon. 2015;92:150–76.

[6] Postiglione G, Natale G, Griffini G, Levi M, Turri S. Conductive 3D microstructures by direct 3D printing of polymer/carbon nanotube nanocomposites via liquid deposition modeling. Composites A Appl Sci Manufact. 2015;76:110–4.

[7] McWilliams A. Advanced materials for 3D printing: Technologies and global markets. BCC Res. 2014 .

[8] Stephens B, Azimi P, El Orch Z, Ramos T. Ultrafine particle emissions from desktop 3D printers. Atmos Environ. 2013;79:334–9.

[9] Turner BN, Strong R, Gold SA. A review of melt extrusion additive manufacturing processes: I. Process design and modeling. Rapid Prototyp J. 2014;20(3):192–204.

[10] Gebler M, Uiterkamp A, Visser C. A global sustainability perspective on 3D printing technologies (vol 74, p. 158, 2014). Energy Policy. 2015;85:511.

[11] Garlotta D. A literature review of poly(lactic acid). J Polym Environ. 2001;9(2):63–84.

[12] Al-Itry R, Lamnawar K, Maazouz A. Improvement of thermal stability, rheological and mechanical properties of PLA, PBAT and their blends by reactive extrusion with functionalized epoxy. Polym Degrad Stabil. 2012;97(10):1898–914.

[13] Pelegrini K, Donazzolo I, Brambilla V, Grisa AMC, Piazza D, Zattera AJ, et al. Degradation of PLA and PLA in composites with triacetin and buriti fiber after 600 days in a simulated marine environment. J Appl Polym Sci. 2016;133(15):43290.

[14] Tan BK, Ching YC, Poh SC, Abdullah LC, Gan SN. A review of natural fiber reinforced poly(vinyl alcohol) based composites: Application and opportunity. Polymers. 2015;7(11):2205–22.

[15] Leigh SJ, Bradley RJ, Purssell CP, Billson DR, Hutchins DA. A simple, low-cost conductive composite material for 3D printing of electronic sensors. PLoS One. 2012 ; 7(11).

[16] McCarthy B, Coleman JN, Curran SA, Dalton AB, Davey AP, Konya Z, et al. Observation of site selective binding in a polymer nanotube composite. J Mater Sci Lett. 2000;19(24):2239–41.

[17] Gao YL, Zhai QJ, Barrett R, Dalal NS, Kroto HW, Acquah SFA. Piezoelectric enhanced cross-linked multi-walled carbon nanotube paper. Carbon. 2013;64:544–7.

[18] Steven E, Saleh WR, Lebedev V, Acquah SFA, Laukhin V, Alamo RG, et al. Carbon nanotubes on a spider silk scaffold. Nat Commun. 2013 ;4.

[19] Ventura DN, Stone RA, Chen KS, Hariri HH, Riddle KA, Fellers TJ, et al. Assembly of cross-linked multi-walled carbon nanotube mats. Carbon. 2010;48(4):987–94.

[20] Bauhofer W, Kovacs JZ. A review and analysis of electrical percolation in carbon nanotube polymer composites. Compos Sci Technol. 2009;69(10):1486–98.

[21] Celzard A, McRae E, Deleuze C, Dufort M, Furdin G, Marêché JF. Critical concentration in percolating systems containing a high-aspect-ratio filler. Phys Rev B. 1996;53(10): 6209–14.

[22] Bai JB, Allaoui A. Effect of the length and the aggregate size of MWNTs on the improvement efficiency of the mechanical and electrical properties of nanocomposites —experimental investigation. Compos A Appl Sci Manufact. 2003;34(8):689–94.

[23] Martin CA, Sandler JKW, Shaffer MSP, Schwarz MK, Bauhofer W, Schulte K, et al. Formation of percolating networks in multi-wall carbon-nanotube–epoxy composites. Compos Sci Technol. 2004;64(15):2309–16.

[24] Mahfuz H, Khan MR, Leventouri T, Liarokapis E. Investigation of MWCNT reinforcement on the strain hardening behavior of ultrahigh molecular weight polyethylene. J Nanotechnol. 2011;2011:9.

[25] Smook J, Pennings AJ. Suspension spinning of ultrahigh molecular-weight polyethylene. Polym Bull. 1983;10(7–8):291–7.

[26] Pakhomov PM, Golikova AY, Khizhnyak SD, Shavyrina MA, Galitsin VP, Gribanov SA, et al. The structure of high-strength ultrahigh-molecular-weight polyethylene fibres fabricated by the gel-spinning method. Fibre Chem. 2006;38(3):200–6.

[27] Daver F, Baez E, Shanks RA, Brandt M. Conductive polyolefin-rubber nanocomposites with carbon nanotubes. Compos A Appl Sci Manufact. 2016;80:13–20.

[28] Besco S, Modesti M, Lorenzetti A, Donadi S, McNally T. Effect of modified clay on the morphology and electric properties of PC/ABS-MWCNT composites. J Appl Polym Sci. 2012;124(5):3617–25.

[29] Ji MZ, Deng H, Yan DX, Li XY, Duan LY, Fu Q. Selective localization of multi-walled carbon nanotubes in thermoplastic elastomer blends: An effective method for tunable resistivity-strain sensing behavior. Compos Sci Technol. 2014;92:16–26.

[30] Sun Y, Guo Z-X, Yu J. Effect of ABS rubber content on the localization of MWCNTs in PC/ABS blends and electrical resistivity of the composites. Macromol Mater Eng. 2010;295(3):263–8.

[31] Wu D, Zhang Y, Zhang M, Yu W. Selective localization of multiwalled carbon nanotubes in poly(ε-caprolactone)/polylactide blend. Biomacromolecules. 2009;10(2):417–24.

[32] Pötschke P, Pegel S, Claes M, Bonduel D. A novel strategy to incorporate carbon nanotubes into thermoplastic matrices. Macromol Rapid Commun. 2008;29(3):244–51.

[33] Olalde B, Aizpurua JM, Garcia A, Bustero I, Obieta I, Jurado MJ. Single-walled carbon nanotubes and multiwalled carbon nanotubes functionalized with poly(L-lactic acid): a comparative study. J Phys Chem C. 2008;112(29):10663–7.

[34] Vatani M, Lu Y, Lee K-S, Kim H-C, Choi J-W. Direct-write stretchable sensors using single-walled carbon nanotube/polymer matrix. J Electr Packag. 2013;135(1):011009.

[35] Davenport M. Twists and shouts: A nanotube story. Chem Eng News. 2015;93(23):10–5.

[36] Barletta M, Puopolo M, Tagliaferri V, Vesco S. Graphene-modified poly(lactic acid) for packaging: Material formulation, processing and performance. J Appl Polym Sci. 2016 ; 133(2).

[37] Huang X, Qi XY, Boey F, Zhang H. Graphene-based composites. Chem Soc Rev. 2012;41(2):666–86.

[38] Wei XJ, Li D, Jiang W, Gu ZM, Wang XJ, Zhang ZX, et al. 3D printable graphene composite. Sci Rep. 2015 ;5.

[39] Kim JH, Chang WS, Kim D, Yang JR, Han JT, Lee GW, et al. 3D printing of reduced graphene oxide nanowires. Adv Mater. 2015;27(1):157–61.

[40] Soares BG, Touchaleaume F, Calheiros LF, Barra GMO. Effect of double percolation on the electrical properties and electromagnetic interference shielding effectiveness of carbon-black-loaded polystyrene/ethylene vinyl acetate copolymer blends. J Appl Polym Sci. 2016 ;133(7).

[41] Reboul JP, Moussalli G. About some DC conduction processes in carbon-black filled polymers. Int J Polym Mater. 1976;5(1–2):133–46.f

[42] Balberg I. A comprehensive picture of the electrical phenomena in carbon black–polymer composites. Carbon. 2002;40(2):139–43.

[43] Balberg I. Tunneling and nonuniversal conductivity in composite-materials. Phys Rev Lett. 1987;59(12):1305–8.

[44] Rubin Z, Sunshine SA, Heaney MB, Bloom I, Balberg I. Critical behavior of the electrical transport properties in a tunneling-percolation system. Phys Rev B. 1999;59(19):12196–9.

[45] Sichel EK, Gittleman JI, Sheng P. Transport properties of the composite-material carbon-poly(vinyl chloride). Phys Rev B. 1978;18(10):5712–6.

[46] Ning F, Cong W, Qiu J, Wei J, Wang S. Additive manufacturing of carbon fiber reinforced thermoplastic composites using fused deposition modeling. Compos B Eng. 2015;80:369–78.

[47] Kang H-W, Lee SJ, Ko IK, Kengla C, Yoo JJ, Atala A. A 3D bioprinting system to produce human-scale tissue constructs with structural integrity. Nat Biotech. 2016;34(3):312–9.

[48] Paggi RA, Beal VE, Salmoria GV. Process optimization for PA12/MWCNT nanocomposite manufacturing by selective laser sintering. Int J Adv Manuf Technol. 2013;66(9–12):1977–85.

[49] Athreya SR, Kalaitzidou K, Das S. Microstructure, thermomechanical properties, and electrical conductivity of carbon black-filled nylon-12 nanocomposites prepared by selective laser sintering. Polym Eng Sci. 2012;52(1):12–20.

[50] Makuch A, Trzaska M, Skalski K, Bajkowski M. PA-G composite powder for innovative additive techniques. Compos Theory Pract. 2015;15(3):152–7.

[51] Varotto A. Global markets for 3D printing. BCC Res. 2015. Report No.: IAS102A.

[52] Liu H, Webster TJ. Enhanced biological and mechanical properties of well-dispersed nanophase ceramics in polymer composites: From 2D to 3D printed structures. Mater Sci Eng C. 2011;31(2):77–89.

[53] Penkova AV, Acquah SFA, Dmitrenko ME, Sokolova MP, Mikhailova ME, Polyakov ES, et al. Improvement of pervaporation PVA membranes by the controlled incorporation of fullerenol nanoparticles. Mater Des. 2016;96:416–23.

Chapter 9

Polymer Nanocomposite Artificial Joints

Samy Yousef

Additional information is available at the end of the chapter

http://dx.doi.org/10.5772/62269

Abstract

Artificial joints (AJ) often have a polymeric component to decrease the wear rate and the total weight at the same time and make it more flexible. Ultrahigh molecular weight polyethylene (UHMWPE) is considered as the standard material for these applications. Therefore, UHMWPE has been reinforced by many of the nanomaterials, hoping to improve the tribology characteristic, which is considered as the most important factor in determining the life span of AJ. However, all attempts were in laboratory scale and did not live up to the actual implementation due to the high viscosity of UHMWPE, leading to poor dispersion with bulk components. This chapter aims to explain in detail a novel technique to produce a real UHMWPE nanocomposite (UNC) hip cup using the paraffin oil dispersion technique and tested by an artificial joint simulator (AJS), which was designed by the author. The chapter contains three parts: the first part starts with a brief account of AJ and then illustrates the AJ polymeric components. The wear behavior of the polymeric components is also presented. The second part reviews some previous attempts for the synthesis of UNC and the common nanomaterials [carbon nanofiber (CNF), carbon nanotubes (CNT), and graphene (GA)], which are used as a nanofiller. The problems that had arisen during the mixing process and UNC characterizations are also presented. For the third part, the chapter concludes by explaining a novel technique to produce UHMWPE hip cup reinforced by CNT using the paraffin oil dispersion technique and testing by AJS, which was designed especially for this.

Keywords: artificial joint simulator, artificial joints, carbon nanofiber, carbon nanotubes, graphene, paraffin oil dispersion technique, polymer nanocomposites, polymer nanocomposite hip cup, UHMWPE, wear

1. Artificial joints (AJ)

1.1. Introduction

The human body contains many orthopedic joints to give it more flexibility during movement (**Figure 1**). Usually, the damages in these joints occur due to internal effects and environmental changes, such as weather and pollution, in addition to bad nutrition and especially calcium deficiency, as well as advancing age and the diseases of aging. However, the possibility of damage is significant in the hip, shoulder, knee, and fingers as a result of exposure to overload. Currently, millions of orthopedic patients around the world are suffering from pain and psychological disorders, even after performing joint replacement procedures, due to the short life span for the AJ, and this means the return of pain from thus restoring the implant and additionally the extra cost of the implant [2]. To eliminate pain, increase mobility, and restore the quality of life, AJ was innovated in 1961 when Sir John Charnley performed the procedure of the first surgery to implant joint prostheses successfully in a particular artificial hip joint [3]. At the time, AJ was proven to be an extremely successful and cost-effective means of relieving arthritis pain. After that, AJ has been extended to include shoulder, knee, and finger joint replacement bearings. **Figure 2** shows the AJ radiographs for the hip, shoulder, knee, and finger joint bearings respectively [4–7]. This section presents the AJ components and the common methods used to investigate the wear behavior for AJ polymeric components and its failure shapes.

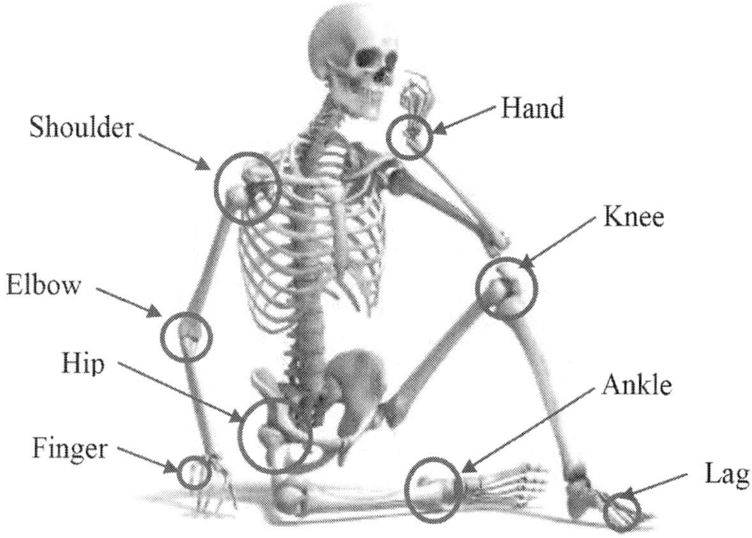

Figure 1. Schematic drawing for human body joints [1].

http://dx.doi.org/10.5772/62269

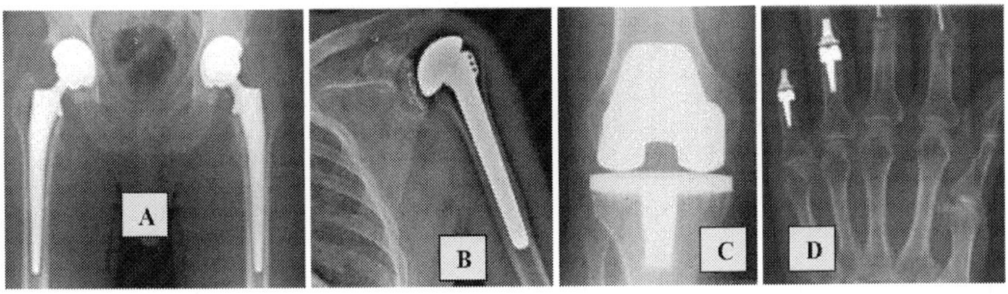

Figure 2. AJ radiographs of (A) hip, (B) shoulder, (C) knee, and (D) finger joints.

1.2. AJ components

The hip, shoulder, knee, and finger AJ prostheses are usually composed of two main components articulating with one another, polymeric against a metal or ceramic component (**Figure 3**). This section focuses on the polymeric material components that consider the weaker part in total AJ components. At the beginning, polytetrafluoroethylene (PTEF) and polyaryletheretherketone (PEEK) polymeric materials have been employed in these applications, but the results were not good due to the large aseptic loosening and osteolysis [8]. Nonetheless, ultrahigh molecular weight polyethylene (UHMWPE) remains the gold standard polymeric material for AJ component due to its many unique properties, such as high wear resistance, strength, modulus, excellent toughness, chemical resistance and impact, low moisture absorption, good wave transmission, and electrical insulation [9]. In addition, the mechanical and wear properties of UHMWPE are considered the most important effective factors, which can be controlled at the service time of the total joint [10].

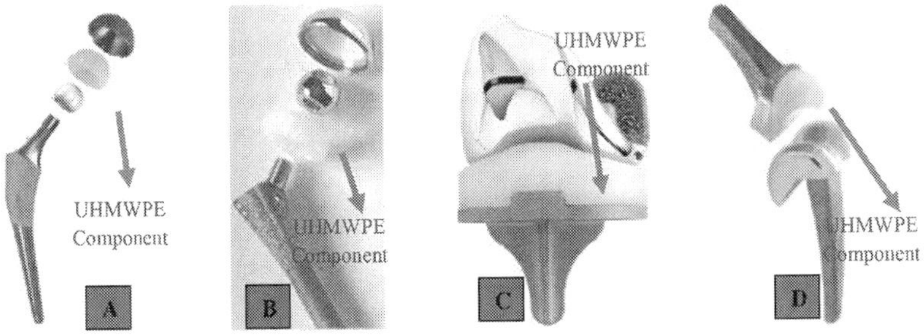

Figure 3. Contemporary AJ prosthesis system components for (A) hip, (B) shoulder, (C) knee, and (D) finger bearing joints [11].

1.3. Investigation of AJ polymeric component wear

In general, the UHMWPE components in the AJ are considered the weaker components, and failure happens due to two kinds of wear failure [sliding and rolling (abrasive and adhesive)] as a result of sliding between polymeric and metal components [12]. The established wear depends on several variables such as load (pressure), speed, contact geometry, state of lubrication, and roughness [13]. Nevertheless, it is difficult to take all these variables into consideration. The pin-on-disc or ball-on-disc machine is considered the leading method to evaluate the wear behavior (in the form of weight loss or volume loss) under constant load conditions [14]. To increase the accuracy of wear investigation and to simulate the real failure condition, many previous test rigs or joint replacement simulators (JRS) have been designed and built to measure the wear rate of hip, shoulder, and finger joint bearings under different conditions. The basic difference between each JRS and another is the loading mechanism (**Figure 4**).

Figure 4. AJ wear simulators of (A) hip, (B) shoulder, (C) knee, and (D) finer joints [15–18].

1.4. Failure of AJ polymeric components

Studies that were conducted on the failure of polymeric components in AJ point to the damages that were featured in the following forms: rim erosion, surface irregularities, component fracture, and wear [19]. Wear failure is considered a key design parameter of AJ and the created wear particles might be the reason for component loosening, leading to pain and the need for

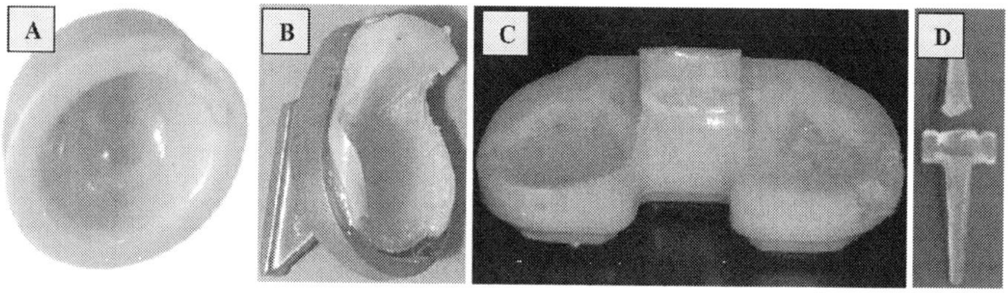

Figure 5. Photograph of failure shapes of (a) hip, (b) shoulder, (c) knee, and (d) finger polymeric components [18–21].

a revision operation [20]. **Figure 5** shows the photograph failure of the polymeric components of the hip, shoulder, knee, and finger, respectively. It is clear that the main failure is located on the contact surface between femoral (metal component) and tested (UHMWPE component) samples. Also, it has seen that failure occurs due to three reasons: scratching, pitting, and delamination or cracking [21].

2. Nanocomposite polymer biomaterials

2.1. Introduction

With the increasing demand for AJ, it has become necessary to improve the mechanical and wear behaviors of UHMWPE component to increase the total life span of joints and also to keep pace with the progress for prosthetic applications. Nanotechnology science is considered the most appropriate alternative at the moment after it that has proven high efficiency in many areas, especially in polymer nanocomposites. Mixing UHMWPE with nanofiller materials, which have a higher surface area to volume ratio, leads to rapid interaction and more mixing between the nanofiller and UHMWPE and thus improvement of the chemical and physical properties [22]. This section reviews the common nanofiller materials, which are used to improve the properties of UHMWPE. In addition, different dispersion methods were used to obtain a uniform dispersion. Finally, the characterization methods for the synthesized UHMWPE nanocomposite (UNC) have been listed.

2.2. Nanofiller materials

There are many types of nanofiller materials that have a good tribological characteristic reinforced by UHMWPE such as carbon nanofibers (CNF), carbon nanotubes (CNT), and graphene (GA) for the synthesis of UNC to have a good wear resistance [23–25]. The results showed that GA and CNT additives have been proven more effective compared to other nanofillers because GA and CNT have unique physical, mechanical, tribological, high aspect ratio, and chemical properties. Also, GA has better biocompatibility compared to other nanofiller materials [25]. Although CNT is widely used, it has to be economic. This chapter focuses on the CNT/UHMWPE synthesis. Therefore, this study focuses on CNT as an economic nanofiller material. CNT is composed of molecular-scale sheets of graphite called GA that roll up to make a tube. CNT can be classified into single-wall nanotubes (SWCNT) and multiwall nanotubes (MWCNT). SWCNT consists of single GA rolls, whereas MWCNT consists of two or more coaxial tubes within a tube [26]. There are three common different methods used for the synthesis of CNT: arc discharge, laser vaporization, and chemical vapor deposition (CVD) [27]. Arc discharge multielectrode and CVD multiquartz tubes were designed and built to produce CNT with high yield and that is more economic (**Figure 6**). Finally, **Figure 7** illustrates the different carbon structures between single-wall and multiwall CNT using transmission electron microscopy (TEM).

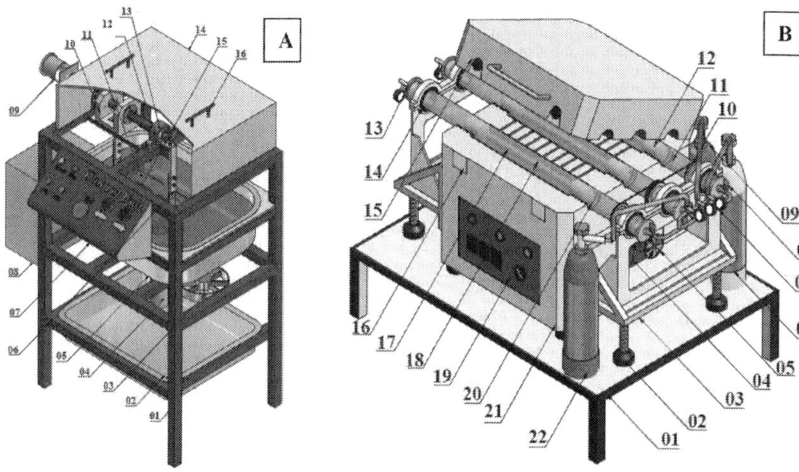

(A) Arc discharge multi-electrodes technique
(1) Base frame, (2) lower tank, (3) fans for cooling, (4) valve, (5) upper tank, (6) carbon holder, (7) control unit, (8) electrical unit, (9) steeper motor, (10) nuts housing, (11) deionized water, (12) power screw, (13) bearing, (14) cover, (15) pure graphite, (16) cover hand.
(B) CVD multi-quartz tubes technique
(1) Base, (2) Set screw, (3) Bearing holder, (4) Motor, (5) Sprocket, (6) Argon cylinder, (7) Pressure gauge, (8) Control valve, (9) Hydrogen cylinder, (10) Chain, (11) Quartz tube, (12) Ceramic sealant, (13) Exhaust gas, (14) Bearing, (15) Furnace movable part, (16) Padlock, (17) Heater, (18) Catalyst position, (19) Control unit, (20) Graphite rings, (21) Furnace fixed part, (22) Methane cylinder.

Figure 6. Fully automatic system for producing CNT using arc discharge multielectrode CVD multiquartz tube design [28,29].

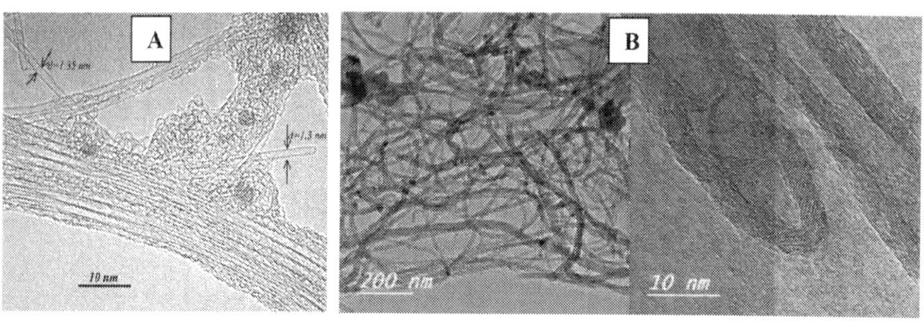

Figure 7. TEM images of (A) SWCNT and (B) MWCNT [29,30].

2.3. Dispersion of UNC polymer

In the dispersion system, UHMWPE and nanofiller particle materials are dispersed in a continuous phase with a different composition. Several methods of mixing techniques were employed to prepare UHNs, such as (a) hot plate and magnetic stir bar, (b) ultrasonic bath, (c) ball milling, (d) twin-screw extrusion, and (e) three roll mill. It is worth mentioning that each technique has different mixing conditions to improve the dispersion of nanofiller inside UHMWPE polymer base. Most mixing conditions are focused on the mixing temperature and mixing time. **Figure 8** shows the dispersion techniques with different mixing parameters. Finally, all the previous dispersion methods obtain UHNs in powder shape expected Min-Lab Extruder given in wire form and then cut into small pieces (pellets), as shown in **Figure 9**.

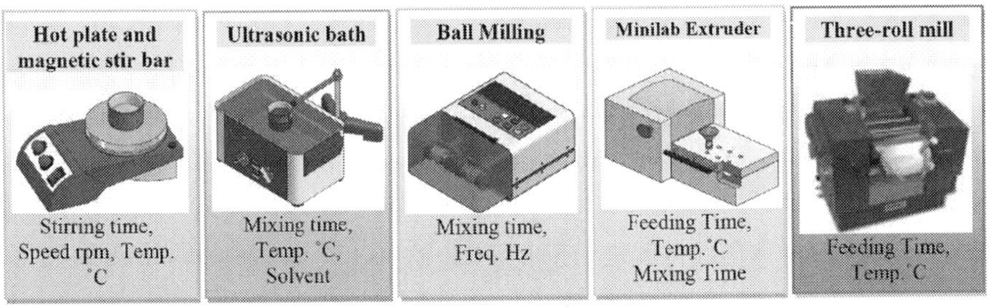

Figure 8. Nanocomposite UHMWPE dispersion [31].

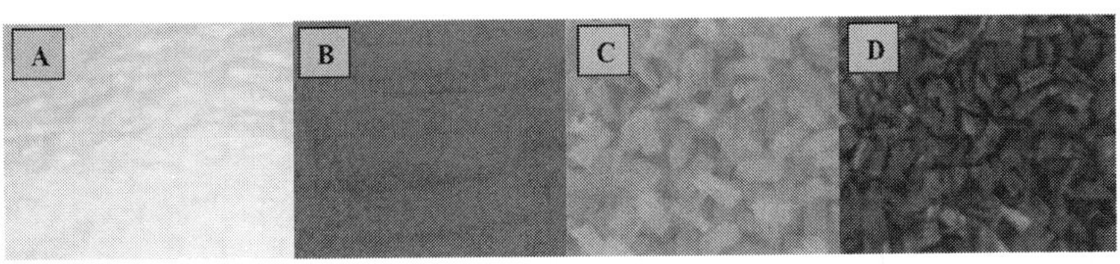

Figure 9. Compression conditions and hot press components [32].

2.4. Synthesis of UNC polymer

To produce the final sheet or film, hot press can be employed to compress the powder or pellet UNC inside copper or aluminum die between two Teflon sheets to produce UNC sheets with very fine surfaces and uniform thickness as shown in **Figure 10**. It is worth mentioning that each type of polymeric material has different compression conditions depending on the melting temperature and viscosity. This section focuses on UHMWPE compression conditions. UNC can be found in two forms: powder or pellets. The powder shape can be produced by magnetic stirring, ultrasonic bath, ball milling, and three roll mill, whereas the pellet shape can be produced by twin-screw extrusion. **Table 1** shows the compression conditions of UNC in powder and pellet shapes at 200°C and a pressure of 200 bar.

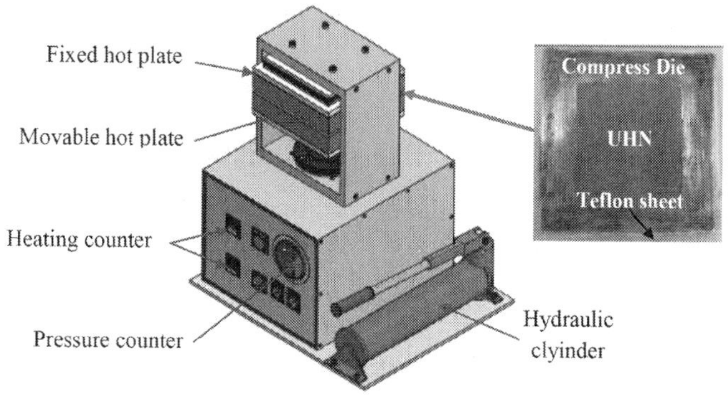

Figure 10. Hot press schematic drawing.

	Powder		Pellets	
Step. no.	Time (min)	Pressure (bar)	Time (min)	Pressure (bar)
1	1	0	10	Without pressure
2	1	50	5	0
3	1	100	4	50
4	1	150	4	100
5	16	200	4	150
6			16	200

Table 1. Compression conditions of UNC [31,32]

2.5. UNC polymer characterizations

After preparing the UNC sheets using the previous steps, the following procedures were used to study the effect of nanofiller adding on UHMWPE characterizations. In this section, the author shows the effect of the feeding ratio of the CNT on the different properties of UHMWPE. Also, other nanofiller additives such as GA and CNF are presented just for comparison.

2.5.1. Scanning electron microscopy (SEM)

SEM was used to examine the dispersion of nanofillers inside UHMWPE. **Figure 11** shows the fracture surfaces of virgin UHMWPE and UHMWPE reinforced by GA, CNT, and CNF. As clearly seen from the SEM images, the incorporation of CNT, GA, and CNF in the UHMWPE matrix resulted in a drastic change in the topography of the fracture surfaces, whereas the fracture surfaces of virgin UHMWPE are smoother.

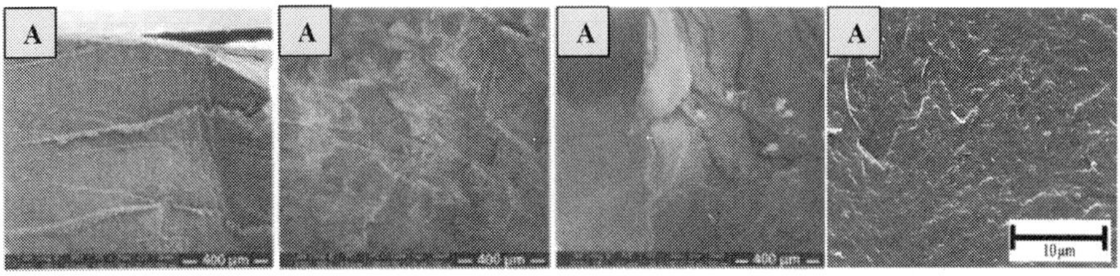

Figure 11. SEM images of fracture surface of (A) UHMWPE, (B) GA/UHMWPE, (C) CNT/UHMWPE, and (d) CNF/UHMWPE [24,33].

2.5.2. Differential scanning calorimetry (DSC)

DSC was used to investigate the thermal properties (crystallization temperature, melting temperature, and lamellar thickness) of UNC. The tests were carried out in nitrogen ambient environment with a heating rate of 10°C/min until 230°C, and then the specimen cools down to room temperature with cold water [34]. The previous results showed that the melting temperature and crystallinity degree increased by adding nanofiller, whereas the lamellar thickness was decreased. This indicated that CNT acted as effective heterogeneous nucleating agents to facilitate the crystallization of UHMWPE [35].

2.5.3. Rheological performance of GUC

UHMWPE has high viscosity (η) representing the main obstacle for blends with the nanofiller materials; more precisely, high η means bad dispersion [36]. The parallel plate rheometry was used to investigate the rheological properties of UNC, particularly viscosity and elastic modulus. The tested samples were cut according to ASTM standard into round shapes having a diameter of 25 mm and thickness of 1 mm. The tests conducted with following data gape 1 mm at stress-controlled rheometer in constant strain mode. The experiment was performed in the linear viscoelastic regime at a temperature up to 180°C and frequency range from 0.005 to 100 Hz and the applied strain was controlled at 0.5% [37]. Unfortunately, all the previous results showed that the viscosity of UHMWPE was increased by the addition of a nanofiller and this led to bad dispersion [38]. To avoid this problem, Galetz et al. and Wood et al. used PO as a solvent and assisted melt material during the mixing process, respectively, and then extracted PO to decrease η of UHMWPE; the result was uniform dispersion approximately [24].

2.5.4. Mechanical properties

The manual press was used to prepare the standard tensile specimens according to ASTM D638-10 standard or any standard. The mechanical properties of UNC standard specimens can be measured by universal testing machine with load cell of 500 N and crosshead speed of 10 mm/min. The results showed that the addition of nanofiller leads to significant improvement in the mechanical properties with varying proportions, and the variance depends on the percentage weight of the filler and dispersion technique. In general, the mechanical properties of UHMWPE improved due to the increase of the crystallinity degree [39].

2.5.5. *Wear behaviors*

JRS are still not widely used to test UNC components due to economic reasons, especially as the shape of the testing samples is more complicated. The pin-on-disk or ball-on-disk test rig is recommended in these applications to overcome the previous problems (**Figure 12**). In this case, the tested UNC sample may be flat or cylindrical shape sliding against stainless steel or ceramic pin. The specific wear rate in this case depends on four parameters: applied normal load, speed, test duration, and type of lubricant. Furthermore, the specific wear rate can be calculated by the following equation.

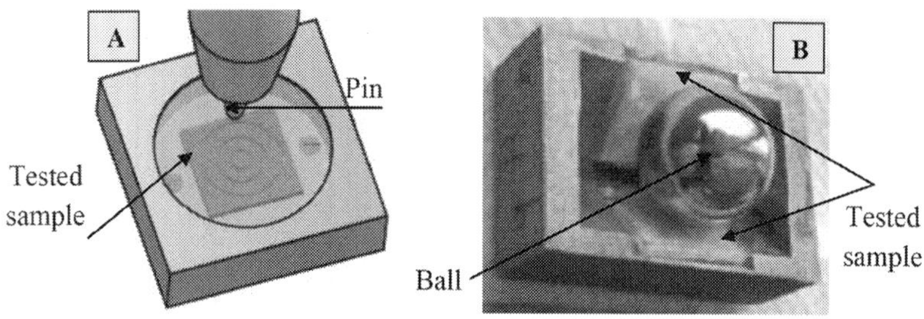

Figure 12. (A) Pin-on-disc test rig model and (B) ball-on-disc model [14,35].

The previous results showed that the wear resistance of UHMWPE has been improved by adding a nanofiller, especially CNT and GA, because CNT and GA have a high aspect ratio. Also, the addition of CNT and GA leads to increase the melting temperature, which considers the obstacle in wear progress [35]. The wear particles are created due to the continuous sliding between UNC and the stainless steel pin, as thermal softening and the melting of the surface layer materials occurred. The surface layer increases over time and causes several scratches inside the wear track. In addition, the wear mechanism can be found in two forms: ductile or brittle fractures. The brittle fracture is safer when compared to ductile fracture, because the amount of materials removed in the form of particles is less.

$$Ws = \frac{\Delta m}{P x Fn x L}$$

Where W_s (10^{-6}mm^3/Nm) is the specific wear rate, Δm (mg) is the mass loss of the specimen and measured by using a high sensitivity electronic weighing balance with accuracy

(10^{-4}gram), p (g/ml) is the density of the specimen, F_n (N) is the normal load and L *(m)* is the total sliding distance.

Finally, **Figure 13** shows the three wear mechanisms of UHMWPE reinforced by CNF, CNT, and GA. It is clear that the amount and length of the removed chips decrease with the addition of CNF (as indicated in the images by circles). While adding CNT and GA, the removed chips are converted to discontinuous chips or small particles, especially with GA, due to the high aspect ratio of GA. Well-dispersed GA in UHMWPE provided a large surface area available for the interaction between UHMWPE molecules and GA, which facilitates a good load transfer to the GA network [24,40,41].

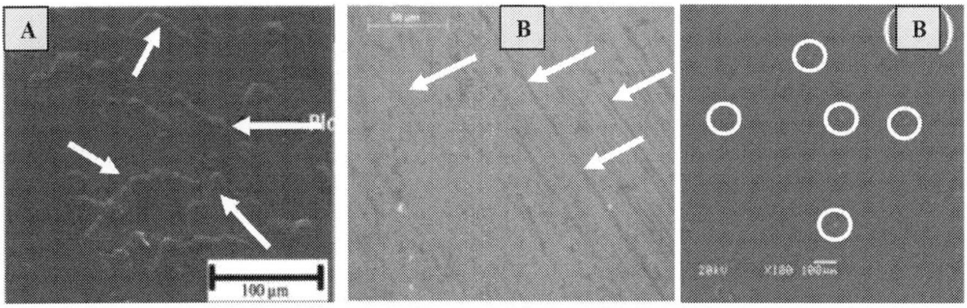

Figure 13. SEM images of worn surfaces of (A) CNF/UHMWPE, (B) CNT/UHMWPE, and (C) GA/UHMWPE.

2.5.6. *Physical properties*

2.5.6.1. *Density*

The density of UNC can be calculated with the Archimedes principle through weighing the sample in air and ethanol as an immersion medium using a high-sensitivity electronic weighing balance with accuracy (10^{-4} g). Previous studies showed that the density of UHMWPE remained the same, because the ratio of nanofiller to the volume of nanocomposite is very small and the densities of the nanofillers are very low [41].

2.5.6.2. *Wettability*

The wettability or contact angle (θ) of UNC can be evaluated using a high-resolution camera as shown in **Figure 14**. A 5 μl drop was deposited on the sample surface and θ can be measured after a few seconds. Any software, such as ImageJ, can be used to capture and analyze the contact angle. The results show that the contact angle increased with the addition of the nanofiller due to the modification of the surface and improvement of the surface roughness by adding nanofills [42].

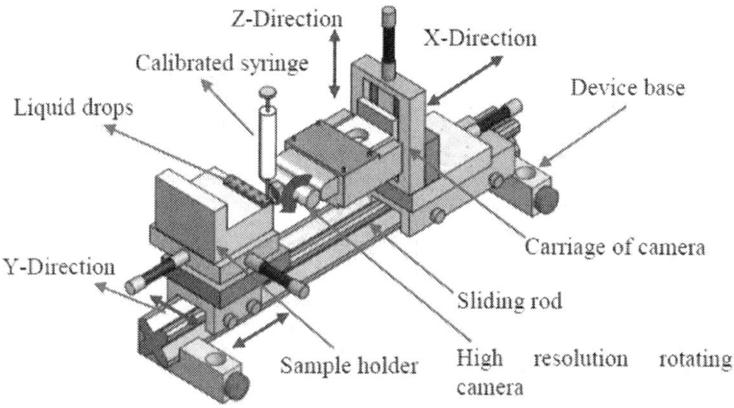

Figure 14. Schematics of contact angle device measurement.

2.5.7. Biocompatibility

Biocompatibility or cytotoxicity studies were conducted to show and evaluate the biocompatibility of the UNC. Ge et al. and Huang et al. studied the effect of adding ultralow molecular weight polyethylene and natural coral particles on the biocompatibility of UHMWPE [43,44]. The results showed that all cells are attached and spread, proliferate nicely, and grow in synthesis sheets. For carbon fillers, the studies reported GA more than CNT with a slight increase [45]. Therefore, the trend now is to use GA as a biocompatibility nanofiller.

2.5.8. Effect of UV radiation on wear resistance

It is supposed that replacement joints are exposed to the influence of some environmental factors during daily life. These factors should be taken into consideration during the design process to determine the extent of their impact on the performance and service life. Weather tester equipment is used for simulating sunlight with a light source of fluorescence. The experiment can be conducted according to the ASTM G154-06 standard. The samples were exposed to UV radiation for a period of 300 h at a temperature of 60°C without water condensation or spraying cycle. The results showed that UV radiation has no effect on the wear durability or any peeling off of the nanocomposite at the low number of cycles, with microvoids appearing with the increase in the number of cycles [46].

2.5.9. Degrading action

In general, AJ works in wet lubricant ambient environment to reduce the wear rate. Therefore, nature's lubricant might affect the degradation of UHMWPE, particularly mechanical and physical properties. In this section, the prepared samples (UHMWPE and its composites) are exposed to hyaluronic acid for different intervals (several months) and then studied for the

http://dx.doi.org/10.5772/62269

effects on the performance. Hyaluronic acid is a simulated synovial liquid. Most previous studies focused on pristine UHMWPE and modified UHMWPE. The results showed that the performance of pure UHMWPE decreased, whereas the modified one had a small change as a result of cross-linking that prevents their sliding in the presence of acid and thus increase the polymeric resistance to the degradation action [47]. Unfortunately, there are no previous studies focused on UHMWPE, except Yousef (under review). Regarding UHMWPE reinforced by carbon nanofiller, the results showed that nanocomposites are more stable.

2.6. Summary

CNT, GA, and CNF are considered the common nanomaterial types used to enhance UHMWPE using many methods for dispersion such as hot plate, magnetic stir bar, ultrasonic bath, ball milling, twin-screw extrusion, and three roll mill. The hot press is used to compress the UNC clay and then produce the sheets. The characterizations of the UNC sheets depend on the percentage weight of the nanofiller and dispersion technique. The previous results showed that the characterizations of UHMWPE improved by adding CNT and GA. Also, the studies reported the biocompatibility of GA/UHMWPE more than CNT/UHMWPE with a slight increase.

3. Nanocomposite polymer CNT/UHMWPE hip cup

This section presents a novel technique to produce a real polymer nanocomposite hip cup using paraffin oil dispersion technique, which has been innovated in the recent years by Yousef et al. [48]. In addition, the wear behavior of the hip can be investigated using an AJ simulator (AJS) built especially for this by Yousef et al. (under research).

3.1. Synthesis of nanocomposite polymer hip cup

Previous studies have focused on the production of UNC fit for use in biomaterial applications within the laboratory scale precisely for UNC thin sheets. These studies did not live up to the actual implementation due to the high viscosity, which is considered as the main obstacle in obtaining a regular dispersion during the mixing process even on the laboratory scale. This was the motivation for Yousef to innovate the paraffin oil dispersion technique to produce polymer nanocomposite bulk components. In this section, the new dispersion technique will be employed after making some modifications to produce the CNT/UHMWPE hip cup as explained in the following steps.

3.2. Materials

Three materials are used in this study for the synthesis of the real nanocomposite hip cup: UHMWPE as a base material, CNT as a nanofiller material, and paraffin liquid (PL) as an assist melt during the injection process.

3.3. Synthesis of CNT/UHMWPE short bars

First, UHMWPE and CNT (0.5–1% wt.) were mixed together using (a) hot plate or magnetic stir bar and (b) ultrasonic bath, ball milling, or three roll mill at different conditions to find the optimum dispersion method and conditions to obtain CNT/UHMWPE clay. With regard to the EX dispersion, it is not preferred in this case because it is more expensive. Second, the CNT/UHMWPE powder is mixed with PL (2% wt.) at 125°C using magnetic stirring bar for 5 to 10 to absorb PL. Third, the injection moulding die, which was designed and manufactured by Yousef et al., was used to synthesize short PNC bars with diameters of 45 mm and length of 80 mm (**Figure 15**). The following steps are followed in the production of the flanges and short bars:

- CNT/UHMWPE clay is poured into a hopper machine.
- An electric heater increases the mixing chamber temperature to 140°C.
- When 140°C is reached, a screw thread begins to rotate to push the melt powder along the heater.
- The liquid is injected into the molded die to form the short bar and then cooled to produce the final shape.

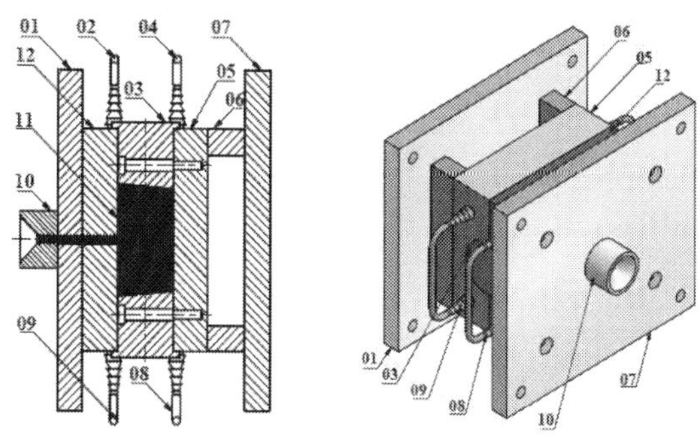

01) Supported movable die, 02-04) cooling water exit, 03) Movable die, 05) Supported fixed die, 06) Fixed die, 07) Injected flange, 08) Die support with extruder head, and 02-04) Cooling water inlet.

Figure 15. Schematic drawing of the short bar injection die [49,50,51].

3.4. Manufacturing of CNT/UHMWPE hip cup

The lathe machine was employed to machining the short bar to realize that the final shape of the hip cup has an inner diameter of 28 mm and thickness of 7 mm. In addition, external flange, which is used to fix the hip cup on the test rig, can be machined by the lathe also.

http://dx.doi.org/10.5772/62269

3.5. AJS design

This section presents a new approach to investigate the wear rate of CNT/UHMWPE hip cup. The new simulator has been designed to address some of the frailties of the other hip test rigs and to make it more flexible by controlling the following parameters: (i) dynamic weight value, (ii) length of leg and thigh, (iii) gait angle, (iv) linear velocity, and (v) dry or wet lubricant conditions. The idea of the new design depends on preventing the upper human body from moving in the front and back directions (X-direction) and allowing to move in the vertical direction (Y-direction) only to overcome the reaction that results during the collision of the foot with the ground (**Figure 16**). It is worth mentioning that the spring is used to generate the dynamic load during the testing process and the length of the vertical movement is equal to the spring deflection. Also, the stepper motor is used to rotate the quick-return mechanism and thus move the ground as an oscillatory motion in the X-direction. The oscillatory motion has been designed to simulate the real movements of the human body (reverse the reality; the ground is fixed and the human body is moveable) and thus decrease the simulator size. In addition, the oscillatory motion is responsible to rotate the leg link around the knee joint and then the resulting rotational motion transfer to the thigh arm to begin the contact between the

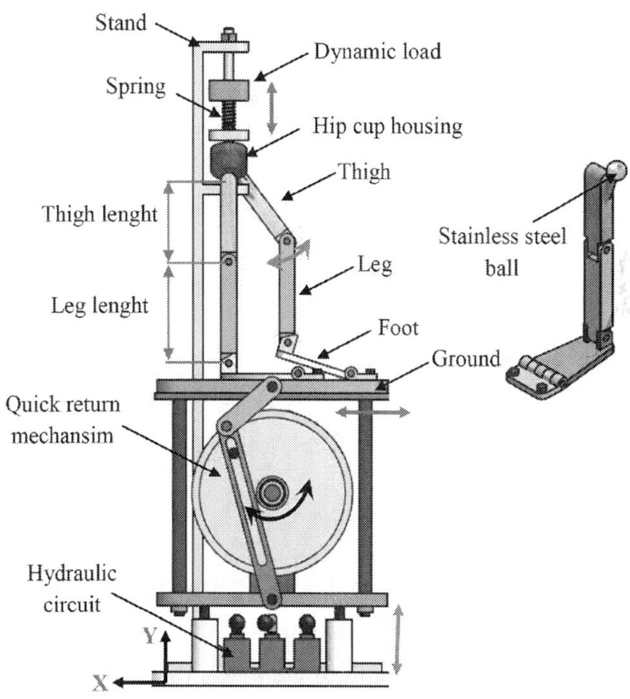

Figure 16. Joint Replacement Simulator.

metal and polymer components. Furthermore, loosening occurs and two types of wear failure mechanisms (sliding and rolling) appear. Finally, the simulator has the possibility to investigate the wear under dry and wet conditions by adding the lubricant inside the hip cup housing.

3.6. Wear hip cup investigation

The wear rate in the new design depends on six variables, such as the length of the femur, length of the tibia, gait angle, applied load, and metallic material (stainless steel or ceramic material), in addition to the lubricant conditions. To simulate the operating environment, usually the wear of the hip cup is investigated in dry or wet ambient environment. Most previous studies use distilled water or synovial fluid (SF) as a wet lubricant. SF can be defined as a natural lubricant and considered as the standard lubricant of these applications, because SF consists of many biological molecules such as proteins, lipids, and polysaccharides leading to improve the surface properties [51]. Distilled water is used as a simulated lubricant due to it being more abundant compared to SF. Furthermore, it is difficult to take all the previous variables into account during the testing process, but there are priorities that can be chosen depending on the human age and environmental effects. The wear behavior in this case can be evaluated by the weight or volume loss (in the form of mass loss) by weighing the tested hip cup on a high-sensitivity balance with high accuracy (10^{-4} g) after and before every test. The test can be repeated many times to determine the optimal percentage weight ratio of the CNT in the UHMWPE matrix. To investigate the effect of CNT added on the wear mechanism of the UHMWPE hip cup, optical microscopy is recommended to examine the worn surfaces. Finally, it is expected that these variables will affect the service life and give the most accurate values. Also, wear resistance will increase with the addition of CNT using the new dispersion technique and obstruct wear progress due to the following causes:

- The addition of paraffin oil during the injection process to UHMWPE led to a decrease in viscosity and then resulted in a good mixing (dispersed) with the CNT.
- Paraffin oil can be used also in this case as a lubricating oil. Similarly, CNT is used as a solid lubricant even in the event of agglomeration.
- Previous studies showed that the melting temperature of UHMWPE grows when mixed with the CNT. This process leads to the destruction of the surface layer because thermally softening occurs early, thus, decreasing the wear rate.

4. Conclusion

In this chapter, the author reviewed an overview of the AJ, including its components, most common polymeric materials used for these applications, reasons of failures, and characterization methods. Then, the author touched the methods of improving the performance of polymeric materials through mixing it with several types of nanofiller materials using many methods of dispersion. The chapter overview is focused on UHMWPE as a polymer base and CNT as a nanofiller. In addition, some nanofiller materials such as GA and CNF are presented also for comparison. The author presented a novel technique to produce a real UNC hip cup that has high wear resistance using the paraffin oil dispersion technique. Finally, to investigate the wear behavior of the new cup, AJS has been designed to be comparable to the human body movement approximately to increase the accuracy of the results. Finally, regarding the advantages of the polymer nanocomposite AJ compared to the traditional AJ, the wear resistance of the new joints will increase and thus increase the service life. The side effect summarized in the traditional AJ has biocompatibility slightly better than the nanocomposite.

Author details

Samy Yousef

Address all correspondence to: ahcann@hotmail.com

Department of Production Engineering and Printing Technology, Akhbar Elyom Academy, Giza, Egypt

References

[1] http://www.clipartof.com/portfolio/pmrk/illustration/3d-human-male-skeleton-sitting-and-thinking-1073684.html.

[2] http://www.niams.nih.gov/health_info/joint_replacement/#2.

[3] Charnley J. Arthroplasty of the hip. A new operation. Lancet 1 (1961) 1129–1132.

[4] http://www.hipandpelvis.com/for-patients/patient-education/patient-education-anterior-approach-tha/.

[5] Brems J. The glenoid component in total shoulder arthroplasty. J. Shoulder Elbow Surg. 2 (1993) 47–54.

[6] https://www.hss.edu/conditions_arthritis-of-the-knee-total-knee-replacement.asp.

[7] http://www.arthritiscentre.org/for-the-professional/services-for-gps-other-professionals/upper-limb-surgery-clinic/.

[8] Bistolfi A., Bellare A.. The relative effects of radiation crosslinking and type of counterface on the wear resistance of ultrahigh-molecular-weight polyethylene. Acta Biomater. 7 (2011) 3398–3403.

[9] Lu S., Liang G., Zhou Z., Li F.. Structure and properties of UHMWPE fiber/carbon fiber hybrid composites. J. Appl. Polym. Sci. 101 (2006) 1880–1884.

[10] Baena J.C., Wu J., Peng Z. Wear performance of UHMWPE and reinforced UHMWPE composites in arthroplasty applications: a review. Lubricants 3 (2015) 413-436.

[11] Hulbert S.F., Megremis S.J.. Effects of UHMWPE wear debris generated from total hip and total knee replacements. 1996 IEEE 0-7803-3 13 1.

[12] Valenza A., Visco A.M., Torrisi L., Campo N.. Characterization of ultra-high-molecular-weight polyethylene (UHMWPE) modified by ion implantation. Polymer 45 (2004) 1707–1715.

[13] Xue Y., Wu W., Jacobs O., Schädel B.. Tribological behaviour of UHMWPE/HDPE blends reinforced with multi-wall carbon nanotubes. Polym. Test. 25 (2006) 221–229.

[14] Saikko V., Shen M.. Wear comparison between a dual mobility total hip prosthesis and a typical modular design using a hip joint simulator. Wear 268 (2010) 617–621.

[15] Geary C., G.E. O'Donnell, Jones E., FitzPatrick D., Birkinshaw C.. Automated in-vitro testing of orthopaedic implants: a case study in shoulder joint replacement. Proc. Inst Mech Eng Vol. 224 Part H: J. Eng. Med.

[16] Saikko V., Ahlroos T., Calonius O.. A three-axis knee wear simulator with ball-on-flat contact. Wear 249 (2001) 310–315.

[17] Bascarevic Z., Vukasinovic Z., Slavkovic N., Dulic B., Trajkovic G., Bascarevic V., Timotijevic S.. Alumina-on-alumina ceramic versus metal-on-highly cross-linked polyethylene bearings in total hip arthroplasty: a comparative study. Int. Orthopaed. (SICOT) 34 (2010) 1129–1135.

[18] van der Pijl A.J., S′wiszkowski W., Bersee H.E.N.. Design of a wear simulator for in vitro shoulder prostheses testing. Technical Application Series, 2004.

[19] Rodriguez J.A. Cross-linked polyethylene in total knee arthroplasty. J. Arthroplast. 23 (2008).

[20] Joyce T.J., Unsworth A.. The design of a finger wear simulator and preliminary results. Proc. Inst. Mech. Eng. Vol. 214 Part H. S. Yousef. Polymer Nanocomposite Components: A Case Study on Gears (Woodhead, Cambridge, in press).

[21] Wannasri S., Panin S.V., Ivanova L.R., Kornienko L.A., Piriyayon S.. Increasing wear resistance of UHMWPE by mechanical activation and chemical modification combined with addition of nanofibers. Procedia Eng. 01 (2009) 67–70.

[22] Maksimkin A.V., Kaloshkin S.D., Kaloshkina M.S., Gorshenkov M.V., Tcherdyntsev V.V., Ergin K.S., Shchetinin I.V.. Ultra-high molecular weight polyethylene reinforced with multi-walled carbon nanotubes: fabrication method and properties. J Alloy Compd. 536 (2012) S538–S540.

[23] Lahiri D., Hec F., Thiesse M., Durygind A., Zhang C., Agarwal A.. Nanotribological behavior of graphene nanoplatelet reinforced ultra high molecular weight polyethylene composites. Tribol. Int. 70 (2014) 165–169.

[24] Wood W.J., Maguire R.G., Zhong W.H.. Improved wear and mechanical properties of UHMWPE-carbon nanofiber composites through an optimized paraffin-assisted melt-mixing process. Composites Pt. B 42 (2011) 584–591.

[25] Kanagaraj S., Mathew M.T., Fonseca A., Oliveira M.S.A., Simoes J.A.O., Rocha L.A., L.A. Tribological characterisation of carbon nanotubes/ultrahigh molecular weight polyethylene composites: the effect of sliding distance. Int. J. Surf. Sci. Eng. 4 (2010) 305–321.

[26] http://www.imatproject.eu/en/technology-111/carbon-nanotubes-124.

[27] Yousef S., Khattab A., Osman T.A., Zaki M.. Fully automatic system for producing carbon nanotubes (CNTs) by using arc-discharge technique multi electrodes. ICIES (2012).

[28] Yousef S., Khattab A., Osman T.A., Zak M.. Effects of increasing electrodes on CNTs yield synthesized by using arc-discharge technique. J. Nanomater. 2013 (2013) 392126.

[29] Mohamed A, Yousef S.. CNT mass production by using CVD multi quartz tubes.

[30] Eklund P.C., Pradhan B.K., Kim U.J., Xiong Q.. Large-scale production of single-walled carbon nanotubes using ultrafast pulses from a free electron laser. Nano Lett. 2 (2002) 561–566.

[31] Visco A.M., Yousef S., Galtieri G., Nocita D., Pistone A., Njuguna J.. Thermal, Mechanical and Rheological Behaviors of Nanocomposites Based on UHMWPE/Paraffin Oil/ Carbon Nanofiller Obtained by Using Different Dispersion Techniques. JOM-Springer 2015. DOI 10.1007/s11837-016-1845-x

[32] Yousef S., Visco A., Galtieri G., Nocita D.. Improved wear resistance of UHMWPE based nanocomposites for prosthetic applications, filled with liquid paraffin and carbon nanofibers. INSTM Conf. Italy (2015).

[33] Golchin A., Wikner A., Emami N.. An investigation into tribological behaviour of multi-walled carbon nanotube/graphene oxide reinforced UHMWPE in water lubricated contacts. Tribol. Int. 95 (2016) 156–161.

[34] Zuo J., Zhu Y., Liu S., Jiang Z., Zhao J.. Preparation of HDPE/UHMWPE/MMWPE blends by two-step processing way and properties of blown films. Polym. Bull. 58 (2007) 711–722.

[35] Yousef S., Visco A.M., Galtieri G., Njuguna J.. Wear characterizations of polyoxymethylene (POM) reinforced with carbon nanotubes (POM/CNTs) using the paraffin oil dispersion technique. JOM-Springer 2015. DOI: 10.1007/s11837-015-1674-3.

[36] Ruan S.L., Gao P., Yang X.G., Yu T.X.. Toughening high performance ultrahigh molecular weight polyethylene using multiwalled carbon nanotubes. Polymer 44 (2003) 5643–5654.

[37] Ma H., Chen X., Hsiao B.S., Chu B.. Improving toughness of ultra-high molecular weight polyethylene with ionic liquid modified carbon nanofiber. Polymer 55 (2014) 160–165.

[38] Chen Y., Zou H., Liang M., Liu P.. Rheological, thermal, and morphological properties of low-density polyethylene/ultra-high-molecular-weight polyethylene and linear low-density polyethylene/ultra-high-molecular-weight polyethylene blends. J. Appl. Polym. Sci. 2013. DOI: 10.1002/APP.38374.

[39] Kumar R.M., Sharma S.K., Kumar B.V.M., Lahiri D.. Effects of carbon nanotube aspect ratio on strengthening and tribological behaviour of ultra high molecular weight polyethylene. Composite 76 (2015) 62–72.

[40] Liu Y., Sinha S.K.. Wear performances and wear mechanism study of bulk UHMWPE composites with nacre and CNT fillers and PFPE overcoat. Wear 300 (2013) 44–54.

[41] Tai Z., Chen Y., An Y., Yan X., Xue Q.. Tribological behavior of UHMWPE reinforced with graphene oxide nanosheets. Tribol. Lett. 46 (2012) 55–63.

[42] Yousef S., Visco A.M., Galtieri G., Njuguna J.. Flexural, Impact, Rheological and physical Characterizations of POM Reinforced by Carbon Nanotubes and Paraffin Oil. Polymers for Advanced Technologies. (In press)

[43] Huang Y.-F., Xu J.-Z., Li J.-S., He B.-X., Xu L., Li Z.-M.. Mechanical properties and biocompatibility of melt processed, self-reinforced ultrahigh molecular weight polyethylene. Biomaterials xxx (2014) 1–11.

[44] Ge S., Wang S., Huang X.. Increasing the wear resistance of UHMWPE acetabular cups by adding natural biocompatible particles. Wear 267 (2009) 770–776.

[45] Chen Y., Qi Y., Tai Z., Yan X., Zhu F., Xue Q.. Preparation, mechanical properties and biocompatibility of graphene oxide/ultrahigh molecular weight polyethylene composites. Eur. Polym. J. 48 (2012) 1026–1033.

[46] Abdul Samad M., Sinha S.K.. Effects of counterface material and UV radiation on the tribological performance of a UHMWPE/CNT nanocomposite coating on steel substrates. Wear 271 (2011) 2759– 2765.

[47] Visco A.M., Campo N., Torrisi L., Cristani M., Trombetta D., Saija A.. Electron beam irradiated UHMWPE: degrading action of air and hyaluronic acid. Bio-Med. Mater. Eng. 18 (2008) 137–148.

[48] Yousef S., Khattab A., Zak M., Osman T.A.. Wear characterization of carbon nanotubes reinforced polymer gears. IEEE Trans. Nanotechnol. 12 (2013) 616–620.

[49] Yousef S., Osman T.A., Khattab M., Bahr A.A., Youssef A.M.. A new design of the universal test rig to measure the wear characterizations of polymer acetal gears (spur, helical, bevel, and worm). Adv. Tribol. DOI: org/10.1155/2015/926918.

[50] Yousef S., Osman T.A., Abdalla A.H., Zohdy G.A.. Wear characterization of carbon nanotubes reinforced acetal spur, helical, bevel and worm gears using a TS universal test rig. JOM-Springer, DOI: 10.1007/s11837-014-1268-5.

[51] Yousef S. Chapter 16: Polymer nanocomposite components: a case study on gears. Elsevier, 2016. Lightweight Composite Structures in Transport,. ISBN: 978-1-78242-325-6. http://dx.doi.org/10.1016/B978-1-78242-325-6.00016-5

[52] Kung M.S., Markantonis J., Nelson S.D., Campbell P.. The synovial lining and synovial fluid properties after joint arthroplasty. Lubricants 3 (2015) 394–412. DOI: 10.3390/lubricants3020394.

Chapter 10

Carbon Nanotube-Based UV-Curable Nanocomposite Coatings

Saeed Bastani and Masoume Kaviani Darani

Additional information is available at the end of the chapter

http://dx.doi.org/10.5772/62507

Abstract

This chapter covers the preparation and properties of ultraviolet (UV)-curable nanocomposite containing carbon nanotubes as fillers. UV-curing technology is of particular interest due to its unique properties such as rapid curing process and solvent-free formulation. Alongside with the advantages of utilizing this curing method, carbon nanotubes undergo benefits including high aspect ratio, high transparency, and good mechanical properties. Carbon nanotubes (CNTs) are hollow cylindrical shaped configuration; consist of one, two, or more walls with an interlayer of non-covalent van der Waals force acting among the carbon atoms of various walls. Besides influencing the UV curing process, the CNTs loaded UV-curable nanocomposites sustain modified surface, thermal, mechanical, physical, and conductive properties which are discussed in this chapter. The health and safety concerns of using these classes of nanocomposite are further discussed.

Keywords: Carbon Nanotubes, UV-Curable Coatings, Nanocomposites, Physical and Mechanical Properties, Ultraviolet Radiation

1. UV-curable nanocomposite coatings

In recent years, polymer nanocomposite coatings have found many applications due to their high strength, light weight, good fatigue, corrosion resistance, and superior optical properties and appearance, controlled anisotropic properties and low costs [1–5].

The addition of inorganic materials to polymers is an important method to produce materials benefiting each individual component properties. Theses composite materials are determined by the components behavior, the degree of dispersion and the interfacial properties [6]. This multi-functionality is important because polymer coatings should not only propose decorative or protective functions, but also should provide other demands such as electronic discharge, superior mechanical properties, large operating temperature range, and good chemical resistance [7–9].

Among nanocomposite preparation methods, ultraviolet (UV)-curing technique is largely employed to produce hybrid materials, especially in form of films, developing polymeric thermoset matrices. UV light allows a fast transformation of liquid monomer to solid films with tailored physicochemical and mechanical properties. This is known as a fast, environmental friendly method. Unlike traditional thermal curing, the substrate should not be heated for curing, so it can be finished at room temperature, no solvent is used and the complete conversion is obtained in seconds [6, 10–17].

Generally, nanocomposites are divided into two main categories: nanocomposites with filler directly incorporated inside, and in-situ production of fillers.

The properties of final UV-cured nanocomposites are dependent on factors such as properties of each component, size and shape of the fillers, morphology of system, and the nature of interface between components. There properties are as follows: high stiffness, high strength, and high dimensional stability, increased strength and toughness, high distortion temperature, good mechanical damping, increased permeability to gasses and liquids, excellent electrical properties, and low manufacturing costs.

In UV-curable nanocomposites, one should consider the absorption, refraction, and reflection of UV light by the reinforcing fillers for not reducing the curing rate and conversion [18].

2. CNT-based UV-curable nanocomposites coatings

Among reinforcing fillers, carbon nanotubes (CNTs), appears as an interesting candidate [11, 12]. CNTs are gaining scientific and industrial interests due to their outstanding characteristics such as high tensile strength and modulus, diffusion and transport properties, antivibration and damping properties, low intensity and excellent electrical and thermal properties [19–23].

CNTs can be considered as graphene sheets rolled to form a tube. Depending on synthesis route and reaction parameters, single-walled carbon nanotubes (SWCNTs), doubled-walled CNTs (DWCNTs), triple-walled CNTs (TWCNTs), and multi-walled CNTs (MWCNTs) can be formed (**Figure 1**). MWCNTs are grown of several independent tubes in concentric circles. The electronic properties are depending on the number of graphene walls [22, 24, 25].

Various methods have been reported for CNTs synthesizes, arc discharge [26], laser ablation [27–29], and chemical vapor deposition [30–32]. Arc discharge is the arc evaporation of graphite in the presence of inert gas, so the CNT is formed on electrode during quenching. The second method, laser ablation, vaporization of graphite is targeted by laser, then, CNTs are formed on receiver during quenching. The chemical deposition method which is the most used consists of decomposition of hydrocarbons over transition metal catalyst to form CNT [33].

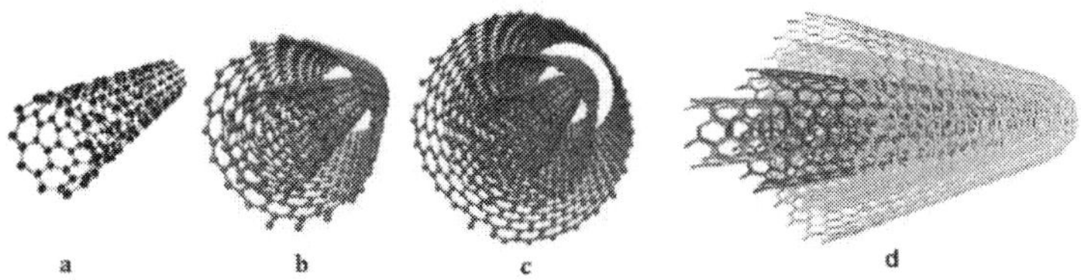

Figure 1. Schematic illustration of (a) SWCNT, (b) DWCNT, (c) TWCNT, and (d) multi-walled carbon nanotube (MCNT) [25].

The diameter of CNTs is in the range of 5–40 nm, which is 2000 times thinner than a human hair. The length is usually about 10 μm and their aspect ratio is in the range of hundreds to thousands and also they have strength s 10–100 times higher than the strongest steel at a fraction of weight and possess electrical current transfer capacity 1000 times greater than copper wire [1, 6, 13].

Among conductive fillers, CNTs offer the highest aspect ratio lead to higher transparency and better mechanical properties than carbon black or metal nanofillers. The only consideration to make is their black color and also the need of lowest possible concentration in formulation [24]. Besides, CNTs bear self-aggregation and poor solubility in water and organic solvents [34, 35]. In order to overcome these drawbacks, several approaches such as non-covalent and covalent functionalization have been reported. Covalent functionalization is grafting macromolecules using both *"grafting to"* and *"grafting from"* methods. These approaches make the CNTs to exhibit hydrophilic, hydrophobic, and amphiphilic properties [20, 21]. Non-covalent functionalization approach includes solution mixing [36–38], melt mixing [39, 40], and in situ polymerization [22]. It should be notice that the electrical properties of CNTs are influenced by their structure and diameter of nanotubes [41].

CNTs-based nanocomposites are used in many application including strain sensors [42], damage sensors [43], gas sensors [44], electromechanical actuators [45], conducting plastics-photovoltaic devices [46], optoelectronics [47], electrostatic dissipation [48], electromagnetic interference shielding [49], optical barriers [50], cost-effective transparent electronics [46], composite mirrors [51], plastics with high thermal dissipation, and biomaterial devices [52, 53]. They also have optical properties that fall in to the following categories: photoluminescence, light emission and photonic properties, optical non-linearity, and optical limiters [54–57].

3. Preparation of CNT-based UV-curable nanocomposites

To have a desirable CNT-based nanocomposite, homogeneously dispersion of CNTs within the polymer matrix is required [58, 59]. It is also very important to stabilize the dispersion in order to prevent aggregation of CNTs [60, 61]. Sonication and mixing are known to result in a good dispersion, but, the most proper method is surface functionalization of CNTs [62–66]. Though several mixing routes are known to disperse CNTs in polymer matrices including dry powder mixing, solution blending, melt mixing, in-situ polymerization and surfactant-assisted mixing [67–71]. **Table 1** represents the advantages and disadvantages of known CNT functionalization methods. By functionalizing CNTs, a direct bonding between the nanotubes and polymer is formed [72]. Researchers have shown that the high dispersion of these CNTs is achieved by sonication [20]. This functionalization is known to enhance the interface and load–transfer in CNTs/polymer nanocomposites. **Figure 2** reveals the functionalization of SWCNTs by two macromolecules.

Method		Principle	Possible damages to CNTs	Interaction with polymer matrix	Reagglomeration f-CNTs in matrix
Chemical method	Side wall	Hybridization of C atoms from Sp^2 to sp^3	*	Strong	*
	Defect	Defect transformation	*	Strong	*
Physical method	Polymer wrapping	Van der Waals forces, π–π stacking	–	Variable	–
	Surfactant adsorption	Physical adsorption	–	Weak	–
	Endohedral method	Capillary effect	–	Weak	*

Table 1. Advantage and disadvantage of various CNT functionalization routes [33].

Commonly, solution method, a combination of dispersing CNTs in a liquid medium by sonication, mechanical, and magnetic stirring mixing by the polymer solution, and finally evaporating solvents is suggested to prepare a CNT-based UV-curable nanocomposite [7, 19, 54].

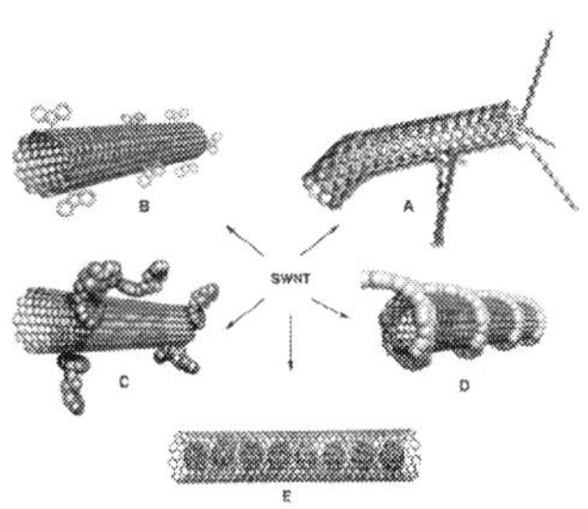

Figure 2. Different functionalized SWCNTs [73].

Sonication is considerable step for breaking aggregates and entanglements of CNTs [20, 74, 75]. Homogeneously dispersed MWCNTs UV-curable resin by solution mixing method can be achieved (**Figure 3**) [6].

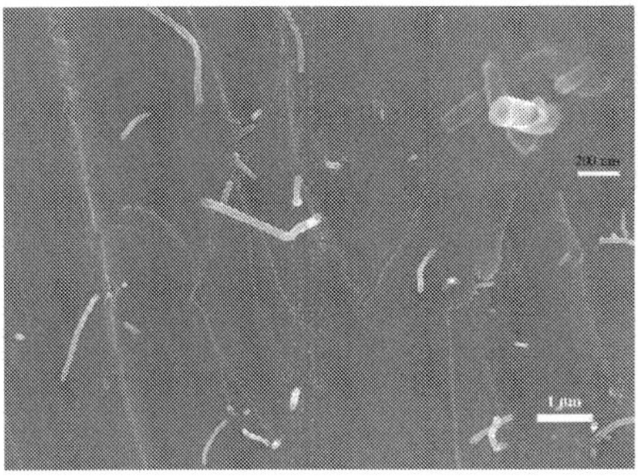

Figure 3. SEM rift image of the morphology of AG/MWCNTs UV-cured nanocomposite [6].

4. Characterization methods of CNT-based UV-curable nanocomposites

To characterize a CNTs/polymer nanocomposite, both the dispersion of nanofillers and the interface between components must be studied. These two factors have massive influence of the final nanocomposite properties.

The morphology investigation by the mean of dispersion is achieved by scanning electron microscopy (SEM) and transmission electron microscopy (TEM) [19, 21, 41].

The photo curing process is perused by photo-differential scanning calorimetry (photo-DSC) [10, 19, 21] and calculating gel content [7]. The kinetic of photopolymerization of nanocomposites is provided by real-time Fourier transform infrared spectroscopy (RT-FTIR).

Viscoelastic characterization, glass transition temperature (T_g), Tan delta (the value of loss factor), and damping values of the cured nanocomposites are carried out by dynamic mechanical thermal analysis (DMTA) and DSC [19, 23].

Thermal stability of the cured nanocomposite and its hardness can be investigated by thermogravimetry [19] and nanoindentation [7, 19], pencil test [13], respectively.

5. Properties of UV-cured nanocomposite coatings based on CNT

In this section, the properties of CNTs based UV-cured nanocomposites are discussed. As mentioned before, the final properties are influenced by the dispersion stated and interface between CNTs and polymer. Having a well-dispersed CNTs within the polymer matrix results in good mechanical, physical, electrical, and thermal properties which individually will be discussed. The addition of these filler to the polymeric matrices, influence the UV-curing process, though enhance the final properties of the loaded composite such as conductivity, surface and tribological properties, thermal and physical-mechanical of the cured nanocomposite.

5.1. Dispersion

The homogeneity and stability of CNTs in nanocomposites plays a significant role in the final product. **Figure 4** represents the TEM images of a CNTs/polymer nanocomposite. It reveals even after sonication, some agglomerates are present. This suggests the use of wetting and dispersing additives alongside with shear forces [10, 24].

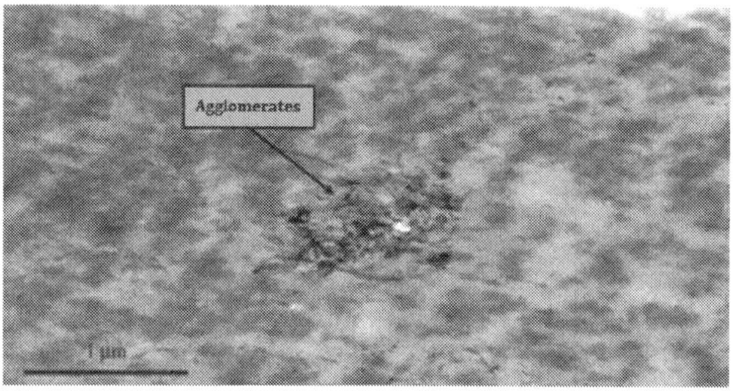

Figure 4. The remaining agglomerates in TEM images even after sonication [19].

A good dispersion of CNTs within the polymer matrix (**Figure 5**) increases the physical and mechanical properties of nanocomposite coatings [21].

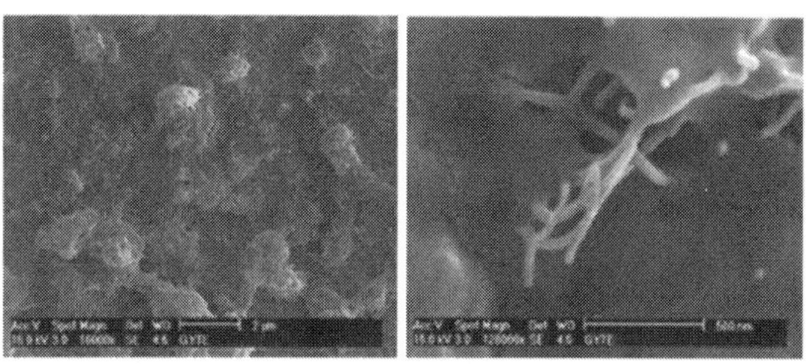

Figure 5. SEM images of CNTs/polymer nanocomposite indicating good dispersion and interaction between nanofiller and polymer matrix [21].

5.2. Curing process

As mentioned in Section 1, the nanofiller may itself absorb the UV light, resulting to a competition between photoinitiator and reinforcing nanofiller which can led to a less effective curing process.

Figure 6 illustrates the absorption spectroscopy results of SWCNTs. As can be seen, CNTs have light absorption in UV region.

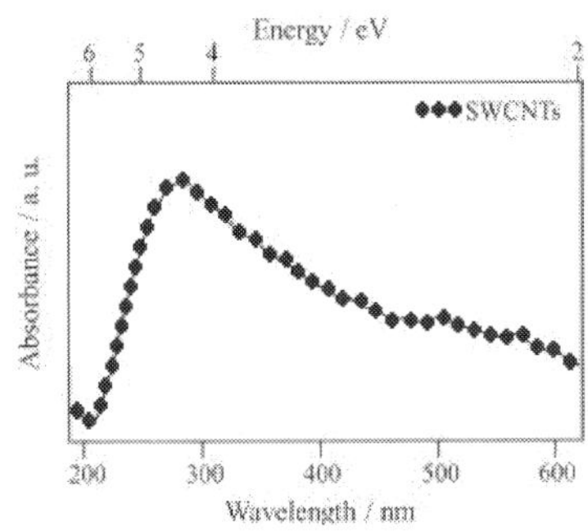

Figure 6. SWCNTs absorption spectroscopy.

To overcome this drawback, one can use higher UV light intensities and longer irradiation times [7, 76]. Another approach to dominate this effect of CNTs. They used a hyper-branched polymer (HP) in their formulation which caused an enhancement in final conversion. Their results (**Figure 7**) reveal by the aid of HP in the CNTs/polymer nanocomposite, a higher conversion in shorter time can be achieved.

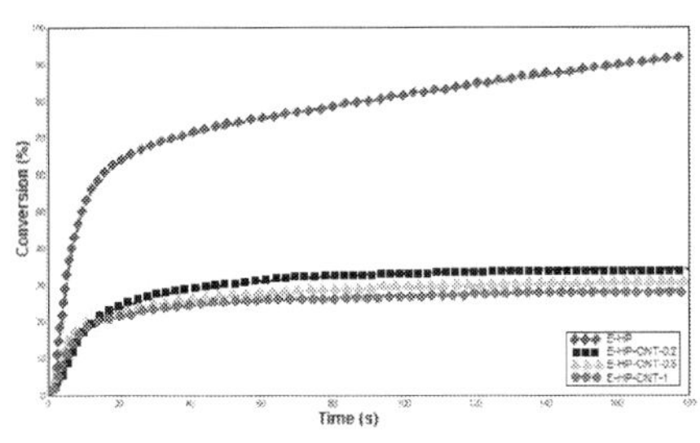

Figure 7. The effect of hyper-branches polymer in a CNT-based UV-curable nanocomposite [77].

Commonly used method for investigating curing process, kinetic of photopolymerization, and curing behavior of UV-curable polymer is by photo-DSC, RT-FTIR and measuring the gel content. In photo-DSC analysis, the heat of the photo-initiated polymerization reaction is measured [4, 21, 77], In RT-FTIR method, the conversion of active groups is followed by monitoring the decrease in of those groups peaks in the conversion curves as a function of radiation time (**Figure 8**) [7]. In this curve, the rate of polymerization (slope of the initial linear segment) and the final value of conversion can be driven. It is perceived by using CNTs, both the polymerization rate and the final conversion are decreased, but utilization of functionalized CNTs, rises the T_g of the system due to the formation of crosslinked polymer network, hampering the mobility of active groups while the reaction progresses result in higher value of conversion. This effect is more pronounced in UV-cured nanocomposite containing oxidized CNTs.

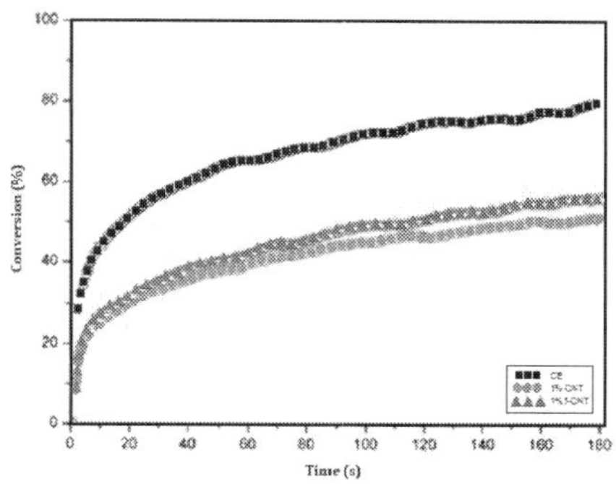

Figure 8. FTIR conversion curves for the neat cationically photocurable epoxy resin (CE) and it's composites at 1 wt% of multi-walled CNTs (1% CNT) and 1 wt% oxidized of multi-walled CNTs (1% f-CNT) content [7].

The gel content values of the cured coatings can be obtained by the weight loss after being adrift in solvent.

Table 2 represents the polymerization rate and gel contents of the neat polymer and the loaded ones. The results suggest that the nanosize filler have great absorption of the UV-light, leading to a strong shielding effect. It is revealed that these fillers reduce the generation of reactive types at the initial of UV-induced reaction [7].

Sample	Polymerization rate[a] (s^{-1})	Conversion[b] (%)	Conversion[c] (%)	Gel content (%)
CE	9.6	80	100	100
1% CNT	5.5	51	98	99
1% f-CNT	5.6	51	88	100

[a]Values of the slope at the initial stages of the conversion-time curves.
[b]Value of the plateau in the conversion curves with a UV-light intensity of 35 mW/cm^2 and an irradiation time of 3 min. Film thickness of 25 μm.
[c]Determined by single spectra taken before and after 5 min of irradiation at 55 mW/cm^2. Film thickness of 25 μm.

Table 2. Conversion values and gel content of the nanocomposites with 1 wt% loading fraction [7].

Furthermore, the addition in content of CNTs, influence the curing behavior of the nanocomposites. The more loading, leads to less curing conversion. This can be attributed to the competition of UV absorption between the fillers and photoinitiators [77]. This reduction in final conversion is shown in **Figure 9**.

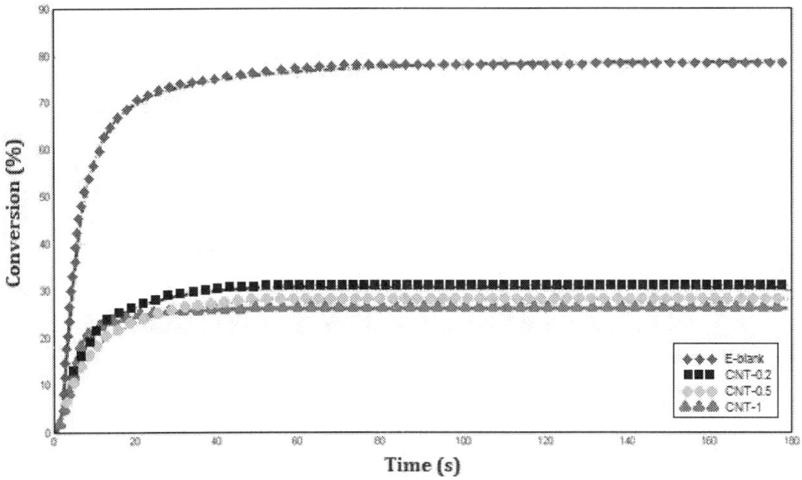

Figure 9. Curing behavior of photocurable epoxy acrylate with different CNTs content [77].

5.3. Surface properties

Surface properties of a UV-cured nanocomposite is divided into surface hardness and its surface chemistry.

CNTs are unsaturated systems owning highly mobile electrons, depending on the polymer chemistry, there would be distinct interactions [24]. Likewise, the functionalization type and method is very important parameter to be controlled.

5.4. Tribological properties

One indication of good dynamic mechanical property of nanocomposite can be the elevated surface hardness [7, 19]. Surface hardness can be measured by nano-/micro-indentation and pencil test [20]. CNTs nanofillers are known to increase this property. These results are shown in **Figure 10**.

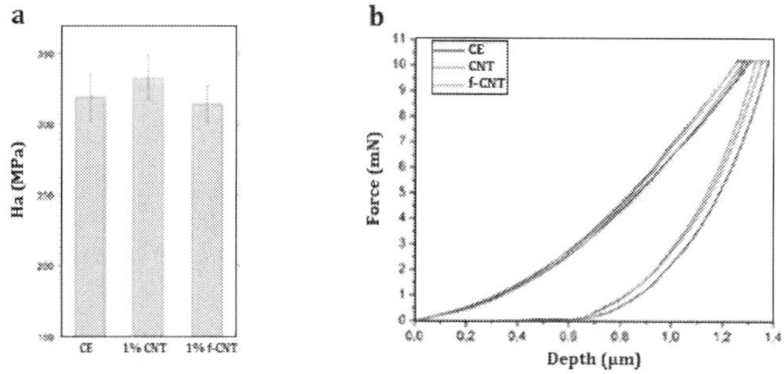

Figure 10. (a) Surface hardness values and (b) load vs depth curve [7].

CNTs have self-lubricating properties. In higher CNTs content in a nanocomposite, agglomerates are higher which increase the friction coefficient. Therefore, the tendency for matrix fragmentation is enhanced, producing loose wear debris at the contact interface [19]. **Figure 11** represents the friction coefficient of a UV-cured nanocomposite as a function of different loadings. It is observed that wear rate is raised by the increasing of CNTs content.

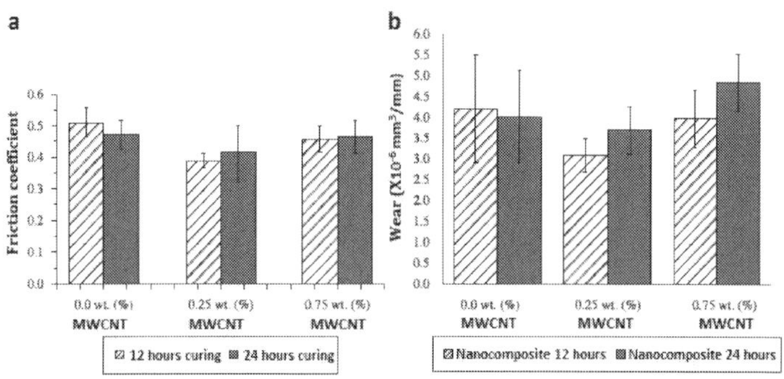

Figure 11. Friction coefficient (a) and wear rate (b) for CNT/polymer nanocomposite as a function of curing time by UV-A irradiation cycles 12 and 24 h [19].

5.5. Conductivity

CNTs embedded in polymer matrix, are protruded from the surface, offer permanent electrical, surface and volume conductivity [24]. In a CNT/polymer nanocomposite, the critical filler content is called the percolation threshold. By formation of a CNTs interconnected tridimensional network, a low-resistance path for the moving charge carriers is provided [54].

The geometry of nanofillers plays an important role in the formation of conductive pathways. Using CNTs would increase both electrical conductivity and dielectric permittivity of the cured nanocomposite. This enhancement (**Figure 12**) is a proof of the formation of a conductive network in polymer matrix [7].

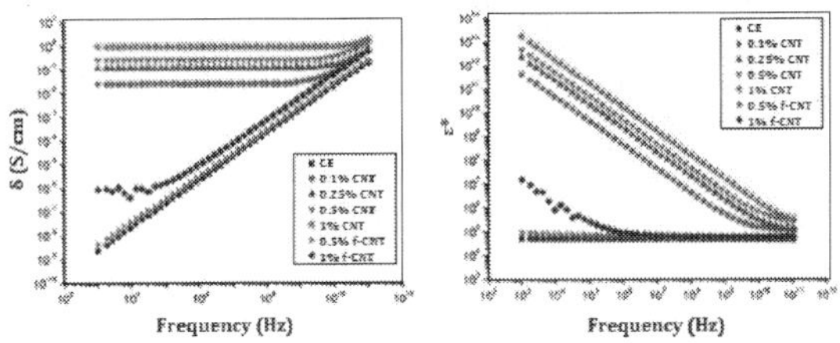

Figure 12. AC conductivity and permittivity vs frequency for nanocomposites at different loading [7].

By extrapolating the broadband AC conductivity, it is possible to obtain DC conductivity values (**Figure 13**). In this figure, the filled areas represent the electrical percolation region for each of nanocomposite differing in filler. It reveals that the most efficient nanofiller is MWCNT [7].

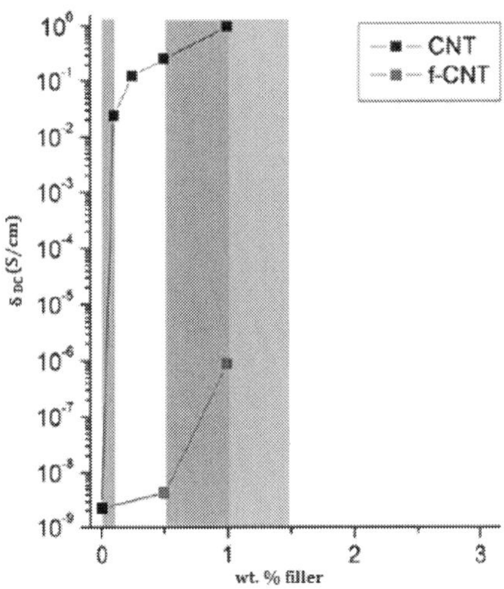

Figure 13. Values of the DC electrical conductivity of nanocomposite as a function of different nanofillers (CNT and oxidized-CNT) [7].

Besides electrical conductivity, industry is interested in advantages of thermally conductive polymer nanocomposite. These nanocomposites are light weight, corrosion resistant and easily processed, which make them better candidate than metal parts. Using CNTs with its superior efficient, thermal conductivity will overcome this demand [54].

5.6. Dynamic mechanical properties

In analogy to neat polymers, filled nanocomposites are known to have higher T_g. **Figure 14** demonstrates the tan delta curves as a function of temperature for prepared nanocomposites. The increased T_g is followed by a shift in the maximum value of tan delta and a reduction in damping. It can be revealed that CNTs would withhold the polymer chain movements led to damping decrement and higher T_g. By increasing the CNTs content, a stronger nanofiller/ polymer is attained, resulting in higher maximum value of tan delta and greater damping effect due to the covalent banding between the components [77].

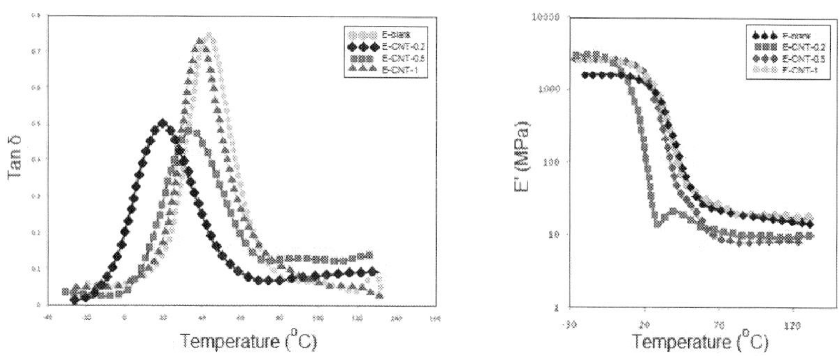

Figure 14. Tan delta and E′ curve obtained by DMTA analysis for nanocomposites containing CNTs and functionalized CNTs (f-CNT) [77].

Meanwhile, the changes in the modulus values driven by DMTA analysis are not significant (**Figure 15**), although a slight increase in rubbery region is detected [7].

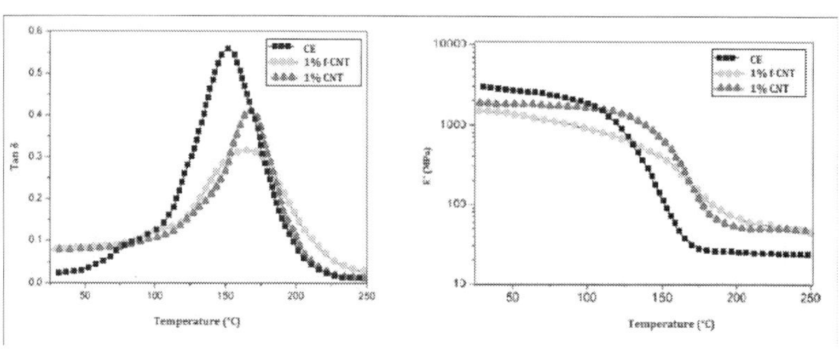

Figure 15. Modulus obtained by DMTA analysis for different nanocomposites containing CNTs and f-CNTs [7].

It is well known that the addition of CNTs will raise the storage modulus and stiffness of the nanocomposite (**Table 3 Figure 16**) [14, 19].

Samples	Storage modulus[a] (MPa)	Loss modulus[a] (MPa)	T_g (°C)
Neat 12 h	428	141	55
0.25 wt% MWCNT 12 h	570	166	58
0.75 wt% MWCNT 12 h	551	154	63
Neat 24 h	288	110	53
0.25 wt% MWCNT 24 h	507	135	58
0.75 wt% MWCNT 24 h	732	152	68

[a]Data taken at 40°C.

Table 3. DMA results for neat polymer and nanocomposites [19].

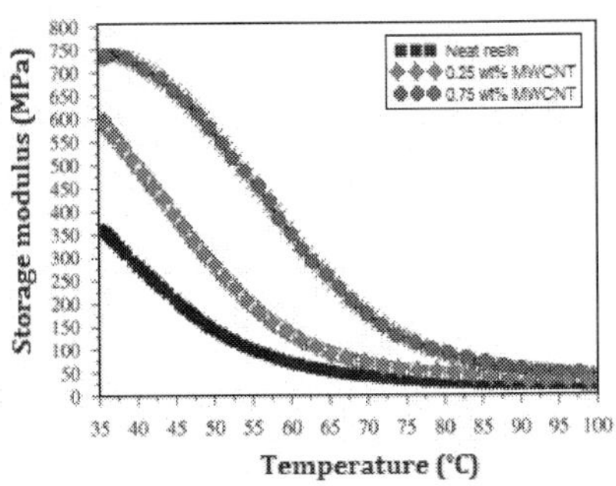

Figure 16. Storage modulus for neat resin and it's nanocomposite at different loadings [19].

In nanocomposites, CNTs frequently turn in their direction within the polymer matrix. This could improve the mechanical properties, likewise fatique resistance or bending strength. Even in the case of no improving in mechanical properties, the good dispersion of CNTs would prevent any negative impression on it [24].

6. Health and safety concerns

Using nanocomposites carrying nanofillers, some consideration must be count from fabrication of CNTs to the final application of the nanocomposite. The consideration for health and safety of humans and environment.

The concern about CNTs is using its powder form and dusts. To prevail this issue, dispersions of polymeric wetting and additives of high molecular weight is suggested [24].

When embedded in polymer matrix, the concerns are about releasing of CNTs to the environment after the polymer degradation (**Figure 17**) [78].

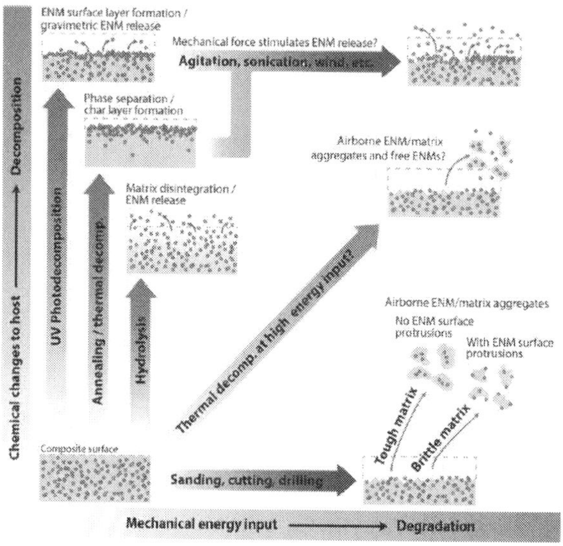

Figure 17. Schematic of releasing engineered nanomaterials (ENM) pathways from nanocomposite due to mechanical degradation or chemical decomposition of the host materials [79].

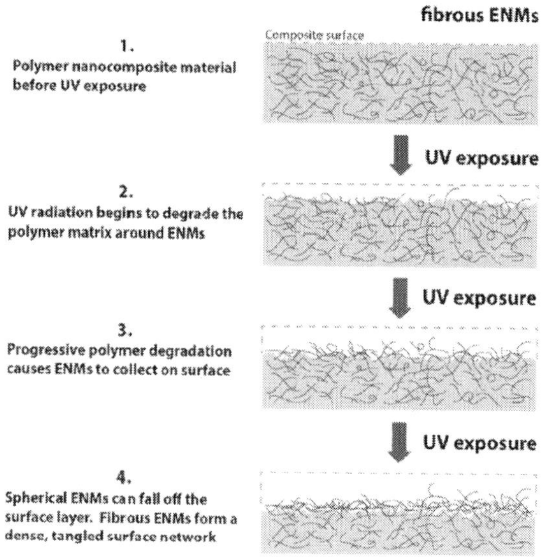

Figure 18. Conceptual model of engineered nanomaterial (ENM) released from UV-degraded polymer composite [79].

Figure 18 represents the schematic of the UV exposure of nanocomposite filled CNTs and its releasing procedure.

Researchers showed a good dispersion of CNTs within the polymer matrix may decrease the releasing of nanofiller to the environment (**Figure 19**). It can be seen a large amount of fragments are released from the bad dispersed material.

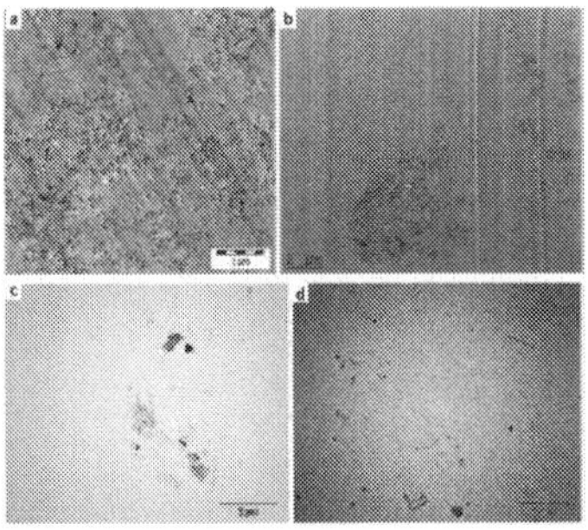

Figure 19. Electron microscope images of CNT/polymer composite with (a) good dispersion and (b) bad dispersion. (c) and (d) show particles released from these material [79].

7. Conclusion

Filled UV-curable nanocomposites have gained attentions due to their low cost, improved properties and being a fast and green process. Among nanofillers, carbon nanotube with their unique properties as a nanofiller have been widely used. A challenge in these classes of nanocomposite is the dispersion of the filler within the polymer matrix. Dispersion and the interface between components play an important role in the final properties of the cured nancomposite. A well-dispersed system led to better curing, enhanced mechanical properties, improved thermal, and electrical behavior. Finally, the degradation of these nanocomposites should take into account for the health of human and environment.

Author details

Saeed Bastani[1,2*] and Masoume Kaviani Darani[1]

*Address all correspondence to: bastani@icrc.ac.ir

1 Surface Coatings and Corrosion Department, Institute for Color Science and Technology, Tehran, Iran

2 Center of Excellence for Color Science and Technology, Tehran, Iran

References

[1] P. S. Gouda, R. Kulkarni, S. Kurbet, and D. Jawali, "Effects of multi walled carbon nanotubes and graphene on the mechanical properties of hybrid polymer composites," *Advanced Materials Letters*, vol. 4, pp. 261–270, 2013.

[2] L. P. Kollár and G. S. Springer, *Mechanics of Composite Structures*. Cambridge University Press, 2003.

[3] V. V. Vasiliev and E. Morozov, *Mechanics and Analysis of Composite materials*. Elsevier, 2001.

[4] M. Mohseni, S. Bastani, and A. Jannesari, "Influence of silane structure on curing behavior and surface properties of sol–gel based UV-curable organic–inorganic hybrid coatings," *Progress in Organic Coatings*, vol. 77, pp. 1191–1199, 2014.

[5] M. Benelmekki and M. Sowwan, "*Handbook of Nano-ceramic and Nano-composite Coatings and Materials*. In: H. Makhlouf and D. Scharnweber eds.," Amsterdam: Elsevier, (at press), 2015.

[6] S. Ceccia, E. Turcato, P. Maffettone, and R. Bongiovanni, "Nanocomposite UV-cured coatings: organoclay intercalation by an epoxy resin," *Progress in Organic Coatings*, vol. 63, pp. 110–115, 2008.

[7] M. Martin-Gallego, M. Hernandez, V. Lorenzo, R. Verdejo, M. Lopez-Manchado, and M. Sangermano, "Cationic photocured epoxy nanocomposites filled with different carbon fillers," *Polymer*, vol. 53, pp. 1831–1838, 2012.

[8] G. Mashouf, M. Ebrahimi, and S. Bastani, "UV curable urethane acrylate coatings formulation: experimental design approach," *Pigment & Resin Technology*, vol. 43, pp. 61–68, 2014.

[9] S. I. Yengejeh, S. A. Kazemi, and A. Öchsner, *A Primer on the Geometry of Carbon Nanotubes and Their Modifications*. Springer, 2015.

[10] R. F. Landel and L. E. Nielsen, *Mechanical Properties of Polymers and Composites*: CRC Press, 1993.

[11] Y. Tong, S. Bohm, and M. Song, "Graphene based materials and their composites as coatings," *Austin Journal of Nanomedicine & Nanotechnology*, vol. 1, p. 1003, 2013.

[12] S. C. Tjong, *Carbon Nanotube Reinforced Composites: Metal and Ceramic Matrices*. John Wiley & Sons, 2009.

[13] G. Kickelbick, "Introduction to hybrid materials," *Hybrid Materials: Synthesis, Characterization, and Applications*, p. 7, 2007.

[14] Y. G. Jeong and J. E. An, "UV-cured epoxy/graphene nanocomposite films: preparation, structure and electric heating performance," *Polymer International*, vol. 63, pp. 1895–1901, 2014.

[15] B. Yu, X. Wang, W. Xing, H. Yang, L. Song, and Y. Hu, "UV-curable functionalized graphene oxide/polyurethane acrylate nanocomposite coatings with enhanced thermal stability and mechanical properties," *Industrial & Engineering Chemistry Research*, vol. 51, pp. 14629–14636, 2012.

[16] B. Yu, X. Wang, W. Xing, H. Yang, X. Wang, L. Song, *et al.*, "Enhanced thermal and mechanical properties of functionalized graphene/thiol-ene systems by photopolymerization technology," *Chemical Engineering Journal*, vol. 228, pp. 318–326, 2013.

[17] M. Sangermano, S. Marchi, L. Valentini, S. B. Bon, and P. Fabbri, "Transparent and conductive graphene oxide/poly (ethylene glycol) diacrylate coatings obtained by photopolymerization," *Macromolecular Materials and Engineering*, vol. 296, pp. 401–407, 2011.

[18] P. Kardar, M. Ebrahimi, and S. Bastani, "Curing behaviour and mechanical properties of pigmented UV-curable epoxy acrylate coatings," *Pigment & Resin Technology*, vol. 43, pp. 177–184, 2014.

[19] M. N. D. Santos, C. V. Opelt, S. H. Pezzin, S. C. Amico, C. E. D. Costa, J. C. Milan, *et al.*, "Nanocomposite of photocurable epoxy-acrylate resin and carbon nanotubes: dynamic-mechanical, thermal and tribological properties," *Materials Research*, vol. 16, pp. 367–374, 2013.

[20] K. S. Wagner, "Investigate methods to increase the usefulness of stereolithography 3D printed objects by adding carbon nanotubes to photo-curable resins," 2014.

[21] Z. Doğruyol, G. Temel, S. K. Doğruyol, Ö. Pekcan, and N. Arsu, "Investigation of PSt-MWCNT concentration on epoxyacrylate photopolymerization and conductivity of polymer films," *Progress in Organic Coatings*, vol. 76, pp. 944–949, 2013.

[22] K. K. Kar, J. K. Pandey, and S. Rana, "*Handbook of Polymer Nanocomposites. Processing, Performance and Application.*"

[23] M. Yoonessi, M. Lebrón-Colón, D. Scheiman, and M. A. Meador, "Carbon Nanotube Epoxy Nanocomposites: the effects of interfacial modifications on the dynamic mechanical properties of the nanocomposites," *ACS Applied Materials & Interfaces*, vol. 6, pp. 16621–16630, 2014.

[24] M. Berkei, "Conductive Coatings Using Carbon Nanotubes: A Fascinating Material for the Coating Producer's Toolbox," *CHEManager Europe*, p. 10, 2011.

[25] P. Mukhopadhyay and R. K. Gupta, *Graphite, Graphene, and Their Polymer Nanocomposites*. CRC Press, 2012.

[26] T. Ebbesen and P. Ajayan, "Large-scale synthesis of carbon nanotubes," *Nature*, vol. 358, pp. 220–222, 1992.

[27] Y. Zhang, H. Gu, and S. Iijima, "Single-wall carbon nanotubes synthesized by laser ablation in a nitrogen atmosphere," *Applied Physics Letters*, vol. 73, pp. 3827–3829, 1998.

[28] J. Chrzanowska, J. Hoffman, A. Małolepszy, M. Mazurkiewicz, T. A. Kowalewski, Z. Szymanski, *et al.*, "Synthesis of carbon nanotubes by the laser ablation method: effect of laser wavelength," *Physica Status Solidi (B)*, 2015.

[29] G. Yang, *Laser Ablation in Liquids: Principles and Applications in the Preparation of Nanomaterials*. CRC Press, 2012.

[30] Y.-L. Li, I. A. Kinloch, and A. H. Windle, "Direct spinning of carbon nanotube fibers from chemical vapor deposition synthesis," *Science*, vol. 304, pp. 276–278, 2004.

[31] G. Allaedini, S. M. Tasirin, and P. Aminayi, "Synthesis of CNTs via chemical vapor deposition of carbon dioxide as a carbon source in the presence of NiMgO," *Journal of Alloys and Compounds*, vol. 647, pp. 809–814, 2015.

[32] S. Santangelo, E. Piperopoulos, M. Lanza, E. Mastronardo, and C. Milone, "Synthesis of three-dimensional macro-porous networks of carbon nanotubes by chemical vapor deposition of methane on Co/Mo/Mg catalyst," *Applied Catalysis A: General*, 2015.

[33] I.-Y. Jeon, D. W. Chang, J.-B. Baek, and N. A. Kumar, *Functionalization of Carbon Nanotubes*: INTECH Open Access Publisher, 2011.

[34] N. Moghtaderi, M. R. Bozorgmehr, and A. Morsali, "The study of self-aggregation behavior of the bilirubin molecules in the presence and absence of carbon nanotubes: molecular dynamics simulation approach," *Journal of Molecular Liquids*, vol. 208, pp. 342–346, 2015.

[35] K. Y. Cho, Y. S. Yeom, H. Y. Seo, Y. H. Park, H. N. Jang, K.-Y. Baek, *et al.*, "Rational design of multiamphiphilic polymer compatibilizers: versatile solubility and hybridization of noncovalently functionalized CNT nanocomposites," *ACS Applied Materials & Interfaces*, vol. 7, pp. 9841–9850, 2015.

[36] F. Bottacchi, L. Petti, F. Späth, I. Namal, G. Tröster, T. Hertel, *et al.*, "Polymer-sorted (6, 5) single-walled carbon nanotubes for solution-processed low-voltage flexible microelectronics," *Applied Physics Letters*, vol. 106, p. 193302, 2015.

[37] M. TabkhPaz, M. Mahmoodi, M. Arjmand, U. Sundararaj, J. Chu, and S. S. Park, "Investigation of chaotic mixing for MWCNT/polymer composites," *Macromolecular Materials and Engineering*, vol. 300, pp. 482–496, 2015.

[38] G. Mittal, V. Dhand, K. Y. Rhee, S.-J. Park, and W. R. Lee, "A review on carbon nanotubes and graphene as fillers in reinforced polymer nanocomposites," *Journal of Industrial and Engineering Chemistry*, vol. 21, pp. 11–25, 2015.

[39] M. T. Müller, P. Pötschke, and B. Voit, "Dispersion of carbon nanotubes into polyethylene by an additive assisted one-step melt mixing approach," *Polymer*, vol. 66, pp. 210–221, 2015.

[40] A. V. Poyekar, A. R. Bhattacharyya, A. S. Panwar, and G. P. Simon, "Evolution of phase morphology and 'network-like'structure of multiwall carbon nanotubes in binary polymer blends during melt-mixing," *Polymer Engineering & Science*, vol. 55, pp. 429–442, 2015.

[41] M. Sangermano, S. Pinneri, P. Calza, and C. Paganini, "Photocatalytic activity of epoxy/CNT nanocomposite films," *Macromolecular Materials and Engineering*, vol. 297, pp. 353–358, 2012.

[42] J. Deignan, J. Foroughi, S. Farajikhah, A. Jeirani, P. Innis, S. Coyle, *et al.*, "Wearable textile strain sensor with integrated carbon nanotubes," 2015.

[43] H. Zhang, Y. Liu, M. Kuwata, E. Bilotti, and T. Peijs, "Improved fracture toughness and integrated damage sensing capability by spray coated CNTs on carbon fibre prepreg," *Composites Part A: Applied Science and Manufacturing*, vol. 70, pp. 102–110, 2015.

[44] A. Karakuscu, L. H. Hu, A. Ponzoni, C. Baratto, R. Ceccato, G. Sberveglieri, *et al.*, "SiOCN functionalized carbon nanotube gas sensors for elevated temperature applications," *Journal of the American Ceramic Society*, vol. 98, pp. 1142–1149, 2015.

[45] P. Chen, S. He, Y. Xu, X. Sun, and H. Peng, "Electromechanical actuator ribbons driven by electrically conducting spring-like fibers," *Advanced Materials*, vol. 27, pp. 4982–4988, 2015.

[46] I. Jeon, K. Cui, T. Chiba, A. Anisimov, A. G. Nasibulin, E. I. Kauppinen, *et al.*, "Direct and dry deposited single-walled carbon nanotube films doped with MoO x as electron-blocking transparent electrodes for flexible organic solar cells," *Journal of the American Chemical Society*, vol. 137, pp. 7982–7985, 2015.

[47] E.-C. Cho, C. Li, J.-H. Huang, K.-C. Lee, and J.-H. Huang, "Three-dimensional conductive nanocomposites based on MWCNT networks and PEDOT: PSS as flexible transparent electrode for optoelectronics," *ACS Applied Materials & Interfaces*, 2015.

[48] C. L. Bertin, "Electrostatic discharge protection circuits using carbon nanotube field effect transistor (CNTFET) devices and methods of making same," Google Patents, 2015.

[49] A. P. Singh, M. Mishra, D. P. Hashim, T. Narayanan, M. G. Hahm, P. Kumar, *et al.*, "Probing the engineered sandwich network of vertically aligned carbon nanotube–reduced graphene oxide composites for high performance electromagnetic interference shielding applications," *Carbon*, vol. 85, pp. 79–88, 2015.

[50] I. Zahou, R. Mlika, R. B. Chaabane, and H. B. Ouada, "Effect of MWCNT loading on the morphological, optical and electrical properties of Nickel phtalocyanine based composites," *Synthetic Metals*, vol. 202, pp. 39–48, 2015.

[51] Z. Zhou, L. Chen, C. Liu, and S. Fan, "A curvature-controllable, convex-mirror actuator based on carbon nanotube film composites," *Carbon*, vol. 96, pp. 672–677, 2016.

[52] M. J. Mitchell, C. A. Castellanos, and M. R. King, "Surfactant functionalization induces robust, differential adhesion of tumor cells and blood cells to charged nanotube-coated biomaterials under flow," *Biomaterials*, vol. 56, pp. 179–186, 2015.

[53] V. S. Basker, T. Furukawa, M. C. Hakey, S. J. Holmes, C. W. Koburger III, and K. V. Singh, "Carbon nanotube structures for enhancement of thermal dissipation from semiconductor modules," Google Patents, 2015.

[54] A. Pantano, G. Modica, and F. Cappello, "Multiwalled carbon nanotube reinforced polymer composites," *Materials Science and Engineering: A*, vol. 486, pp. 222–227, 2008.

[55] X. Ma and H. Htoon, "Tailoring the photophysical properties of carbon nanotubes by photonic nanostructures," *Modern Physics Letters B*, p. 1530004, 2015.

[56] P. V. Jena, Y. Shamay, J. Shah, D. Roxbury, N. Paknejad, and D. A. Heller, "Photoluminescent carbon nanotubes interrogate the permeability of multicellular tumor spheroids," *Carbon*, vol. 97, pp. 99–109, 2016.

[57] S. Park, M. Vosguerichian, and Z. Bao, "A review of fabrication and applications of carbon nanotube film-based flexible electronics," *Nanoscale*, vol. 5, pp. 1727–1752, 2013.

[58] M. Campo, A. Jiménez-Suárez, and A. Ureña, "Effect of type, percentage and dispersion method of multi-walled carbon nanotubes on tribological properties of epoxy composites," *Wear*, vol. 324, pp. 100–108, 2015.

[59] X.-L. Xie, Y.-W. Mai, and X.-P. Zhou, "Dispersion and alignment of carbon nanotubes in polymer matrix: a review," *Materials Science and Engineering: R: Reports*, vol. 49, pp. 89–112, 2005.

[60] L. Vaisman, H. D. Wagner, and G. Marom, "The role of surfactants in dispersion of carbon nanotubes," *Advances in Colloid and Interface Science*, vol. 128, pp. 37–46, 2006.

[61] H. Hyung, J. D. Fortner, J. B. Hughes, and J.-H. Kim, "Natural organic matter stabilizes carbon nanotubes in the aqueous phase," *Environmental Science & Technology*, vol. 41, pp. 179–184, 2007.

[62] H. Chen, O. Jacobs, W. Wu, G. Rüdiger, and B. Schädel, "Effect of dispersion method on tribological properties of carbon nanotube reinforced epoxy resin composites," *Polymer Testing*, vol. 26, pp. 351–360, 2007.

[63] P.-C. Ma, N. A. Siddiqui, G. Marom, and J.-K. Kim, "Dispersion and functionalization of carbon nanotubes for polymer-based nanocomposites: a review," *Composites Part A: Applied Science and Manufacturing*, vol. 41, pp. 1345–1367, 2010.

[64] Z. Spitalsky, D. Tasis, K. Papagelis, and C. Galiotis, "Carbon nanotube–polymer composites: chemistry, processing, mechanical and electrical properties," *Progress in Polymer Science*, vol. 35, pp. 357–401, 2010.

[65] J. Hilding, E. A. Grulke, Z. George Zhang, and F. Lockwood, "Dispersion of carbon nanotubes in liquids," *Journal of Dispersion Science and Technology*, vol. 24, pp. 1–41, 2003.

[66] R. Shvartzman-Cohen, Y. Levi-Kalisman, E. Nativ-Roth, and R. Yerushalmi-Rozen, "Generic approach for dispersing single-walled carbon nanotubes: the strength of a weak interaction," *Langmuir*, vol. 20, pp. 6085–6088, 2004.

[67] V. Choudhary and A. Gupta, "Polymer/carbon nanotube nanocomposites," *Carbon Nanotubes–Polymer Nanocomposites*, pp. 65–90, 2011.

[68] M. Moniruzzaman and K. I. Winey, "Polymer nanocomposites containing carbon nanotubes," *Macromolecules*, vol. 39, pp. 5194–5205, 2006.

[69] C. Park, Z. Ounaies, K. A. Watson, R. E. Crooks, J. Smith, S. E. Lowther, *et al.*, "Dispersion of single wall carbon nanotubes by in situ polymerization under sonication," *Chemical Physics Letters*, vol. 364, pp. 303–308, 2002.

[70] X. Gong, J. Liu, S. Baskaran, R. D. Voise, and J. S. Young, "Surfactant-assisted processing of carbon nanotube/polymer composites," *Chemistry of Materials*, vol. 12, pp. 1049–1052, 2000.

[71] B. Krause, T. Villmow, R. Boldt, M. Mende, G. Petzold, and P. Pötschke, "Influence of dry grinding in a ball mill on the length of multiwalled carbon nanotubes and their dispersion and percolation behaviour in melt mixed polycarbonate composites," *Composites Science and Technology*, vol. 71, pp. 1145–1153, 2011.

[72] S. Campidelli, C. Klumpp, A. Bianco, D. M. Guldi, and M. Prato, "Functionalization of CNT: synthesis and applications in photovoltaics and biology," *Journal of Physical Organic Chemistry*, vol. 19, pp. 531–539, 2006.

[73] S. Yellampalli, "*Carbon Nanotubes–Polymer Nanocomposites*," InTech, Rijeka, Croatia, 2011.

[74] M. M. R. Nayini, S. Bastani, and Z. Ranjbar, "Synthesis and characterization of functionalized carbon nanotubes with different wetting behaviors and their influence on the wetting properties of carbon nanotubes/polymethylmethacrylate coatings," *Progress in Organic Coatings*, vol. 77, pp. 1007–1014, 2014.

[75] N. Martin and J.-F. Nierengarten, *Supramolecular Chemistry of Fullerenes and Carbon Nanotubes*. John Wiley & Sons, 2012.

[76] P. Kardar, M. Ebrahimi, and S. Bastani, "Influence of temperature and light intensity on the photocuring process and kinetics parameters of a pigmented UV curable system," *Journal of Thermal Analysis and Calorimetry*, vol. 118, pp. 541–549, 2014.

[77] F. Mirshahi, "Investigation of nano-hyper branched polymer on final properties of UV-curable nanocomposites containing graphene and CNT," 2013.

[78] T. V. Duncan and K. Pillai, "Release of engineered nanomaterials from polymer nanocomposites: diffusion, dissolution, and desorption," *ACS Applied Materials & Interfaces*, vol. 7, pp. 2–19, 2014.

[79] T. V. Duncan, "Release of engineered nanomaterials from polymer nanocomposites: the effect of matrix degradation," *ACS Applied Materials & Interfaces*, vol. 7, pp. 20–39, 2014.

Chapter 11

Carbon Nanotube Composites as Electromagnetic Shielding Materials in GHz Range

Marta González, Guillermo Mokry, María de Nicolás,
Juan Baselga and Javier Pozuelo

Additional information is available at the end of the chapter

http://dx.doi.org/10.5772/62508

Abstract

Following the development of the new electronic systems and communication networks, the levels of electromagnetic contamination have risen dramatically in the recent years. Every day, new studies appear searching for a way to mitigate the electromagnetic interferences (EMI). At the same time, the rapid evolution of technology forces the field to search for lighter and more efficient materials. The composites using carbon allotropes (such as carbon nanotubes) and polymers as reinforcement are gaining importance, due to the many advantages they exhibit in comparison to the materials that were used until now. A great number of applications require absorption to be the main electromagnetic shielding mechanism, thereby making this review necessary as a way to summarize the latest studies on CNT/polymer composites and how to improve the absorption mechanism by changing the morphology and composition of CNTs.

Keywords: Carbon nanotube, Electromagnetic shielding materials, Electromagnetic characterization, Nanocomposite, Electromagnetic absorber

1. Introduction

Electromagnetic interferences (EMIs) occur when electromagnetic signals are unintentionally transmitted from an emitter to another element by radiation and/or conduction, causing it to behave in an unexpected way [1]. The problem arises from high-frequency signals coupled to the main one that will radiate as they are conducted through and along the power wire. The wire will behave as an antenna that will pick up other signals and transfer them to the circuit elements of the device [2]. EMIs, though part of our modern world, affect all electrical and electronic devices, which are more indispensable everyday for the society. EMI shielding is thus needed to suppress or attenuate electromagnetic radiation from emitters with materials that are able to interact with those signals. Thus, one of the present needs is to find broad-band shields, able to neutralize electromagnetic radiation in the GHz range. This requirement arises from the fast development of electronics, which has led to microprocessors with greatly enhanced data transfer speeds that operate at higher frequencies. Furthermore, the miniaturization and manufacturing of such components demand high-performance and lightweight materials. Due to the wide impact of telecommunications, several financially strong industrial sectors are expectant about the progress of EMI shielding materials. These technological fields demand not only efficient shields but also materials that meet specific criteria for each engineered design. For example, chemical and corrosion resistance, lightweight, flexibility, tuneable morphology, processing easiness and inexpensiveness are requirements that the materials must fulfil in order to be applicable in flexible electronics (e.g., personal computers and mobile phones), aerospace (e.g., satellite and aircraft´s manufacture) and automotive (e.g., integrated circuits) industries [3]. In the field of aerospace engineering, a challenging sector is focused on military stealth. For this specific purpose, radar-absorbing materials are needed. Much effort has been done in this direction although absorption of electromagnetic energy remains challenging, due to the need of accomplishing both low-reflection and high-absorption losses [2].

The exploration of new functional materials that enable to effectively block or filter broad-band electromagnetic energy is an active field of research nowadays. Polymers offer several advantages over traditional metals and ceramics used for EMI shielding. They can be easily shaped; it is possible to prepare a variety of configurations and formulations and they are substantially lighter. Although polymers are electromagnetically transparent, different strategies are available to convert them into active electromagnetic shields. Graphite, carbon black and carbon fibres were the first to be combined with polymers for the fabrication of EMI shields [3–5]. The attention soon shifted to nanocarbons, since with lower-weight fractions, more conductive composites could be obtained. In addition, to enable the entire cross-section of a filler to be active in shielding, the dimensions of the conductive filler should be less than the length of penetration of the incident radiation [6]. This penetration length is usually very small for good conductors at high frequencies. Thereby, nanocarbon fillers are adequate to shield in the GHz range. In this context, carbon nanofibres, nanotubes and graphene, which have higher specific surface area and aspect ratio than their microscale analogues, are promising candidates for the preparation of efficient EMI shielding composites. Despite having been studied extensively in recent years, several factors affecting the performance of nanocarbon/polymer composites as high-frequency electromagnetic shields remain unexplored or not completely understood up to date. Indeed, there are few reported reviews on this application of nanocarbons [7–10], which show the need for new and reproducible experimental data to facilitate the construction of realistic models to improve our understanding.

http://dx.doi.org/10.5772/62508

2. Electromagnetic shielding mechanism

When an electromagnetic wave (E_I) impacts on a material (**Figure 1**), two waves are created on the surface: a reflected wave (E_R) and a transmitted wave into the material ($E_{I\text{-}R}$). Inside the material, a fraction of the wave ($E_{I\text{-}R}$) may be dissipated as heat (E_A) until it reaches the second surface of the material. At this point, two new waves appear: one that is transmitted through the surface (E_T) and a new wave that is reflected into the material. This process is repeated successively until it meets the criteria stated in equation 1:

$$E_I = \sum E_R + \sum E_A + \sum E_T \tag{1}$$

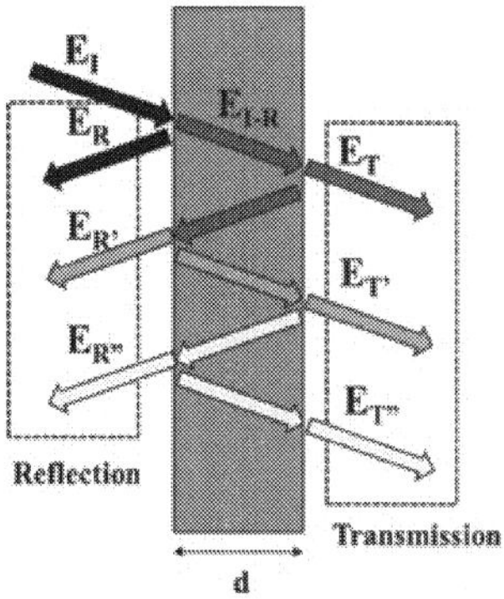

Figure 1. Mechanisms of attenuation of the incident EM power when it strikes a finite-dimensional media.

Both in the reflection and transmission processes, waves generated at each step may cause constructive and destructive interferences depending on the sample thickness and frequency. The reflection process on each plane of the material is what is called multiple reflections. Therefore, the electromagnetic shielding efficiency, *SE*, of a material can be quantified as the sum of three contributions: reflection, absorption and multiple reflections.

$$SE = SE_R + SE_A + SE_{MR} \tag{2}$$

The electromagnetic shielding efficiency can be expressed as a function of: impedance mismatch between the medium and the material (η_0 and η), sample thickness (d) and skin thickness (δ):

$$SE = 20log\left(\frac{\eta}{4\eta_0}\right) + 20log\left(exp\left(\frac{d}{\delta}\right)\right) + 20log\left(1 - exp\left(\frac{2d}{\delta}\right)\right) \tag{3}$$

where the skin thickness is defined as the depth at which the field decreases to $1/e$ from its initial value and is a function of frequency (f), permeability (μ) and conductivity (σ):

$$\delta = (\pi f \mu \sigma)^{-1/2} \tag{4}$$

The impedance can be expressed as the relation between the electric and the magnetic fields, where j accounts for an imaginary value and ε for the complex permittivity:

$$\eta = \frac{|E|}{|H|} = \left(\frac{j2\pi f \mu}{\sigma + j2\pi f \varepsilon} \right)^{1/2} \tag{5}$$

If the conductivity of the propagation medium is zero ($\sigma = 0$) and the conductivity in the conductive material is($\sigma \gg 2\pi f\varepsilon$), equation 5 can be simplified as:

$$\eta_0 = \left(\frac{\mu_0}{\varepsilon_0} \right)^{1/2} \quad and \quad \eta_0 = \left(\frac{2\pi f \mu}{\sigma} \right)^{1/2} \tag{6}$$

where the 0 index stands for the values in free space. Multiple reflections can be neglected when the sample thickness is bigger than the skin thickness. In this case, the absorbed radiation is high enough so that those constructive and destructive interferences cannot be produced, and only two mechanisms are possible in the electromagnetic shielding process: reflection and absorption. Taking this into account, and using equations 4 and 6, the electromagnetic shielding (equation 3) can be expressed as:

$$SE = \left[39.5 + 10 log \left(\frac{\sigma}{2\pi f \mu} \right) \right] + \left[8.7d (\pi f \mu \sigma)^{1/2} \right] \tag{7}$$

Electromagnetic shielding can be determined from the reflection (R), absorption (A) and transmission (T) coefficients as a function of the reflected, transmitted and incident powers (P_R, P_T and P_I, respectively), and therefore:

$$R = \frac{P_R}{P_I}; T = \frac{P_T}{P_I}; A = 1 - (R + T) \tag{8}$$

In this way, the electromagnetic shielding can be expressed as:

$$SE = 10log\left(\frac{1}{T}\right); SE_R = 10log\left(\frac{1}{1-R}\right); SE_A = 10log\left(\frac{1-R}{T}\right); \quad (9)$$

3. Measurement techniques

With a "Network Analyser," it is possible to study the properties of electrical networks, especially those associated with reflection and transmission of electrical signals, known as scattering parameters, from a few MHz to 100 GHz. A two-ports network analyser emits electromagnetic radiation (*I*) in the required frequency range from each side, and analyses the reflected (*R*) and transmitted radiation (*T*) through the studied material. The three most common configurations for measuring solid samples are waveguide, coaxial line and free space arrangements (**Figure 2**).

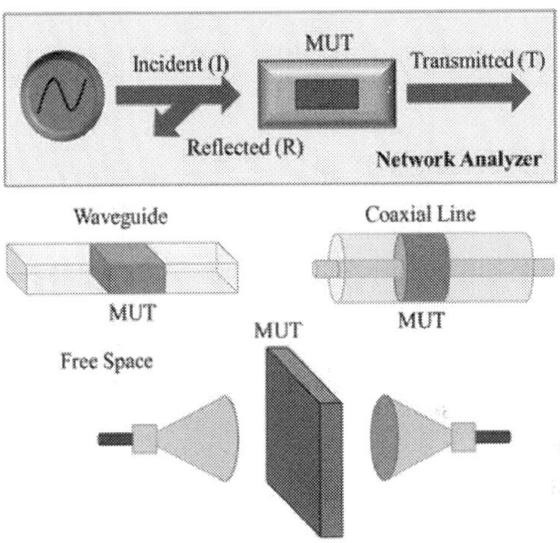

Figure 2. Configurations for EMI measuring solid samples.

Waveguide. Usually, it has a rectangular section. The sample is introduced at a precise distance from the waveguide end. The size of the waveguide depends on the frequency range for which it is prepared, and size decreases with increasing frequencies. The advantage of this system is the easiness of the sample preparation. The disadvantage is the narrow range of frequencies that can be measured. This causes that several waveguides are necessary to measure at high-frequency ranges.

Coaxial line. The sample is introduced between the inner and the outer conductors at an exact distance from the ends, and must be prepared as a rectangular toroid. For example, a coaxial line for measuring frequencies from 0.5 to 18 GHz has an inner conductor of 3.04 mm and a 7.00 mm external conductor. The main advantage is that it is possible to measure at high-frequency ranges on the same sample.

Free Space. It is a non-contact measurement based on two opposing antennae, flanking the sample. With this system, it is possible to measure a wide range of frequencies and to modify the angle of incidence of radiation on the samples. The most important disadvantage of this system is that the size of the samples must be considerably bigger than the previous techniques: Dimension range is from 30–40 cm for a few GHz to 10–15 cm for frequencies near 20 GHz.

With a two-ports network analyser, the voltage ratio between the generated and the returning electromagnetic waves for both ports is obtained (**Figure 3**).

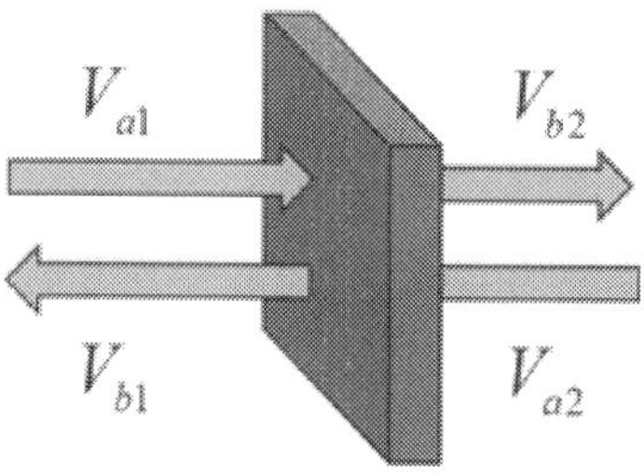

Figure 3. Voltage generated and returning in a two-ports network analyser.

where V_{a1} y V_{a2} are voltages that come out of ports 1 and 2, respectively, and V_{b1} y V_{b2} are voltages that return to ports 1 and 2, respectively. These voltages are related to the scattering parameters as:

$$\begin{pmatrix} V_{b1} \\ V_{b2} \end{pmatrix} = \begin{pmatrix} S_{11} & S_{12} \\ S_{21} & S_{22} \end{pmatrix} \begin{pmatrix} V_{a1} \\ V_{a2} \end{pmatrix} \tag{10}$$

Therefore,

$$V_{b1} = S_{11}V_{a1} + S_{12}V_{a2} \text{ and } V_{b2} = S_{21}V_{a1} + S_{22}V_{a2} \tag{11}$$

According to the maximum transfer theorem, if the opposite port is charged with an identical voltage to the system impedance, then scattering parameters will be:

$$S_{11} = \frac{V_{b1}}{V_{a1}};S_{22} = \frac{V_{b2}}{V_{a2}};S_{12} = \frac{V_{b1}}{V_{a2}};S_{21} = \frac{V_{b2}}{V_{a1}} \tag{12}$$

If the material is homogeneous, then $S_{11} = S_{22}$ and $S_{12} = S_{21}$. Therefore, S_{11} is the reflection produced in the sample and S_{21} is the transmission through the sample. The coefficients of reflection, absorption and transmission powers will therefore be expressed as:

$$R = \frac{P_R}{P_I} = |s_{11}|^2 \tag{13}$$

$$T = \frac{P_T}{P_I} = |s_{21}|^2 \tag{14}$$

$$A = 1 - (R - T) = 1 - \left(|S_{11}|^2 + |S_{21}|^2\right) \tag{15}$$

And therefore the shielding effectiveness is:

$$SE_T = 10log\left(\frac{1}{T}\right) = 10log\left(\frac{1}{|s_{21}|^2}\right) \tag{16}$$

$$SE_R = 10log\left(\frac{1}{1-R}\right) = 10log\left(\frac{1}{1-|s_{11}|^2}\right) \tag{17}$$

$$SE_A = 10log\left(\frac{1-R}{T}\right) = 10log\left(\frac{1-|s_{11}|^2}{|s_{21}|^2}\right) \tag{18}$$

Shielding effectiveness gives information about the preferred deactivation mechanism, while coefficients provide information on the fractions of the radiation lost by reflection and absorption, when multiple reflections are neglected. Multiple reflections produced by the interference of the reflected radiation in the first plane of incidence, and reflection produced in the last plane of the material, originate constructive and destructive interference dependant on the frequency. According to the Schelkunoff's theory [11–13], multiple reflections can be neglected when the sample thickness is larger than the skin depth, as mentioned above.

Figure 4 shows the skin depth as a function of frequency for samples with conductivities of 20, 30, 40 and 50 S.m^{-1}. In samples with a thickness of 4 mm, multiple reflections may be negligible above 5 GHz. In **Figure 4b**, the effect of multiple reflections on measurements of two different samples with the same thickness is shown. We can observe that the red sample gives a signal close to 2 GHz while the black sample gives three signals close to 2, 5 and 9 GHz, mainly due to the lower conductivity of the black sample. In order to avoid the sample thickness causing misleading reflections due to multiple reflections, the minimum frequency for a shielding study must be 5 GHz for the sample in red and 11 GHz for the sample in black.

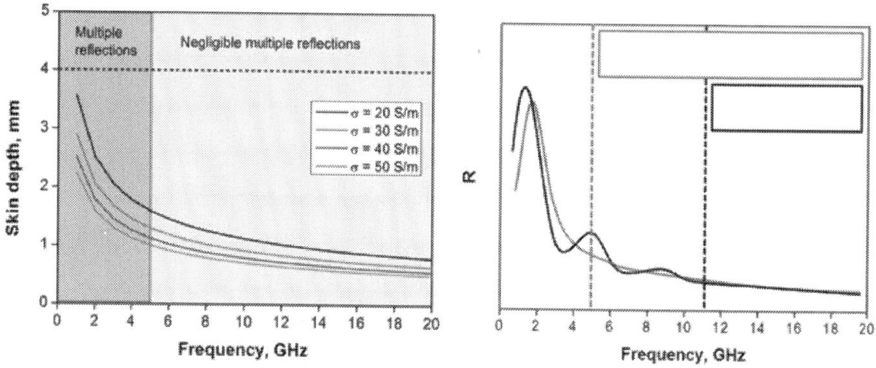

Figure 4. (a) Skin depth as a function of frequency for samples with different conductivities. (b) Effect of multiple reflections of two samples with the same thickness and different conductivity. Colored rectangles show the frequency range where multiple reflections are negligible.

4. Electrical properties of CNT composites

Electromagnetic shielding is closely related to the electrical properties of the composite. Although several conducting particles have been studied, such as carbon black particles [14], carbon fibres [15] or metallic fillers [16], CNTs have clearly demonstrated better properties due to their high aspect ratio (*L*/*d*), higher strength and flexibility and lower density, making them ideal as fillers.

There is a point, depending on the filler concentration in the composite, at which a conductive path is formed. This point has also been defined as the least concentration at which a composite with conductive inclusions is capable of conducting direct current. This is known as "percolation threshold," the point at which conductivity increases greatly. Although a broad range of polymers has been employed, there is a clear trend showing that with very small quantities of carbon nanotubes, it is possible to achieve the required percolation threshold. Below the percolation threshold, polymers behave as insulators for which conductivity, that increase with temperature, is generally determined by thermally assisted hopping or charge tunnelling between the charged particles [17]. On the other hand, when the percolation threshold is reached, conduction through the polymer occurs. A schematic diagram of how percolation threshold occurs is shown in **Figure 5**. Numerous studies have shown that the percolation threshold depends strongly on the polymer type, synthesis method used, aspect ratio of the CNT, how agglomerated these CNTs are in the polymer matrix and their degree of alignment [18].

http://dx.doi.org/10.5772/62508

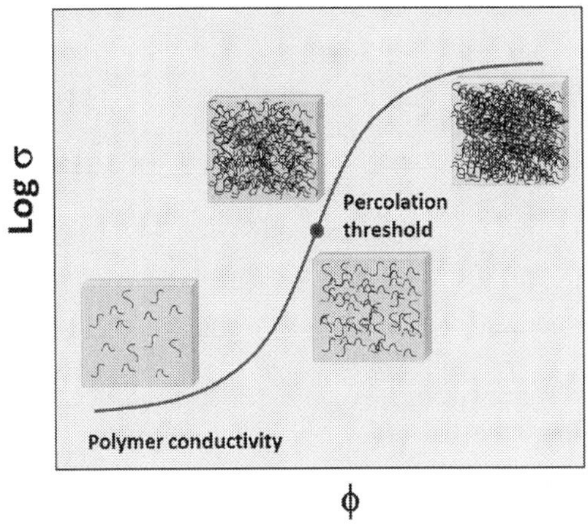

Figure 5. Schematic variation of electric conductivity with filler volume fraction for conductive nanocomposites.

The conductivity in these composites can have different values depending on the kind of conductivity that is measured. Conductivity can be measured either by direct current measurements (DC) or by alternating currents (AC). Determining the conductivity of a composite with direct current is quite straight forward, and there are several paths to achieve this. For a sample with length L and cross-section A, conductivity is given by:

$$\sigma_{DC} = \frac{L}{AR} \tag{19}$$

where R is the resistance which is measured applying a direct current and measuring a value for the voltage through the use of Ohm's Law. But this is not the only method available to calculate DC conductivity. The DC conductivity of the composites can also be determined from the frequency dependency of the AC conductivity graphs, as the value obtained for low frequencies where the conductivity is independent of the frequency used. The value of this plateau is the value of the DC conductivity. If the DC conductivity is known, then the percolation threshold can be easily calculated by using the known percolation theory, which defines an insulator-conductor transition and is stated as $\sigma_{DC}=\sigma_0(p-p_c)^t$, where σ_0 is a scaling factor, p is the CNT weight fraction, p_c is the percolation critical concentration and t is a critical exponent that governs the scaling law when close to the percolation threshold. This critical exponent is expected to depend on the system dimensionality with calculated values of 1.33 for two dimensions and a value of 2 for three dimensions [19]. The data can then be fitted to the equation in order to obtain the best-fitting values for percolation threshold and critical exponent for each case [20].

AC conductivity can be calculated as:

$$\sigma_{ac} = 2\pi f \varepsilon_0 \varepsilon'' \quad (20)$$

where f is the frequency at which the measurement is done, ε_0 is the dielectric constant in free space, ε_r'' is the imaginary part of the complex permittivity. It is therefore obvious that this conductivity depends on the frequency used. This is the reason why DC conductivity remains as a constant value when a sweep in frequency is done, whilst AC conductivity varies. To understand fully how the changes in current affects the medium, it is important to notice that carbon nanotubes in a dielectric medium can be compared to many capacitors inside the composite.

The permittivity of a medium represents how much flux (or electric field) is generated per unit charge in the medium. More electric flux is generated in a medium with lower permittivity due to polarization effects. An efficient dielectric supports a varying charge with minimal dissipation of energy in the form of heat. The real part of the complex permittivity (ε') is related to the stored energy in the capacitor when a polarization in the dielectric occurs. It can be explained as the ability of a material to polarize as a response to an applied electric field. So, the greater the polarization developed by a material in an applied electric field, the greater the value of the dielectric constant. However, in the presence of an alternating current, this polarization is different. As the alternating current switches, the direction of the field also switches and the polarization of the dielectric needs to change in order to align with the new direction. For this new orientation to occur, a small time known as relaxation time is needed and this value is usually close to 10–11 seconds. However, if the frequency of the AC current is higher than the relaxation time, the orientation of the dipole would not be able to keep up with this change and hence will cease to contribute to the polarization of the dielectric. Therefore, as the frequency of the AC current is increased, the value of the dielectric constant drops, until it becomes equal to 1, which is the same as the dielectric constant of vacuum. Dielectrics usually loose energy through two main mechanisms and this is known as the dielectric loss parameter (ε''); in conduction loss, a flow of charge through the material causes energy dissipation, and the dielectric loss usually occurs due to the movement of the charges in an alternating magnetic field as polarization switches direction. Dielectric loss is usually highest around the relaxation or resonance frequencies of the polarization mechanisms as the polarization starts to lag behind the applied field, which causes interactions between the field and the dielectric polarization that result in heating. This must be taken into account when calculating the AC conductivity, and thus the reason why it appears in the formula mentioned above [21, 22].

5. Electromagnetic shielding of CNT composites

A wide range of values for EMI SE and conductivities have been reported in the literature over the past few decades, depending greatly on the processing method, polymer matrix and carbon nanotube type (**Table 1**). This part of the chapter is designed to be a comprehensive source for polymer composite research including fundamental relationships between the CNT structure, percolation threshold values, processing techniques, EMI SE and conductivity values. The

http://dx.doi.org/10.5772/62508

results in most cases show an enhancement of the electrical conductivity by several orders of magnitude, with the addition of CNT to the matrix.

Matrix	CNT content	Thickness (mm)	σ S/m	SET (dB)	Frequency (GHz)	Reference
PU	22 wt %	0.1	5	20	8–12	[23]
PU	10 wt% MWCNTs	>0.2	12.4	29	X-band	[24]
PU	5 wt.% SWCNTs	2	100	22	X-band	[25]
PU	10 wt% MWCNT	2.5	790	41.6	X-band	[26]
PU/PEDOT	30 wt% MWCNT	2.5	275	45	12.4	[27]
PU	76 wt%	1	2100	80	X-band	[28]
Cellulose	9.1 wt %	0.2	375	20	15–40	[29]
Cellulose	0.45 vol%	0.01–0.02	1.80	20.8	8–12	[30]
PS	15 wt%	N/A	0.1	19	8–12	[31]
PAN	2 wt%	N/A	0.006	20	0.3–3	[32]
PMMA	40 wt% MWCNT	0.06–0.165	3000	27	0.05–13.5	[33, 34]
PMMA	10 vol% MWCNT	2.1	150	40	X-band	[35]
Epoxy	15 wt% SWCNT	1.5	15	30	X-band	[36]
PC	15 wt% MWCNT	6	1000	28	X-band	[37]
PTT	10 wt% MWCNT	1.5–2	30	42	Ku-band	[38]
PVDF	3.5 wt%	1.1	100	17.7	X-band	[39]
UHMWPE	10 wt%	1	100	50	8–12.5	[40]
PEDOT	15 wt%	2.8	1935	58	12–18	[41]
HDPE	18 wt% MWCNT	3	1000	58	0.5–1.5	[42]
PE	5 wt%	2.1	80	46.4	8–12	[43]
PCL	0.25 vol%	2	2.5	80	X-band	[44]

Table 1. Electromagnetic shielding of some CNT polymer composites.

As a general trend, it can be observed that the addition of carbon nanotubes increases the conductivity and electromagnetic shielding efficiency in the composites at the same frequency value. Jia et al. [42] stated that to obtain 20 dB (the minimum commercially acceptable EMI SE), a minimum conductivity of at least 1 $S.m^{-1}$ is necessary. It is worth mentioning that the thickness of the sample is an important factor to consider [41], as thicker samples would display higher EMI SE, as shown in equation 3. According to Pande et al. [33], multiple laminated composites exhibit higher absorption capacities than bulk composites, due to internal reflections taking place between the boundaries of each layer. This effect can be compared to the effect found on Salisbury screens.

EMI SE and conductivity are affected greatly by the processing techniques used. Al-Saleh et al. [39] compares various processing techniques and concludes that the best processing method in order to obtain the best EMI SE is by melt mixing [45]. This technique is better than solution processing [12], and at the same time solution processing is better than wet mixing [39].

When good EMI SE properties are required, melt mixing is the main technique used due to several factors; first, the dispersion of the carbon nanotubes obtained with this technique is much better than the one obtained with other techniques. Secondly, the nanotubes do not break with melt mixing, unlike with techniques such as ball milling or co-precipitation, which gives the composite a better conductivity overall and therefore enhances the electromagnetic shielding properties. Ultimately, melt mixing does not use any solvents, making this technique environmental-friendly unlike in-situ polymerization or solvent casting [46].

Some studies have shown that the percolation threshold depends strongly on the alignment and aspect ratio of the filler used. On this regard, rod-like fillers such as carbon nanotubes exhibit higher aspect ratios than spherical particles, allowing the percolation threshold to be achieved with lower filler concentration. If the fillers stick together, in a process known as agglomeration, the percolation threshold rises due to the aspect ratio being smaller as compared to well-dispersed fillers. Again, continuing with the importance of aspect ratio, Gupta et al. [26] and Huang et al. [35] demonstrated that the longer the CNTs the better the conductivity and therefore the electromagnetic shielding efficiency. Qin et al. [10] and Du et al. [18] proved that a secondary mechanism affecting the percolation threshold exists: alignment percolation. It was shown that aligned fillers tend to have higher percolation thresholds than anisotropic oriented fillers. This is due to the probability of a conductive path forming in the matrix being higher, than in oriented filler composites. Also, aligned fillers have worse conductivity, as fewer contacts between the tubes exist [45].

The type of carbon nanotube used, being single-walled carbon nanotubes (SWCNT) and multi-walled carbon nanotubes (MWCNT) the main types, also seems to have an effect on the EMI SE parameters. Comparing SWCNTs to MWCNTs, it is possible to notice the greater number of defects present in the MWCNT, which also causes higher permittivity. In a series of papers, it has been studied that the defects and impurities of CNTs have become impediment to properly describe the polarization mechanisms and ohmic losses of the composite materials [41]. This makes MWCNT have higher EMI SE, with the main shielding mechanism being absorption. However, other references analyse the differences in conductivity caused by the type of CNT used, and it seems that the impact on conductivity depending on CNT type is very small [42].

It is widely accepted that altering the carbon nanotube structure by functionalizing them disrupts the conductivity of isolated nanotubes. However, there are exceptions to this rule, as demonstrated by Tamburri et al. (2005) [22] where functionalizing the SWCNT with carboxylic groups enhanced the conductivity in the composites by a factor of 140, as compared with the unfunctionalized nanotubes which only showed an increase of 20 times the conductivity. This functionalization also improved the dispersion in the matrix, outweighing the advantages as compared to untreated carbon nanotubes. Apart from good EMI SE, composites must fulfil some important requirements, such as enhanced mechanical properties [27], light weight [40], good processability [23–26], low cost [38] and environmental-friendly [29, 44]. This is the reason why several polymer matrices have been studied in order to suit each and any of the desired applications, in particular EMI SE.

Dielectric properties of the polymer used are important, as polymers with high permittivity result in higher conductivities. The main objective is to search for conductive polymers (such as PEDOT), in order to block the EM radiation. However, some polymers are not conductive and alternatives for achieving high conductivities are searched. There are two main categories of the polymers used to achieve conductive composites: intrinsically conductive polymers (ICPs) and polymer composites with conductive particles. Polypyrrole and polyaniline are the most frequently used ICPs. Non-conductive polymers have been added to these composites to enhance the mechanical integrity of the mixture, as ICPs are normally brittle. Farukh et al. [27] used polyurethane with the previously mentioned PEDOT, with the objective to obtain a final composite with better elasticity, high impact strength and elongation properties. The drawback in most cases is the high load of ICP that is needed to attain high conductivities, which is detrimental for the processability of the blends. The alternative like mentioned above is to insert conductive fillers in the polymer matrix. The fillers do not depend only on the intrinsic electrical properties of the individual particles but also depend on the interparticle interactions. On this regard, carbonaceous materials such as CNTs and graphene have been used as conductive fillers, as they possess large conductivities with very low densities as compared to metallic particles, providing the composite with very good electrical properties and maintaining low weights.

From a theoretical point of view, four composites with a thickness of 4 mm and varying CNT compositions, which can originate DC conductivities of 20, 30, 40 and 50 S.m^{-1}, are studied. If the reflection and absorption shielding are calculated (**Figure 6**), it will be observed that increasing the conductivity increases both shielding mechanisms.

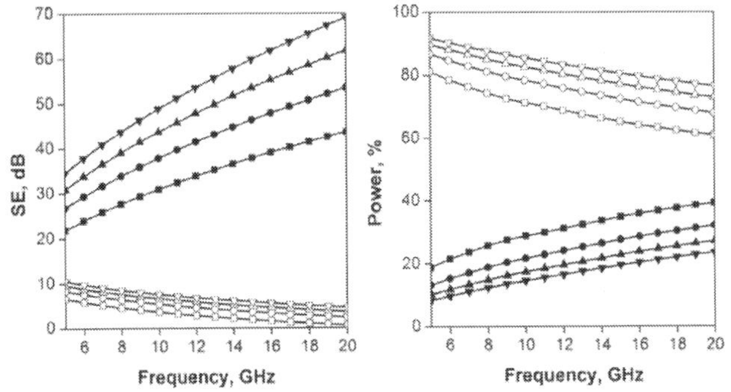

Figure 6. Reflection shielding and reflection coefficient (R) of specimens with 20 (□), 30 (○), 40 (Δ) and 50 S.m^{-1} (∇). Absorption shielding and transmission coefficient through the first plane of incidence ($1 - R$) of specimens with 20 (■), 30 (●), 40 (▲) and 50 (▼) S.m^{-1}.

The absorption shielding is the preferred mechanism in all cases with values higher than 20 dB for samples with only 4 mm thickness, whereas the reflection is always less than 10 dB. This would suggest that the predominant process is absorption. However, if the reflection coefficient from equation 8 is calculated, the conclusions are not the same.

Figure 6 shows the reflection coefficient (R) and the transmission coefficient through the first plane of incidence ($1 - R$). If these values are analysed, it is observed that between 60 and 90% of the incident radiation is reflected, while the transmission coefficient is never over 40%. This means that while the preferred shielding mechanism is the absorption, most of the radiation will not be able to enter the material and therefore will be unable to absorb it.

We can conclude, from the previous explanation, that both reflection and absorption increase with the conductivity of the material. High conductivity promotes both processes: absorption increases but reflection is also favoured, so that just a small part of radiation is able to penetrate into the material and behaves as a reflector. Low conductivity worsens both reflection but mainly absorption, as more radiation is able to penetrate into the material and thus be absorbed. Nevertheless, when conductivity is low, the multiple reflection process increases and the material behaves as a reflector. Therefore, is high or low conductivity necessary? "That is the question".

Reflection process is located in the incidence plane, while absorption process depends on the material thickness. Modification of the first incidence plane to decrease the impedance mismatch between the medium and the material is therefore necessary.

In an attempt to minimize the reflection from a surface, it is useful to consider the physical equations that represent the reflection process. This equation describes the reflection coefficient at an interface:

$$r = \frac{\eta_M - \eta_o}{\eta_M + \eta_o} = \frac{Z_M - Z_o}{Z_M + Z_o} \tag{22}$$

where r is the reflection coefficient and η the admittance of the propagating medium (subscript o for incident medium or air and M for the substrate). The admittance in this equation can be replaced by the intrinsic impedance ($Z = 1/\eta$). The reflection coefficient falls to zero when $\eta_M = \eta_o$, or in other words the material in the layer is impedance matched to the incident medium.

Absorbers can be classified into impedance matching and resonant absorbers (Salisbury screens). Impedance matching (pyramidal) absorbers are shielding materials that reduce the impedance step between the incident and the absorbing media. In resonant materials, the impedance is not matched between incident and absorbing media and the material is thin so that the power is not completely absorbed. Pyramidal absorbers [45] are typically thick materials with pyramidal or cone structures extending perpendicular to the surface in a regularly spaced pattern. The interface of these structures presents a gradual transition in impedance from air to that of the absorber. The height and periodicity of the pyramids tend to be on the order of one wavelength. The disadvantage of pyramidal absorbers is their thickness and fragility. They are usually used for anechoic chambers.

It is well known that a "Salisbury screen" is based on a sandwich structure of a dielectric between two conductive sheets. When the electromagnetic wave is incident upon the first layer, a portion of the radiation is reflected, while another portion penetrates into the dielectric sheet towards the second conductive sheet. In the second plane of incidence, the reflection of radiation occurs in the direction of the first plane of incidence and part of this is able to go through this conductive sheet and towards the exterior. The reflected radiation in the first and the second planes can produce constructive and destructive interference. Destructive interference occurs when the distance between the conductive sheets is a multiple of a quarter of the wavelength (see **Figure 7**).

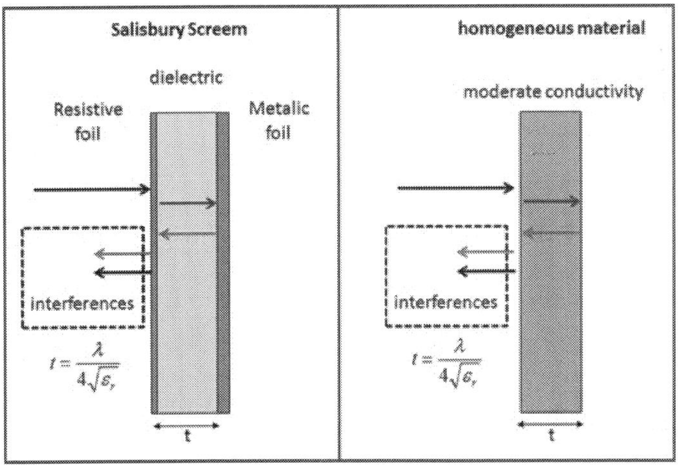

Figure 7. Interferences between reflections produced over the first and second planes in a Salisbury screen and a homogeneous conductive material.

The process is equivalent in homogeneous material with moderate conductivity. In both cases, a maximum electromagnetic shielding occurs at a certain frequency. In **Figure 7**, a qualitative behaviour of (SE) with the permittivity in samples with thickness of 4 mm is shown. It is possible to observe how a maximum displacement of the shielding occurs by increasing the permittivity of the dielectric. In homogeneous materials with moderate conductivity, processes of energy dissipation can exist. These processes that arise from the first to the second plane, back and forth propagation of the waves, dissipate some of the radiation and therefore interferences that should have been produced in the first plane are lower.

Another form to minimize the impedance mismatch between the medium and the sample is the use of porous materials. To address the problem of the impedance mismatch when using highly conductive carbonaceous nanoparticles, special focus on their microscopic organization should be taken. Optimally, a highly porous outer surface with a pore size similar to the incident radiation would be required. In principle, when radiation impinges on the surface of a porous conductive material, it becomes distributed among the pores and their walls. The major portion of the reflected radiation will arise from reflections with the pore walls, being reflections from the inner part of the pores a much smaller contribution. Therefore, the fraction and size of the pores created on the surface of a conductive compound may be good tools for modulating the phenomenon of electromagnetic reflection. If the process of reflection is decreased, a larger fraction of the radiation will penetrate into the material, being able to remove it by another mechanism (absorption). Different studies support this effect. For example, Zhao et al. [47] studied the shielding behaviour of carbon fibre epoxy composites, varying the thickness of the fibres and the size of the grid; the authors observed that when the ratio grid size/thickness of fibre increased, the reflection of electromagnetic radiation decreased. These data confirm the possibility to decrease the impedances mismatch for minimizing the reflection losses using porous structures in such a way that the walls are formed by conductive fibres for increasing the absorption in the material.

An important consideration for the designer of absorbing materials concerns the absorption efficiency with respect to weight. CNTs absorbing properties originate either from polarization, ohmic losses or multiple scattering due to the large specific surface area. There are differences in absorbing properties depending on the type of CNTs [48]. As we have commented previously, compared with single-walled CNTs (SWCNTs), multi-walled CNTs (MWCNTs) have more defects due to their complicated structure and higher permittivity, and their absorption is due to dielectric relaxation. The absorption properties of CNT composites depend on multiple factors [4, 10]: interfacial polarization, particle geometry, fibre concentration and particle dispersion.

Composite materials with conducting fibres have a frequency-dependent effective permittivity. To characterize the electromagnetic properties of composite media, it is important to know the electromagnetic parameters of a host (matrix, base) material and inclusions.

6. Trend in electromagnetic shielding materials

The development of new materials with the purpose of electromagnetic shielding has evolved in the past few years towards the study of carbon structures. The three principal types of carbon architectures employed for this purpose are carbon nanotubes and graphene.

6.1. Carbon nanotubes

Carbon nanotubes, in the form of cylindrical carbon molecules, exhibit high strength and count also with high electrical and thermal conductivity due to their symmetry. These allotropes can be formed by a single (SWNT) or several walls (MWNT), and the variation in their diameter can vary greatly their electrical properties [46, 49]. As studied in previous sections, their aspect ratio has a great effect in electromagnetic shielding applications. Carbon nanotubes have been used in several investigations as reinforcements in host matrices, in order to accomplish a 3D shielding structure. In particular, Li et al. merged single-walled carbon nanotubes with a polymeric matrix (15%), and a EMI SE of 15–20 dB was obtained for 1.5 GHz of frequency, a frequency range dominated by reflection [50]. Increasing their weight percentage in the composite enhances the shielding efficiency of the material but decreases its mechanical properties and processability, so a compromise between these properties for each application must be obtained. **Figure 8** shows CNT foams synthesized by chemical vapour deposition (CVD) with SE_T of to 22 dB in almost the entire frequency window (1–18 GHz) for the 2.38 mm thick CNT-foam slab [11]. CNT foams possess extremely low densities lower than 0.02 g·cm^{-3} and the higher specific shielding efficiency obtained of 1100 dB·cm^3·g^{-1}.

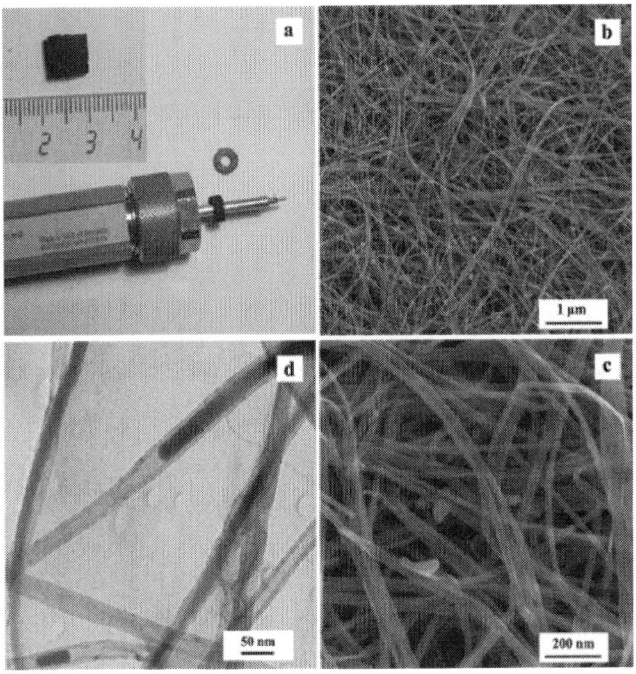

Figure 8. (a) Specimens and coaxial airline used for EMI shielding measurements. (b,c) SEM images of sponges. (d) TEM image of CNTs that are forming the sponges [13].

6.2. Graphene

Graphene can be regarded as the current state-of-the-art material in the study of EMI shielding. It is a two-dimensional hexagonally-packed carbon structure, of sp^2 hybridization, where the free carbon electrons align perpendicular to the plane, thus forming an out-of-plane π bond. Graphene counts with high mechanical properties, charge carriers mobility, thermal conductivity and light weight and quantum Hall effect at room temperature. Its ability to form three-dimensional structures is the main advantage that differentiates it from carbon fibres and nanotubes, enhancing this carbon hybrid as a suitable candidate for the ES of structures, such as radars.

Conductivity of graphene decreases as the layer increases, approaching to that of graphite. This conductivity is determinant for the losses in reflection and conduction; on the other hand, polarization loss is enhanced by the functional groups and defects. The decreased thickness of graphene enhances the conduction loss due to the increase in conductive paths. Polarization loss is the reason why chemically fabricated graphene is normally used for EMI shielding, whose conductivity is, in fact, lower.

Over time, ultra-thin graphene composites arose, with which more than 20 dB of EMI SE could be achieved. This mixture combines the excellent EMI shielding properties of graphene, with the mechanical strength of the matrix material. Graphene has been blended with multiple polymers, such as epoxy [51], polystyrene (PS) [52], polyethylene (PE) [53] and polyurethane (PU) [54, 55], among others [56].

More specifically, as mentioned before, the possibility to form three-dimensional (3D) porous foams with it have been carefully studied in recent investigations. As porous carbon structures, they count with excellent conductivity, and a high specific surface area can be achieved. With these holes, reflection can be lowered, due to the decrease in the external surface of the structure, and absorption can be enhanced, thanks to the multi-reflection inside the pores. **Figure 6** shows a pore structure inside a three-dimensional CNT structure.

In an investigation of Zhang et al. [57], a three-dimensional graphene porous structure was fabricated by hydrothermal polymerization of a carbon source (a mixture of phenol and formaldehyde) with GO (graphene oxide) in order to get these three-dimensional structures, which were later activated chemically—with KOH—to achieve the conductivity and desired surface specific area values. It is worth mentioning that surface specific area increases when GO is combined with a carbon source: the carbon source or GO by themselves had much lower values of specific surface area. The obtained material, few-nanometres long, had a specific surface area of 3523 $m^2 \cdot g^{-1}$ and a conductivity of 303 $S.m^{-1}$.

Another type of graphene three-dimensional structures were the graphene sponge (GS) scaffolds infiltrated in epoxy resin to form a GS/epoxy composite, built and studied by Li et al. [58]. There are two main problems of manufacturing graphene-polymer composites. The first is the reduced dispersion of graphene in a polymeric matrix, as a result of the graphene sheets' strong intermolecular connections and their high specific surface area; the second one arises when low filler content is added, so that the graphene sheets are surrounded by polymer chains, which affects the thermal and electrical conductivity of the composite material. It is clear then that an optimum ratio of polymer/graphene is needed in order to avoid these problems. In this study, the assembly into a three-dimensional sponge of GO sheets was performed using a hydrothermal plus freeze-drying treatment, followed by a vacuum infusion process to obtain the final composite. This method permits ease in the production, saving resources (low cost) and can be applied at different scales.

GO suspension was synthesized using the modified Hummers' approach. The GS obtained had a density of 0.022 g.cm^{-3}, an ultra-low value for this property. The epoxy resin could be infused into the GS because of its open pore structure. The size and of the GS did not change after the vacuum process, showing that no shrinkage occurs. It could be stated that the conductivity of the GS/epoxy composite was comparable to that of the GS. It is worth to mention that the conductivity of the graphene sponges is isotropic; an effect accredited to the random distribution of the reduced graphene sheets.

The thickness of these graphene structures is also of vital importance for the electromagnetic shielding efficiency. Song et al. [59] proved in their investigation that the SE of graphene-based structures increased with increasing thickness of the sponges: 28 dB for 2mm and 36 dB for 3 mm of thickness.

6.3. Graphene-CNT hybrid structures

In recent investigations, the possibility to blend large surface area, one-dimensional carbon nanotubes within a high charge density, two-dimensional graphene matrix has been studied. Generally, CNTs tend to agglomerate in organic dispersion; therefore, numerous efforts were developed to disperse the CNT by using micelles, ionic liquids, surfactants, polymer wrapping and other chemical functionalization approaches. It has been proved that GO could be a better dispersant to form a stable dispersion of CNT and the resulted dispersion is a novel hybrid named as graphene oxide-CNT (GO-CNT) [60]. Studies proved that GO-CNT and graphene-CNT hybrid nanomaterials exhibit higher electrical conductivities, large specific area and catalytic properties compared with either pristine CNTs or GO/graphene [61]. The strong π–π stacking interaction operating between graphene and CNT makes a three-dimensional network for the hybrid material and provide exceptional stability [62]. The CNTs act as conducting wires inside the already conducting graphene structure, thus promoting the conductivity of the hybrid material. In particular, Mani et al. [63] compared the suitability of graphene oxide (GO) structures merged with CNTs. GO was obtained by chemical oxidation of graphite into graphite oxide, and subsequent exfoliation into monolayer graphene. It could be stated that GO was more beneficial to obtain a stable dispersion of CNTs in its three-dimensional structure. The forces between these two architectures were found to be π-π interactions. In another study, Chen et al. [64] used graphene-MMCNTs structures to study the EMI shielding of these hybrid structures. A shield of 80-90% of the power density could be achieved with these sponges.

7. Conclusions

The latest techniques and strategies to develop new CNT/polymer composites with electromagnetic shielding properties have been reviewed. In composites with CNTs dispersed into a polymeric matrix, better EMI SE results are obtained with the increase of sample thickness and filler conductivity. Thus, it is possible to conclude that multi-walled carbon nanotubes exhibit better shielding properties than single-walled carbon nanotubes. At the same time, having longer or better dispersed carbon nanotubes translates into lower volume fractions of filler needed in order to reach the percolation threshold in the composite. At the same time, the possibility of minimizing the reflection mechanisms in electromagnetic shielding leading to purely (or mostly) absorbent materials has been proved. Equilibrium between the morphology and the conductivity of the material must be searched, being this one of the reasons why foams are gathering most of the attention in the field, as they allow the radiation to enter through the porous surface and is then absorbed in the interior. At the same time, it possesses the enormous advantages of being light-weight and easy to manufacture, opening the field of these foams for new application. A theoretical introduction to the problem of electromagnetic shielding has been introduced, showing an insight into the ways to calculate it and the measuring equipment used at present for this purpose.

Acknowledgements

This work was supported by the grant Nanoarq (MAT2014-57557-R) from the Spanish Ministerio de Ciencia e Inovación.

Author details

Marta González, Guillermo Mokry, María de Nicolás, Juan Baselga and Javier Pozuelo*

*Address all correspondence to: jpozue@ing.uc3m.es

Materials Science and Engineering and Chemical Engineering (IAAB), Carlos III University of Madrid, Spain

References

[1] Christopoulos C. Principles and Techniques of Electromagnetic Compatibility. CRC Press; 2007, 1–4.

[2] Clayton R.P. Frontmatter. In: Introduction to Electromagnetic Compatibility. 2nd ed. Hoboken, NJ: John Wiley & Sons, Inc.; 2005, 1–48.

[3] Tong X.C. Advanced Materials and Design for Electromagnetic Interference Shielding. CRC Press; 2008, 1–35.

[4] Yusoff A.N., Abdullah M.H., Ahmad S.H., Jusoh S.F., Mansor A.A., Hamid S.A.A. Electromagnetic and absorption properties of some microwave absorbers. J. Appl. Phys. 2002;92:876.

[5] Chung D.D.L. Electromagnetic interference shielding effectiveness of carbon materials. Carbon 2001;39:279–285.

[6] Huang J.C. EMI shielding plastics: A review. Adv. Polym. Technol. 1995;14:137–150.

[7] Das N.C., Khastgir D., Chaki T.K., Chakraborty A. Electromagnetic interference shielding effectiveness of carbon black and carbon fibre filled EVA and NR based composites. Compos. Part A Appl. Sci. Manuf. 2000;31:1069–1081.

[8] Vovchenko L., Perets Y., Ovsienko I., Matzui L., Oliynyk V., Launetz V. Shielding coatings based on carbon–polymer composites. Surf. Coatings Technol. 2012;211:196–199.

[9] Ren F., Yu H., Wang L., Saleem M., Tiana Z., Rena P. Current progress on the modification of carbon nanotubes and their application in electromagnetic wave absorption. RSC Adv. 2014;4:14419–14431.

[10] Qin F., Brosseau C. A review and analysis of microwave absorption in polymer composites filled with carbonaceous particles. J. Appl. Phys. 2012;111:061301.

[11] Gonschorek K.H., Vick R. Electromagnetic Compatibility for Device Design and System Integration, Chap. 17: Skin Effect and Shielding Theory of Schelkunoff. Berlin, Heidelberg: Springer; 2009, 377–392.

[12] Al-Saleh M.H., Sadaah W.H., Sundararaj U. EMI shielding effectiveness of carbon based nanostructured polymeric materials: a comparative study. Carbon 2013;60:146–156.

[13] Crespo M., González M., Elías A.L., Rajukumar L.P., Baselga J., Terrones M., Pozuelo J. Ultra-light carbon nanotube sponge as an efficient electromagnetic shielding material in the GHz range. Phys. Status Solidi RRL 2014;8:698–704.

[14] Zois H., Apekis L., OmastováM. Electrical properties of carbon black-filled polymer composites. Macromol. Symposia 2001;170:249–256.

[15] Ezquerra T.A., Connor M.T., Roy S., Kulescza M., Fernandes-Nascimento J., Baltá-Calleja F.J. Alternating-current electrical properties of graphite, carbon-black and carbon-fiber polymeric composites. Compos. Sci. Technol. 2001;61:903–909.

[16] Kortschot M., Woodhams R. Computer simulation of the electrical conductivity of polymer composites containing metallic fillers. Poly. Compos. 1988;9:60–71.

[17] Barrau S., Demont P., Peigney A., Laurent C., Lacabanne C. DC and AC conductivity of carbon nanotubes–polyepoxy composites. Macromolecules 2003;36:5187–5194.

[18] Du F., Fischer J.E., Winey K.I. Effect of nanotube alignment on percolation conductivity in carbon nanotube/polymer composites. Phys. Rev. B Condens. Matter Mater. Phys. 2005;72:1–4.

[19] Stauffer D., Aharony A. Introduction to Percolation Theory. CRC Press; 1994; pp. 1–92.

[20] Bauhofer W., Kovacs J.Z. A review and analysis of electrical percolation in carbon nanotube polymer composites. Compos. Sci. Technol. 2009;69: 1486–1498.

[21] Du J., Hsieh Y.L. Cellulose/chitosan hybrid nanofibers from electrospinning of their ester derivatives. Cellulose 2009;16:247–260.

[22] Tamburri E., Orlanducci S., Terranova M.L., Valentini F., Palleschi G., Curulli A., Brunetti F., Passeri D., Alippi A., Rossi M. Modulation of electrical properties in single-walled carbon nanotube/conducting polymer composites. Carbon 2005;43:1213–1221.

[23] Hoang A.S. Electrical conductivity and electromagnetic interference shielding characteristics of multiwalled carbon nanotube filled polyurethane composite films. Adv. Nat. Sci. Nanosci. Nanotechnol. 2011;2:025007/025001-025007/025005.

[24] Gupta T.K., Singh B.P., Dhakate S.R., Singh V.N., Mathur R.B. Improved nanoindentation and microwave shielding properties of modified MWCNT reinforced polyurethane composites. J. Mat. Chem. A 2013;1:78.

[25] Liu Z., Bai G., Huang Y., Li F., Ma Y., Guo T., He X., Lin X., Gao H., Chen Y. Microwave absorption of single-walled carbon nanotubes/soluble cross-linked polyurethane composites. J. Phys. Chem. C 2007;111:13696.

[26] Gupta T.K., Singh B.P., Teotia S., Katyal V., Dhakate S.R., Mathur R.B. Designing of multiwalled carbon nanotubes reinforced polyurethane composites as electromagnetic interference shielding materials. J. Polym. Res. 2013;20:169.

[27] Farukh M., Dhawan R., Singh B.P., Dhawan S.K. Sandwich composites of polyurethane reinforced with poly(3,4-ethylene dioxythiophene) coated multiwalled carbon nanotubes with exceptional electromagnetic interference shielding properties. RSC Adv. 2015;5:75229–75238.

[28] Zeng Z., Chen M., Jin H., Li W., Xue X., Zhou L., Pei Y., Zhang H., Zhang Z. Thin and flexible multi-walled carbon nanotube/waterborne polyurethane composites with high-performance electromagnetic interference shielding. Carbon 2016;96:768–777.

[29] Imai M., Akiyama K., Tanaka T., Sano E. Highly strong and conductive carbon nanotube/cellulose composite paper. Compos. Sci. Technol. 2010;70:1564–1570.

[30] Yang Y., Gupta M.C., Dudley K.L., Lawrence R.W. Conductive carbon nanofiber-polymer foam structures. J. Nanosci. Nanotechnol. 2005;5:927–931.

[31] Chen I.H., Wang C.C., Chen C.Y. Fabrication and structural characterization of polyacrylonitrile and carbon nanofibers containing plasma-modified carbon nanotubes by electrospinning. J. Phys. Chem. C 2010;114:23532–13539.

[32] Kim H.M., Kim K., Lee C.Y., Joo J., Cho S.J., Yoon H.S., Pejakovic D.A., Yoo J.W., Epstein A.J. Electrical conductivity and electromagnetic interference shielding of multiwalled carbon nanotube composites containing Fe catalyst. J. Appl. Phys. Lett. 2004;84:589.

[33] Pande S., Singh B.P., Mathur R.B., Dhami T.L., Saini P., Dhawan S.K. Improved electromagnetic interference shielding properties of MWCNT–PMMA composites using layered Structures. Nanoscale Res. Lett. 2009;4:327.

[34] Kim H.M., Kim K., Lee S.J., Joo J., Yoon H.S., Cho S.J., Lyu S.C., Lee C.J. Charge transport properties of composites of multiwalled carbon nanotube with metal catalyst and polymer: application to electromagnetic interference shielding. Curr. Appl. Phys. 2004;4:577–580.

[35] Huang Y., Du F., He X., Lin X., Gao H., Ma Y., Li F., Chen Y., Eklund P.C. Electromagnetic interference (EMI) shielding of single-walled carbon nanotube epoxy composites. Nano Lett. 2006;6:1141.

[36] Singh A.P., Gupta B.K., Mishra M., Govind M., Chandra A., Mathur R.B., Dhawan S.K. Multiwalled carbon nanotube/cement composites with exceptional electromagnetic interference shielding properties. Carbon 2013;56:455.

[37] Gupta A., Choudhary V. Electromagnetic interference shielding behaviour of poly(trimethylene terephthalate)/multi-walled carbon nanotube composites. Compos. Sci. Technol. 2011;71:1563–1568.

[38] Arjmand M., Sundararaj U. Electromagnetic interference shielding of nitrogen-doped and undoped carbon nanotube/polyvinylidene fluoride nanocomposites: a comparative study. Compos. Sci. Technol. 2015;118:257–263.

[39] Al-Saleh M.H. Influence of conductive network structure on the EMI shielding and electrical percolation of carbon nanotube/polymer nanocomposites. Synth. Metals 2015;205:78–84.

[40] Farukh M., Singh A.P., Dhawan S.K. Enhanced electromagnetic shielding behavior of multi-walled carbon nanotube entrenched poly (3,4-ethylenedioxythiophene) nanocomposites. Compos. Sci. Technol. 2015;114:94–102.

[41] Yim Y.J., Park S.J. Electromagnetic interference shielding effectiveness of high-density polyethylene composites reinforced with multi-walled carbon nanotubes. J. Indus. Eng. Chem. 2015;21:155–157.

[42] Jia L.C., Yan D.X., Cui C.H., Jiang X., Ji X., Li Z.M. Electrically conductive and electromagnetic interference shielding of polyethylene composites with dev carbon nanotube networks. J. Mater. Chem. C 2015;00:1–8.

[43] Huang H.D., Liu C.Y., Zhou D., Jiang X., Zhong G.J., Yan D.Y., Li Z.M. Cellulose composite aerogel for highly efficient electromagnetic interference shielding. J. Mater. Chem. A 2015;3:4983–4991.

[44] Thomassin J.M., Pagnoulle C., Bednarz L., Huynen I., Jerome R., Detrembleur C. Foams of polycaprolactone/MWNT nanocomposites for efficient EMI reduction. J. Mater Chem. 2008;18:792–796.

[45] Arjmand M., Apperley T., Okoniewski M., Sundararaj U. Comparative study of electromagnetic interference shielding properties of injection molded versus compression molded multi-walled carbon nanotubes/polystyrene composites. Carbon 2012;50(14):5126–5134.

[46] Thomassin J.M., Jérôme C., Pardoen T., Bailly C., Huynen I., Detrembleur C. Polymer/carbon based composites as electromagnetic interference (EMI) shielding materials. Mater. Sci. Eng. R Rep. 2013;74:211–232.

[47] Zhao N., Zou T., Shi C., Li J., Guo W. Microwave absorbing properties of activated carbon-fiber felt screens (vertical-arranged carbon)/epoxy resin composites. Mater. Sci. Eng. B. 2006;127:207–211.

[48] Al-Saleh M.H., Sundararaj U. Electromagnetic interference shielding mechanisms of CNT/polymer composites. Carbon 2009;47:1738–1746.

[49] Celozzi S., Araneo R., Lovat G. Electromagnetic Shielding. IEEE Press; 2008.

[50] Li N., Huang Y., Du F., He X., Lin X., Gao H., Ma L.F., Chen Y., Eklund P.C. Electromagnetic interference (EMI) shielding of single-walled carbon nanotube epoxy composites. Nano Lett. 2006;6(6):1141–1145.

[51] Jiang J.J., Wang Y., Liang J.J., Wang Y., Huang Y., Ma Y.F., Liu Z.F., Cai J.M., Zhang C.D., Gao H.J., Chen Y.S. Electromagnetic interference shielding of graphene/epoxy composites. Carbon 2009;47:922–925.

[52] Yan D.X., Pang H., Li B., Vajtai R., Xu L., Ren P.G., Wuang J.H., Li Z.M. Structured reduced graphene oxide/polymer composites for ultra-efficient electromagnetic interference shielding. Adv. Funct. Mater. 2015;25:559–566.

[53] Yan D.X., Pang H., Xu L., Bao Y., Ren P.G., Lei J., Li Z.M. Electromagnetic interference shielding of segregated polymer composite with an ultralow loading of *in situ* thermally reduced graphene oxide. Nanotechnology 2014;25:145705.

[54] Hsiao S.T., Ma C.C.M., Tien H.W., Liao W.H., Wang Y.S., Li S.M., Huang Y.C. Using a non-covalent modification to prepare a high electromagnetic interference shielding performance graphene nanosheet/water-borne polyurethane composite. Carbon 2013;60:57–66.

[55] Hsiao T., Ma C.C.M., Tien H.W., Liao W.H., Wang Y.S., Li S.M., Yang C.Y., Lin S.C., Yang R.B. Effect of covalent modification of graphene nanosheets on the electrical property and electromagnetic interference shielding performance of a water-borne polyurethane composite. Acs. Appl. Mater. Inter. 2015;7:2817–2826.

[56] Chen Z.P., Xu C., Ma C.Q., Ren W.C., Cheng H.M. Lightweight and flexible graphene foam composites for high-performance electromagnetic interference shielding. Adv. Mater. 2013;25:1296–1300.

[57] Zhang L., Zhang F., Yang X., Long G., Wu Y., Zhang T., Leng K., Huang Y., Ma Y., Yu A., Chen Y. Porous 3D graphene-based bulk materials with exceptional high surface area and excellent conductivity for supercapacitors. Sci. Rep. 2013;3:1408.

[58] Li Y., Samad Y.A., Polychronopoulou K., Alhassan S.M., Liao K. Highly electrically conductive nanocomposites based on polymer-infused graphene sponges. Sci. Rep. 2014;4:4652.

[59] Song W.L., Guan X.T., Fan L.Z., Cao W.Q., Wang C.Y., Cao M.S. Tuning three-dimensional textures with graphene aerogels for ultra-light flexible graphene/texture composites of effective electromagnetic shielding. Carbon 2015;93:151–160.

[60] Cao M.S., Wang X.X., Cao W.Q., Yuan J. Ultrathin graphene: electrical properties and highly efficient electromagnetic interference shielding. J. Mat. Chem. C 2015;3:6589–6599.

[61] Mani V., Chen S.M., Lou B.S. Three dimensional graphene oxide-carbon nanotubes and graphene- carbon nanotubes hybrids. Int. J. Electrochem. Sci. 2013;2013:11641–11660.

[62] Kachoosangi R.T., Musameh M.M., Abu-Yousef I. Carbon nanotube-ionic liquid composite sensors and biosensors. Anal Chem. 2009;81(1):435–442.

[63] Mani V., Devadas B., Chen S.M. Direct electrochemistry of glucose oxidase at electrochemically reduced graphene oxide-multiwalled carbon nanotubes hybrid material modified electrode for glucose biosensor. Biosens. Bioelectron. 2013;41:309–315.

[64] Chen Y.J., Li Y., Chu B.T.T., Kuo I.T., Yip M., Tai N. Porous composited coated with hybrid nano carbon materials perform excellent electromagnetic interference shielding. Composit. B 2015;70:231–237.

Chapter 12

Safer Production of Water Dispersible Carbon Nanotubes and Nanotube/Cotton Composite Materials

Mohammad Jellur Rahman and Tetsu Mieno

Additional information is available at the end of the chapter

http://dx.doi.org/10.5772/62880

Abstract

Water-dispersible carbon nanotubes (WD-CNTs) have great importance in the fields of biotechnology, microelectronics, and composite materials. Sidewall functionalization is a popular method of enhancing their dispersibility in a solvent, which is usually achieved by strong acidic treatment. But, treatment under such harsh conditions deviates from green chemistry and degrades the structure and valuable properties of CNTs. Alternative safer and easier plasma method is discussed to produce functionalized CNTs (*f*-CNTs). The *f*-CNTs remain dispersed in water for more than 1 month owing to the attachment of a large number of carboxyl groups onto their surfaces. The WD-CNTs are applied to produce conductive cotton textile for the next generation textile technologies. Nonconducting cotton textile becomes electroconductive by repeatedly dipping into the *f*-CNT-ink and drying in air. The *f*-CNTs uniformly and strongly cover the individual cotton fibers. After several cycle of dipping into the *f*-CNT-ink, the textile becomes conductive enough to be used as wire in lighting up an LED. As a demonstration of practical use, the textile is shown as a conductive textile heater, where the textile can produce uniformly up to *ca.* 80°C within *ca.* 5 min by applying an electric power of *ca.* 0.1 W/cm^2.

Keywords: carbon nanotubes, plasma-functionalization, water dispersibility, electroconductive cotton, nanocomposites

1. Introduction

Carbon nanotubes (CNTs) possess a unique place in nanoscience owing to their exceptional electrical, thermal, and mechanical properties [1]. They have found applications in areas diverse as composite materials, energy storage and conversion, sensors, drug delivery, field emission devices, and nanoscale electronic components [2]. Water-dispersible CNTs (WD-CNTs) have great importance in the fields of biotechnology, microelectronics, and composite materials [2–5]. However, the stable dispersion of CNTs in solvent without changing their physical properties is a significant challenge and a prerequisite for their applications [6–8]. Sidewall functionalization is a popular method of enhancing the dispersibility of CNTs, which is achieved usually by oxidizing CNTs by strong acids or oxidative gases [6, 9–10]. But, treatment under such harsh conditions deviates from green chemistry, and effect in the change of the CNT structure [9–11], which markedly degrade their basic properties [6, 12]. To overcome these problems, alternative safer and easier functionalization methods should be considered. In this chapter, we will discuss about different functionalization methods, especially, the method we have developed in our laboratory to functionalize the CNTs to enhance their water dispersibility.

Possible applications of WD-CNTs will also be discussed, where we will demonstrate the applications of the WD-CNTs to produce conductive cotton textile for the next generation textile technologies. Integration of electronics is one of the smart applications of the textile, which covers the applications in high-performance sportswear, wearable displays, new classes of portable power, and embedded health monitoring devices [13–20]. Recently, interests on the preparation of lightweight and flexible electrothermal materials have been increased for the aviation and aerospace industries, microreactor technologies, and different kinetic systems [14–20]. Resistive wires made of metal alloys have been used as the heat source in many appliances, but in those cases flexibility is poor, and the heat is localized at the wires [20]. If flexible cotton textile can be used as a heating element, it would offer a spectrum of advantages over these traditional materials [13, 20–23].

2. Potential applications of CNTs

The properties of CNTs have caused researchers and companies to consider using them in several fields [2, 24].

2.1. In composite technology

Because CNTs have the highest strength to weight ratio of any known material, combining them with other materials into composites can be used to build lightweight spacecraft, windmill blades to increase the amount of electricity generated, stronger bicycle components made by adding CNTs to a matrix of carbon fibers, cables strong enough to be used for the space elevator to drastically reduce the cost of lifting people, and materials into orbit. In addition, new materials combined with nanosensors and nanorobots could improve the performance of spaceships, spacesuits, and the equipment used to explore planets and moons.

2.2. In biotechnology

CNTs can easily penetrate membranes such as cell walls [25]. The long and narrow shape makes them look like miniature needles, so it makes sense that they can function like a needle at the cellular level [25]. Medical researchers are using this property by attaching molecules to CNTs that are attracted to cancer cells to deliver drugs directly to the diseased cells. Another

http://dx.doi.org/10.5772/62880

interesting property of CNTs is that their electrical resistance changes significantly when other molecules are attached to the carbon atoms [26]. This property is utilized to develop sensors that can detect chemical vapors such as carbon monoxide or biological molecules [26]. They are also used to improve the healing process for broken bones by providing a CNT scaffold for new bone material to grow on.

2.3. In electronics

CNTs can be used to increase the capabilities of electronics devices while reducing their weight, size, and power consumption, for example display screens on electronics devices or highly dense memory chips with a projected density of one terabyte of memory per square inch or greater. CNT ink is used in inkjet printers for printable electronics devices.

2.4. In environmental issue

CNTs are being used in several applications to improve the environment. These include cleaning up existing pollution, improving manufacturing methods to reduce the generation of new pollution, and making alternative energy sources more cost effective. Inexpensive CNT-based sensor can detects bacteria in drinking water. Because of the small size of CNTs with high surface area, a few gas molecules are sufficient to change the electrical properties of the sensing elements. This allows the detection of a very low concentration of chemical vapors.

2.5. In energy

Use of CNT in solar cells can reduce manufacturing costs as a result of using a low temperature process instead of the high temperature vacuum deposition process typically used to produce conventional cells made with crystalline semiconductor material [27]. They can reduce installation costs by producing flexible rolls instead of rigid crystalline panels, and therefore can be installed as a coating on windows or other building materials as integrated photovoltaic [27]. CNTs can decrease the power needed to run reverse osmosis desalination plants because water molecules pass through CNTs more easily than through other types of nanopores. They are used to make current collecting layer for the cathode in batteries and as electrodes in thermocells that generate electricity from waste heat. Combining CNTs with buckyballs and polymers inexpensive solar cells can be produced by simply painting on a surface. CNT-based supercapacitors do even better than batteries in hybrid cars by significantly reducing the weight of the batteries needed to provide adequate power, increasing the available power, and decreasing the time required to recharge a battery.

2.6. In consumer products

CNT has already found its way into lots of consumer products such as fabric, sporting goods, cleaning products, food, building materials, and skin care. The composite fabric with CNTs allows improvement of fabric properties without a significant increase in weight, thickness, or stiffness.

3. Functionalization of CNTs

CNTs in all their forms are difficult to disperse and dissolve in water or organic media [24, 28]. They are extremely resistant to wetting, which is very important for different applications. A suitable functionalization of the CNTs, i.e., the attachment of chemical functionalities represents a strategy for overcoming these barriers, and thus become an attractive target for synthetic chemists and materials scientists. Functionalization can improve dispersibility [29] and processibility, and will allow combination of the unique properties of CNTs with those of other types of materials. Chemical bonds might be used to tailor the interaction of the CNTs with other entities, such as a solvent, polymer and biopolymer matrices, and other nanotubes. Functionalized CNTs might have mechanical or electrical properties that are different from those of the unfunctionalized CNTs, and thus may be utilized for fine-tuning the chemistry and physics of CNTs.

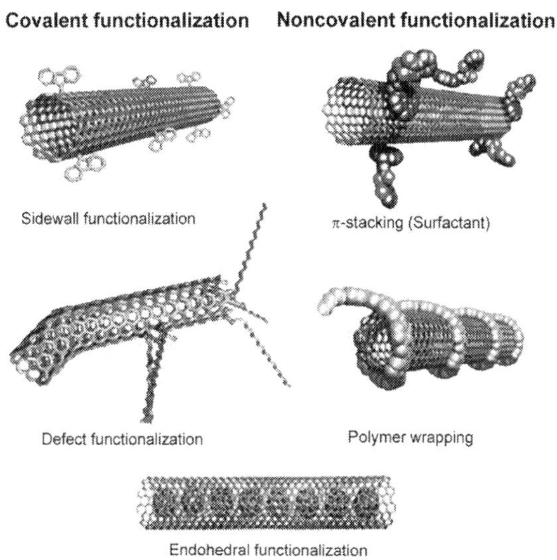

Figure 1. Different possibilities of the CNT functionalization. (Reproduced with permission of from ©2002, John Wiley and Sons [30]).

Both covalent and noncovalent functionalizations of CNTs are possible [6, 24, 31]. Different possibilities of these are the sidewall functionalization, defect-group functionalization, noncovalent exohedral functionalization with molecules through π-stacking, noncovalent exohedral functionalization with polymers, and endohedral functionalization as shown in **Figure 1**. Covalent functionalization is based on covalent linkage of functional entities onto the nanotube's carbon scaffold. It can be performed at the termini of the tubes or at their sidewalls. Direct covalent sidewall functionalization is associated with a change of hybridization from sp^2 to sp^3 and a simultaneous loss of conjugation. Defect functionalization takes advantage of chemical transformations of defect sites already present. Defect sites can be the open ends and holes in the sidewalls, which are terminated by carboxylic groups, and pentagon and heptagon irregularities in the graphene framework. Oxygenated sites, formed through oxidative purification, have also been considered as defects. A noncovalent functionalization is mainly based on supramolecular complexation using various adsorption forces, such as van der Waals' and π-stacking interactions. All these functionalizations are exohedral derivatizations. A special case is the endohedral functionalization of CNTs, i.e., filling of the tubes with atoms or small molecules [6, 32]

3.1. Functionalization to increase water dispersibility

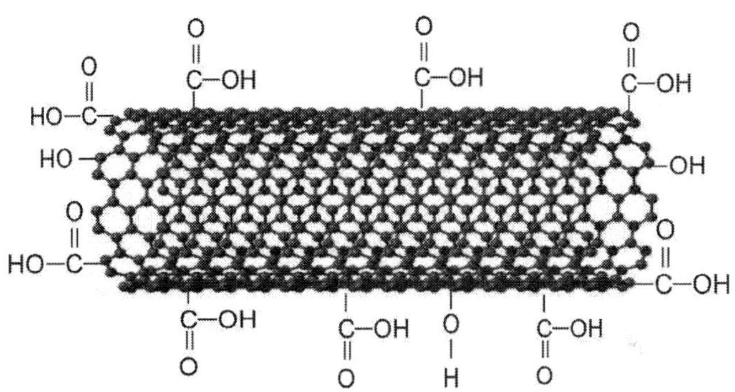

Figure 2. Section of an oxidized CNT, reflecting terminal and sidewall oxidation.

One of the most common functionalization techniques is the oxidative treatment of CNTs by liquid-phase or gas-phase oxidation, introducing carboxylic (–COOH) groups and some other oxygen-bearing functionalities such as hydroxyl, carbonyl, ester, and nitro groups into the tubes. In this process, CNTs are treated by strong acids, such as refluxing in a mixture of sulfuric acid and nitric acid [9, 10], "piranha" solution (sulfuric acid-hydrogen peroxide) [11], boiling in nitric acid [12], or treating with oxidative gases, such as ozone [6, 33]. Upon oxidative treatment the introduction of –COOH groups and other oxygen-bearing groups at the end of the tubes and at defect sites is promoted, decorating the tubes with a somewhat indeterminate number of oxygenated functionalities. However, mainly because of the large aspect ratio of CNTs, considerable sidewall functionalization takes place (**Figure 2**) [34]. However, treatment under such harsh conditions clearly deviates from green chemistry, and results in the opening of the tube tips [9], shortening of the tubes [11], and fragmentation of the sidewalls [12]. These markedly degrade their basic properties [6, 35]. Since reactivity is a function of curvature [36], the oxidative stability also depends on the tubes' diameter and on the production process responsible for the tubes' dimensions [37].

The surface modification of CNTs can be carried out through a wide range of plasma processes, which provide a cost effective and environmentally friendly alternative to other processes, related to environmental issues [38] and biomedical applications [39]. Compared with other chemical modification methods, the plasma-induced functionalization presents interesting properties, is solvent-free and time efficient process. Moreover, this treatment allows the grafting of a wide range of different functional groups depending on the plasma parameters such as power, gas used, duration of treatment, and pressure [40]. In addition, the amount of functional groups can also be tailored. This is important since having saturation of these groups on the surface can alter the electronic conductivity of CNTs. The most common plasma treatment of CNTs is the low pressure RF cold plasma, which is successfully used to bind oxygen [41], hydrogen [42], and fluorine groups [43]. It has been observed that a complete purification of CNTs can also be reached out after their treatment in glow discharges (RF or MW) [44]. However, it was also observed that the average diameter of CNTs decreases with treatment duration. Therefore, the nature of the plasma gas is important, because oxygenated ions or radicals are very reactive in the etching processes. However, destruction of CNT sidewalls is also observed for other less reactive plasma gas such as CF_4 or Ar [45, 46]. Moreover, it was shown that UV photons promote the defunctionalization of moieties grafted on the CNTs [47]. A probable solution could be the reduction of the power supplied to the plasma or duration of the treatment to limit destruction of the sidewalls. In this work, RF plasma is used to functionalization the CNTs using a parallel plate capacitively coupled reactor.

4. Methodology of the research

Here, an environmentally friendly approach to functionalizing CNTs has been described, which is developed to attach –COOH groups onto their surfaces, and carried out under a wet condition using citric acid solution in RF (13.56 MHz) oxygen plasma [48, 49]. CNTs are first pretreated supersonically in ethanol. Then they are wetted with citric acid solution and subsequently treated using oxygen plasma including citric acid and water. This method is safer than the methods available in the literature, as no hazardous reagents are used here. The surfaces of the CNTs are chemically functionalized with –COOH groups, and they can be easily dispersed in water. To achieve the main objective of avoiding the destruction of the structure, which could change the valuable properties of CNTs, the functionalize conditions are optimized in each step. The functionalized CNTs can be used as a multifunctional coating material in improved electronic applications [3], in energy storage devices [50], as well as in the pharmaceutical industry, particularly in the area of drug delivery or as components of biosensors [4, 51]. They are also highly suitable as a filler component for water-soluble polymer composites [4].

4.1. Functionalization of CNTs

A flow chart of the functionalization process is shown in **Figure 3** and the setup of the plasma reactor, indicating the dissociation of oxygen, water, and citric acid molecules is shown schematically in **Figure 4**. 20–30 mg of CNT powder (Sigma-Aldrich, outer diameter = 10–30 nm, inner diameter = 3–10 nm, length = 1–10 μm, purity >90%) is added to 20 mL of pure ethanol (Wako Pure Chemicals Co., purity >95%) and sonicated at room temperature using a supersonic homogenizer (Sonics Vibra cell, VC 130, Sonic & Materials Inc., f = 20 kHz, 6.0 mm ϕ probe) at an input power of 15 W for 60 min.

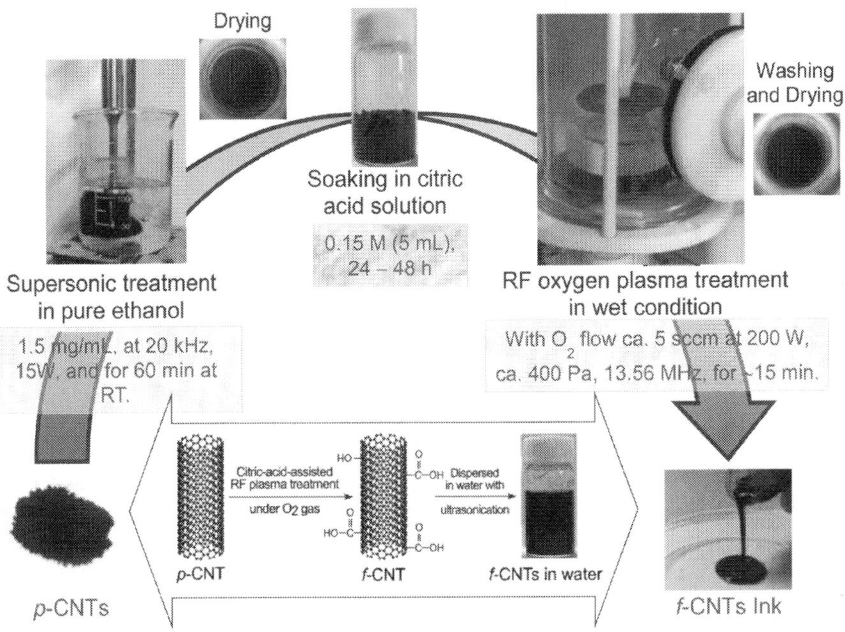

Figure 3. The flow chart of the functionalization process with the optimum treatment conditions and the schematic of the functional group attachment.

The suspension is dried under reduced pressure and soaked in 0.15 mole (5 ml) of citric acid (Wako Pure Chemicals Co., assay >98%) solution for more than 24 h. The CNTs in the solution are then placed on the lower electrode (SUS, 50 mmϕ) of a plasma reactor as shown in **Figure 4**, which is evacuated to ca. 400 Pa using a rotary pump at a very slow rate. When the wet phase starts to disappear, oxygen gas is introduced into the reactor at a rate of 5 sccm and the background chamber pressure is kept at about 400 Pa. Though the water molecules, and part of the citric acid molecules evaporate, it is considered that they remain inside the chamber and in the gas container connected to the chamber, and contribute to the functionalization process. Then the plasma reaction is carried out for about 15 min by an RF input power of P_{rf} = 200 W,

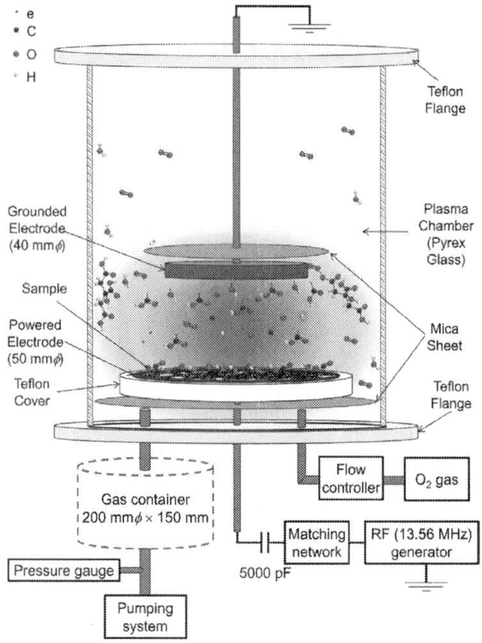

Figure 4. Schematic diagram of the plasma reactor.

f = 13.56 MHz. The reflected RF power is minimized (< 20 W) by controlling the matching network during the plasma reaction. It is noted that when the samples become fully dried before starting the plasma reaction, they are less reactive for the oxygen plasma. However, when the plasma started in the wet phase, the water molecules and part of the citric acid molecules evaporate with the processing time, and considered to remain inside the chamber and in the gas container connected to the chamber, contributing to the functionalization process. After the treatment, the CNTs are washed at least three times using pure water (Wako Pure Chemicals Co., distilled water) and dried under reduced pressure at room temperature. It is observed that approximately 13% of the CNTs are lost by the oxygen plasma, which is measured from the mass difference of the samples before placing in the citric acid and after the final washing process.

4.2. Characterizations

The dispersibility and dispersion stability of CNTs are primarily observed by mixing 20 mg of CNTs in 10 mL of pure water with bath sonication for 2 min, then keeping the mixture undisturbed for more than 20 days. To confirm the dispersion stability of the pristine, and plasma-treated CNTs (hereinafter denoted as *p*-CNTs and *f*-CNTs, respectively ca. 1.0 mg of the dried sample is mixed with 8.0 mL of pure water with brief sonication for 2 min in a bath sonicator (Iuchi Japan, US-1, 110 W, 38 kHz) so that the they are dispersed homogeneously. Then the sample is placed in a quartz cuvette (1.0 × 1.0 × 4.5 cm^3) and the change in the absorbance (*Abs*) at a wavelength of 250 nm is observed for 4 h using a UV-visible spectrometer (JASCO V-630). Here, the wavelength is chosen corresponding to the maximum absorbance region of the UV-visible spectra, because according to Beer's law, the relationship between the absorbance and the concentration of the absorbing particles remain linear up to higher concentration in this region as compared to other region [52].

FT-IR spectroscopy is used to identify the chemical groups attached onto the CNTs. Approximately 0.5 mg of the dried sample is dispersed in 1.0 mL of propanol, and the mixture is uniformly coated on a CaF_2 substrate (Sigma Koki Co., 20 mm diameter and 1 mm thickness), dried, and measured using an FT-IR spectrometer (Shimadzu Co., 8700, 100 scans averaged). The spectra in this thesis are presented after baseline correction.

The dispersibility of the CNTs is also observed by a TEM (JEOL JEM-1400 Plus, acceleration voltage of 120 kV), and their structural quality is measured by a Raman spectrometer (JASCO Co., NR-1800, λ = 532 nm). For thermal analyses a thermogravimetric analyzer (Rigaku Thermo plus TG8120) is used, which is operated under air ambient at a heating rate of 10 K/min from room temperature to 973 K [48].

4.3. Preparation of the cotton nano-composites

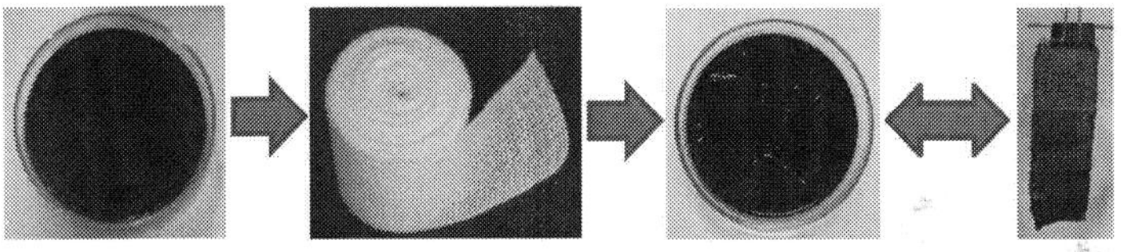

Figure 5. Coating scheme of the f-CNTs/cotton textile.

To obtain the cotton composite, *f*-CNTs are dispersed into water to produce stable [49]. A piece of 100% cotton (*p*-cotton) textile of ca. 24 mesh/inch with thread diameter of ca. 0.25 mm is repeatedly dipped into the 0.25–1.0 wt% of the *f*-CNT ink and dried at room temperature. The steps of the *f*-CNT loading process in the cotton textile are shown in **Figure 5**. The textile sheet is dipped and dried repeatedly until the electrical resistance becomes of the order of several kΩ for an area of 5 × 5 cm^2. To calculate the amount of the loaded *f*-CNTs (W_{nt}), the cotton textile is weighed before dipping in the *f*-CNT ink and after drying it sufficiently.

4.4. Measurements

Electrical resistance, $R = V/I$, of the textile is obtained from the current, I, in the circuit, and potential drop, V, across the sample measured by a digital multimeter (Iwatsu VOAC 87, Japan). A dc voltage, V_{in}, is applied by a regulated power supply (Kikusui Electronics PAB 350–0.2, Japan) between two Nickel electrodes (thickness of 0.15 mm, 40 mesh, Nilaco Co.,

Japan) of width D placed on the coated cotton textile at a distance L. The sheet resistance (R_s) of the textile is then calculated using the formula: $R_s = R/(D \times L)$. The coating uniformity of CNT on the cotton fibers is confirmed by the SEM images taken by a JEOL JSM 6510LV SEM at an operating voltage of 20 kV, and the thermal stability of the textile is observed by a TG/DTA (Rigaku Thermo Plus TG 8120). The temperature of the *f*-CNTs/cotton textile is measured using a thermometer (AS ONE TM-301) and an infrared camera (FLIR i3, emissivity 0.95, focus length = 6.8 mm). A standard washing test is also performed on the coated textile using a standard launder meter to study the toughness of the *f*-CNTs on the cotton fibers.

5. Results and discussions

5.1. Functionalization of CNTs

The enhancement of dispersion stability is confirmed from visual observation of the mixer of CNTs in pure water and measuring the settling speed for a dispersion using the absorbance data of the UV-visible spectroscopy as shown in Figures 6(a) and (b), respectively [48]. Attachment of the functional groups onto the CNTs is studied by a Fourier transform infrared (FT-IR) spectrometer (Shimadzu Co., 8700, 100 scans averaged). The FT-IR spectra of the *p*-CNTs and *f*-CNTs are presented in **Figure 6(c)** after the base line correction. The *f*-CNT spectrum shows additional peak compared to that of the *p*-CNTs at *ca.* 1713 cm^{-1}, which corresponds to the C=O stretching bonds, and the broad band at *ca.* 3200 cm^{-1} corresponding to the O–H/–OH bonds [48]. Therefore, carboxyl groups are considered to be attached on the CNTs after functionalization.

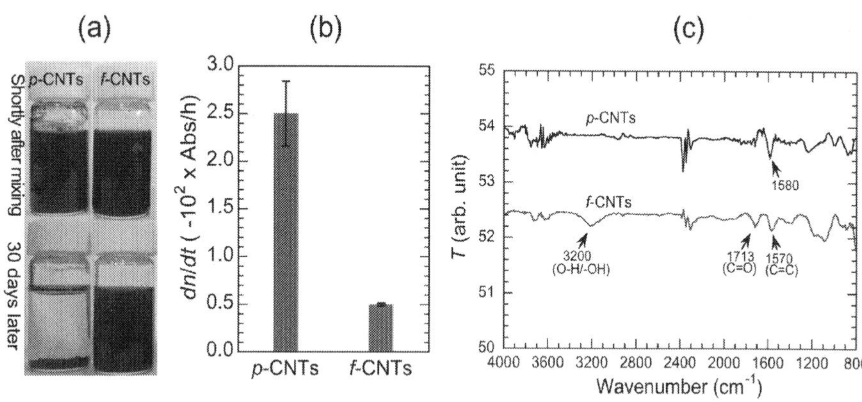

Figure 6. (a) Photographs of the dispersion of 20 mg of *p*-CNTs and *f*-CNTs in 10 mL of pure water using 2 min of bath sonication. (b) Settling speed, *dn/dt*, calculated from the *Abs* (250 nm) versus *time* graphs for the *p*-CNTs and *f*-CNTs. (c) FT-IR spectra of the *p*-CNTs and *f*-CNTs.

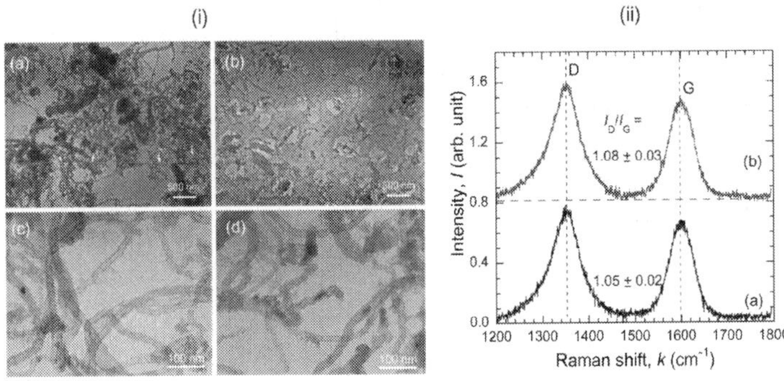

Figure 7. (i) TEM images of the (a, c) *p*-CNTs, and (b, d) *f*-CNTs in two different magnifications. (ii) Raman spectra of the (a) *p*-CNTs, and (c) *f*-CNTs.

The improved dispersion of the *f*-CNTs is further confirmed from the transmission electron microscope (TEM; HITACHI High Technology Co., H-7500, acceleration voltage of 100 kV) images as shown in **Figure 7(i)**. Compared to the *p*-CNTs, the dispersibility of the *f*-CNTs is enhanced clearly as shown in **Figure 3(a) and 3(b)**. In the higher magnification images, it is observed that the structural quality of the CNTs remain similar after the citric-acid-assisted RF plasma functionalization, which is also confirmed from the Raman spectroscopy as shown in **Figure 7(ii)**, which are normalized and bodily shifted in the intensity axis. [48]. Because, Raman spectrum is considered to be an important tool to study the CNT structure [53, 54] and is widely used to assess the amount of defects [55, 56]. The defect-induced D band appears at ca. 1350 cm^{-1}, which indicates the amount of disordered carbon in the CNT structure, and its intensity, I_D, corresponds to the degree of disorderness. The G band ca. 1600 cm^{-1} corresponds to the graphitic (ordered) carbon, and its intensity, I_G, corresponds to the amount of ordered carbon. Therefore, the ratio, I_D/I_G, is used to estimate the change in structural quality of CNTs after functionalization in different processes. Defect density corresponding to I_D/I_G for the *p*-CNTs and *f*-CNTs are observed to be 1.05 ± 0.02, and 1.08 ± 0.03, respectively. The changes are very small, and also no distinct changes are observed in terms of the Raman shift. These suggest that the CNT structure and the chemical composition of the interior of the CNTs are almost unaffected by the ultrasonic and plasma treatments.

On the basis of the above results, the basic functionalization scheme of the CNTs by the citric-acid-assisted oxygen plasma treatment is summarized in **Figure 8**. CNTs are long, web-like, and remain strongly aggregated. When they are dispersed in ethanol by the supersonic treatment, ethanol molecules enter the aggregated parts of the CNTs and weaken the attractive forces between them. When the sonicated CNTs are placed in the citric acid solution, the citrate and hydronium ions attack their weak parts. During the plasma reaction, oxygen, water, and citric acid molecules or ions are fragmented to generate oxygen containing ions, radicals, and CO or CO_2, which react with the defect sites [56–58]. The CO and CO_2 are oxidized to form –COOH groups and attach to the CNT surfaces [59]. Also, the attached –OH groups are further oxidized to form –COOH groups [58, 60]. These functional groups enable the CNTs to readily

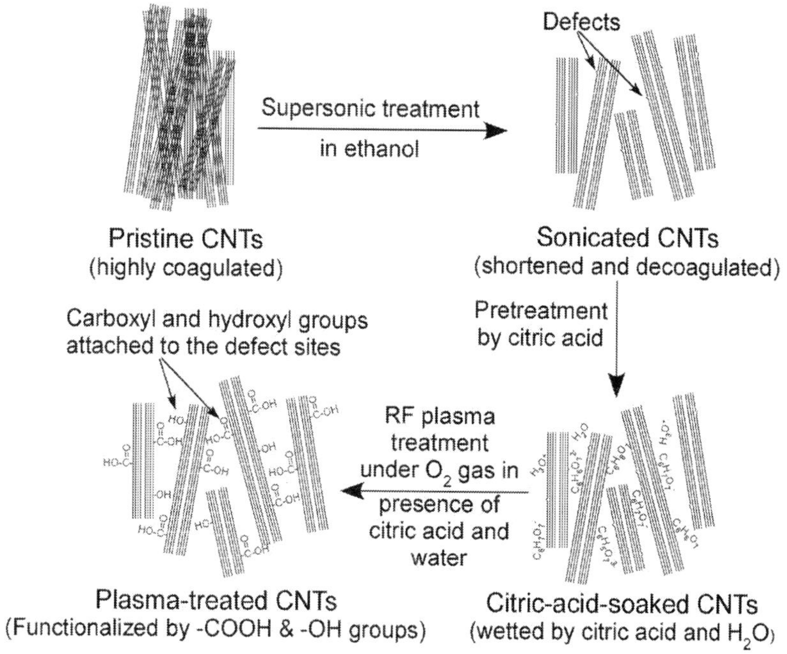

Figure 8. A model of the functionalization process of CNTs [48].

disperse in water due to hydrogen bonds formed between the carboxylic acid groups and water molecules [61]. The negatively charged surfaces of the CNTs repel each other and prevent them from coagulating. The polar interactions of the functional groups with the water molecules reduce the settling speed of the *f*-CNTs [56].

Hussain et al. functionalized CNTs by H_2O plasma treatment under controlled environment. Through the controlled functionalization process the electrochemical properties of the CNTs were modified, expanding the range of potential applications of the *f*-CNTs in the field of energy and environment [62]. The different oxygen containing groups attached on to the surfaces of the nanotubes change the physicochemical properties of the CNTs. After the plasma treatment, the specific capacitance of the CNTs increased from 23 up to 68 Fg^{-1}. Hordy et al. [63] functionalized the grown CNTs via an $Ar/O_2/C_2H_6$ capacitively coupled RF plasma discharge. CNTs are grown directly from a stainless steel mesh and are subsequently plasma functionalized in the same reactor. The functionalized CNTs removed from the substrate are found to remain stable for extended periods of time. But in this case acidic pre-treatment step were followed and the process needed some complicated steps. Chen et al. [56] worked on attaching functional groups other than oxygen containing groups on to the CNTs, such as amine groups by using a microwave-excited NH_3/Ar surface-wave plasma. This functional group improved the hydrophilicity of CNTs, though the atomic composition and structural properties of the *f*-CNTs were compromised.

5.2. Properties of the cotton nanocomposites

When the *p*-cotton textile is dipped into the *f*-CNT ink it becomes easily coated with the *f*-CNTs. After sufficient amount of loading (>5.0 wt%), the coated textile shows measurable resistance, *R*. The schematic setup of the resistance measurement is shown in **Figure 9(a)**. This is owing to the formation of conducting path on the cotton fibers. *R* values of the coated textile are observed to vary linearly with *L*. As W_{nt} increases with the dipping-drying cycles, R_s of the *f*-CNTs/cotton textile decreases gradually as shown in **Figure 9(b)**. The *f*-CNTs/cotton textile becomes conductive with R_s *ca*. 9.65 kΩ/sq, when W_{nt} is *ca*. 18.5 wt%. After first and second dipping cycles, only parts of the textile become poorly conductive. However, after 5-6 cycles of dipping, the variation of R_s along the width or length becomes very small indicating the homogenous coating of the cotton textile. Comparing the R_s of the *f*-CNTs/cotton textiles (*ca*. 10^2 kΩ/sq) with those of the *p*-CNTs/cotton textiles (*ca*. 10^5 kΩ/sq) for the same W_{nt}, it is realized that homogenous dispersion of the CNTs is very important in the coating process. When thickness of the *f*-CNTs/cotton textile is, $b \approx 0.5$ mm, the bulk resistivity, $\rho = [R\ (D \times b)]/L$, is calculated to be *ca*. 2.3 kΩ.m, and the conductivity, $\sigma = 1/\rho$, is *ca*. 0.43 S/m.

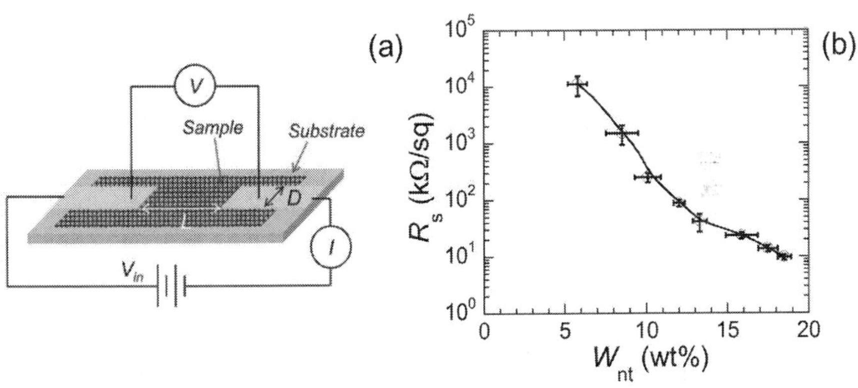

Figure 9. (a) Schematic setup of the sheet resistance (R_s) measurement. (b) The R_s versus the loaded *f*-CNTs, W_{nt}, in the cotton textile [49].

SEM images of the *p*-cotton and *f*-CNTs/cotton textiles at different magnifications are presented in **Figure 10**. It is observed that the *f*-CNTs coat the cotton fibers uniformly, and are firmly attached to them as shown in the cross sectional view of the fiber in **Figure 10(c)**. Cotton has hydroxyl groups in the cellulose fibers [13, 22, 64], which enable the *f*-CNTs to form strong hydrogen bonds with the surfaces of the fibers, resulting in the high density network coating [13, 65, 66].

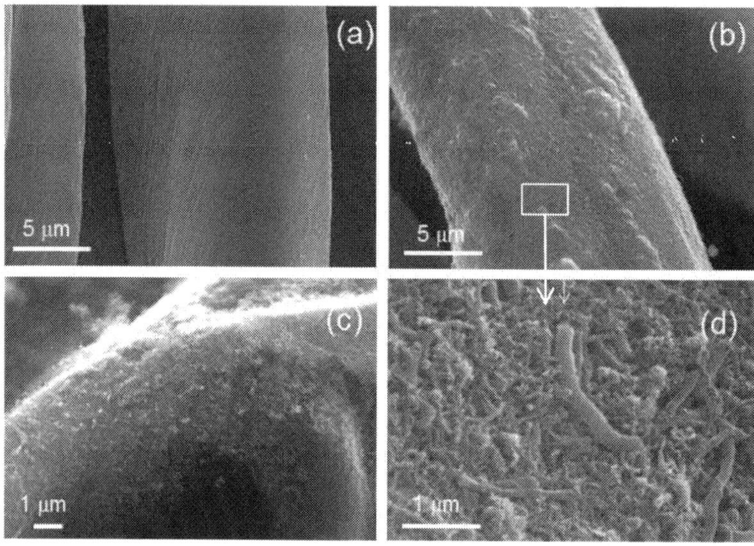

Figure 10. SEM images of the (a) *p*-cotton, (b–d) *f*-CNTs/cotton textile.

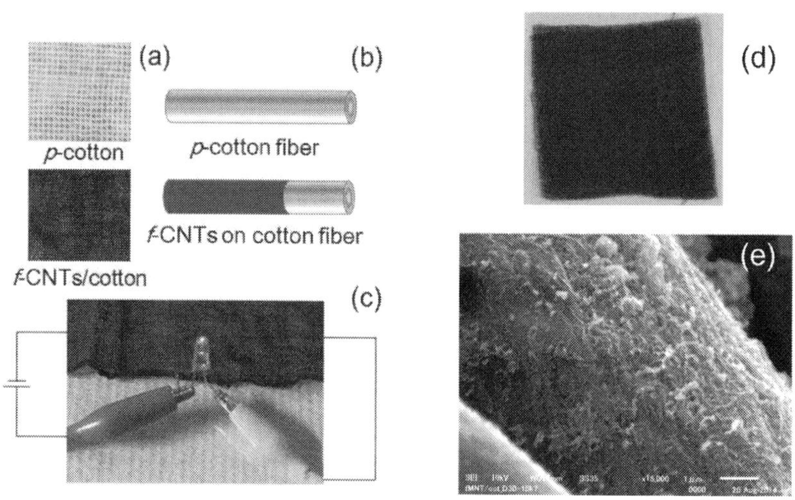

Figure 11. (a) Comparison between the *p*-cotton (upper) and *f*-CNTs/cotton (lower) textiles. (b) Schematic of the fiber coating. (c) The *f*-CNTs/cotton textile acting as a conducting path in the emission of an LED. (d) The textile after a standard washing test for 30 min, (e) SEM image of the cotton fiber after the washing test.

After dipping in the *f*-CNT ink, the *p*-cotton textile becomes black as shown in **Figure 11(a)**, which is owing to the formation of CNT network armor on each cotton fiber as shown schematically in **Figure 11(b)**. After 10 cycles of dipping in 0.5 wt% *f*-CNT ink, the resulting *f*-CNTs/cotton textile becomes conductive with σ more than *ca*. 0.20 S/m. As a demonstration of the conductivity, an LED device connected to a dc source can be easily powered through the prepared conducting textile as shown in **Figure 11(c)**. It is observed that the *f*-CNTs cannot be removed from the coated textile after normal washing, and the *f*-CNTs/cotton textile remain as black as it is just coated. To study the toughness of the *f*-CNTs on the cotton fiber, a piece of *f*-CNTs/cotton textile is washed in a standard launder meter with detergent/DI water (0.5 g/L) at a temperature of 40 ± 2°C with a revolution rate of 0.67 Hz for 30 min. It is observed that only a very little portion of the loaded *f*-CNTs are removed from the textile and lots of CNTs still remain on the fiber as shown in **Figure 11(d) and (e)**, suggesting the high level of toughness of the *f*-CNTs on the textile.

Thermal conductivity of the *f*-CNTs/cotton textile is observed to be *ca*. 0.045 W/mK, which suggests *ca*. 70% improvement. It is conjectured that strong bonding between the *f*-CNTs and the cotton fibers ensure better heat transfer in the textile [22, 66]. From the thermogravimetric study, it is observed that the final combustion of the *f*-CNTs/cotton textile is delayed by *ca*. 100°C as compared to *p*-cotton, signifying the improved thermal stability of the textile. This improvement in thermal stability is due to the uniform coating of the cotton fibers by the *f*-CNTs, because thermal stable CNTs act as armor for the cotton fibers protecting them from burning in a relatively lower temperature. The combustion processes of the *p*-cotton and *f*-CNTs/cotton textiles are studied by hanging them on a metal supporter, and then igniting simultaneously by a gas flame. It is observed that the *p*-cotton textile catches fire immediately and burns quickly to ashes completely. However, the *f*-CNTs/cotton textile becomes charred only, which provides direct evidence of improved flame retardancy.

5.2.1 The textile as a low powered flexible heater

The above results strongly suggest that the *f*-CNTs/cotton textile can be used as an electro-thermal heating element. The electric heating behavior of the patterned textiles is evaluated by measuring the changes in the resistance and temperature under V_{in} = 10 × 60 V. When few layers of such *f*-CNTs/ cotton textiles (5 × 5 cm^2) is stacked together, the resultant R values become low enough to produce heat applying small voltages based on the Joule heating from electric power [67, 68]. When V_{in} is applied through a four layered such textile of R_s ca. 1.67 kΩ/sq as shown schematically in **Figure 12(a)**, it is observed that heat starts to radiate from the textile indicating the increase in temperature with time as presented in **Figure 12(b)**. The little difference between the thermocouple-based and infrared-based measurements shown in **Figure 12(c)** validates the data obtained from the measurement, which is because of the slow response of the thermocouple owing to the slow heat propagation [69]. Temperature of the f-CNTs/cotton textile can be increased above 60°C within 5 min depending on the applied voltage.

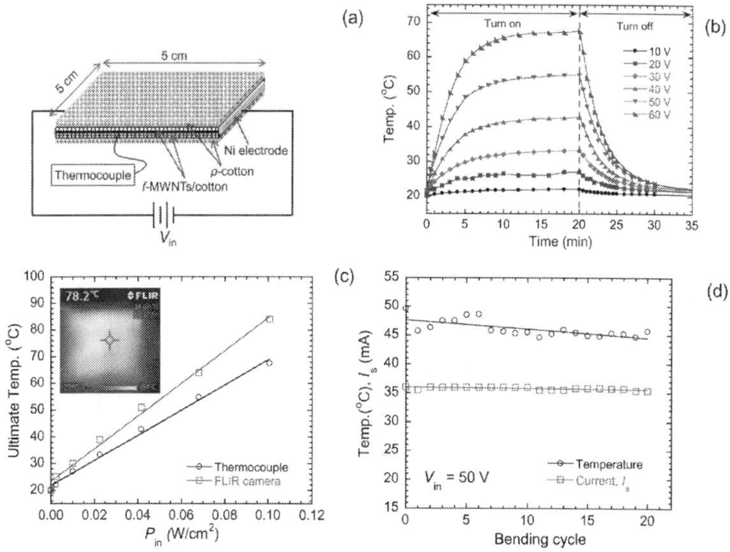

Figure 12. Heating and cooling temperatures of the *f*-CNTs/ cotton textile of R_s = 1.67 kΩ/sq using a dc power supply. (a) Schematic of the heat measurement setup. (b) Heating and cooling temperatures at the center of the textile with time measured by a thermocouple. (c) The increase in ultimate temperature of the *f*-CNTs/cotton textile with input power (inset: thermal image at V_{in} = 60 V showing the uniform heating of the textile). (d) Changes in the temperature and current after repeated bending.

When an electric power of approximately 0.1 W/cm^2 is applied to the textile its temperature increases uniformly to ca. 84°C, which is shown in the IR image of **Figure 12(c)**. The white color indicates the area over which the temperature reaches more than 90% of the maximum temperature. Temperature homogeneity of this textile as heating element is better than those made with only stainless steel yarns, because in the latter case, heat is produced and localized only at the conductive yarns [21, 70]. Uniform distribution and dissipation of heat allow the heating element to be located in close proximity to the heated area in order to maximize warmth/heat production/output, to minimize response time, and to eliminate hot spots. The samples with four layers show a temperature increase of ca. 40°C (ca. 0.1 W/cm^2) within 0.5 min. The heat releasing ability is also high for the coated textile, which is observed from the exponential decrease of the temperature to the room temperature within few minutes. The temperature of the water could be raised more than 80°C inserting the textile heater into it.

The flexibility of the textile as heating element is studied by measuring the changes in the temperatures within 20 cycles of bending over a time frame of 80 min. It is observed that with repeated bending, the current conducting through the heating textile does not change significantly, and the decrease in temperature is very small as shown in **Figure 12(d)**. These indicate that the flexibility of the textile is high, which along with the temperature homogeneity suggests the use of the *f*-CNTs/cotton textile as a flexible heating element. These lightweight textiles can be configured and cut into any size or any shape, and are useful for portable heating system. For heating up to ca. 90°C, a power of 0.1 W/cm^2 is required, which is much lower than that of the conventional wired heater (ca. 0.3 W/cm^2) [71]. This allows more efficient heating with less energy, which will find potential applications in garments and in transportation to improve thermal comfort.

Mattana et al. [72] employed nanoscale modification of natural cotton fibers with conformal coatings of gold nanoparticles, deposition of thin layers of the conductive polymer poly(3,4-ethylenedioxithiophene) and a combination of these two processes to obtain conductive cotton from plain cotton yarns. The electrical and mechanical properties of these yarns are improved to be successfully used as conductors, in order to bias electronic devices. They demonstrated the possibility of realizing a fully textile circuit, including passive and active elements, and paves the way for a future complete integration between electronics and textiles. Kotov groups in their communication showed the coating of cotton yarn with CNTs and polyelectrolytes [19]. Their method provides a fast, simple, robust, low-cost, and readily scalable process for making e-textiles. Even though the cotton yarn became slightly harder after being coated with SWNTs, it is still very flexible and soft, both of which are important for the wearability of electronic fabric.

It was informed that CNTs have toxic effects. Isolated CNTs in human body would make damage to the organic cells. In order to avoid these effects the textile can be covered with some additional fabric or by other means before using it in different applications, by which direct contact to the human body or diffusion into the air could be avoided.

6. Summary

Functionalization of CNTs is very important to realize their applications in modern technological advancements. A safe method has been developed to functionalize CNTs. In the sequence of treatments, CNTs are pretreated in pure ethanol using a supersonic homogenizer, wetted using citric acid solution, and plasma treated using RF oxygen plasma. By the plasma reaction in the presence of water vapor, O_2 and citric acid, plasma species interact with them to create many kinds of ions and radicals. They attack the CNT surfaces and activate a large number of sites to enhance the attachment of –COOH groups onto their surfaces. These attached groups significantly enhance the dispersion stability of the CNTs in water. Therefore, we are able to produce highly stable dispersed *f*-CNT ink, where the structural integrity of the *f*-CNTs is conjectured to be preserved, which is verified by the Raman and TEM measurements.

Nonconducting cotton textile becomes electroconductive by repeatedly dipping into the stable *f*-CNT ink and drying in air. The *f*-CNTs uniformly and strongly cover the individual cotton fibers, which remain attached even after 30 min of standard washing. After several cycle of dipping into *f*-CNT ink, the textile becomes conductive enough to be used as wire in lighting up an LED, and its conductivity becomes more than 0.20 S/m depending on the loaded content of *f*-CNTs in it. The thermal conductivity of the textile is enhanced by *ca*. 70%, and the thermal stability and flame retardancy are also improved. As a demonstration of practical use, the textile is shown as a conductive textile heater designed with parallel electrodes. The temperature of the *f*-CNTs/cotton textile can be increased homogeneously to *ca.* 80°C within *ca.* 5 min by applying an electric power of *ca.* 0.1 W/cm^2, which is much less than the power required for conventional wired heater. In our process, the dispersion of the CNTs are achieved by functionalizing them with a safe and surfactant-free method of plasma treatment, which helps to avoid impurities as well as to preserve the properties of the CNTs to be incorporated in the cotton fibers.

7. Acknowledgments

This study was supported by the Promotion of Nano-Biotechnology Research to Support Aging, Welfare Society from Ministry of Education, Culture, Sports, Science & Technology, Japan. We used the TEM and TG/DTA at Research Institute of Green Science & Technology, Shizuoka University. We would like to thank Dr. C. Sawatari of Shizuoka University for her sincere help to do the washing test.

Author details

Mohammad Jellur Rahman[1,2] and Tetsu Mieno[2,3*]

*Address all correspondence to: sptmien@ipc.shizuoka.ac.jp

1 Department of Physics, Bangladesh University of Engineering and Technology, Dhaka, Bangladesh

2 Graduate School of Science & Technology, Shizuoka University, Shizuoka, Japan

3 Department of Physics, Shizuoka University, Shizuoka, Japan

References

[1] Yamamoto T, Watanabe K, Hernández ER. Mechanical Properties, Thermal Stability and Heat Transport in Carbon Nanotubes. In: Jorio A, Dresselhaus G, Dresselhaus MS, editors. Carbon Nanotubes-Topics in Applied Physics. Heidelberg: Springer; 2008. pp. 165–195.

[2] Endo M, Strano MS, Ajayan PM. Potential Applications of Carbon Nanotubes. In: Jorio A, Dresselhaus G, Dresselhaus MS, editors. Carbon Nanotubes-Topics in Applied Physics. Heidelberg: Springer; 2008. pp. 13–62.

[3] Murakami H, Nakashima N: Soluble Carbon Nanotubes and Their Applications. Journal of Nanoscience and Nanotechnology. 2006; 6:16–27.

[4] De Volder MFL, Tawfick SH, Baughman RH, Hart AJ: Carbon Nanotubes: Present and Future Commercial Applications. Science. 2013; 339:535–539. DOI: 10.1126/science.1222453

[5] Baughman RH, Zakhidov AA, de Heer WA: Carbon Nanotubes—the Route Toward Applications. Science. 2002; 297:787–792. DOI: 10.1126/science.1060928

[6] Hirsch A, Vostrowsky O: Functionalization of Carbon Nanotubes. Top. Curr. Chem. 2005; 245: 193–237. DOI 10.1007/b98169

[7] Tasis D, Tagmatarchis N, Bianco A, Prato M: Chemistry of Carbon Nanotubes. Chemical Review. 2006; 106:1105–1136. DOI: 10.1021/cr050569o

[8] Grossiord N, Regev O, Loos J, Meuldijk J, Koning CE: Time-Dependent Study of the Exfoliation Process of Carbon Nanotubes in Aqueous Dispersions by Using UV-Visible Spectroscopy. Analytical Chemistry. 2005; 777:5135–5139. DOI: 10.1021/ac050358j

[9] Liu J, Rinzler AG, Dai H, Hafner JH, et al.: Fullerene Pipes. Science. 1998; 280:1253–1256.

[10] Pavese M, Musso S, Bianco S, Giorcelli M, Pugno N: An analysis of carbon nanotube structure wettability before and after oxidation treatment. Journal of Physics: Condensed Matter, 2008; 20:474206. DOI:10.1088/0953-8984/20/47/474206

[11] Ziegler KJ, Gu Z, Peng H, Flor EL, Hauge RH, Smalley RE: Controlled oxidative cutting of single-walled carbon nanotubes. Journal of American Chemical Society. 2005; 127:1541–1547. DOI: 10.1021/ja044537e

[12] Worsley KA, Kalinina I, Bekyarova E, Haddon RC: Functionalization and Dissolution of Nitric Acid Treated Single-Walled Carbon Nanotubes. Journal of American Chemical Society. 2009; 131:18153–18158. DOI: 10.1021/ja906267g

[13] Hu LB, Pasta M, Mantia FL, Cui LF, et al.: Stretchable, Porous, and Conductive Energy Textiles. Nano Letters. 2010; 10: 708–714. DOI: 10.1021/nl903949m

[14] Lukowicz P, Kirstein T, Troster G: Wearable Systems for Health Care Applications. Methods of Information in Medicine. 2004; 43: 232–238.

[15] Park S, Jayaraman S: Smart Textiles_Wearable Electronic Systems. MRS Bulletin. 2003; 28: 585–591.

[16] Han J-W, Kim B, Li J, Meyyappan M: A carbon nanotube based ammonia sensor on cotton textile. Applied Physics Letter. 2013; 102: 193104. DOI: 10.1063/1.4805025

[17] Wang K, Zhao P, Zhou X, Wua H, Wei Z: Flexible supercapacitors based on cloth-supported electrodes of conducting polymer nanowire array/SWCNT composites. Journal of Materials Chemistry. 2011; 21:16373–16378. DOI: 10.1039/C1JM13722K

[18] Yuan C, Hou L, Li D, Shen L, Zhang F, Zhang X: Synthesis of flexible and porous cobalt hydroxide/conductive cotton textile sheet and its application in electrochemical capacitors. Electrochimica Acta. 2011; 56: 6683–6687. DOI:10.1016/j.electacta.2011.05.050

[19] Shim BS, Chen W, Doty C, Xu C, Kotov NA: Smart Electronic Yarns and Wearable Fabrics for Human Biomonitoring made by Carbon Nanotube Coating with Polyelectrolytes. Nano Letters. 2008; 8:4151–4157. DOI: 10.1021/nl801495p

[20] Negru D, Buda CT, Avram D: Electrical Conductivity of Woven Fabrics Coated with Carbon Black Particles. Fibres & Textiles in Eastern Europe. 2012; 20: 53–56.

[21] Janas D, Koziol KK: A review of production methods of carbon nanotube and graphene thin films for electrothermal applications. Nanoscale. 2014; 6: 3037. DOI: 10.1039/c3nr05636h

[22] Abbas A, Zhao Y, Zhou J, Wang X, Lin T: Improving Thermal Conductivity of Cotton Fabrics Using Composite Coatings Containing Graphene, Multiwall Carbon Nanotube or Boron Nitride Fine Particles. Fibers and Polymers. 2013; 14:1641–1649. DOI: 10.1007/s12221-013-1641-y

[23] Liu Y, Wang X, Qi K, Xin JH: Functionalization of cotton with carbon nanotubes. Journal of Materials Chemistry. 2008; 18: 3454–3460. DOI: 10.1039/B801849A

[24] Krueger A. Carbon Nanotubes. In: Carbon Materials and Nanotechnology. Weinheim: Wiley-VCH; 2010. p. 123–281. DOI: 10.1002/9783527629602.ch3

[25] Properties of Carbon Fibre [Internet]. Available from: http://www.smithersrapra.com/SmithersRapra/media/Sample-Chapters/Update-on-Carbon-Fibre.pdf [Accessed: 2016-02-21]

[26] Nanotubes: A Carbon-based Nanoparticle [Internet]. Available from: Boysen E, Muir NC: Nanotechnology for Dummies. http://www.dummies.com/how-to/content/nanotubes-a-carbonbased-nanoparticle.html [Accessed: February 21, 2016]

[27] Nanotechnology in Solar Cells [Internet]. Available from: Compiled by Earl Boysen of Hawk's Perch Technical Writing, LLC and UnderstandingNano.com. http://www.understandingnano.com/solarcells.html [Accessed: 2016-02-21]

[28] von Helden G, Hsu MT, Gotts NG, Bowers MT: Carbon Cluster Cations with up to 84 Atoms: Structures, Formation Mechanism, and Reactivity. Journal of Physical Chemistry. 1993; 97:8182–8192. DOI: 10.1021/j100133a011

[29] Bahr JL, Mickelson ET, Bronikowski MJ, Smalley RE, Tour JM: Dissolution of small diameter single-wall carbon nanotubes in organic solvents. Chemical Communications. 2001; 2:193–194. DOI: 10.1039/b008042j

[30] Hirsch A: Functionalization of Single-Walled Carbon Nanotubes. Angewandte Chemie International Edition. 2002; 41 :1853–1859. DOI: 10.1002/1521-3773(20020603)41

[31] Vigolo B, Hérold C. Processing Carbon Nanotubes. In: Yellampalli S, editor. Carbon Nanotubes-Synthesis, Characterization, Applications. Croatia: InTech Open; 2011. p. 3–28. DOI: 10.5772/17773

[32] Spizzirri UG, Curcio M, Cirillo G, Spataro T, et al.: Recent Advances in the Synthesis and Biomedical Applications of Nanocomposite Hydrogels. Pharmaceutics. 2015; 7: 413–437. DOI: 10.3390/pharmaceutics7040413

[33] Peng K, Liu L -Q, Li H, Meyer H, Zhang Z: Room temperature functionalization of carbon nanotubes using an ozone/water vapor mixture. Carbon. 2011; 49: 70–76. DOI: 10.1016/j.carbon.2010.08.043

[34] Hirsch A, Vostrowsky O. Functionalization of Carbon Nanotubes. In: Müller TJJ, Bunz UHF, editors. Functional Organic Materials: Syntheses, Strategies and Applications. Weinheim, Germany: Wiley-VCH Verlag; 2007, p. 1–57. DOI: 10.1002/9783527610266.ch1

[35] Zhao B, Hu H, Yu A, Perea D, Haddon RC: Synthesis and Characterization of Water Soluble Single-Walled Carbon Nanotube Graft Copolymers. Journal of American Chemical Society. 2005; 127: 8197–8203. DOI: 10.1021/ja042924i.

[36] Chen Y, Haddon RC, Fang S, Rao AM, et al.: Chemical Attachment of Organic Functional Groups to Single-walled Carbon Nanotube Material. Journal of Materials Research. (1998) 2423–2431. DOI: http://dx.doi.org.sci-hub.cc/10.1557/JMR.1998.0337

[37] Zhou W, Ooi YH, R Russo, Papanek P, et al.: Structural characterization and diameter-dependent oxidative stability of single wall carbon nanotubes synthesized by the catalytic decomposition of CO. Chemical Physics Letters. 2001; 350: 6–14. DOI: 10.1016/S0009-2614(01)01237-4

[38] Hammer T: Application of Plasma Technology in Environmental Techniques. Contributions to Plasma Physics. 1999; 39: 441–462. DOI: 10.1002/ctpp.2150390507

[39] Frauchiger VM, Schlottig F, Gasser B, Textor M: Anodic plasma-chemical treatment of CP titanium surfaces for biomedical applications. Biomaterials. 2004; 25 : 593–606. DOI: 10.1016/S0142-9612(03)00560-X

[40] Meyer-Plath AA, Finke B, Schröder K, Ohl A: Pulsed and cw microwave plasma excitation for surface functionalization in nitrogen-containing gases. Surface and Coatings Technology. 2003; 174–175: 877–881. DOI: 10.1016/S0257-8972(03)00563-2

[41] Felten A, Bittencourt C, Pireaux J-J, Van Lier G, Charlier J-C: Radio-frequency plasma functionalization of carbon nanotubes surface O_2, NH_3, and CF_4 treatments. Journal of Applied Physics. 2005; 98: 074308. DOI: 10.1063/1.2071455

[42] Zhang J, Feng T, Yu W, Liu X, Wang X, Li Q: Enhancement of field emission from hydrogen plasma processed carbon nanotubes. Diamond and Related Materials. 2004; 13: 54–59. DOI: 10.1016/j.diamond.2003.08.018

[43] Valentini L, Puglia D, Armentano, Kenny JM: Sidewall functionalization of single-walled carbon nanotubes through CF_4 plasma treatment and subsequent reaction with aliphatic amines. Chemical Physics Letters. 2005; 403: 385–389. DOI: 10.1016/j.cplett.2005.01.042

[44] Xu T, Yang J, Liu J, Fu Q: Surface modification of multi-walled carbon nanotubes by O_2 plasma. Applied Surface Science. 2007; 253: 8945. DOI: 10.1016/j.apsusc.2007.05.028

[45] Kalita G, Adhikari S, Aryal HR, Ghimre DC, et al.: Fluorination of multi-walled carbon nanotubes (MWNTs) via surface wave microwave (SW-MW) plasma treatment. Physica E. 2008; 41: 299–303. DOI: 10.1016/j.physe.2008.07.015

[46] Qin Y, Hu M: Effects of microwave plasma treatment on the field emission properties of printed carbon nanotubes/Ag nano-particles films. Applied Surface Science. 2008; 254:1757–1762.

[47] Khare BN, Meyyappan M, Cassell AM, Nguyen CV, Han J: Functionalization of Carbon Nanotubes Using Atomic Hydrogen from a Glow Discharge. Nano Letters. 2002; 2: 73–77.

[48] Rahman MJ, Mieno T: Water-dispersible multiwalled carbon nanotubes obtained from citric-acid-assisted oxygen plasma functionalization. Journal of Nanomaterials. 2014; 2014:508192. DOI: http://dx.doi.org/10.1155/2014/508192

[49] Rahman MJ, Mieno T: Conductive cotton textile from safely functionalized carbon nanotubes. Journal of Nanomaterials. 2015; 2015: 978484. DOI: http://dx.doi.org/10.1155/2015/978484

[50] Dai L, Chang DW, Baek J -B, Lu W: Carbon nanomaterials for advanced energy conversion and storage. Small 2012; 8:1130. DOI: 10.1002/smll.201101594

[51] Kurkina T, Vlandas A, Ahmad A, Kern K, Balasubramanian K: Label-free detection of few copies of DNA with carbon nanotube impedance biosensors. Angewandte Chemie International Edition. 2011; 50: 3710.

[52] Owen T: Fundamentals of Modern UV-visible Spectroscopy. Germany: Hewlett-Packard Co. 2000. p.1.

[53] Dresselhaus MS, Dresselhaus G, Saito R, Jorio A: Raman spectroscopy of carbon nanotubes. Physics Reports. 2005; 409:47–99. DOI: 10.1016/j.physrep.2004.10.006

[54] Graupner R: Raman spectroscopy of covalently functionalized single-wall carbon nanotubes. Journal of Raman Spectroscopy., 2007; 38: 673–683. DOI: 10.1002/jrs.1694

[55] Tian R, Wang X, Li M, Hu H, Chen R, Liu F, Zheng H, Wan L: An efficient route to functionalize singe-walled carbon nanotubes using alcohols. Applied Surface Science. 2008; 255:3294–3299. DOI: 10.1016/j.apsusc.2008.09.040

[56] Chen C, Liang B, Lu D, Ogino A, Wang X, Nagatsu M: Amino group introduction onto multiwall carbon nanotubes by NH_3/Ar plasma treatment: Carbon. 2010; 48: 939–948. DOI: 10.1016/j.carbon.2009.10.033

[57] Zhang X, Lei L, Xia B, Zhang Y, Fu J: Oxidization of carbon nanotubes through hydroxyl radical induced by pulsed O_2 plasma and its application for O_2 reduction in electro-Fenton. Electrochimica Acta. 2009; 54: 2810–2817. DOI: 10.1016/j.electacta.2008.11.029

[58] Imasaka K, Suehiro J, Kanatake Y, Kato Y, Hara M: Preparation of water-soluble carbon nanotubes using a pulsed streamer discharge in water. Nanotechnology. 2006; 17:3421–3427. DOI :10.1088/0957-4484/17/14/013

[59] Naseh MV, Khodadadi A. A, Mortazavi Y, Pourfayaz F, Alizadeh O, Maghrebi M. Fast and clean functionalization of carbon nanotubes by dielectric barrier discharge plasma in air compared to acid treatment. Carbon. 2010; 48:1369–1379. DOI:10.1016/j.carbon. 2009.12.027

[60] Tong DG, Luo YY, Chu W, Guo YC, Tian W. Cutting of carbon nanotubes via solution plasma processing. Plasma Chemistry and Plasma Processing. 2010; 30: 897–905. DOI: 10.1007/s11090-010-9262-3

[61] Kolacyak D, Ihde J, Merten C, Hartwig A, Lommatzsch U. Fast functionalization of multi-walled carbon nanotubes by an atmospheric pressure plasma jet. Journal of Colloid and Interface Science. 2011; 359: 311–317. DOI: 10.1016/j.jcis.2011.03.069

[62] Hussain S, Amade R, Jover E, Bertran E.: Functionalization of carbon nanotubes by water plasma. Nanotechnology. 2012; 23: 385604. DOI: 10.1088/0957-4484/23/38/385604

[63] Hordy N, Coulombe S, Meunier J-L.: Plasma functionalization of carbon nanotubes for the synthesis of stable aqueous nanofluids and poly(vinyl alcohol) nanocomposites. Plasma Processes and Polymers. 2013; 10: 110–118. DOI: 10.1002/ppap.201200075

[64] Gonçalves AG, Jarrais B, Pereira C, Morgado J, Freire C, Pereira MFR: Functionalization of textiles with multi-walled carbon nanotubes by a novel dyeing-like process. Journal of Materials Science. 2012; 47:5263–5275. DOI: 10.1007/s10853-012-6412-4

[65] Yu G, Hu L, Vosgueritchian M, Wang H, Xie X, McDonough JR, Cui X, Cui Y, Z Bao: Solution-processed graphene/MnO_2 nanostructured textiles for high-performance electrochemical capacitors. Nano Letters. 2011; 11: 2905–2911. DOI: dx.doi.org/10.1021/ nl2013828

[66] Hu S, Rajamani R, Yu X: Flexible solid-state paper based carbon nanotube supercapacitor. Applied Physics Letters. 2012; 100:104103. DOI: http://dx.doi.org/ 10.1063/1.3691948

[67] Marconnet AM, Panzer MA, Goodso KE: Thermal conduction phenomena in carbon nanotubes and related nanostructured materials. Reviews of Modern Physics. 2013; 85: 1295–1326. DOI: 10.1103/RevModPhys.85.1295

[68] Ragab T, Basaran C: Joule heating in single-walled carbon nanotubes. Journal of Applied Physics. 2009; 106: 063705. DOI: 10.1063/1.3204971

[69] Kang J, Kim H, Kim KS, Lee SK, et. al: High-performance graphene-based transparent flexible heaters. Nano Letters. 2011; 11: 5154–5158. DOI: dx.doi.org/10.1021/nl202311v

[70] Chu K, Kim D, Sohn Y, Lee S, Moon C, Park S: Electrical and thermal properties of carbon-nanotube composite for flexible electric heating-unit applications. IEEE Electron Device Letters. 2013; 34: 668–670. DOI: 10.1109/LED.2013.2249493

[71] Markevicius T, Olsson N, Furferi R, Meyer H: Flexible mild heaters in structural conservation of paintings: state of the art and conceptual design of a new carbon nanotubes-based heater. Journal of Applied Sciences. 2012; 12: 211–220. DOI: 10.3923/jas.2012.211.220

[72] Mattana G, Cosseddu P, Beatrice Fraboni B, George G. Malliaras GG, et al.: Organic electronics on natural cotton fibres. Organic Electronics. 2011; 12: 2033–2039. DOI: 10.1016/j.orgel.2011.09.001

Chapter 13

Functionalization of Carbon Nanotubes with Stimuli-Responsive Molecules and Polymers

Li Wang and Yuming Zhao

Additional information is available at the end of the chapter

http://dx.doi.org/10.5772/64078

Abstract

"Smartly" functionalized carbon nanotubes (CNTs) constitute an actively pursued research topic in the fields of nanomaterials and nanotechnology. The development of highly efficient and selective methodologies for dispersing CNTs in the liquid phase has not only made efficient separation and purification of CNTs possible, but also opened the doors to many fascinating material and biological applications. Very recently, the development of CNT hybrid systems with controlled stimuli-responsiveness has achieved significant breakthroughs. This chapter outlines the state of the art within this vibrant research area, and examples from the most recent literature are selected to demonstrate progress in the preparation of CNT composites, the physical properties of which can be readily switched by various external stimuli (*e.g.*, pH, photoirradiation, solvent, temperature, etc.).

Keywords: carbon nanotubes, chemical functionalization, stimuli-responsiveness, reversibility, nanomaterials, supramolecular chemistry

1. Introduction

Since the first synthesis of carbon nanotubes (CNTs) by Ijima in 1991 [1], CNT-based materials have attracted tremendous interest from both academia and industry. Indeed, CNTs along with other nano-sized carbon materials (*e.g.*, fullerenes and graphenes) have taken the central stage of the present research of nanoscience and nanotechnology, and it is easily foreseeable that research efforts dedicated to these topics will continue to grow in the future [2]. For decades, the extraordinary structural and physical properties of CNTs have promoted extensive investigation aimed at their synthesis, processing, and functionalization, which in turn create a wide range of applications in electronics [3, 4], optics [5], environment [6], biology, and medicinal science [7, 8].

CNTs are carbon allotropes in a tubular shape with nanoscale diameter (*ca.* 1–2 nm) and high aspect ratio. Generally speaking, CNTs can be classified as single-walled carbon nanotubes (SWNTs), double-walled carbon nanotubes (DWNTs), and multi-walled carbon nanotubes (MWNTs). A SWNT can be viewed as a single layer of graphene sheet rolled up in a cylindrical shape, while the DWNTs and MWNTs are similar carbon nanostructures but made of two or more layers of SWNTs, in which the smaller diameter tube(s) coaxially nested in the larger one(s). The exact structures of SWNTs differ by the ways of rolling the graphene sheet, which can be described using their specific chiral indices (n,m) and chiral angles $\theta(n,m)$. As illustrated in **Figure 1**, SWNTs with chiral indices of $m = 0$ or $n = m$ are achiral and often referred to as *zigzag* and *armchair* tubes, respectively. Other combinations of n and m define the family of so-called chiral nanotubes. The electronic properties of SWNTs significantly depend on their chiral indices. When $|n - m| = 3q$ where q is an integer, the nanotube is metallic or semi-metallic, whereas for all other cases the nanotubes are semiconducting in nature.

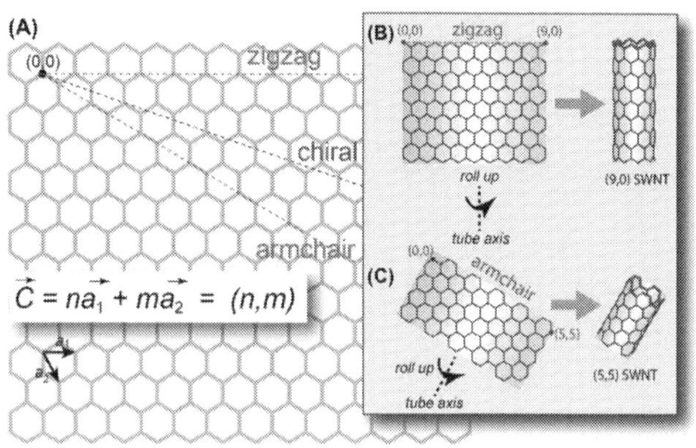

Figure 1. (A)–(C) Illustration of the formation of SWNTs with various chiral indices by rolling graphene sheets.

The synthesis of CNTs can be done in several ways, for example, arc discharge, laser ablation, chemical vapor deposition (CVD), and others [9]. However, all the currently existing processes for CNT production generate mixtures of various CNTs, amorphous carbon, and/or residual metal species. For this reason, post-treatment of as-produced CNT products becomes a critically important and indispensable step toward the application of CNTs [10], since removal of the unwanted impurities can give rise to considerably improved performances for CNT-based electronic and optoelectronic devices, while the purity of CNTs plays a key role in reducing or averting undesired toxicity and side effects in biological and medicinal applications. Another significant technical challenge encountered in the application of CNTs is the extremely low solubility and hence very poor processability of pristine CNTs in common solvents. The very large hydrophobic π-surface of CNTs makes them prone to aggregation

http://dx.doi.org/10.5772/64078

(forming bundles) through π-stacking and van der Waals interactions. Breaking the bundles into individual tubes is an energetically costly process that usually relies on chemical modifications to overcome the strong inter-tube attractions. There are two general approaches for preparation of functionalized CNTs (*f*-CNTs), namely covalent [11] and non-covalent [12, 13]. The covalent method involves the attachment of certain molecular functionalities to the sidewall and/or the defects/ends of CNTs through covalent bonds. As such, not only very high solubility or dispersity in the solution phase become readily achievable, but new chemical and physical properties can be introduced to the *f*-CNTs. One significant limitation of the covalent functionalization is that it inevitably converts a number of sp^2 carbons into sp^3 on the sidewall of CNTs, which is an unwanted scenario for applications where the integrity of graphene-type backbone of CNTs must be retained to keep their pristine electronic and mechanical characteristics. In this respect, the non-covalent approach has been deemed as an intrinsically better alternative due to its non-destructive nature.

The generally employed strategy for non-covalent functionalization of CNTs relies on the use of certain dispersing agents (*i.e.*, dispersants) to break CNT bundles. Application of external forces, commonly ultrasonication or mechanical stirring, allows the dispersants to be assembled on the surface of CNTs via non-covalent binding forces (*e.g.*, π-π stacking, CH-π, van der Waals, and charge transfer interactions). As such, CNT bundles could be exfoliated, resulting in individualized (or dispersed) tubes encapsulated or wrapped with dispersants to form stable suspension. Up to now, a vast array of chemical species, ranging from conjugated molecules, synthetic polymers, surfactants, peptides, to biopolymers and DNA molecules, has been demonstrated to show effectiveness at dispersing CNTs in various solvents [10, 12–14]. The scope of this topic is rather broad and encompasses an enormous body of literature. Generally speaking, the purposes of dispersing CNTs are multi-fold and significantly dependent on the requirements of specific applications. For instance, effective exfoliation and dispersion of CNTs in organic solvents allow CNTs to be readily blended with other functional polymers to form new composite materials with enhanced mechanical and electrical properties [13, 15]. Dispersion of CNTs into aqueous media with biocompatible matrices opens the doors to exciting biological and medicinal applications of CNT hybrids [8, 16]. Many initial studies on CNT dispersion were primarily aimed at enhanced dispersity and compatibility with various solvents and matrix materials. As the research moved on, it was observed that certain dispersants exhibited selectivity in binding with CNTs of certain electronic types, diameters, and chirality. The ramification of such behavior was quickly recognized and exploited for various purposes. As mentioned above, as-produced CNTs are structurally heterogeneous and contain significant amounts of impurities. In the past decade, a great deal of research efforts has been dedicated to the purification of CNTs through selective non-covalent functionalization, particularly the studies of sorting SWNTs have been greatly advanced [17, 18]. Very recently, remarkable breakthroughs have also been achieved in the fabrication of high-performance nanodevices using purified CNTs [19–23], which in turn fuels up the development of CNT processing/purification methods featuring excellent efficiency and cost-effectiveness.

In short, the present knowledge and techniques for dispersed CNT systems and subsequent applications have been considerably expanded, thanks to the continued efforts of material chemists on the design and synthesis of novel molecular and macromolecular systems as functional dispersants or agents for CNT modifications. Indeed, the current design of CNT dispersants and development of related processing methods have already surpassed being merely a single-parameter issue of improving dispersity. Rather, the objectives of research have turned out to be more and more diversified and application-oriented. In the following sections, our discussions are firstly focused on a newly emerging topic—reversible dispersion and release of CNTs regulated by rationally designed stimuli-responsive molecular and supramolecular systems—which has attracted rapidly growing attention in recent years. In the second part, the recent progress in polymer-functionalized CNT composites that show controlled responsiveness to one or more environmental stimuli is described.

2. Dispersion and release of CNTs: a microscopic and supramolecular perspective view

Effective dispersion of CNTs requires sufficient energy input (*e.g.*, mechanical stirring, shear mixing, ultrasonication) to disrupt the inter-tube attractions, with the amount of dispersion energy dependent on the length and diameter of CNTs as well as solvent nature and temperature. It has been found that stable dispersion of CNTs could be formed with certain solvents, such as 1,2-dichlorobenzene [24], CS_2 [25], and particular organic amides [26, 27]. For the majority of commonly used solvents, however, the surface of CNTs dispersed in them must be subsequently modified to prevent re-aggregation of CNTs from occurring, and this can be realized by having the CNTs' outer surface adsorbed or encapsulated with solvophilic layers. In this sense, effective CNT dispersants should contain the unit(s) having sufficient affinity for the graphitic surface of CNTs (henceforth referred to as the *graphenophile*). To the graphenophiles various solubilizing groups can be attached, the presence of which dictates the dispersity of the functionalized CNTs in different solvents or matrices.

In the literature, a large array of molecular and macromolecular motifs has been reported to give satisfactory graphenophilic performance. In the category of small molecules, planar structurally rigid polycyclic aromatic hydrocarbons (PAHs) have been known to easily adsorb on the surface of CNTs through a combination of hydrophobic and π-π interactions [28, 29]. Examples of such PAHs include pyrene, naphthalene, and phenathrene, which carry the fused aromatic backbones, mapping out the segmental hexagonal arrangement on the sidewall of CNTs. It is worth noting that since the initial study by Dai and co-workers [30], pyrene has become a popularly employed anchoring group allowing facile assembly of CNT supermolecular constructs imparted with various optoelectronic properties [31–34] and biologic activities [35–38] By the same token, highly π-extended heterocyclic aromatic compounds, such as phorphyrins, [39], phthalocyanines [40], and tetrathiafulvalenes [41, 42], have been found to show strong non-covalent interactions with CNTs, rendering the resulting hybrid systems intriguing photophysical and optoelectronic properties. Besides the small molecule-based systems, π-conjugated polymers have been widely used in CNT dispersion as well [15].

http://dx.doi.org/10.5772/64078

In these conjugated polymers, arene groups are embedded in the repeat units, which collectively engender strong binding to CNTs via π-π stacking. Experimental observations and molecular dynamic (MD) simulations have demonstrated that the polymer backbones with a certain degree of flexibility would favor wrapping around the CNTs in different fashions [43, 44]. DNA or RNA molecules can interact with the CNT sidewall via π-π stacking. In particular, single-strand DNA (ssDNA) oligomers and polymers have been extensively explored as SWNT dispersants and the interactions were found to be sequence-dependent [45–51]. Such properties have enabled fascinating applications including precise sorting of SWNTs, DNA sequencing, and chemical sensing applications [52]. Indeed, for many CNT dispersants developed so far, aromatic functional groups (*i.e.*, graphenophiles) are the essential structural components, and the resulting CNT supramolecular assemblies are usually stable and do not dissociate easily. On the other hand, the dispersion of CNTs can be realized by utilizing other types of non-covalent binding forces. For example, the hydrophobic interactions between surfactants and CNTs in water. Typical surfactants such as SDS are known to adsorb on the surface of CNTs with their hydrophobic tails (*e.g.*, alkyl chains), while their hydrophilic cationic or anionic heads directly interact with water molecules. Recent experimental and theoretical modeling studies [53–55] have shown that surfactant molecules can assemble into various arrangements of micellar structures, which effectively encapsulate individual CNTs to form well-dispersed stable suspensions in aqueous media.

With the wide range of CNT dispersants studied over the past decade, fundamental understanding of their non-covalent interactions with CNTs at the molecular and supramolecular levels has been continually developed by state-of-the-art spectroscopic and microscopic analyses as well as high-level theoretical simulations. Knowledge in this field is highly instructive to the design of more sophisticated CNT-dispersing methods that are transferrable to the application of CNT-related materials in science and technology. One topic receiving much attention in the recent research of CNTs is the reversible dispersion and release of CNTs under the control of external stimuli, such as irradiation, chemical reactions, and solvation effects. Basically, to generate stable CNT dispersion in the solution phase, the CNT surface needs to be functionalized or encapsulated with dispersants to a sufficient degree. This in turns makes the dissociation of CNTs from dispersants not an easy task; for instance, many polyaromatics and π-conjugated polymers are known to irreversibly adsorb onto the surface of CNTs, which makes it extremely difficult to remove them from CNTs by physical means (*e.g.*, solvent rinsing). An efficient approach to release CNTs out of a well-stabilized dispersion is to reverse the attractive forces between the dispersants of CNTs into repulsive interactions, such as the strategy schematically illustrated in **Figure 2**. Generally speaking, a chemical process (*e.g.*, acid-base interactions, redox reactions, and photodegradation) that leads to the reversal of the dispersants from being graphenophilic to graphenophobic would be useful for triggering the release or re-aggregation of CNTs. On the other hand, various supramolecular means (*e.g.*, metal coordination, conformational changes) can also offer sufficient driving forces to induce CNT release. For all the methods perceivable, it is essential for the CNT dispersants to show responsiveness to external inputs one way or another. The following section provides an overview of the recent progress in this field, and the detailed discussion is organized according to the type and mechanism of stimuli-responsiveness.

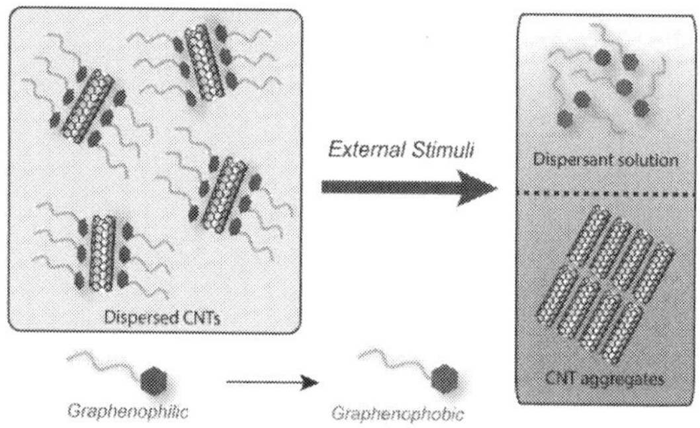

Figure 2. Schematic illustration of release of CNTs by changing the dispersants from being graphenophilic to graphenophobic.

3. Recent advances in reversible dispersion and release of CNTs with stimuli-responsive dispersants

3.1. By acid-base interactions

Acid-base interactions are probably the most straightforward ways to drastically alter the chemical and physical properties, for instance, changing from neutral to cationic/anionic, from organic solubility to water solubility. For this reason, this approach has become the actively pursued one in this field. Das and co-workers [56] recently developed a series of cholesterol-based amino acid carboxylates **1** and dipeptide carboxylates **2** (**Figure 3**) as dispersant for SWNTs as well as for graphene oxide (GO). Herein, the dispersants are amphiphilic, with the structurally rigid, hydrophobic cholesterol unit acting as the graphenophile to interact with the sidewall of SWNTs and the carboxylate-appended tails being hydrophilic. All the dispersants were able to disperse SWNTs in water, forming stable colloidal solutions. Compounds **2** with dipeptide moieties were found to bring about better dispersity than the mono amino acid-containing compounds **1**. Upon lowering the pH of the SWNT-**2c** colloidal suspension by adding 1N HCl, the carboxylate anions were converted into carboxylic acids, which in turn reduced the stability of the colloidal suspension as evidenced by zeta potential analysis. As a consequence of this, SWNTs were found to precipitate out of the aqueous solution. Addition of NaOH to increase the pH of the mixture converted the carboxylic acid groups back to hydrophilic carboxylate anions and re-generated the SWNT dispersion in water. Moreover, the SWNT dispersions engendered using compounds **1-2** as dispersants were tested to show cytocompatibility and potential biological applications. Later on, Das and co-workers reported another class of cholesterol-dipeptide amphiphiles **3** (**Figure 3**), which were designed to be highly sensitive and responsive in the pH range of tumorogenic environment (*ca*. pH 5–6) [57]. With these dispersants, SWNTs were well suspended in PBS buffer at pH 8. The SWNT-**3** assemblies could be further loaded with an anticancer drug DOX. At pH 6.0–6.5, precipitation

http://dx.doi.org/10.5772/64078

of SWNTs occurred and concomitantly the drug DOX was released. This performance is useful for specific drug delivery to cancer cells.

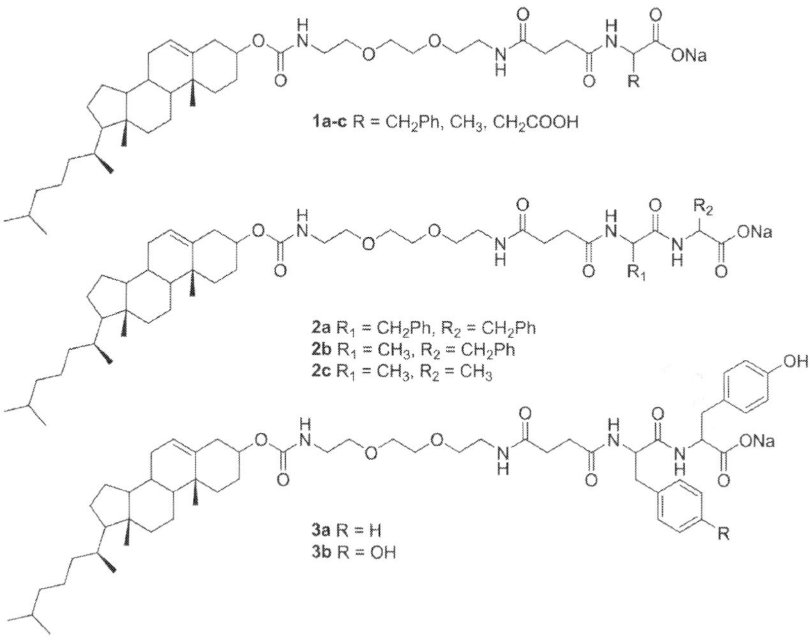

Figure 3. Cholesterol-based carboxylates amphiphiles **1–4** enabling reversible SWNT dispersion in water under pH control.

Based on a similar acid-base exchange concept, Huang and co-workers [58] recently synthesized a pH-responsive pillar[6]arene **4**, the backbone of which was decorated with water-solubilizing carboxylate groups (**Figure 4**). In their work, a pyrene compound **5** was first anchored to the sidewall of MWNTs via π-π stacking. The resulting hybrids were insoluble in water; however, addition of pillar[6]arene **4** to the functionalized MWNTs under sonication resulted in guest-host complexation taking place on the CNT sidewall. Of great interest was that the water solubility of resulting MWNT supramolecular assemblies could be reversibly switched by pH control. Under basic conditions, the carboxylic groups were deprotonated to form carboxylate anions, making the complexed MWNTs water-soluble. Upon acidification, the MWNTs were found to precipitate out of water. This work provided an elegant example of using pH-sensitive guest-host chemistry to achieve reversible dispersion and release of CNTs. Nevertheless, it is worth remarking that this work did not clearly address whether the complexation between **4** and **5** was altered on the CNT sidewall with changing pH, and such an issue may deserve further investigation.

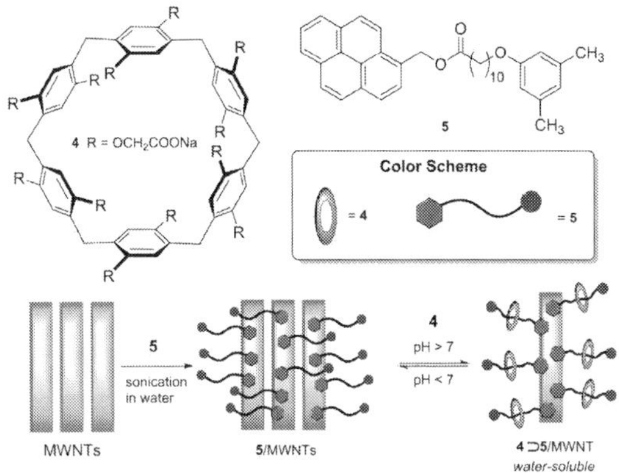

Figure 4. Reversible dispersion and release of MWNTs by pH-sensitive guest-host complexation between pillar[6]arene **4** and pyrene derivative **5**.

In 2015, Bryce and Lambert [59] synthesized a total of 13 amphiphilic surfactants, the structures of which were made of a pyrene head (hydrophobic) and various hydrophilic tails ending with carboxylate groups. Similar to the design of numerous other CNT dispersants, the pyrene group here was employed to act as a strong graphenophile to irreversibly link the surfactant molecules to the surface of CNTs. The performances of these pyrene-based surfactants in terms of MWNT dispersion in aqueous media were assessed. In particular, the authors examined the pH responsiveness of two of the surfactants **6** and **7** (**Figure 5**) in aqueous NaCl solution, and found that the corresponding MWNT-surfactant complexes could be reversibly dispersed and re-aggregated under basic and acidic conditions. A working mechanism for this acid-base-triggered MWNT dispersion was proposed based on the switching between hydrophobicity and hydrophilicity in different pH environments (see **Figure 5**).

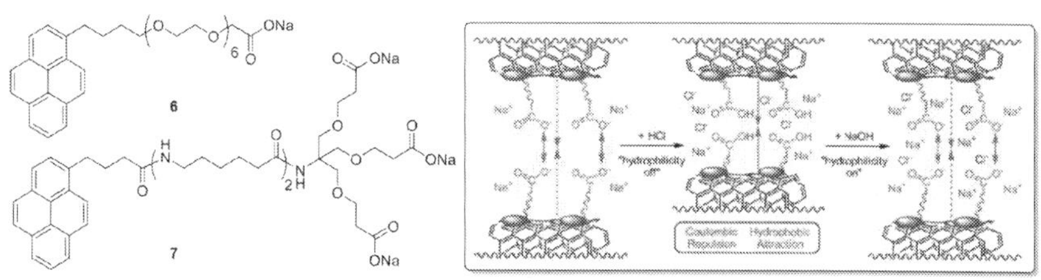

Figure 5. Pyrene-based surfactants **6** and **7** that show pH-responsiveness in dispersing MWNTs. The inset scheme was adopted from reference [59] with permission.

A novel dumbbell-shaped 2-ureido-6[1H]-pyrimidinone (UPy)-based fluorene derivative **8** (**Figure 6**) was recently devised by the Bao group [60] for selective and reversible dispersion of semiconducting SWNTs. In their design, fluorenyl group was chosen as the graphenophile to promote selectivity for semiconducting SWNTs, while the two UPy terminal groups enabled **8** to self-associate into supramolecular polymers through H-bonding interactions at high concentration. As illustrated in **Figure 6**, the supramolecular polymers of **8** selectively dispersed semiconducting SWNTs allowing for the separation of metallic and semiconducting tubes from as-synthesized SWNTs. Moreover, semiconducting SWNTs could be quantitatively released from the suspension by addition of 1% TFA solution, which induced the disassembly of the supramolecular polymers by breaking the H-bonds within them. Compared with many covalent polymer dispersants, this supramolecular polymer approach offers advantages in terms of releasing dispersant-free SWNTs and easiness in reusing the fluorene-UPy monomer **8**.

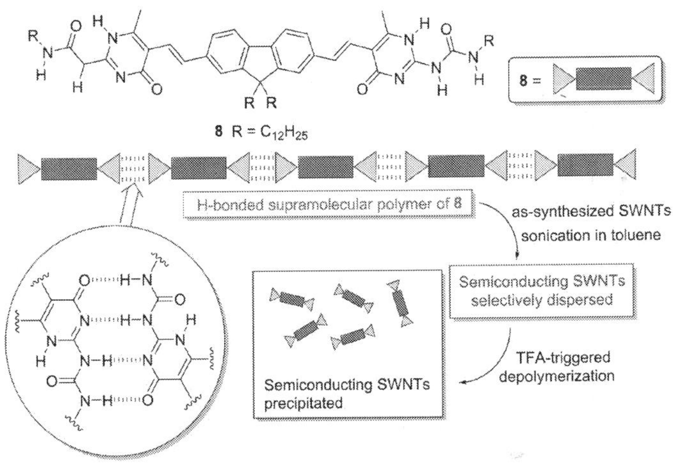

Figure 6. Selective dispersion and quantitative release of semiconducting SWNTs using a pH-sensitive H-bonded supramolecular polymer.

In addition to the above-mentioned rationally designed CNT dispersant systems, some other pH-sensitive molecules and biopolymers have also been reported to induce reversible dispersion/precipitation of CNTs in aqueous media. For example, Sun and co-workers [61] reported that 1-pyreneacetic acid after deprotonation under basic conditions could be non-covalently functionalized on nitric acid-treated as-produced SWNTs to form stable dispersion in water. Upon acidification, 1-pyreneacetic acid was quantitatively removed to yield purified SWNTs. The easy recovery and reuse of dispersants make this method potentially useful for large-scale CNT processing and production. Bhattacharya *et al.* [62] in 2014 reported that the dispersion of SWNTs with *ss*-DNA oligomers was pH-sensitive. At pH 7, the *ss*-DNA oligomers wrapped them around SWNTs to effect debundling and dispersion in water. Under acidic conditions (*e.g.*, at pH 3.5), however, the *ss*-DNA oligomers underwent a reversible structural change leading to unwrapping and precipitation of SWNTs. This method shows the potential in separating metallic and semiconducting SWNTs. Wang and Chen [63] discovered that poly-

L-lysine was capable of dispersing SWNTs in water at pH < 9.7, presumably driven by hydrophobic and cation-π interactions. When the basicity was increased to pH 9.7, poly-L-lysine adopted an α-helix conformation and the ammonium groups were deprotonated into amino, and these changes led to precipitation of SWNTs.

Apart from typical protic acids, CO_2 has been utilized as a form of acid to trigger reversible release of CNTs from corresponding CNT-dispersant assemblies in solution. In 2010, Zhang and co-workers [64] reported the use of *N,N*-dimethyl-*N'*-(pyren-1-ylmethyl)acetimidamidinium ($PyAH^+$) as a CO_2-responsive dispersant to control the dispersion and aggregation of SWNTs. The $PyAH^+$ system was functionalized on the surface of SWNTs through π-π stacking and the hydrophilic amidinium group rendered the SWNT-$PyAH^+$ assemblies water soluble. Reversible interconversion between amidinium and amidine could be done by bubbling the solution with CO_2 or argon, and this chemical behavior allowed the water solubility of $PyAH^+$ functionalized SWNTs to be turned on and off by alternated bubbling of CO_2 and argon. Following this strategy, Feng and co-workers [65] in 2013 designed and synthesized a class of functionalized polystyrenes **9** (**Figure 7**) using reversible addition-fragmentation chain transfer (RAFT) polymerization and click reactions. To the backbone of polymers **9** were grafted pyrene and amidine groups. The pyrenes induced binding of the polymers with SWNTs, while the amidine units gave rise to CO_2-sensitivity. With these polymers, reversible dispersion and aggregation of SWNTs were achieved both in water and in a mixed solvent of water and methylene chloride by simply bubbling CO_2 or N_2 into the SWNT/polymer suspension.

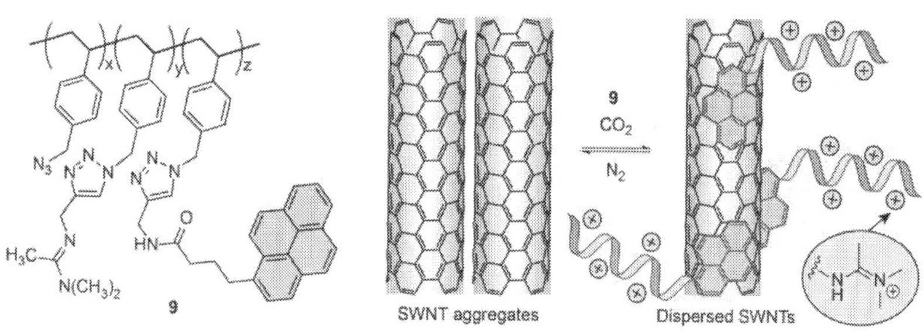

Figure 7. Reversible dispersion and aggregation of SWNTs using CO_2-responsive polymers **9** as dispersants.

3.3. By photoirradiation

Photoirradiation offers an effective way to trigger conformational and bonding changes in molecules, and therefore photo-responsive systems have found wide applications in molecular switches and photochromic devices [66]. In the field of CNT dispersion, photo-responsive dispersant systems have been explored, but not yet to a very large extent.

In 2008, Zhang and co-workers [67] developed poly(ethylene glycol) **10**, the structure of which contains a photo-reactive terminal moiety of malachite green derivative (PEG-MG). It was found that SWNTs could be dispersed in water with the aid of **10** under sonication. Upon irradiation of the SWNT-**10** suspension with UV light, the MG moiety underwent a C-C bond cleavage reaction to form PEG-MG$^+$ cation **11**. After standing in air for a few hours, the PEG-MG$^+$ cation was gradually dissociated from SWNTs resulting in the precipitation of SWNTs out of water. This work demonstrated that re-aggregation of non-covalently functionalized SWNTs can be achievable by light control; however, the involvement of an irreversible photo-induced bond breaking step would not make the dispersion/release sequence repeatable in multicycles. Indeed, the authors noted that the dispersion and release of SWNTs could *"only stand one recycle probably because of the fatigue of the photo reaction."* Kappes and Mayor in 2011 developed a fluorene-based polymer, the structure of which bears photo-cleavable *o*-nitro-benzylether moieties [68]. The presence of fluorene units made the polymer quite selective for dispersing semiconducting SWNTs in toluene. Upon photoirradiation of the resulting suspension for a short period of time, SWNT precipitation was observed as a result of photo-induced depolymerization reactions. This method presents an easy way to selectively disperse SWNTs and cleanly remove the dispersants afterwards. Nevertheless, like the previous method reported by Zhang [67], the use of photo-cleavage reactions hinders multicycle reversibility (**Figure 8**).

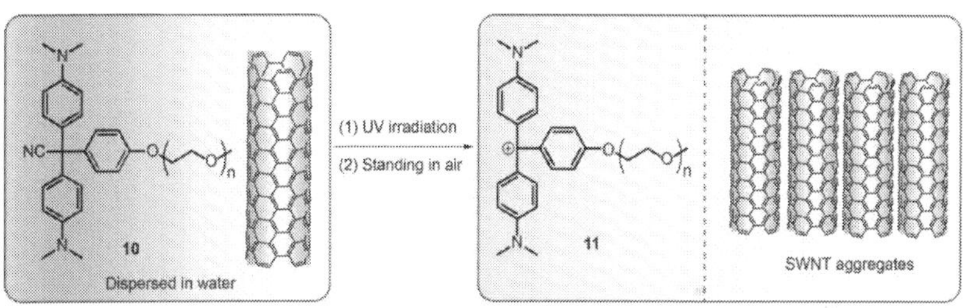

Figure 8. Dispersion and re-aggregation of SWNTs in water by a photo-reactive PEG-MG dispersant 10.

In a very recent report, Feng and co-workers [93] devised a photo-controlled method to achieve reversible dispersion and re-aggregation of SWNTs by means of photoswitchable guest-host chemistry. In their work, a pyrene-attached cyclodextrin **12** was used in combination with an azobenzene-terminated poly(ethylene glycol) **13** as the dispersant system (**Figure 9**). In the *trans* form, azobenzene-PEG **13** readily formed a supramolecular guest-host complex with cyclodextrin **12**, while the pyrene head group of **12** anchored the complex to the sidewall of SWNTs to effect good dispersion in water. Upon photoirradiation with UV light, the azobenzene unit underwent a *trans*-to-*cis* isomerization, which consequently broke the complexation of **12** and **13**. The decomplexation caused precipitation of SWNTs from the aqueous phase. The azobenzene could be further isomerized back to the *trans* form after exposure to sunlight for 24 hours. Sonication for about 10 min then led to the re-formation of SWNT dispersion in water. The authors reported that the dispersion and re-aggregation of SWNTs could be repeated in multicycles under the control of UV and sunlight, which testifies to a very good reversibility.

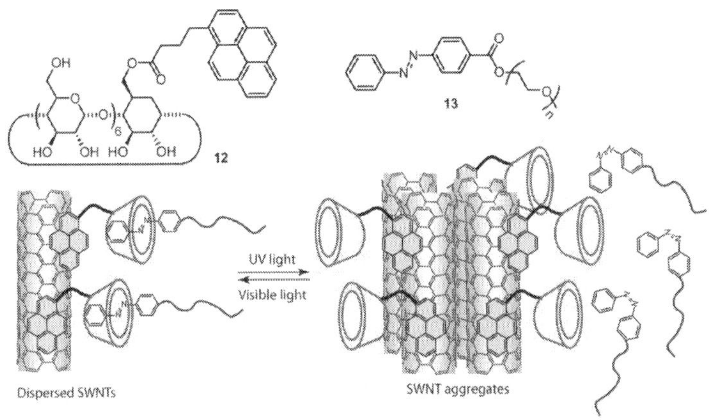

Figure 9. Reversible dispersion and release of SWNTs in water using photo-regulated supramolecular guest-host complexation.

The easy synthesis and controllable photo-switchability of azobenzene have made it a popular building block in the design of photo-responsive CNT dispersants, but one important issue related to it warrants particular attention. It has been noted in some previous studies that when the azobenzene unit is tightly bound to the surface of CNT, the *cis*-to-*trans* photoisomerization behavior may vanish due to the quenching effect of CNTs on the excited state of azobenzene [69, 70]. It was reported that the effect of photoisomerization in combination with other types of controls (*e.g.*, thermal) [71] could effectively remove the dispersants out of the surface of CNTs.

3.3. By redox chemistry

There are many redox-active systems known to undergo facile reversible electron transfers under the controls of either chemical or electrochemical means, and often a reversible redox reaction is associated with a dramatic conformational change in the substrate. Such properties can be utilized to exert control over the dispersion and release of CNTs, if the dispersants are rationally designed to carry certain redox-active units. In 2008, Ikeda and co-workers [72] developed a Cu-based redox-active complex **14** (**Figure 10**) as CNT dispersant. When the oxidation state of **14** was Cu(II), the complex was found to be able to disperse CoMoCAT SWNTs in chloroform. When ascorbic acid was added to the suspension of SWNT/**14**, Cu(II) was reduced into Cu(I) species. Accompanying this reduction, precipitation of SWNTs immediately occurred because the Cu(I) complex did not interact with SWNTs as strongly as the Cu(II) complex did. Re-dispersion of SWNTs could be achieved by bubbling air through the mixture, which converted the Cu(I) back to Cu(II). As such, SWNTs could be reversibly dispersed and released in chloroform by switching the oxidation state of the Cu center in complex **14**.

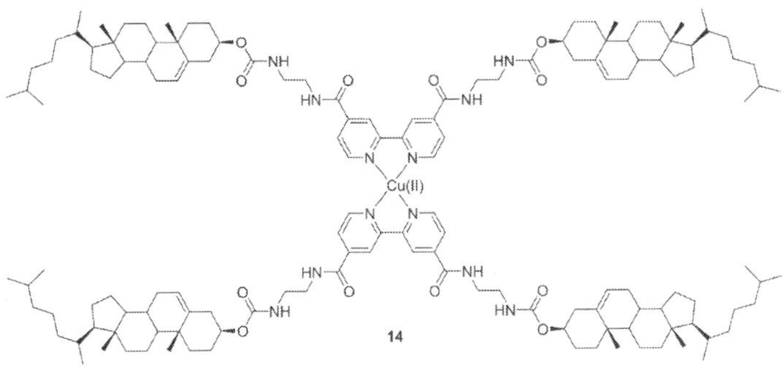

Figure 10. Cu complex **14** developed by Ikeda [72] for reversible dispersion and release SWNTs via redox control.

A class of highly electron-donating organic compounds, namely tetrathiafulvalene vinylogoues (TTFV), has been investigated by Zhao and co-workers as redox-responsive units to be integrated in redox-regulated "smart" molecular and polymer systems [73]. In general, a TTFV unit can undergo a simultaneous two-electron transfer to form a dication in the presence of a chemical oxidant (*e.g.*, iodine) or under controlled electrochemical conditions. As shown in **Figure 10**, this redox process is reversible and associated with a dramatic conformational change from *pseudo cis* (neutral) to *trans* (dication). Taking advantage of such properties, a series of TTFV-based conjugated polymers **15a-c** was synthesized by the Zhao group [74, 75]. The presence of electron-donating TTFV units here not only enabled redox-control to be readily exerted, but also endowed these polymers with high efficiency in dispersing SWNTs in various organic solvents, such as chloroform, toluene, and THF. The SWNT/**15** suspension in organic solvents was very stable. Upon addition of iodine as oxidant, an immediate color change to dark green was observed, which is indicative of the formation of TTFV dication. Accompanying this oxidation, SWNT precipitation was formed. In this method, excess iodine was needed, which complicated the separation and recovery procedure of the TTFV polymers. The addition of trifluoroacetic acid (TFA) was discovered to be not only very effective in inducing the precipitation of SWNT, but more efficient in recovering the polymers by simple neutralization with base. Herein, the exact effect of acidification on TTFV has not been completely clarified; however, it is believed that TTFV would undergo dramatic conformational changes after protonation, given that tetrathiafuvalene (TTF), the parent structure of TTFV, was known to form cation and radical cation species by protonation [76]. In 2014, Adronov and co-workers [77] prepared a TTFV-fluorene copolymer **16** (**Figure 11**). The inclusion of fluorene in this polymer brought about good selectivity for small-diameter semiconducting SWNTs to be dispersed in organic solvents. Triggered by TFA addition, the selectively dispersed SWNTs were efficiently released from the suspension and collected as dispersant-free pristine nanotubes. The polymer dispersant **16** was easily recovered after neutralization and could be reused in multicycles. This method provides an efficient way to sort nanotubes with specific chiral indices out of as-produced SWNTs.

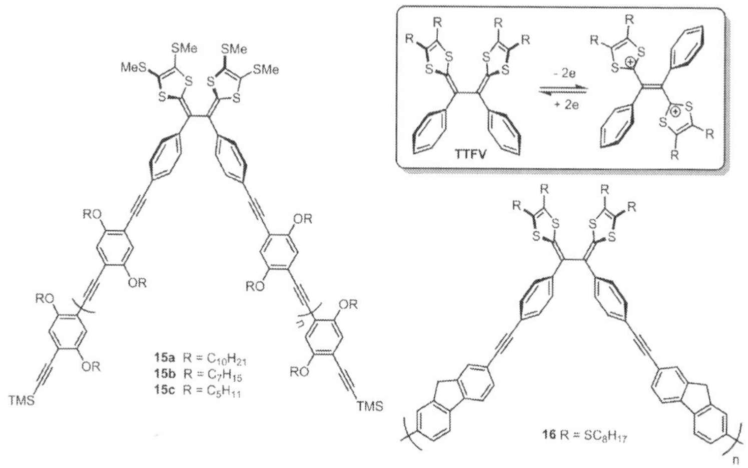

Figure 11. Conjugated polymers **15** and **16** containing redox-active TTFV units as CNT dispersants.

3.4. By temperature control

The control over dispersion and release of CNTs can also be achieved using temperature-sensitive dispersants. For example, Wang and Chen in 2007 investigated the dispersion of SWNTs with temperature-responsive poly(*N*-isopropylacrylamide) [63]. SWNTs dispersed with this polymer were observed to precipitate when heated at a temperature higher than the lower critical solution temperature (LCST), while re-dispersion of SWNTs could be done by cooling and sonication. Later, Theato and Grunlan [78] prepared a pyrene-functionalized poly(*N*-cyclopropylacrylamide) which could disperse SWNTs in water at temperature lower than LCST. When the temperature was increased above LCST, the polymers surrounding the SWNTs underwent a conformational change from coil to a globule-like shape. The authors proposed that such a transformation reduced the steric layer thickness that hinders SWNT aggregation.

Most recently, a family of pyrene-based non-ionic surfactants was synthesized and studied by Bryce and Lambert [79]. Of these compounds, two surfactants **17** and **18** (**Figure 12**) were found to exhibit very good performance in reversible dispersion and aggregation of MWNTs in 0.6 M aqueous NaCl solution. Upon heating the MWNTs dispersion with either of these surfactants at 85 °C for 15 minutes, precipitation of MWNTs was observed. The resulting precipitate was stable for several hours after cooling. With gentle shaking, dispersion of MWNTs was readily attained. The temperature-controlled dispersion and precipitation of MWNTs could be repeatedly executed, testifying to the excellent reversibility of this method. The remarkable performance in terms of reversibility was attributed to two factors: (i) The pyrene group strongly anchored the surfactant molecules to the surface of MWNTs, which prevented the re-bundling of precipitated MWNTs. (ii) The LCST transition of the surfactants switched the surface of the surfactant-functionalized MWNTs from being hydrophilic to hydrophobic.

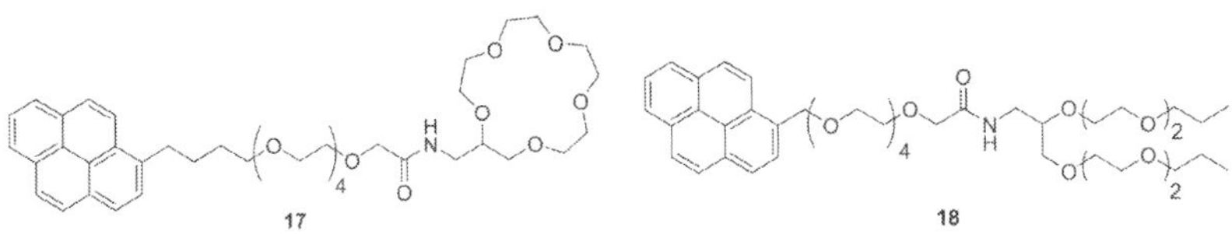

Figure 12. Temperature-sensitive surfactants **17** and **18** that allowed reversible dispersion and precipitation of MWNTs under thermal controls.

3.5. By solvent control

Compared with the aforementioned types of stimuli-responsiveness, the tuning of solvent properties is a much easier way as it neither involves multiple steps of addition and separation of chemical species, nor requires significant energy inputs (*e.g.*, light, heat). In practice, the design of systems with specific solvent-responsiveness at the molecular level is not a trivial task, since solvation is a rather complex issue to be thoroughly understood. Inspired by the type of organic oligomers (*i.e.*, foldamers) [80, 81] that change their conformations in different solvent environments, Moore and Zang in 2010 prepared a 13-mer of olig(*m*-phenylene ethynylene) **19** [82]. In a nonpolar solvent, chloroform, oligomer **19** adopted a flexible non-folded conformation, which allowed it to wrap around SWNTs through π-π stacking, resulting in a stable dispersion of nanotubes. Increasing the solvent polarity by addition of acetonitrile caused the oligomer to self-aggregate into a rigid, helical structure. As a result of this conformational change, the oligomer was unwrapped from the SWNTs, releasing the SWNTs as precipitate (**Figure 13**).

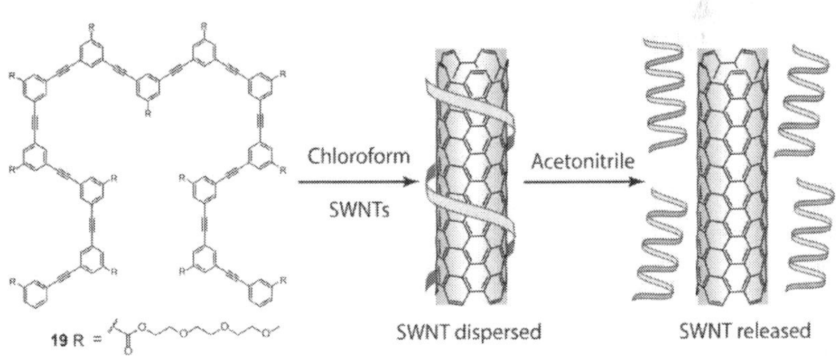

Figure 13. Solvent-controlled wrapping and unwrapping of SWNTs with oligo(*m*-phenylene ethynylene) **19**.

In 2013, Mulla and Zhao [83] reported the synthesis and properties of a series of linear and Z-shaped π-conjugated oligomers with redox-active dithiafulvenyl (DTF) groups attached to the terminal positions. Some of the oligomers showed very good performance in dispersing SWNTs in organic solvents. Of great interest was a *Z*-shaped oligomer **20** (**Figure 14**) that showed high efficiency of dispersing SWNTs in chloroform (*ca.* 0.29 mg nanotube/mL). Addition of an equal volume of hexanes to the chloroform suspension resulted in immediate precipitation of SWNTs. The good dispersity and solvent-dependent effect were attributed to the presence of DTF groups, as the parent oligomer without DTF terminal groups was found to give no dispersing effect at all. By the same token, Zhao and co-workers later designed and synthesized poly(phenylene butadiynylene) **21** (**Figure 14**), the backbone of which was click-functionalized with DTF-terminated side chains[84]. This polymer showed a high selectivity in dispersing (6,5), (7,5) and (7,6) semiconducting SWNTs in organic solvents (*e.g.*, toluene, chloroform). Addition of hexanes to the suspension of SWNT-**21** in chloroform also led to precipitation of SWNTs. Thermogravimetric analysis (TGA) showed that after addition of hexanes, most of the dispersant **21** had been stripped off the surface of SWNTs and only 7.7% wt of dispersant still remained on the released SWNTs. The solvent-controlled SWNT dispersity exhibited by the DTF-functionalized π-conjugated oligomers and polymers can be easily applied to large-scale SWNT processing and purification. In the meantime, fundamental studies on the solvent-regulated SWNT-polymer adsorption-desorption equilibrium are warranted to devise better solvent control over efficient and selective SWNT dispersion and release by this method.

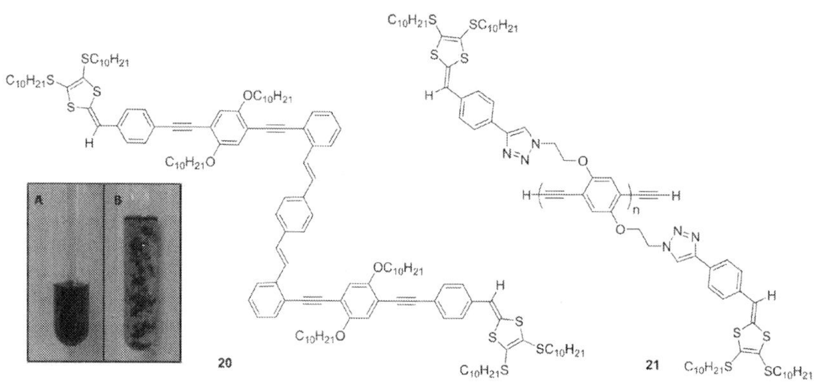

Figure 14. Dithiafulvenyl-functionalized conjugated oligomer **20** and polymer **21**. Inset photographic images: (A) SWNTs dispersed in chloroform with 20, (B) SWNTs released after addition of hexanes.

Bonifazi and co-workers [85] in 2011 reported a strategy of solvent-controlled hydrogen bonding interactions to achieve reversible dispersion and release of MWNTs. In their work, H-bonding supramolecular polymers were respectively assembled by complementary H-bonding recognition between di(acetylamino)pyridine-terminated molecules (**22** and **23**) and uracil-terminated compounds (**24–26**). The supramolecular polymers were able to strongly interact with MWNTs, yielding stable dispersion in nonpolar solvents. Addition of H-bond

http://dx.doi.org/10.5772/64078

breaking solvents (*e.g.*, MeOH, DMSO) to the MWNT/H-bonding polymer hybrids induced depolymerization and hence released MWNTs from the solution (**Figure 15**).

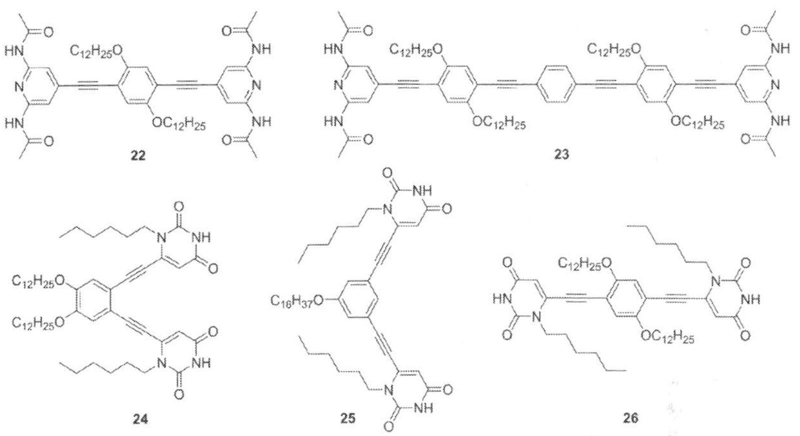

Figure 15. Molecular building blocks **22–26** that formed supramolecular polymers for solvent-controlled dispersion and release of MWNTs.

4. CNT-polymer composites responsive to single or multiple external stimuli

Stimuli-responsive polymers show the intriguing behavior that their shapes, physical, electrical, and optical properties can be significantly changed in response to small variations of environmental conditions, such as pH, temperature, electrical field, ionic strength, solvent, and so on [86, 87]. As such, stimuli-responsive polymers have been widely used as active building blocks to develop advanced nanomaterials and molecular devices. Investigations on functionalization of CNTs, either covalently or non-convalently, with stimuli-responsive polymers have been actively carried out over the past decade. In 2008, Pan and Hong [88] contributed a review article outlining the progress in designing *f*-CNTs that showed responsiveness to a range of external stimuli and potential applications in biosensing. A prominent advantage of synthetic polymers lies in the versatility of structural tuning and modifications to bring about synergistic effects and/or multiple functions in one system. Most recently, there has been a growing effort in developing "smart" polymer materials that are responsive to single or multiple stimuli so as to achieve more sophisticated applications. The following section hence highlights the most recent progress in the design and application of stimuli-responsive CNT-polymer composites.

In 2014, Luo *et al.* prepared CNT composites by mixing MWNTs with shape memory polyurethane (SMPU) through a transfer method [89]. The composites showed good electrical conductivity and improved hydrophilicity. Of significant interest, the CNT-SMPU composites were found to exhibit dual-stimuli responsive shape memory behavior upon exposure to water and electrical stimulation. The authors proposed that the presence of CNT in the composites facilitated the water diffusion to accelerate shape recovery, while the formation of CNT networks in the composites provided a conductive pathway enabling Joule heat to stimulate the shape recovery. Overall, the incorporation of CNTs in the composites led to enhanced shape memory performance, while the responses to electrical stimulation resulted in much faster shape recovery than water.

In 2012, Yuan and co-workers prepared a copolymer through free radical polymerization of *N*-isopropylacrylamide (NIPAM) and an ionic liquid monomer, 1-ethyl-3-vinylimidazolium bromide (EVImBr) [90]. The NIPAM moieties imparted the copolymer with thermo-sensitivity, while the EVImBr units assisted in the dispersion of CNTs by strong polarization interactions and responded to ionic strength in solution. **Figure 16** schematically illustrates the unique dual stimuli-responsiveness of MWNTs dispersed by this copolymer. Aqueous dispersion of MWNTs with this copolymer was found to retain stability when heated to 60 °C or mixed with excess KBr. Only the combination of heating and KBr addition led to the precipitation of MWNTs. The authors claimed that this method was advantageous in terms of allowing precise and designable control of CNT dispersion over a wide temperature range and at a desired temperature.

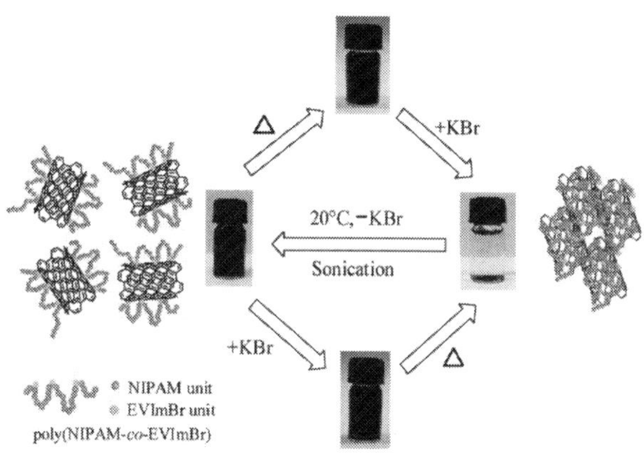

Figure 16. Dispersion and precipitation of MWNTs in aqueous solution of a dual stimuli-responsive copolymer of NIPAM and EVImBr. Adopted from reference [90] with permission. Copyright 2012 American Chemical Society.

In 2016, Mandal and co-workers used RAFT polymerization to synthesize a type of cationic poly(ionic liquid) **22**, namely poly(triphenyl-4-vinylbenzylphosphonium chloride)s (P[VBTP][Cl]s) (**Figure 17A**) [91]. These polymers showed double responsiveness towards halide ions and temperature (upper critical solution temperature-type) in aqueous solution. The authors reported that one of the polymers with $M_n = 40{,}000$ g mol^{-1} could effectively debundle MWNTs in water to form well-dispersed suspension (**Figure 17C**). In response to the addition of NaCl and temperature, P[VBTP][Cl] **22** acted as a "smart" dispersant to control the dispersion and precipitation of MWNTs in the aqueous phase (see **Figure 17B**).

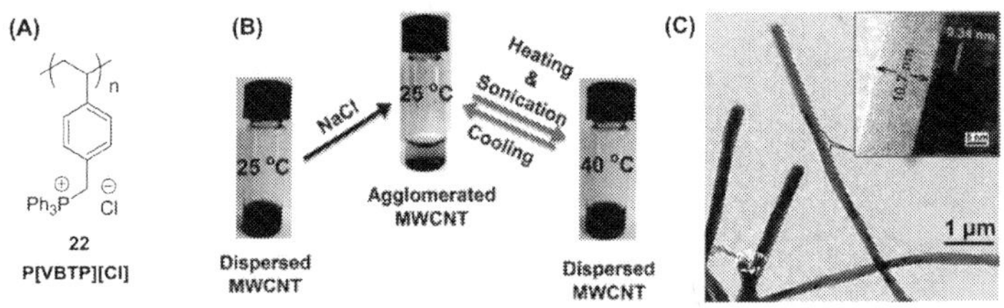

Figure 17. (A) Structure of P[VBTP][Cl] **22**. (B) Photographic images showing the dispersion and precipitation of MWNTs in an aqueous solution of **22** in response to addition of NaCl and temperature. (C) TEM image of MWNTs coated with **22**. Adopted from reference [91] with permission.

Barner-Kowollik and co-workers recently synthesized cyclopentadienyl end-capped poly(*N*-isopropylacrylamide) (PNIPAM-Cp, **23**, **Figure 18**) by RAFT polymerization and Cu-catalyzed alkyne-azide coupling [92]. This polymer was then covalently attached to the sidewall of SWNTs via the Diels-Alder reaction at different temperatures. At low temperature (cooled by an ice bath), stable dispersion of the functionalized SWNTs was attained in water by sonication. As the temperature increased, the polymer chains responded by collapsing onto the CNT surface. As a result, the CNT dispersion became destabilized, leading to the aggregation of the functionalized SWNTs. This work demonstrates that covalent functionalization can be a method of choice to effectively modify/switch the physical properties (*e.g.*, dispersity) of CNT-based nanocomposites.

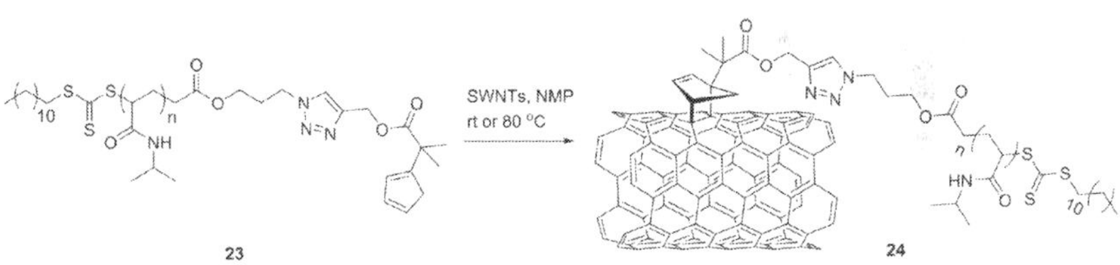

Figure 18. Synthesis of thermo-responsive functionalized SWNTs **24** via the Diels-Alder reaction.

5. Conclusions and perspectives

The literature survey discussed above has demonstrated that stimuli-responsive molecular and macromolecular systems can be successfully applied to attain reversible dispersity of CNTs in solutions as well as sensitive changes in other physical properties. Each type of the external stimuli aforementioned exhibits certain advantages and offers promising opportunities for the preparation of "intelligent" nanohybrids with improved properties and enhanced performance than conventional CNT-based materials. On the other hand, significant challenges are still present, which require continued research efforts to address both the fundamental and practical aspects. For the systems based on chemical or photochemical stimuli, controllability and tenability of the reversibility of the chemical and supramolecular reactions involved are the key issues to investigate. Besides the currently used methods (*e.g.*, acid/base, photoisomerism, redox, hydrogen bonding), new design concepts and ideas can be developed from related stimuli-responsive molecular and biological systems already established in other fields. For methods relying on relatively simple physical inputs (*e.g.*, temperature, solvent polarity), molecular modeling studies are of great value for gaining in-depth mechanistic understanding and more reliable and predictive theoretical models for the design of task-specific and better performing stimuli-responsive CNT-based systems. In this light, recent advances in molecular dynamic (MD) simulations of various CNT systems have paved a way for achieving this goal. Overall, the major drivers of technological advancements in CNTs are their wide-ranging applications, while new research thrusts are expected to emerge from synergistic efforts by the theoretical, synthetic, materials, and engineering communities.

Author details

Li Wang[1] and Yuming Zhao[2*]

*Address all correspondence to: yuming@mun.ca

1 Institute of Environmental Science, Shanxi University, Taiyuan, Shanxi, China

2 Department of Chemistry, Memorial University, St. John's, Newfoundland and Labrador, Canada

References

[1] Iijima S. Helical Microtubules of Graphitic Carbon. Nature. 1991;354(6348):56–8.

[2] Zhang Q, Huang J-Q, Qian W-Z, Zhang Y-Y, Wei F. The Road for Nanomaterials Industry: A Review of Carbon Nanotube Production, Post-Treatment, and Bulk Applications for Composites and Energy Storage. Small. 2013;9(8):1237–65.

[3] Park S, Vosguerichian M, Bao Z. A Review of Fabrication and Applications of Carbon Nanotube Film-based Flexible Electronics. Nanoscale. 2013;5(5):1727–52.

[4] Sgobba V, Guldi DM. Carbon Nanotubes-Electronic/Electrochemical Properties and Application for Nanoelectronics and Photonics. Chem Soc Rev. 2009;38(1):165–84.

[5] Avouris P, Freitag M, Perebeinos V. Carbon-Nanotube Photonics and Optoelectronics. Nat Photon. 2008;2(6):341–50.

[6] Liu X, Wang M, Zhang S, Pan B. Application Potential of Carbon Nanotubes in Water Treatment: A Review. J Environ Sci. 2013;25(7):1263–80.

[7] Heister E, Brunner EW, Dieckmann GR, Jurewicz I, Dalton AB. Are Carbon Nanotubes a Natural Solution? Applications in Biology and Medicine. ACS Appl Mater Interf. 2013;5(6):1870–91.

[8] Prato M, Kostarelos K, Bianco A. Functionalized Carbon Nanotubes in Drug Design and Discovery. Acc Chem Res. 2008;41(1):60–8.

[9] Prasek J, Drbohlavova J, Chomoucka J, Hubalek J, Jasek O, Adam V, et al. Methods for Carbon Nanotubes Synthesis-Review. J Mater Chem. 2011;21(40):15872–84.

[10] Karousis N, Tagmatarchis N, Tasis D. Current Progress on the Chemical Modification of Carbon Nanotubes. Chem Commun. 2010;110(9):5366–97.

[11] Dyke CA, Tour JM. Covalent Functionalization of Single-Walled Carbon Nanotubes for Materials Applications. J Phys Chem A. 2004;108(51):11151–9.

[12] Di Crescenzo A, Ettorre V, Fontana A. Non-covalent and Reversible Functionalization of Carbon Nanotubes. Beilstein J Nanotechnol. 2014;5:1675–90.

[13] Bilalis P, Katsigiannopoulos D, Avgeropoulos A, Sakellariou G. Non-covalent Functionalization of Carbon Nanotubes with Polymers. RSC Adv. 2014;4(6):2911–34.

[14] Zhao Y-L, Stoddart JF. Noncovalent Functionalization of Single-Walled Carbon Nanotubes. Acc Chem Res. 2009;42(8):1161–71.

[15] Samanta SK, Fritsch M, Scherf U, Gomulya W, Bisri SZ, Loi MA. Conjugated Polymer-Assisted Dispersion of Single-Wall Carbon Nanotubes: The Power of Polymer Wrapping. Acc Chem Res. 2014;47(8):2446–56.

[16] Kharissova OV, Kharisov BI, de Casas Ortiz EG. Dispersion of Carbon Nanotubes in Water and Non-aqueous Solvents. RSC Adv. 2013;3(47):24812–52.

[17] Yang H, Bezugly V, Kunstmann J, Filoramo A, Cuniberti G. Diameter-Selective Dispersion of Carbon Nanotubes via Polymers: A Competition between Adsorption and Bundling. ACS Nano. 2015;9(9):9012–9.

[18] Hersam MC. Progress Towards Monodisperse Single-walled Carbon Nanotubes. Nat Nanotechnol. 2008;3(7):387–94.

[19] Sun D-m, Timmermans MY, Tian Y, Nasibulin AG, Kauppinen EI, Kishimoto S, et al. Flexible High-Performance Carbon Nanotube Integrated Circuits. Nat Nanotechnol. 2011;6(3):156–61.

[20] Yun D-J, Jeong YJ, Ra H, Kim J-M, An TK, Seol M, et al. Fabrication of High-Performance Composite Electrodes Composed of Multiwalled Carbon Nanotubes and Glycerol-doped Poly(3,4-ethylenedioxythiophene):Polystyrene Sulfonate for Use in Organic Devices. J Mater Chem C. 2015;3(28):7325–35.

[21] Wang F, Kozawa D, Miyauchi Y, Hiraoka K, Mouri S, Ohno Y, et al. Considerably Improved Photovoltaic Performance of Carbon Nanotube-based Solar Cells Using Metal Oxide Layers. Nat Commun. 2015;6.

[22] Jung Y, Li X, Rajan NK, Taylor AD, Reed MA. Record High Efficiency Single-Walled Carbon Nanotube/Silicon p–n Junction Solar Cells. Nano Lett. 2013;13(1):95–9.

[23] Hata K, Futaba DN, Mizuno K, Namai T, Yumura M, Iijima S. Water-Assisted Highly Efficient Synthesis of Impurity-Free Single-Walled Carbon Nanotubes. Science. 2004;306(5700):1362–4.

[24] Bahr JL, Mickelson ET, Bronikowski MJ, Smalley RE, Tour JM. Dissolution of Small Diameter Single-Wall Carbon Nanotubes in Organic Solvents? Chem Commun. 2001(2):193–4.

[25] Zhou B, Lin Y, Li H, Huang W, Connell JW, Allard LF, et al. Absorptivity of Functionalized Single-Walled Carbon Nanotubes in Solution. J Phys Chem B. 2003;107(49): 13588–92.

[26] Landi BJ, Ruf HJ, Worman JJ, Raffaelle RP. Effects of Alkyl Amide Solvents on the Dispersion of Single-Wall Carbon Nanotubes. J Phys Chem B. 2004;108(44):17089–95.

[27] Krupke R, Hennrich F, Hampe O, Kappes MM. Near-Infrared Absorbance of Single-Walled Carbon Nanotubes Dispersed in Dimethylformamide. J Phys Chem B. 2003;107(24):5667–9.

[28] Kah M, Zhang X, Jonker MTO, Hofmann T. Measuring and Modeling Adsorption of PAHs to Carbon Nanotubes Over a Six Order of Magnitude Wide Concentration Range. Environ SciTechnol. 2011;45(14):6011–7.

[29] Yang K, Xing B. Adsorption of Organic Compounds by Carbon Nanomaterials in Aqueous Phase: Polanyi Theory and Its Application. Chem Rev. 2010;110(10):5989–6008.

[30] Chen RJ, Zhang Y, Wang D, Dai H. Noncovalent Sidewall Functionalization of Single-Walled Carbon Nanotubes for Protein Immobilization. J Am Chem Soc. 2001;123(16): 3838–9.

[31] Maligaspe E, Sandanayaka ASD, Hasobe T, Ito O, D'Souza F. Sensitive Efficiency of Photoinduced Electron Transfer to Band Gaps of Semiconductive Single-Walled Carbon Nanotubes with Supramolecularly Attached Zinc Porphyrin Bearing Pyrene Glues. J Am Chem Soc. 2010;132(23):8158–64.

[32] Perez EM, Illescas BM, Herranz MA, Martin N. Supramolecular Chemistry of π-Extended Analogues of TTF and Carbon Nanostructures. New J Chem. 2009;33(2):228–34.

[33] Bottari G, de la Torre G, Torres T. Phthalocyanine–Nanocarbon Ensembles: From Discrete Molecular and Supramolecular Systems to Hybrid Nanomaterials. Acc Chem Res. 2015;48(4):900–10.

[34] Bartelmess J, Frasconi M, Balakrishnan PB, Signorelli A, Echegoyen L, Pellegrino T, et al. Non-covalent Functionalization of Carbon Nano-Onions with Pyrene-BODIPY Dyads for Biological Imaging. RSC Adv. 2015;5(62):50253–8.

[35] Hong G, Diao S, Antaris AL, Dai H. Carbon Nanomaterials for Biological Imaging and Nanomedicinal Therapy. Chem Rev. 2015;115(19):10816–906.

[36] Sudibya HG, Ma J, Dong X, Ng S, Li L-J, Liu X-W, et al. Interfacing Glycosylated Carbon-Nanotube-Network Devices with Living Cells to Detect Dynamic Secretion of Biomolecules. Angew Chem Int Ed. 2009;48(15):2723–6.

[37] Li C, Mezzenga R. Functionalization of Multiwalled Carbon Nanotubes and Their pH-Responsive Hydrogels with Amyloid Fibrils. Langmuir. 2012;28(27):10142–6.

[38] Ogoshi T, Takashima Y, Yamaguchi H, Harada A. Chemically-Responsive Sol–Gel Transition of Supramolecular Single-Walled Carbon Nanotubes (SWNTs) Hydrogel Made by Hybrids of SWNTs and Cyclodextrins. J Am Chem Soc. 2007;129(16):4878–9.

[39] Zhong Q, Diev VV, Roberts ST, Antunez PD, Brutchey RL, Bradforth SE, et al. Fused Porphyrin–Single-Walled Carbon Nanotube Hybrids: Efficient Formation and Photophysical Characterization. ACS Nano. 2013;7(4):3466–75.

[40] Wang Y, Hu N, Zhou Z, Xu D, Wang Z, Yang Z, et al. Single-walled Carbon Nanotube/Cobalt Phthalocyanine Derivative Hybrid Material: Preparation, Characterization and Its Gas Sensing Properties. J Mater Chem. 2011;21(11):3779–87.

[41] Romero-Nieto C, García R, Herranz MÁ, Ehli C, Ruppert M, Hirsch A, et al. Tetrathiafulvalene-Based Nanotweezers—Noncovalent Binding of Carbon Nanotubes in Aqueous Media with Charge Transfer Implications. J Am Chem Soc. 2012;134(22):9183–92.

[42] Ehli C, Guldi DM, Angeles Herranz M, Martin N, Campidelli S, Prato M. Pyrene-Tetrathiafulvalene Supramolecular Assembly with Different Types of Carbon Nanotubes. J Mater Chem. 2008;18(13):1498–503.

[43] Didenko VV, Moore VC, Baskin DS, Smalley RE. Visualization of Individual Single-Walled Carbon Nanotubes by Fluorescent Polymer Wrapping. Nano Lett. 2005;5(8):1563–7.

[44] Von Bargen CD, MacDermaid CM, Lee O-S, Deria P, Therien MJ, Saven JG. Origins of the Helical Wrapping of Phenyleneethynylene Polymers about Single-Walled Carbon Nanotubes. J Phys Chem B. 2013;117(42):12953–65.

[45] Xiao Z, Wang X, Xu X, Zhang H, Li Y, Wang Y. Base- and Structure-Dependent DNA Dinucleotide–Carbon Nanotube Interactions: Molecular Dynamics Simulations and Thermodynamic Analysis. J Phys Chem C. 2011;115(44):21546–58.

[46] Gao H, Kong Y. Simulation of DNA-Nanotube Interactions. Annu Rev Mater Res. 2004;34(1):123–50.

[47] Kato Y, Inoue A, Niidome Y, Nakashima N. Thermodynamics on Soluble Carbon Nanotubes: How Do DNA Molecules Replace Surfactants on Carbon Nanotubes? Sci Rep. 2012;2:733.

[48] Landry MP, Vuković L, Kruss S, Bisker G, Landry AM, Islam S, et al. Comparative Dynamics and Sequence Dependence of DNA and RNA Binding to Single Walled Carbon Nanotubes. J Phys Chem C. 2015;119(18):10048–58.

[49] Manohar S, Tang T, Jagota A. Structure of Homopolymer DNA−CNT Hybrids†. J Phys Chem C. 2007;111(48):17835–45.

[50] Karachevtsev MV, Karachevtsev VA. Peculiarities of Homooligonucleotides Wrapping Around Carbon Nanotubes: Molecular Dynamics Modeling. J Phys Chem B. 2011;115(29):9271–9.

[51] Johnson RR, Johnson ATC, Klein ML. Probing the Structure of DNA−Carbon Nanotube Hybrids with Molecular Dynamics. Nano Lett. 2008;8(1):69–75.

[52] Zhang S, Wang X, Li T, Liu L, Wu H-C, Luo M, et al. Sensitive Detection of a Modified Base in Single-Stranded DNA by a Single-Walled Carbon Nanotube. Langmuir. 2015;31(36):10094–9.

[53] Sohrabi B, Poorgholami-Bejarpasi N, Nayeri N. Dispersion of Carbon Nanotubes Using Mixed Surfactants: Experimental and Molecular Dynamics Simulation Studies. J Phys Chem B. 2014;118(11):3094–103.

[54] Uddin NM, Capaldi FM, Farouk B. Molecular Dynamics Simulations of Carbon Nanotube Dispersions in Water: Effects of Nanotube Length, Diameter, Chirality and Surfactant Structures. Comput Mater Sci. 2012;53(1):133–44.

[55] Tummala NR, Morrow BH, Resasco DE, Striolo A. Stabilization of Aqueous Carbon Nanotube Dispersions Using Surfactants: Insights from Molecular Dynamics Simulations. ACS Nano. 2010;4(12):7193–204.

[56] Dutta S, Kar T, Brahmachari S, Das PK. pH-responsive Reversible Dispersion of Biocompatible SWNT/Graphene-Amphiphile Hybrids. J Mater Chem. 2012;22(14):6623–31.

[57] Ghosh M, Brahmachari S, Das PK. pH-Responsive Single Walled Carbon Nanotube Dispersion for Target Specific Release of Doxorubicin to Cancer Cells. Macromol Biosci. 2014;14(12):1795–806.

[58] Yu G, Xue M, Zhang Z, Li J, Han C, Huang F. A Water-Soluble Pillar[6]arene: Synthesis, Host–Guest Chemistry, and Its Application in Dispersion of Multiwalled Carbon Nanotubes in Water. J Am Chem Soc. 2012;134(32):13248–51.

[59] Welsh DJ, O'Driscoll LJ, Bailey SWD, Visontai D, Howes K, Frampton H, et al. Key Role of the Linker in Pyrene-Linker-Carboxylate Surfactants for the Efficient Aqueous Dispersion of Multiwalled Carbon Nanotubes. RSC Adv. 2015;5(115):95360–8.

[60] Pochorovski I, Wang H, Feldblyum JI, Zhang X, Antaris AL, Bao Z. H-Bonded Supramolecular Polymer for the Selective Dispersion and Subsequent Release of Large-Diameter Semiconducting Single-Walled Carbon Nanotubes. J Am Chem Soc. 2015;137(13):4328–31.

[61] Lu F, Wang X, Meziani MJ, Cao L, Tian L, Bloodgood MA, et al. Effective Purification of Single-Walled Carbon Nanotubes with Reversible Noncovalent Functionalization. Langmuir. 2010;26(10):7561–4.

[62] Maji B, Samanta SK, Bhattacharya S. Role of pH Controlled DNA Secondary Structures in the Reversible Dispersion/Precipitation and Separation of Metallic and Semiconducting Single-Walled Carbon Nanotubes. Nanoscale. 2014;6(7):3721–30.

[63] Wang D, Chen L. Temperature and pH-Responsive Single-Walled Carbon Nanotube Dispersions. Nano Lett. 2007;7(6):1480–4.

[64] Ding Y, Chen S, Xu H, Wang Z, Zhang X, Ngo TH, et al. Reversible Dispersion of Single-Walled Carbon Nanotubes Based on a CO_2-Responsive Dispersant. Langmuir. 2010;26(22):16667–71.

[65] Guo Z, Feng Y, He S, Qu M, Chen H, Liu H, et al. CO_2-Responsive "Smart" Single-Walled Carbon Nanotubes. Adv Mater. 2013;25(4):584–90.

[66] Natali M, Giordani S. Molecular Switches as Photocontrollable "Smart" Receptors. Chem Soc Rev. 2012;41(10):4010–29.

[67] Chen S, Jiang Y, Wang Z, Zhang X, Dai L, Smet M. Light-Controlled Single-Walled Carbon Nanotube Dispersions in Aqueous Solution. Langmuir. 2008;24(17):9233–6.

[68] Lemasson F, Tittmann J, Hennrich F, Sturzl N, Malik S, Kappes MM, et al. Debundling, Selection and Release of SWNTs Using Fluorene-based Photocleavable Polymers. Chem Commun. 2011;47(26):7428–30.

[69] Imin P, Imit M, Adronov A. Supramolecular Functionalization of Single-Walled Carbon Nanotubes (SWNTs) with a Photoisomerizable Conjugated Polymer. Macromolecules. 2012;45(12):5045–50.

[70] Maggini L, Marangoni T, Georges B, Malicka JM, Yoosaf K, Minoia A, et al. Azobenzene-based Supramolecular Polymers for Processing MWCNTs. Nanoscale. 2013;5(2):634–45.

[71] Umeyama T, Kawabata K, Tezuka N, Matano Y, Miyato Y, Matsushige K, et al. Dispersion of Carbon Nanotubes by Photo- and Thermal-responsive Polymers Containing Azobenzene Unit in the Backbone. Chem Commun. 2010;46(32):5969–71.

[72] Nobusawa K, Ikeda A, Kikuchi J-i, Kawano S-i, Fujita N, Shinkai S. Reversible Solubilization and Precipitation of Carbon Nanotubes through Oxidation–Reduction Reactions of a Solubilizing Agent. Angew Chem Int Ed. 2008;47(24):4577–80.

[73] Zhao Y, Chen G, Mulla K, Mahmud I, Liang S, Dongare P, et al. Tetrathiafulvalene Vinylogues as Versatile Building Blocks for New Organic Materials. Pure Appl Chem. 2012;84(4):1005–25.

[74] Liang S, Chen G, Zhao Y. Conformationally Switchable TTFV-Phenylacetylene Polymers: Synthesis, Properties, and Supramolecular Interactions with Single-Walled Carbon Nanotubes. J Mater Chem C. 2013;1(35):5477–90.

[75] Liang S, Chen G, Peddle J, Zhao Y. Reversible Dispersion and Releasing of Single-Walled Carbon Nanotubes by a Stimuli-responsive TTFV-Phenylacetylene Polymer. Chem Commun. 2012;48(25):3100–2.

[76] Adeel SM, Li Q, Nafady A, Zhao C, Siriwardana AI, Bond AM, et al. A Systematic Study of the Variation of Tetrathiafulvalene (TTF), TTF^{+} and TTF^{2+} Reaction Pathways with Water in the Presence and Absence of Light. RSC Adv. 2014;4(91):49789–95.

[77] Liang S, Zhao Y, Adronov A. Selective and Reversible Noncovalent Functionalization of Single-Walled Carbon Nanotubes by a pH-Responsive Vinylogous Tetrathiafulvalene–Fluorene Copolymer. J Am Chem Soc. 2014;136(3):970–7.

[78] Etika KC, Jochum FD, Theato P, Grunlan JC. Temperature Controlled Dispersion of Carbon Nanotubes in Water with Pyrene-Functionalized Poly(N-cyclopropylacrylamide). J Am Chem Soc. 2009;131(38):13598–9.

[79] O'Driscoll LJ, Welsh DJ, Bailey SWD, Visontai D, Frampton H, Bryce MR, et al. Reversible Thermal Switching of Aqueous Dispersibility of Multiwalled Carbon Nanotubes. Chem-Eur J. 2015;21(10):3891–4.

[80] Hill DJ, Mio MJ, Prince RB, Hughes TS, Moore JS. A Field Guide to Foldamers. Chem Rev. 2001;101(12):3893–4012.

[81] Guichard G, Huc I. Synthetic Foldamers. Chem Commun. 2011;47(21):5933–41.

[82] Zhang Z, Che Y, Smaldone RA, Xu M, Bunes BR, Moore JS, et al. Reversible Dispersion and Release of Carbon Nanotubes Using Foldable Oligomers. J Am Chem Soc. 2010;132(40):14113–7.

[83] Mulla K, Zhao Y. When Dithiafulvenyl Functionalized π-Conjugated Oligomers Meet Fullerenes and Single-Walled Carbon Nanotubes. J Mater Chem C. 2013;1(33):5116–27.

[84] Mulla K, Liang S, Shaik H, Younes EA, Adronov A, Zhao Y. Dithiafulvenyl-grafted Phenylene Ethynylene Polymers as Selective and Reversible Dispersants for Single-Walled Carbon Nanotubes. Chem Commun. 2015;51(1):149–52.

[85] Llanes-Pallas A, Yoosaf K, Traboulsi H, Mohanraj J, Seldrum T, Dumont J, et al. Modular Engineering of H-Bonded Supramolecular Polymers for Reversible Functionalization of Carbon Nanotubes. J Am Chem Soc. 2011;133(39):15412–24.

[86] Schattling P, Jochum FD, Theato P. Multi-Stimuli Responsive Polymers – The All-in-One Talents. Polym Chem. 2014;5(1):25–36.

[87] Stuart MAC, Huck WTS, Genzer J, Muller M, Ober C, Stamm M, et al. Emerging Applications of Stimuli-Responsive Polymer Materials. Nat Mater. 2010;9(2):101–13.

[88] Hong CY, Pang CY. Functionalized Carbon Nanotubes Responsive to Environmental Stimuli. J Mater Chem. 2008;18(16):1831–6.

[89] Luo H, Li Z, Yi G, Zu X, Wang H, Huang H, et al. Multi-Stimuli Responsive Carbon Nanotube–Shape Memory Polymeric Composites. Mater Lett. 2014;137:385–8.

[90] Soll S, Antonietti M, Yuan J. Double Stimuli-Responsive Copolymer Stabilizers for Multiwalled Carbon Nanotubes. ACS Macro Lett. 2012;1(1):84–7.

[91] Biswas Y, Maji T, Dule M, Mandal TK. Tunable Doubly Responsive UCST-Type Phosphonium Poly(ionic liquid): A Thermosensitive Dispersant for Carbon Nanotubes. Polym Chem. 2016;7(4):867–77.

[92] Zydziak N, Hubner C, Bruns M, Vogt AP, Barner-Kowollik C. Modular Ambient Temperature Functionalization of Carbon Nanotubes with Stimuli-Responsive Polymer Strands. Polym Chem. 2013;4(5):1525–37.

[93] Guo Z, Feng Y, Zhu D, He S, Liu H, Shi X, et al. Light-Switchable Single-Walled Carbon Nanotubes Based on Host–Guest Chemistry. Adv Funct Mater. 2013;23(40):5010–8.

Chapter 14

Application of Aligned Carbon Nanotube-Reinforced Polymer Composite to Electrothermal Actuator

Keiichi Shirasu, Go Yamamoto and Toshiyuki Hashida

Additional information is available at the end of the chapter

http://dx.doi.org/10.5772/62509

Abstract

Electrothermal bimorph actuators have been widely researched, comprising two layers with asymmetric expansion that generate a bending displacement. Actuation performance greatly relies upon the difference of the coefficients of thermal expansion (CTE) between the two material layers. Since traditionally used bimorph materials have positive CTE values, the generated displacements are restricted because of their relatively low CTE difference. Currently, the synthesis and characterization of carbon nanotube (CNT)/polymer composite actuators are topics of intense research activity. CNTs have been attracting much interest because of their superior electrical, thermal and mechanical properties. In addition, the negative CTE value of CNTs in the axial direction has been investigated analytically, leading one to expect that the CTE of the composites in a direction parallel to the CNT alignment will drastically decrease by containing the aligned CNTs into polymer materials. In this chapter, an experimental method for determining the CTE of a CNT in the axial direction is discussed. Based on this result, we demonstrate an electrothermal bimorph actuator having a large bending displacement and high force output using an aligned CNT-reinforced epoxy composite and thin aluminum foil. Performance characteristics including power and work output per unit volume versus frequency are also reviewed.

Keywords: Carbon nanotube, Composite, Young's modulus, Coefficient of thermal expansion, Actuator

1. Introduction

Electrothermal actuators, which have a displacement/force output at low-voltage operation conditions and a simple fabrication process, are good for applications such as precise-tracking positioning devices, artificial muscles and manipulators [1–3]. One of the electrothermal actuation schemes is the so-called bimorph effect [4], where two materials with different thermal expansions are combined in a bimorph cantilever. The thermal expansion mismatch between the two layers of the cantilever can produce a bending displacement when current is passed through the component. Traditionally used bimorph materials include metals, metal oxides and silicon [1, 5–7], most of which have a positive coefficient of thermal expansion (CTE). Therefore, the generated displacements of these bimorph materials are restricted because of the relatively low CTE difference between them.

Currently, the synthesis and characterization of carbon nanotube (CNT)/polymer composite actuators are topics of intense research activity [8–10]. CNTs have been attracting much interest because of their potential applications as a next-generation electronic material. In particular, their superior electrical, thermal and mechanical properties, including high electrical and thermal conductivity [11, 12] and extremely high mechanical strength exceeding 100 GPa [13], make them a candidate material for nano- and microscale actuators, composites and electronic devices. In addition, the negative CTE of CNTs in the axial direction has been investigated analytically [14–17], and is found to be much lower than that of polymer materials. Thus, it is expected that the CTE of the composites in the direction parallel to CNT alignment will drastically decrease with the addition of aligned CNTs into the polymer material. However, there is no experimental study on the axial CTE of CNTs. Therefore, for such applications, knowledge of the CTE of CNTs is crucial.

Recently, Zhang et al. [18]have developed continuous multiwalled CNT (MWCNT) sheets by directly drawing MWCNTs from super-aligned MWCNT arrays. The MWCNT sheet fabrication technology has allowed aligned MWCNT-reinforced polymer composites to be prepared, where the resultant composites possess a high MWCNT volume fraction and achieve a high Young's modulus [19–23]. In addition, because these kind of MWCNTs have high aspect ratios and are well-aligned in the composites, it can be expected that an evaluation of the CTE of the composites in the MWCNT alignment direction makes it possible to evaluate the axial CTE of the MWCNTs.

In this chapter, to determine the axial CTE of MWCNTs, the CTE of aligned MWCNT-reinforced epoxy composites in the MWCNT alignment direction is measured and the axial CTE of the MWCNTs is estimated using the rule of mixtures. We have found that the MWCNTs in the axial direction possess a negative CTE, and that the CTE of the composites in the MWCNT alignment direction became negative with the addition of more than 10 vol.% MWCNTs. Based on this result, we demonstrate an electrothermal bimorph actuator with a large bending displacement and a high force output by using an aligned MWCNT-reinforced epoxy composite and thin aluminum foil. Because the thermal expansion mismatch between the composite layer and the aluminum layer is enormous, the same temperature rise leads to a larger bending actuation of the structure compared with the conventional electrothermal bimorph actuators. Furthermore, because the Young's modulus of the composite is expected to be enhanced by including aligned MWCNTs in the polymer matrix as mentioned above, the force output of the actuator comprising the composite is also expected to increase.

2. Experimental procedure

2.1. Sample preparation

The MWCNTs were grown vertically on an oxidized silicon wafer substrate with chemical vapor deposition using C_2H_2 and $FeCl_2$ as the base material and the catalyst, respectively. Hereafter, the vertically aligned MWCNTs grown on a substrate are referred to as MWCNT arrays. The detailed procedure for the fabrication of MWCNT arrays has been reported elsewhere [24]. The average diameter and length of the MWCNTs were 39 nm (15–57 nm) and >600 μm, respectively. The MWCNT monolithic sheets were drawn out of the MWCNT arrays and wound onto a rotating plate (**Figure 1a**). In this study, five kinds of stacked MWCNT monolithic sheets (50–250 layers) were prepared.

Figure 1. (a) Photograph and (b) SEM image of a MWCNT array and aligned MWCNT sheet.

Aligned MWCNT/epoxy composites were prepared by a hot-melt prepreg method [21], wherein the MWCNT monolithic sheet was pre-impregnated with an epoxy matrix. A partially cured epoxy resin (B-stage epoxy) with a release paper was used as the starting materials, where the epoxy resin comprised bisphenol-A type epoxy, novolac-type epoxy and an aromatic diamine curing agent. A stacked MWCNT monolithic sheet about 20 mm wide and about 45 mm in length was placed on a polytetrafluoroethylene sheet and covered with the epoxy resin film with the release paper. The epoxy resin was then impregnated into the MWCNT monolithic sheet at 90°C for 3 min between the steel plates of the hot press (AS ONE AH-4015, Japan). After peeling off the release paper from the MWCNT sheet now impregnated with the epoxy resin (prepreg sheet), the prepreg sheet was cured at 130°C for 1.5 h at the pressure of 1 MPa using the hot press, forming a film specimen. To evaluate the actuator properties, an aligned MWCNT-reinforced epoxy composite/aluminum laminate was prepared. The prepreg sheet was prepared under the same processing condition as mentioned above. Subsequently, after peeling off the release paper from the prepreg sheet, aluminum foil was stacked on the prepreg sheet and cured at 130°C for 1.5 h at the pressure of 1 MPa using the hot press.

2.2. Evaluation of mechanical properties and CTE

Tensile tests of the aligned MWCNT/epoxy composites were performed using a tensile test apparatus (Instron Model 5965, USA) with a 50 N load cell. Strain was measured by a laser displacement meter (Keyence LS-7600, Japan) with a resolution of 0.1 μm, whereupon the Young's modulus was calculated from the slope of the stress-strain curve. The thickness and width dimensions of the tensile testing samples were measured by a scanning electron microscope (SEM). The gage length was about 18 mm and the testing speed was 0.2 mm/min, and at least three samples were tested from each batch of composites. Uniaxial tensile tests of individual MWCNT were carried out with a manipulator inside the vacuum chamber of the SEM [25, 26]. Further details of the experimental procedure are described elsewhere [26]. In this study, the diameter of the individual MWCNTs was measured from transmission electron microscope (TEM) images, wherein 23 MWCNTs were observed and measured. The in-plane MWCNT alignment distribution for the composites was observed using an SEM (JEOL JSM6510, Japan) and a TEM (JEOL JEM-2100F, Japan). A thin sample for TEM observations was prepared using a focused ion beam milling machine (Hitachi FB2200, Japan), with a machined area that was approximately 20 μm wide, 4 μm deep and 0.3 μm thick.

We measured the CTE of the aligned MWCNT/epoxy composite films using the experimental setup schematically shown in **Figure 2**. The dimensions of the composite film sample were 19 × 2 × 0.019 – 0.030 mm^3 (length × width × thickness), where the length is along the direction parallel to the MWCNT alignment. The sample was placed on a hot plate and the temperature measured by a thermocouple placed in contact with the composite sample. The length change was then measured by the laser displacement meter in the temperature range 30–60°C and with a heating rate in the range of 1.8–79.0 K/min. In this study, we evaluated the average CTE of the composites in the above-mentioned temperature range. The formula for the CTE of the composites, α_c, was given by

$$\alpha_c = \frac{\Delta L/L}{\Delta T}, \tag{1}$$

where L is the initial length, ΔL is the change in length and ΔT is the change in temperature (30 K) of the composite sample. The slope of the curve $\Delta L/L$ vs. ΔT over the temperature range was obtained using experimental data to determine α_c.

http://dx.doi.org/10.5772/62509

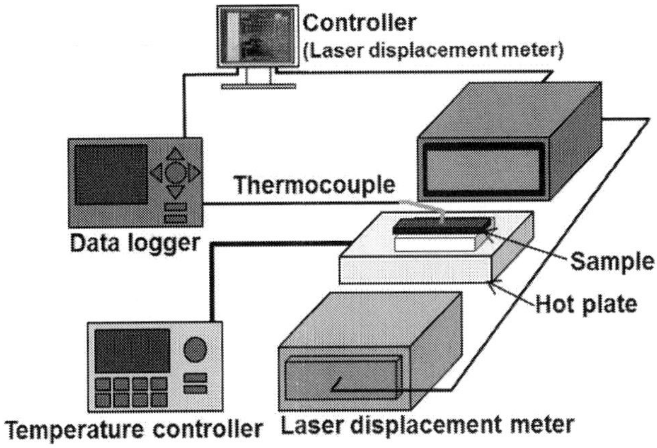

Figure 2. Schematic illustration of experimental setup for determining the CTE of MWCNT/epoxy composites.

2.3. Evaluation of actuator properties

The experimental setup used to characterize the actuator performance is shown in **Figure 3**. To evaluate the actuator property, an electrothermal actuator was prepared using the composite/aluminum laminate, where the MWCNT volume fraction in the composite layer was 27 vol.%. As shown in **Figure 3**a, a U-shaped actuator was formed by cutting out the middle part of the composite/aluminum laminate, where the CNT-aligned direction is parallel to the length direction of the U-shaped actuator. The dimension of the entire U-shaped actuator was 20 × 5 × 0.038 mm^3 (length × width × thickness), wherein the width of each beam was around 2 mm. A gold coating was sputtered onto the surface of the composite layer to decrease the contact resistance between the composite layer and the copper electrodes and to enhance the conductivity in the composite layer. The edge of the sample was masked during deposition by a polyimide tape to prevent electrical shorting on the sidewalls of the sample by the gold layer. The top end of the sample was sandwiched between glass plates and the composite layer was then attached to two copper electrodes. The free length of the actuator was 16 mm. The sample was suspended vertically in a glove box chamber and a DC voltage was applied using a power supply (KEITHLEY 2400, USA). The square waveform input voltage on the composite layer was controlled using a Labview program, and the bending of the sample was captured by the laser displacement meter. The sample surface temperature during actuation was measured using infrared thermography (Apiste FSV-1200, Japan), where the temperature data were obtained from the bare surface of the composite layer. Calibration of the thermography was performed by also heating the sample on a hot plate and using a thermocouple placed on the composite layer to measure the sample temperature. The emissivity of the composite was determined based on the thermocouple temperature measurements. The temperature, displacement and applied voltage values were automatically recorded every 200 ms to a text file using data logger (Graphtec GL220, Japan).

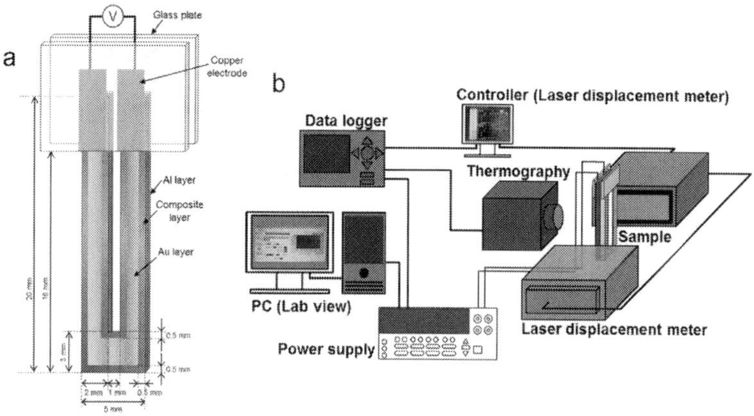

Figure 3. Schematic illustration of (a) the U-shaped actuator and (b) experimental setup for determining bending displacement of the U-shaped actuator.

3. Results and discussion

3.1. Mechanical properties

A MWCNT array prepared in this study is shown in **Figure 1**b. Because of their high areal density and the strong interaction between MWCNTs, well-aligned MWCNT monolithic sheets are produced easily from the MWCNT array by pulling [24]. To evaluate the Young's modulus of individual MWCNTs, uniaxial tensile tests of single MWCNTs were carried out using a nanomanipulator system. The Young's modulus averaged from the values obtained from 23 MWCNTs was 210 GPa.

A SEM image of the MWCNT distribution in the composite is shown in **Figure 4**a. The majority of the MWCNTs are aligned, although some MWCNTs are inclined with respect to the alignment direction. A TEM image shown in **Figure 4**b shows the morphology of the internal structure of the composites, where it is seen that the epoxy resin penetrates thoroughly between the MWCNTs. The TEM image indicates that the densely aligned MWCNT composites are successfully fabricated using the present processing method with only a limited amount of pores, which are marked by black arrows (**Figure 4**b).

The dependence of the Young's modulus of the composites upon the MWCNT volume fraction is shown in **Figure 5**, where it is seen to increase linearly with the increasing MWCNT volume fraction. The Young's modulus of the composite containing 27 vol.% MWCNTs reaches 56.8 ± 3.9 GPa, which is one order of magnitude higher than that of randomly oriented CNT/polymer composites [27]. In these MWCNT/epoxy composites, the MWCNTs possessed a high aspect ratio (length/diameter >15,000) and were aligned along the same direction. This suggests that the composites prepared in this study can be modeled as continuous fibers aligned in parallel within the epoxy matrix [28]. Therefore, the Young's modulus of the composites in the direction of the MWCNT alignment, E_c, may be expressed using the rule of mixtures [28] such that

$$E_c = (1 - V_f)E_m + V_f E_f, \tag{2}$$

where E_m and E_f are the Young's moduli of the epoxy and the MWCNTs, respectively, and V_f is the MWCNT volume fraction. Using Eq. (2), the Young's modulus of the MWCNTs is calculated to be 200 GPa, which is close to the Young's modulus of the individual MWCNTs measured by the uniaxial tensile tests (210 GPa). The solid line in **Figure 5** is a regression line provided by the least-squares regression analysis (the regression coefficient R^2 is calculated to be 0.97). These results suggest that the Young's modulus of the composites in the direction of the MWCNT alignment can be evaluated by the rule of mixtures.

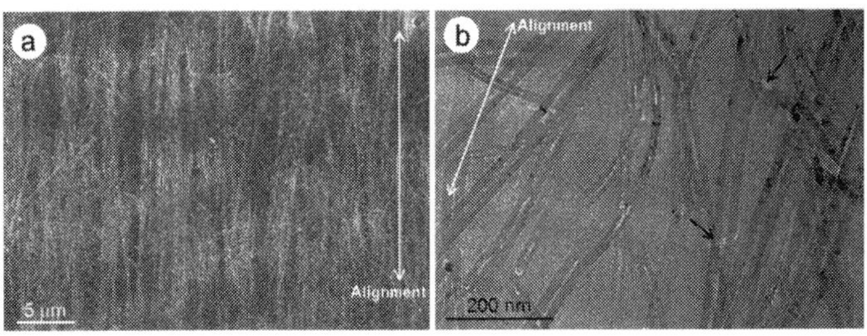

Figure 4. Aligned MWCNT/epoxy composites showing in-plane MWCNT distribution acquired by (a) SEM and (b) TEM. The MWCNT volume fraction is (a) 27 and (b) 22 vol.%.

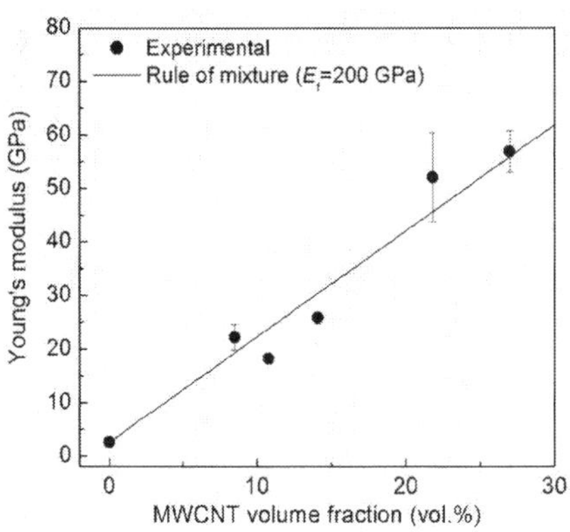

Figure 5. Young's modulus of the aligned MWCNT-reinforced epoxy composites as a function of MWCNT volume fraction.

3.2. Coefficient of thermal expansion

We evaluate the CTE of the aligned MWCNT/epoxy composites. **Figure 6**a shows the variation of the thermal strain of the composites in the axial direction of the MWCNT alignment and of pure epoxy as a function of temperature change. The thermal strain of the epoxy increases with the increasing temperature, and the CTE of the epoxy is 7.0×10^{-5} K^{-1}. The CTE dramatically decreases, however, with the addition of 9 vol.% MWCNTs, and thermal contraction can be observed in composites containing more than 11 vol.% MWCNTs. The dependence of the CTE upon the MWCNT volume fraction is shown in **Figure 6**b, where it can be seen that, though some variations are observed, the CTE tends to decrease with the increasing MWCNT volume fraction. Ultimately, the CTE of the composite containing 27 vol.% MWCNTs is -0.7×10^{-5} K^{-1}. As mentioned above, the composites prepared in this study can be modeled as continuous fibers aligned in parallel within the epoxy matrix. Thus, the CTE of the composites in the direction of the MWCNT alignment, α_c, may be expressed using the rule of mixtures [28] such that

$$\alpha_c = \frac{\alpha_m (1-V_f)E_m + \alpha_f V_f E_f}{(1-V_f)E_m + V_f E_f}, \tag{3}$$

where α_m and α_f are the CTEs of the epoxy and MWCNTs, respectively. As mentioned above, the epoxy resin penetrates thoroughly between individual MWCNTs (**Figure 4**b). Thus, it may be reasonable to substitute the Young's modulus of the individual MWCNTs for E_f in the rule of mixtures. Substituting E_m = 2.5 GPa, E_f = 210 GPa and $\alpha_m = 7.0 \times 10^{-5}$ K^{-1} into Eq. (3) and performing the least square regression analyses for the data shown in **Figure 6**b, the value of α_f is calculated to be -1.1×10^{-5} K^{-1}. The solid line in **Figure 6**b is a regression line provided by the least-squares regression analysis (the regression coefficient R^2 is calculated to be 0.98).

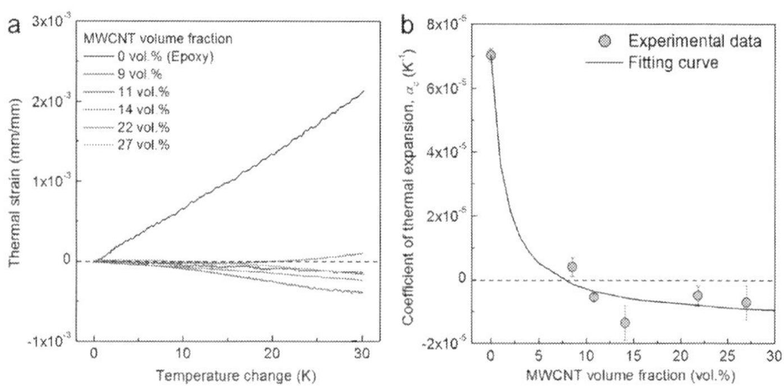

Figure 6. (a) Thermal strain of composites as a function of the temperature change. (b) Coefficient of thermal expansion of the composites as a function of the MWCNT volume fraction.

Some numerical and theoretical studies have been performed on the CTE in the axial direction of the CNTs. The axial CTE values of single-walled CNTs (SWCNTs) and MWCNTs reported in previous studies [14–17, 29, 30] are summarized in **Table 1**, where the majority of the previous studies have predicted that the CNTs contract axially below the temperature of 400 K. The thermal expansion of solids is well understood in terms of Grüneisen theory [31], and whether a solid expands or contracts upon heating depends upon the balance between phonon modes with positive and negative Grüneisen parameters. At low temperatures, transverse acoustic modes that correspond to the out-of-plane atomic vibrations of the crystal lattice may exhibit negative Grüneisen parameters. Schelling et al. [14] have reported that the CTE in the axial direction for SWCNTs is -0.9×10^{-6} K^{-1} at room temperature, and this fact is associated with the negative Grüneisen parameters. The axial CTE value of the MWCNTs measured in this study is in reasonable agreement with those of the CNTs reported previously [14–17].

Material	CTE (K^{-1})	Temperature (K)	Method	Ref.
(10,10) SWCNT	-0.9×10^{-6}	RT	Theory	14
(10,10) SWCNT	2.5×10^{-6}	High	Theory	14
(10,10) SWCNT	-1.2×10^{-5}	400	Simulation	15
(5,5) SWCNT	$-0.2 - 3.6 \times 10^{-6}$	250–1600	Theory	16
(9,0) SWCNT	$-0.4 - 3.3 \times 10^{-6}$	250–1600	Theory	16
(100,100) SWCNT	-1.6×10^{-5}	300	Simulation	17
MWCNT	$0 - -1.0 \times 10^{-6}$	300	Simulation	17
(5,5) SWCNT	3.9×10^{-6}	300–800	Simulation	29
(10,10) SWCNT	2.4×10^{-6}	300–800	Simulation	29
(10,10) SWCNT	$0–2.1 \times 10^{-6}$	0–2000	Simulation	30

Table 1. Full list of the axial CTE of CNTs. Shown are the CNT type, axial CTE of the CNTs, temperature range and method.

3.3. Actuator properties

The cross-sectional view of the laminate comprising the composite containing 27 vol.% MWCNTs and aluminum thin foil is shown in **Figure 7**. The SEM image indicates that the two layers are tightly bonded to each other using the present processing method without delamination in the laminate. The composite layer and aluminum layer have thicknesses of about 30 and 8 μm, respectively. We now evaluate the bending actuation of the U-shaped actuator. **Figure 8** shows the bending actuation performance and temperature variation of the U-shaped actuator. When a DC voltage of 3.0 V was applied, the actuator began to bend immediately and the free-end displacement of the actuator reached 3.0 mm. The free end returned to its initial position after the power source was cut off. The bending displacement was almost identical to the temperature variation during the actuation process.

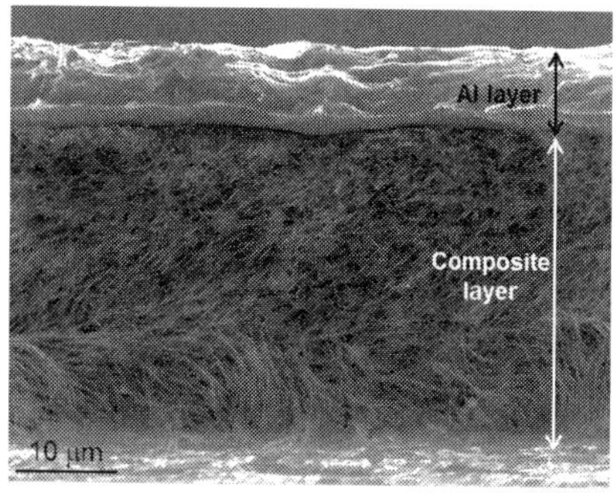

Figure 7. SEM image showing the cross-section of the composite/aluminum laminate.

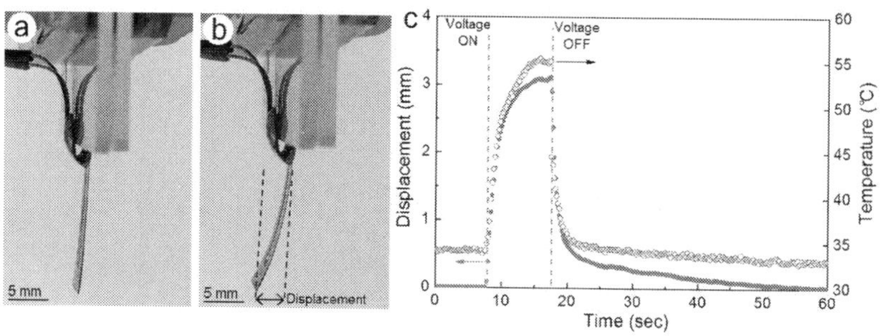

Figure 8. Photographs of an actuator (a) without and (b) with an applied DC voltage of 3.0 V. (c) Bending displacement and temperature variation of the actuator with an applied DC voltage of 3.0 V.

The dependences of the displacement upon the applied voltage and temperature are shown in **Figure 9**a and b. The glass transition temperature of the epoxy used in this study was 130°C, so we evaluated the bending displacement of the actuator at a temperature range of 30–120°C, below the glass transition temperature. The bending displacement is proportional to the temperature change and to the square of the applied voltage, and reaches 7.6 mm under a DC voltage of 5.2 V, which suggests that the actuation mechanism is owing to the Joule heating. The CTE of the composite in the MWCNT alignment direction has been measured to be -0.7×10^{-6} K^{-1} (**Figure 6**). Thus, upon applying an electric voltage on the actuator, the composite layer is directly heated and shrinks along its length. The aluminum layer, however, is heated up by the heat that diffuses from the composite layer and results in a thermal expansion. The vast thermal expansion mismatch between the composite layer and the aluminum layer is expected to cause a large bending displacement for the actuator under electric stimulation. The temperature dependence of the force output is shown in **Figure 9**c. The force output, F, is calculated by the following equation while assuming that the bending displacement is equal to that of a cantilever model subject to a tip concentrated load:

http://dx.doi.org/10.5772/62509

$$F = \frac{3\delta(E_1 I_1 + E_2 I_2)}{L^3}, \tag{4}$$

where δ is the bending displacement, E is the Young's modulus, I is the second moment of inertia and L is the length of the actuator. The subscripts 1 and 2 for E and I refer to the composite layer and aluminum layer, respectively. **Figure 9**c shows that the force output is nearly in line with the temperature change, and reaches 9.0 mN at 5.2 V. **Figure 10** shows the vibration amplitude and calculated force output of the actuator at the square wave voltage with different frequencies. With an increasing frequency higher than 0.25 Hz, both the vibration amplitude and the force output gradually decrease. According to the thermal actuation mechanism, the maximal vibration frequency of the actuator is determined by its heat generation and dissipation rate. For the electrical-induced thermal actuator, the heating and cooling rate will lag the rate of the current charge at a frequency higher than 0.25 Hz, which leads to the decrease of the vibration amplitude and force output.

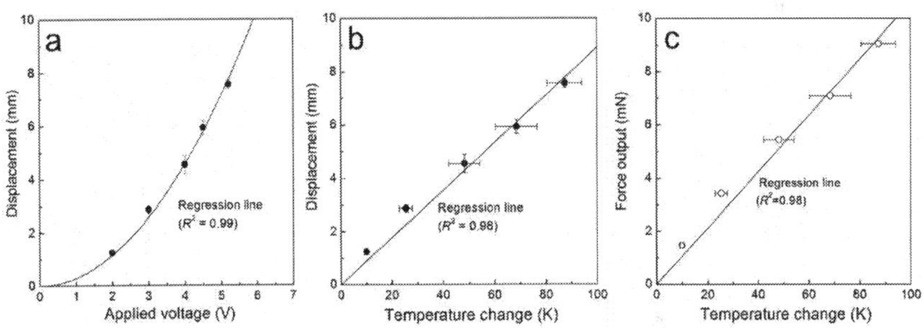

Figure 9. Bending displacement of the actuator as functions of (a) applied voltage and (b) temperature change. (c) Calculated force output as a function of temperature change.

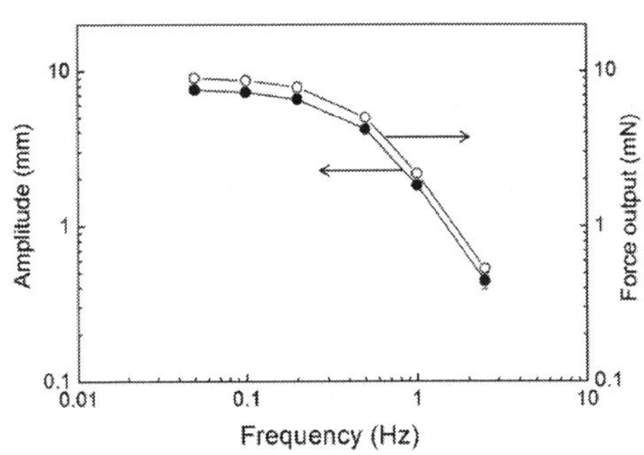

Figure 10. Vibration amplitude and force output as a function of frequency of the applied 5.2 V square voltage.

Where an actuator must operate cyclically, considerations of frequency and power become relevant. The maximum power output per unit volume, P_{max}, is defined by

$$P_{max} = W_{max} \Big/ \frac{t}{2} = 2fW_{max}, \tag{5}$$

where t is the period of time, f is the frequency and W_{max} is the maximum work output per unit volume. The W_{max} is calculated by

$$W_{max} = \frac{F_{max}\delta_{max}}{8wL(h_1 + h_2)}, \tag{6}$$

where w and h are the width and thickness of the actuator, respectively. Because the actuator prepared in this study was stimulated with a square wave voltage, the maximum work output per unit volume was divided by one-half of the period of time to calculate the maximum power output per unit volume. **Figure 11** allows for a comparison of the frequency and power output per unit volume, wherein the dashed lines link actuators that can produce equal W_{max} in each cycle. In addition to the experimental results of this study, **Figure 11** gives some literature data for the previously reported CNT composite electrothermal actuators [8, 9], conventional electrothermal microactuators consisting of metals, ceramics and silicon [1, 5] and other kinds of bending actuators [32–34]. Chen et al. [8] have fabricated a 30-mm long aligned MWCNT-based polydimethylsiloxane (PDMS) composite electrothermal bimorph actuator (the thickness of the composite layer and PDMS layer was 20 and 750 μm, respectively) and achieved a bending displacement of 9.5 mm at a DC voltage of 40 V. Seo et al. [9] have fabricated a 33-mm long electrothermal actuator based on a PDMS slab sandwiched by upper and lower active layers of SWCNT/PDMS composites (the thickness of the composite layer and PDMS layer was about 0.5 and 500 μm, respectively). They reported a large bending displacement of 3.5 mm at a DC voltage of 60 V. Wang et al. [32] have prepared and evaluated a RAINBOW (reduced and internally biased oxide wafer) actuator consisting of a reduced electromechanically passive layer and an unreduced piezoelectric lead zirconate titanate (PZT) layer. Wang et al. [33] have investigated sulfonated poly(styrene-ran-ethylene) (SPSE) as a new ion-change membrane for use in ionomeric polymer-metal composite (IPMC) actuators. Cottinet et al. [34] have studied an IPMC actuator using a Nafion membrane and MWCNT buckypapers. The Young's moduli of the silicon, silicon oxide, silicon nitride, titanium tungsten, SPSE membrane, Nafion membrane, MWCNT buckypaper and PZT used in the above previous studies were estimated based on the literature values [1, 5, 32–34]. On the other hand, no information of the Young's moduli of the CNT composites and PDMS has been given in the literature [8, 9]. Thus, we estimated the Young's modulus of the composites using the rule of mixtures. We substituted $E_m = 0.85$ MPa and $E_f = 480$ GPa into Eq. (2) to estimate the Young's modulus of the aligned MWCNT/PDMS composite prepared by Chen et al. [8]. These values have been reported in the literature [26, 35]. In case of the actuators prepared by Seo et al. [9], the SWCNTs were randomly oriented in the composite layers. Therefore, we estimated the upper and lower limits

http://dx.doi.org/10.5772/62509

of the composite's Young's modulus using Eq. (1) and following Eq. (7), respectively [28], such that

$$E_c = \left[\frac{V_f}{E_f} + \frac{(1-V_f)}{E_m}\right]^{-1}. \tag{7}$$

The Young's moduli of PDMS (0.85 MPa) and SWCNTs (1200 GPa) have been reported in the literature [35, 36]. As shown in **Figure 11**, the actuator prepared in this study has higher values of power and work output per unit volume than that of the CNT composite actuators [8, 9] and IPMC actuators [33, 34]. This is mainly owing to the high Young's modulus of the actuator's constituents and the large bending displacement of the actuator. Although the actuators prepared by Chen et al. [8] and Seo et al. [9] can produce a large bending displacement, the thick pure PDMS layer may limit its force output. On the other hand, the actuator prepared in this study exhibited its large bending displacement and high force output because of the aligned MWCNT-reinforced epoxy composite and thin aluminum foil. The RAINBOW actuator [32] (7.0 × 1.02 × 40.0 mm³ (length × width × thickness)) has a large power output per unit volume under the frequency of 250 Hz even though the bending displacement is in the range of 600 μm. This is mainly because the piezoelectric actuator provides the high force output at high operating frequency. On the other hand, the RAINBOW actuator may provide a power output per unit volume comparable to that of the IPMC actuators [33, 34] with a frequency between 0.05 and 2.5 Hz. In addition, its work output per unit volume is ≤ 10^{-4} mJ/mm³, which is one order of magnitude lower than that of the actuator prepared in this study. The dimensions of the actuators prepared by Yang et al. [1] and Boutchich et al. [5] were 500–550 × 40–400 × 3.5 μm³ (length × width × thickness). Even though the dimensions of the actuators prepared in this study are much larger than those of the actuators prepared by Yang et al. [1] and Boutchich et al. [5], it exhibits comparable or higher power and work output per unit volume under the frequency of 1.0 Hz. Furthermore, the applied voltage is 1–2 orders of magnitude smaller than that of the CNT composite electrothermal actuators [8, 9] and the RAINBOW actuator [32], which is almost identical to that of the IPMC actuators and electrothermal microactuators [1, 33, 34]. By employing microfabrication technology including photolithography, the dimensions of the electrothermal actuator can be decreased to the microscale with a length of hundreds of micrometers. Thus, the rate of heat dissipation as well as the frequency of the generated actuation can be greatly enhanced owing to the enhancement of the large surface-to-volume ratio. In addition, it is expected that the applied voltage may be potentially reduced because of the decrease in the electrical resistance of the actuator. In view of the significance of the Young's modulus and thermal expansion mismatch in these actuators, further studies should be carried out to miniaturize the CNT composite actuator. By developing CNT composite microactuators, a wide range of micro- and nanoscale applications can be envisioned where mechanical motion is needed at high displacement, high force and high speed, such as micromanipulation, optomechanical and electromechanical switches and mirrors, microfluidic valving and pumping, heat regulation and artificial muscles.

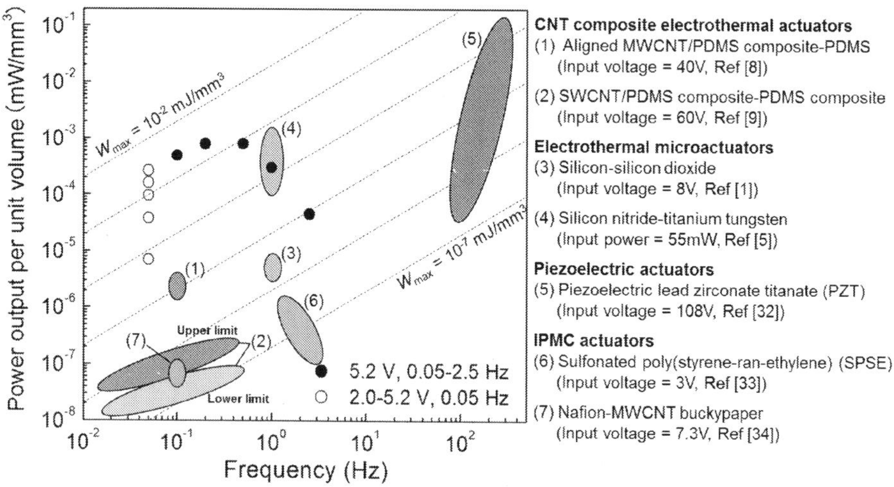

Figure 11. Actuator power output per unit volume as a function of frequency for this work (data points) and published reports (colored areas). The dashed lines link actuators that can produce equal maximum work output per unit volume in each cycle.

4. Conclusions

Combining an aligned MWCNT-reinforced polymer composite possessing a negative CTE in the MWCNT alignment direction with aluminum foil, we created a composite/aluminum bimorph that affords unique opportunities for the development of novel electrothermal actuators. The bending displacement and force output of the actuator, comprising an epoxy composite containing 27 vol.% MWCNTs and thin aluminum foil with a free length of 16 mm, reached 7.6 mm and 9.0 mN under a DC voltage of 5.2 V, respectively. Furthermore, the actuator fabricated in this study exhibited higher values of power and work output per unit volume than those of the actuators reported in previous studies with frequencies between 0.05 and 0.5 Hz. This was mainly owing to the high Young's moduli of both the composite and the aluminum layers as well as the huge mismatch of the CTEs in the composite/aluminum bimorph. The Young's modulus of the composites increased linearly with the increasing MWCNT volume fraction. The composite containing 27 vol.% MWCNTs produced the highest Young's modulus, with a value of 56.8 ± 3.9 GPa. We have also shown that the axial CTE of the MWCNTs was deduced to be -1.1×10^{-5} K^{-1} from the CTE values measured on the aligned MWCNT/epoxy composites using the rule of mixtures, and the CTE of the composite containing 27 vol.% MWCNTs was -0.7×10^{-5} K^{-1}. In this study, we indicated contributions from three sources toward the increased bending displacement and force output in the actuator: (i) preparing the MWCNT/polymer composite to have a high Young's modulus and a negative CTE with the aid of an aligned MWCNT monolithic sheet; (ii) choosing materials with a high

http://dx.doi.org/10.5772/62509

Young's modulus and a large CTE as the second layer in the laminate; and (iii) designing the dimensional parameters of the actuator.

Acknowledgements

The authors thank Prof. Y. Inoue and Prof. Y. Shimamura of Shizuoka University, Prof. T. Ogasawara of Tokyo University of Agriculture and Technology and Prof. T. Ono of Tohoku University for their useful guidance. The authors acknowledge Dr. T. Miyazaki of Technical Division, School of Engineering, Tohoku University, for technical assistance in the TEM analysis. The authors thank our colleagues, Mr. A. Nakamura and Mr. I. Tamaki of the Fracture and Reliability Research Institute (FRRI), Tohoku University, for their helpful discussions. This research was supported in part by the Japan Society for the Promotion of Science (JSPS) Core-to-Core Program. This research was partially supported by the Grant-in-Aid for JSPS 243582, the Grant-in-Aid for Young Scientists (A) 15H05502, the Grant-in-Aid for Research Activity Start-up 15H06025 and the Japan Science and Technology Agency through the Advanced Low Carbon Technology Research and Development Program (ALCA).

Author details

Keiichi Shirasu*, Go Yamamoto and Toshiyuki Hashida

*Address all correspondence to: keiichi.shirasu@rift.mech.tohoku.ac.jp

Fracture and Reliability Research Institute, Tohoku University, Sendai, Japan

References

[1] Yang JP, Deng XC, Chong TC . An electro-thermal bimorph-based microactuator for precise track-positioning of optical disk drives. Journal of Micromechanics and Microengineering. 2005;15:958–965.

[2] Zhu Y, Espinosa HD. An electromechanical material testing system for in situ electron microscopy and applications. Proceedings of the National Academy of Sciences of the United States of America. 2005;102:14503–14508.

[3] Lima MD, Li N, Jung de Andrade M, Fang S, Oh J, Spinks GM, Kozlov ME, Haines CS, Suh D, Foroughi J, Kim SJ, Chen Y, Ware T, Shin MK, Machado LD, Fonseca AF, Madden JDW, Voit WE, Galvão DS, Baughman RH. Electrically, chemically, and photonically powered torsional and tensile actuation of hybrid carbon nanotube yarn muscles. Science. 2012;338:928–932.

[4] Timoshenko S. Analysis of bi-metal thermostats. Journal of the Optical Society of America. 1925;11:233–255.

[5] Boutchich M, Mamtora TJ, McShane GJ, Haneef I, Moore DF, Williams JA. Force measurements on U-shaped electrothermal microactuators: applications to packaging. Proceedings of the Institution of Mechanical Engineers Part C-Journal of Mechanical Engineering Science. 2008;222:87–96.

[6] LeMieux MC, McConney ME, Lin YH, Singamaneni S, Jiang H, Bunning TJ, Tsukruk VV. Polymeric nanolayers as actuators for ultrasensitive thermal bimorphs. Nano Letters. 2006;6:730–734.

[7] Liu K, Cheng C, Cheng Z, Wang K, Ramesh R, Wu J. Giant-amplitude, high-work density microactuators with phase transition activated nanolayer bimorphs. Nano Letters. 2012;12:6302–6308.

[8] Chen L, Liu C, Liu K, Meng C, Hu C, Wang J, Fan S. High-performance, low-voltage, and easy-operable bending actuator based on aligned carbon nanotube/polymer composites. ACS Nano. 2011;5:1588–1593.

[9] Seo DK, Kang TJ, Kim DW, Kim YH. Twistable and bendable actuator: a CNT/polymer sandwich structure driven by thermal gradient. Nanotechnology. 2012;23:075501 (7pp).

[10] Zeng Z, Jin H, Zhang L, Zhang H, Chen Z, Gao F, Zhang Z. Low-voltage and high-performance electrothermal actuator based on multi-walled carbon nanotube/polymer composites. Carbon. 2015;84:327–334.

[11] Ruoff RS, Lorents DC. Mechanical and thermal properties of carbon nanotubes. Carbon. 1995;33:925–930.

[12] Ebbesen TW, Lezec HJ, Hiura H, Bennett JW, Ghaemi HF, Thio T. Electrical conductivity of individual carbon nanotubes. Nature. 1996;382:54–56.

[13] Peng B, Locascio M, Zapol P, Li S, Mielke SL, Schatz GC, Espinosa HD. Measurements of near-ultimate strength for multiwalled carbon nanotubes and irradiation-induced crosslinking improvements. Nature Nanotechnology. 2008;3:626–631.

[14] Schelling PK, Keblinski P. Thermal expansion of carbon structures. Physical Review B. 2003;68:035425 (7pp).

[15] Kwon YK, Berber S, Tománek D. Thermal contraction of carbon fullerenes and nanotubes. Physical Review Letters. 2004;92:015901 (4pp).

[16] Jiang H, Liu B, Huang Y, Hwang KC. Thermal expansion of single wall carbon nanotubes. Journal of Engineering Materials and Technology. 2004;126:265–270.

[17] Alamusi, Hu N, Jia B, Arai M, Yan C, Li J, Liu Y, Atobe S, Fukunaga H. Prediction of thermal expansion properties of carbon nanotubes using molecular dynamics simulations. Computational Materials Science. 2012;54:249–254.

[18] Zhang M, Fang S, Zakhidov AA, Lee SB, Aliev AE, Williams CD, Atkinson KR, Baughman RH. Strong, transparent, multifunctional, carbon nanotube sheets. Science. 2005;309:1215-1219.

[19] Cheng Q, Wang J, Jiang K, Li Q, Fan S. Fabrication and properties of aligned multi-walled carbon nanotube-reinforced epoxy composites. Journal of Materials Research. 2008;23:2975–2983.

[20] Bradford PD, Wang X, Zhao H, Maria JP, Jia Q, Zhu YT. A novel approach to fabricate high volume fraction nanocomposites with long aligned carbon nanotubes. Composites Science and Technology. 2010;70:1980–1985.

[21] Ogasawara T, Moon SY, Inoue Y, Shimamura Y. Mechanical properties of aligned multi-walled carbon nanotube/epoxy composites processed using a hot-melt prepreg method. Composites Science and Technology. 2011;71:1826–1833.

[22] Liu W, Zhang X, Xu G, Bradford PD, Wang X, Zhao H, Zhang Y, Jia Q, Yuan FG, Li Q, Qiu Y, Zhu Y. Producing superior composites by winding carbon nanotubes onto a mandrel under a poly(vinyl alcohol) spray. Carbon. 2011;49:4786–4791.

[23] Wang X, Yong ZZ, Li QW, Bradford PD, Liu W, Tucker DS, Cai W, Wang H, Yuan FG, Zhu YT. Ultrastrong, stiff and multifunctional carbon nanotube composites. Materials Research Letters. 2013;1:19–25.

[24] Inoue Y, Kakihata K, Hirono Y, Horie T, Ishida A, Mimura H. One-step grown aligned bulk carbon nanotubes by chloride mediated chemical vapor deposition. Applied Physics Letters. 2008;92:213113 (3pp).

[25] Yamamoto G, Suk JW, An J, Piner RD, Hashida T, Takagi T, Ruoff RD. The influence of nanoscale defects on the fracture of multi-walled carbon nanotubes under tensile loading. Diamond and Related Materials. 2010;19:748–751.

[26] Yamamoto G, Shirasu K, Nozaka Y, Sato Y, Takagi T, Hashida T. Structure-property relationships in thermally-annealed multi-walled carbon nanotubes. Carbon. 2014;66:219–226.

[27] Spitalsky Z, Tasis D, Papagelis K, Galiotis C. Carbon nanotube–polymer composites: Chemistry, processing, mechanical and electrical properties. Progress in Polymer Science. 2010;35:357–401.

[28] Hull D, Cryne TW. An Introduction to Composite Materials. 2nd ed. Cambridge: Cambridge University Press; 1996.

[29] Raravikar NR, Keblinski P, Rao AM, Dresselhaus MS, Schadler LS, Ajayan PM. Temperature dependence of radial breathing mode Raman frequency of single-walled carbon nanotubes. Physical Review B. 2002;66:235424 (9pp).

[30] Li C, Chou TW. Axial and radial thermal expansions of single-walled carbon nanotubes. Physical Review B. 2005;71:235414 (6pp).

[31] Ashcroft NW, Mermin ND. Solid State Physics. New York: Holt, Reinhart, and Winston; 1976.

[32] Wang QM, Zhang Q, Xu B, Liu R, Cross LE. Nonlinear piezoelectric behavior of ceramic bending mode actuators under strong electric fields. Journal of Applied Physics. 1999;86:3352–3360.

[33] Wang XL, Oh IK, Cheng TH. Electro-active polymer actuators employing sulfonated poly(styrene-ran-ethylene) as ionic membranes. Polymer International. 2010;59:305–312.

[34] Cottinet PJ, Souders C, Labrador D, Porter S, Liang Z, Wang B, Zhang C. Nonlinear strain–electric field relationship of carbon nanotube buckypaper/Nafion actuators. Sensors and Actuators A: Physical. 2011;170:164–171.

[35] Chen LZ, Liu CH, Hu CH, Fan SS. Electrothermal actuation based on carbon nanotube network in silicone elastomer. Applied Physics Letters. 2008;92:263104 (3pp).

[36] Tombler TW, Zhou C, Alexseyev L, Kong J, Dai H, Liu L, Jayanthi CS, Tang M, Wu SY. Reversible electromechanical characteristics of carbon nanotubes under local-probe manipulation. Nature. 2000;405:769–772.

Chapter 15

Carbon Nanotube-Conducting Polymer Composites as Electrode Material in Electroanalytical Applications

Şükriye Ulubay Karabiberoğlu,
Çağrı Ceylan Koçak and Zekerya Dursun

Additional information is available at the end of the chapter

http://dx.doi.org/10.5772/62882

Abstract

This chapter gives a brief overview of the preparation, characterization, and analytical applications for combinations of polymers and carbon nanotubes (CNTs) that have been prepared in different ways, which are used as an electrode material. For this purpose, multiwalled or single-walled CNTs are composed of different types of conductive polymers. The preparation of CNT-conducting polymer composite electrodes was explained by their deposition order. Chemical and morphological surface characterizations of composite electrodes were presented by scanning electron microscopy, transmission electron microscopy, atomic force microscopy, cyclic voltammetry, and electrochemical impedance spectroscopy. In addition, CNT-based polymer composite electrode usage in sensor applications for trace organic/inorganic compounds and energy applications is discussed in the last part of this chapter.

Keywords: Carbon nanotubes, conducting polymers, composite electrodes, metal nanoparticels scanning, electron microscopy

1. Introduction

Recently, considerable attention focused on nanometer-sized structures, which show great catalytic activity due to their high surface area to volume ratio and their effective properties such as magnetic, optical, electronic, and catalytic activity. Carbon nanotubes (CNTs) are one of the most attracted materials in this field. The historical roots of CNTs started in the 1960s when Roger Bacon produced graphene sheets rolled into scrolls, which were supported by microscopic and diffraction results. CNTs were actually discovered by Sumio Iijima in 1991, who published CNTs as "helical microtubules of graphitic carbon," and these soon are known as multiwalled CNTs (MWCNTs). Two years later, Iijima published another type of 1.37-nm-diameter single graphene cylinder, which today are known as single-walled CNT (SWCNTs). Whereas MWCNTs consisted of more than one rolled graphene that is telescoped, SWCNTs only formed from a single graphene sheet. The wrapping of the graphene sheet to form cylindrical nanotubes can occur with different angels. CNTs are sorted by this angle values as zigzag, chiral, and armchair, which affect their property that resulted in the metallic (armchair) or semiconducting (zigzag) CNTs [1].

CNTs have excellent thermal and electrical conductivity and large surface area compared to the other carbon-based materials. Moreover, CNTs formed with sp^2 bonded carbons that are stronger than sp and sp^3 bonds that provide the improved properties, such as flexibility and tensile strength [2, 3]. CNTs are insoluble in water due to their nonpolar nature, although they can be covalently or noncovalently functionalized, which make them preferable to incorporate with different compounds to obtain multifunctional materials as electrode modifiers [4, 5]. The special properties of CNTs, such as easy functionalization ability, high surface area, and unique thermal, mechanical, and electrical properties, make them an appealing component for composite materials, which resulted in the attraction of great interest on CNT-based composites [6, 7].

A composite material is a mixture of different components, which is homogeneous when seen in macroscopic scale and heterogeneous in microscopic scale and shows better physical and chemical properties than those of the individual components used alone. In the case of using CNTs as a conductive component, conducting polymers (CP) attracted great interest among all other secondary components due to their high conductivity to weight ratio and special optical and mechanical properties, which were first reported in 1994 by Ajayan et al. [8, 9]. CNT-polymer composites not only have taken all the advantages of the individual components, such as mechanical and optical properties, electrocatalytic activity, electrical conductivity, and charge density, but also have improved them by a synergistic effect. For this reason, these composite materials can achieve an efficient electrocatalysis [10, 11]. CNTs and polymers can be bound to each other covalently or noncovalently. Conjugated polymers or free electron pair containing heteroatoms in organic polymers can interact with CNTs via van der Waals, π-stacking, or hydrophobic forces, which correspond to noncovalent bonding. Another way is called covalent chemical bonding or grafting that puts together CNTs and polymers with strong chemical bonds by "grafting to" or "grafting from" approaches [12].

CNT-CP composites found a wide range of application field, such as biomedical applications [13], orthopedic implants [14], treatment of periodontal diseases in dentistry [15], and detector for volatile organic compounds [16]. In addition, due to their easily functionalized characteristic and improved electrical and mechanical properties, CNT-CPs have attracted great attention as electrode material for sensing and energy applications [17–20].

http://dx.doi.org/10.5772/62882

2. Preparation of CNT-CP composite electrodes

2.1. Synthesis, purification, and dispersion of CNTs

CNTs can be produced mainly by three different methods: laser ablation, electric arc discharge, and chemical vapor deposition (CVD). CVD provides the synthesis of aligned SWCNTs or MWCNTs with more controllable diameters and lengths. The purification protocol of CNT is an important issue that changes the catalytic activity of CNTs, such as heating the CNTs at temperature under decomposition, chemical treatments with concentrated HNO_3 reflux with H_2SO_4 or HCl, or hydrogen peroxide (H_2O_2) reflux. Dispersing nanotubes at the individual nanotube level is critical for the improved performance of CNTs in most applications, which mainly are sorted as water-soluble dispersions of CNTs with adsorbed surfactants, such as sodium dodecyl sulfate (SDS), dodecyl-benzene sodium sulfonate, and cetyltrimethylammonium bromide, and dispersions of CNTs in organic solvents, such as dimethyl formamide, dimethyl acetamide, tetrahydrofuren, and dimethyl pyrrolidone [21, 22].

2.2. Preparation pathways of CNT-CP composite electrodes

Different procedures for immobilizing the CNT-CP composites onto electrode (E) surface have been described as follows: (i) electropolymerization of monomers after CNT modification on bare electrode surface (CP/CNT/E), (ii) CNT modification after monomer electropolymerization on bare electrode surface (CNT/CP/E), and (iii) CP and CNT can be modified together on bare electrode surface (CNT-CP/E). These CNT-CP preparation pathways are shown with a diagram in (**Figure 1**). The activity and synergistic effect of the prepared CNT-CP composite electrodes towards organic or inorganic compounds can be changed by a preparation pathway that also affects the thickness of the polymer on bare electrode or CNT surface, polymer growth, and surface porosity of the resulting composite. In most of the studies, CNT-CP composites were prepared by following the first pathway [23].

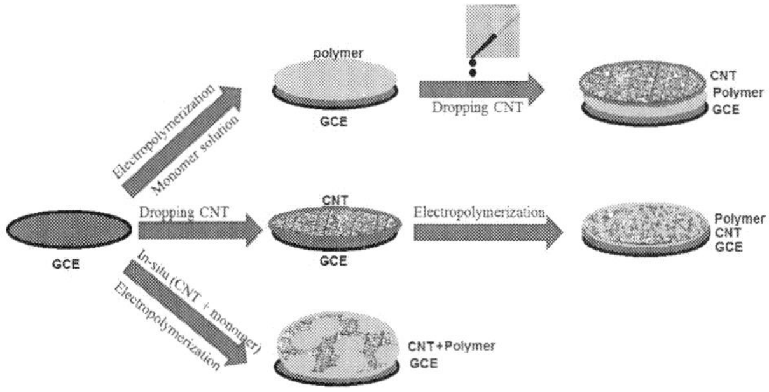

Figure 1. Schematic illustration for preparation pathways of CNT-CP composite electrodes.

2.2.1. CP/CNT/E

The first step for the preparation of CP/CNT-modified electrode is to modify CNTs on bare electrode surfaces, which can be achieved by mainly two ways: dropping the appropriate volume of CNT dispersion on the bulk electrode surface and remains until dispersant evaporation or using the CNT paste electrode that was prepared by mixing CNT and mineral oil. The second step involves the electrochemical deposition of CPs on CNT-modified electrodes, which can be achieved by following the one of the two routes: electropolymerization of CP on CNT/E by applying repetitive potential cycles with cyclic voltammetry (CV) or by applying constant potential with certain time.

Pyrrole (Py), which is one of the most commonly used CPs, was electropolymerized on CNT/E surface mostly by CV [24–26]. Polypyrrole (PPy) was formed on single-stranded DNA (ssDNA)/MWCNT paste electrode in the range 0 to +0.75 V versus saturated calomel electrode (SCE) with 50 mV s^{-1} for six cycles in 0.1 M PBS (pH 7.0) containing 0.05 M Py and $2.0{\times}10^{-5}$ M 32-mer probe oligonucleotide for the electrochemical detection of DNA hybridization [27]. In another study, electrodeposition of PPy film on CNT/glassy carbon electrode (GCE) was carried out from an aqueous solution containing 0.02 M tiron and 0.01 M Py by potential cycling between 0.0 and +0.8 V (vs. Ag/AgCl) at a scan rate of 50 mV s^{-1} for a total of five scans [28].

Poly(aniline) (PANI) and poly(flavin adenine dinucleotide) (FAD)-modified CNT composite electrodes were prepared by CV, which was performed using the mixture of 50 mM aniline and 1 mM FAD in pH 2.28 aqueous solution that was polymerized with consecutive cycles over a suitable potential region from −0.35 to 1.0 V at 50 mV s^{-1} [29]. In other study, the electropolymerization of thiophene and derivatives was carried out using different techniques, such as CV and chronoamperometry. The electropolymerization via CV was carried out with varied potential cycles between 0 and −1.7 V versus Ag/AgCl at 50 mV s^{-1} in sodium perchlorate, acetonitrile, or H_2SO_4 [30–33]. The polymerization with chronoamperometric technique was achieved with the deposition of poly(3,4-ethylenedioxythiophene) (PEDOT) at 1.1 V for 90 s by immersing the β-cyclodextrin (β-CD)-SWCNT/GCE in aqueous solution containing 0.01 M EDOT and 0.02 M $LiClO_4$ [34].

2.2.2. CNT/CP/E

The first step for the preparation of CNT/CP/E is the electropolymerization of CP on bulk electrode surface using CV [35] and chronoamperometry [36, 37]. PEDOT/GCE was obtained individually by one-step electropolymerization of 10 mM monomer on GCE surface in 50 mM $LiClO_4$ by applying a constant potential of 1.1 versus SCE with 90 s deposition time, which was followed by the modification of CNTs on PEDOT/GCE surface by the drop-dry technique [38].

2.2.3. CNT-CP/E

The codeposition of CNT and CP has several advantages, such as using fewer amounts of chemicals and obtaining more uniform surfaces. The homogeneous existence of both components in composite films provides high catalytic activity and sensitivity [39]. CNT-CP/E is generally prepared by the immersion of bulk electrode into the supporting electrolyte containing both CNT and monomer, which is attached to the electrode surface by CV [40–45] or chronoamperometry [46]. CNT-CP electrodes that contain PPy as CP were prepared by applying eight potential cycles from −0.2 to 0.8 V at 0.1 V s^{-1} scan rate in 0.1 M PPy+0.2 mg/mL SWCNTs+0.1 M SDS solution [47]. In another study, CNT-CP composite electrode was prepared by chronoamperometry, which used PEDOT as CP. PEDOT-SWCNT composite on the platinum (Pt) disk electrode surface was prepared by immersion of Pt disk electrode in 0.05 M phosphate buffer solution (pH 6.5) containing 0.02 M EDOT, 1 mg mL^{-1} SWCNT, 0.02 M sodium *N*-lauroylsarcosinate, and 0.02 M $LiClO_4$ followed by the constant potential polymerization at 1.1 V versus SCE for 60 s [48].

3. Characterization of CNT-CP composite electrodes

The developments in composite materials have increased demand for characterization techniques to improve the function and quality of structures. The characterization of CNT-CP composite properties is important in many research fields of material science, including heterogeneous catalysis, semiconductor thin-film technology, corrosion resistance, and studies of the behavior and functions of biological membranes. For this purpose, CNT-CP composite electrodes are characterized by electrochemical impedance spectroscopy (EIS), X-ray photoelectron spectroscopy (XPS), scanning electron microscopy (SEM), transmission electron microscopy (TEM), and atomic force microscopy (AFM).

EIS is generally used to improve the understanding of multistep reactions, such as diffusion impedance, charge transfer, double-layer capacitance, and solution resistance. The charge transfer resistance (Rct) value was obtained from a semicircle diameter of Nyquist plot at high frequencies that were related to dielectric and insulating characteristics of the electrode/electrolyte interface [49]. In Nyquist diagrams, a straight line with a slope of approximately 45° was related to a mass transport process via electroactive compound diffusion [50].

SWCNTs are a well-defined system in terms of electronic properties compared to the MWCNTs that Zhang et al. composed—poly(styrene sulfonic acid) sodium salt (PSS) with SWCNT on GCE surface to form the CNT-CP composite electrode. They showed the EIS results of bare GCE, SWCNT-modified GCE, and PSS film-modified GCE SWCNT/PSS-modified GCE in the presence of 5.0 mM $K_3[Fe(CN)_6]/K_4[Fe(CN)_6]$ and 0.1 M KCl to compare the activity of the surfaces. The EIS of GCE- and SWCNT-modified electrode has close Rct values of 121 and 253 Ω, respectively, which are shown as a line that corresponds to the diffusion-controlled electrode process. The negative PSS film and the probe ion $[Fe(CN)_6]^{3-/4-}$ result in a larger repulsion effect, where PSS/GCE shows a much higher interfacial Rct (7.21×10^3 Ω), whereas a lower Rct (4.13×10^3 Ω) value was obtained in the case of PSS film-modified GCE. This behavior is attributed to the higher conductivity and larger surface area of SWCNTs in PSS film that facilitates the electron transfer [51].

XPS, also known as electron spectroscopy for chemical analysis, is generally used to identify the chemistry of solid surfaces. The sample is irradiated with monoenergetic X-rays causing photoelectrons to be emitted from the sample surface. An electron energy analyzer determines the binding energy of the photoelectrons. The energy of photoelectrons is characteristic of the target material, and the measurement of the energy spectrum (number of count vs. kinetic/binding energy) provides valuable information about the top 2 to 20 atomic layers depending on the material studied. From the binding energy and intensity of a photoelectron peak, the elemental identity, chemical state, and quantity of all elements, except helium and hydrogen, in the periodic table are determined. Karim et al. synthesized and characterized the SWCNT-poly(thiophene) (PTh) composites. XPS analysis enlightened the chemical composition of SWCNT-PTh. Although there is no s2s or s2p spectrum observed in the case of SWCNTs, the s2p core level is formed in the spectrum by the addition of PTh to nanotubes. The s2p core-level spectrum of PTh and SWCNT-PTh can be divided into at least two spin-orbit-split doublet (s2p3/2 and s2p1/2) peaks at approximately 163.6 and 163.7 eV, which were related to the sulfur atoms [52].

SEM is a useful technique to obtain information about the surface morphology of electrode surface. TEM is also used for the same purpose, which has the ability to show the more detailed image about surface in atomic size with narrow focusing areas. Both techniques use electron beams for getting information about the material surface but collect different electrons for imaging.

Lin et al. studied MWCNT and PEDOT to form CNT-CP composite electrodes, which also contain FAD. They identified the surface morphologies of PEDOT, PEDOT-FAD/MWCNT, and MWCNT/PEDOT-FAD-modified indium tin oxide (ITO) electrodes with SEM, where the last two correspond to the electrode modification with MWCNT before and after the electro-codeposition of PEDOT and FAD. SEM images obviously represent the importance of modifier order, which are exactly different from each other. These modified electrodes show a globular shape, except for MWCNT/PEDOT-FAD, and this can be attributed to the coverage of all MWCNT surface by polymer film. The location of the MWCNTs after the electrocodeposition of PEDOT and FAD is more compact in the composite electrode [31].

Au nanoparticle-modified bromocresol purple (BCP) monomers were combined with CNTs on GCE surface, which was prepared by Kocak et al. [17] for the investigation of the electro-chemical behavior of hydrazine oxidation. The evaluation of surface morphology of polymer-CNT composite electrodes was carried out with the SEM technique. (**Figure 2a**) proves that the uniform surface was obtained after modified GCE surface with acid-treated CNT. The web-like appearance of poly(BCP) film that was linked on the walls of the CNTs was smooth and homogeneous **(Figure 2b)**

.

The morphological characteristics of the copolymers of PANI and PPy, which are poly(aniline-co-pyrrole) [poly(An-co-Py)], copper chloride ($CuCl_2$)-doped poly(An-co-Py) [poly(An-co-Py)Cu], and $CuCl_2$-doped poly(An-co-Py)/MWCNTs [poly(An-co-Py)Cu CNT] nanocomposites, were investigated by TEM analysis by Dhibar et al. The image of poly(An-co-Py) Cu composite represents the uniformly dispersed PANI and PPy on the electrode surface. In the case of poly(An-co-Py)Cu CNT nanocomposite, the surface of MWCNTs was uniformly coated with $CuCl_2$-doped poly(An-co-Py). This uniform coating is attributed by the

http://dx.doi.org/10.5772/62882

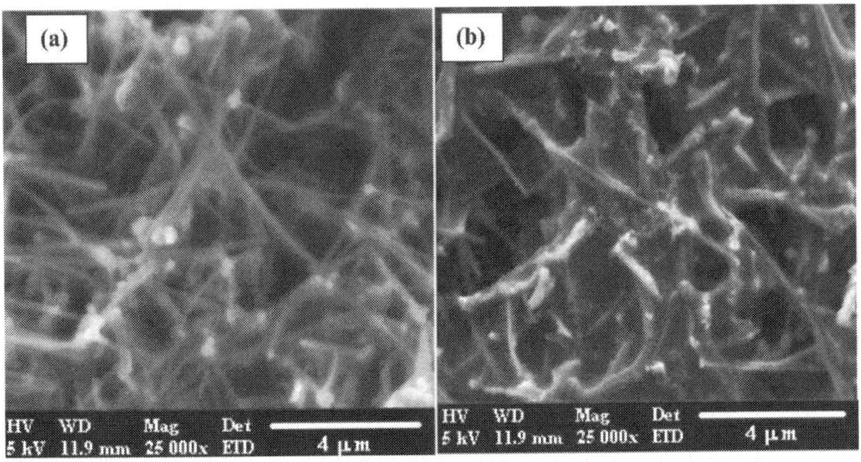

Figure 2. SEM images of (a) CNT/GCE, (b) poly(BCP)/CNT/GCE. Adapted by permission of author [17].

researchers to not only the π-π electron interaction with MWCNTs and polymers but also hydrogen bonding interaction between the amino group of aniline and carboxyl groups of the MWCNTs [53].

AFM is a probe microscopic technique that has the ability to produce atomic-scale images of the surface. Interatomic forces between the tip and the sample are measured, wherein the conductivity of the material is not important [54]. Umasankar et al. used the AFM technique to investigate the surface morphology of CNT-CP composite electrodes, which were prepared by the composition of MWCNTs, Nafion (NF), and poly(malachite green) (PMG) on GCE, gold, and ITO electrodes by the potentiodynamic method. The NF-PMG film shows smaller beads of NF and PMG deposited on the electrode surface. However, there were no bead formations if only NF was coated over the electrode; instead, a porous NF film was formed. MWCNT-NF-PMG had higher thickness than the other two films. These AFM results reveal the coexistence of MWCNT-NF and PMG in the composite film [55].

4. Electroanalytical applications of CNT-CP composite electrodes

This section gives an overview about the applications of CNT-CP composite electrodes in sensing applications. The electroanalysis of organic and inorganic molecules in real samples is an important area. Therefore, the preparation of an efficient composite electrode surface is a challenging issue for sensing applications, as a lot of studies focus on the development of new active materials. One of these enhanced materials is the CNT-CP composite, which is used for various electroanalytical applications. These analytical applications are summarized in **Table 1**.

Analyte	Electrode	Preparation pathway	Polymerization media; technique	Linear range (μM)	LOD (μM)	References
DA	DNA/PAMAM /MWCNT-Chit /Au	CP/CNT/E	–	0.2–10.0, 10.0–100.0	0.03	[46]
	PSS/SWCNT	CP/CNT/E	–	0.016–600.0	0.008	[51]
	ssDNA/SWCNT /PABA	CP/CNT/E	0.05 M 3-aminophenylboronic acid monomer, 0.04 M KF, 0.5 M H_2SO_4; CV	0.001–0.01	0.0006	[58]
	PPy/SWCNT /DM/GCE	CP/CNT/E	0.02 M tiron and 0.01 M Py; CV	0.02–100.0	0.003	[25]
	PGA-SWCNT	CP/CNT/E	PBS (pH 7.2) containing 5 mM glutamic acid; CV	3.3–26.6	0.38	[59]
	PEDOT-FAD /MWCNT	CP/CNT/E, CNT/CP/E	H_2SO_4 (pH 1.5) containing 0.01 M EDOT monomer or 0.01 M EDOT+0.1 mM FAD; CV	6.0–75.0	6.0	[31]
	NF/SWCNT/ P3MT/GCE	CNT/CP/E	0.1 M 3MT and 0.1 M $NaClO_4$ dissolved in acetonitrile; CV and the potentiostatic mode	0.02–0.10, 0.10–1.00, 1.00–6.00	0.005	[35]
	NF/SWCNT/ PMT/GCE	CNT/CP/E	0.1 M MT and 0.1 M tetrabuthylammonium perchlorate dissolved in acetonitrile; the potentiostatic mode	1.5–20.0, 20.0–120.0	–	[37]
EP	PBCB-MWCNTs-DHP	CP/CNT/E	0.1 M phosphate buffer (pH 7.4) containing 0.1 M $NaNO_3$; CV	0.05–10.0	0.01	[39]
	PMG/MWCNT /GCE	CP/CNT/E	10 mM PMG containing 0.5 M $NaNO_3$ and 0.025 M NaH_2PO_4 -Na_2HPO_4 (pH 6.0); CV	0.1–100	0.0820	[60]
	SWCNTs–PMB	CP/CNT/E	1.0 mM MB in PBS;	150.0–3750.0	96.0	[61]

Analyte	Electrode	Preparation pathway	Polymerization media; technique	Linear range (μM)	LOD (μM)	References
			CV			
5-HT	P(EDOP-SWCNTs)/GCE	CP/CNT/E	1.0 mM EDOP, 1.0 mg SWCNTs, and 0.05 M TBAP/MeCN	0.1–10.0	0.005	[41]
L-Trp	PGA/CNTPE	CP/CNT/E	0.1 M PBS (pH 7.0) containing 0.05 M glutamic acid; CV	0.05–100.0	0.01	[68]
CAT	f-SWCNT/PEDOTM/GCE	CNT/CP/E	10 mM monomers on GCE surface in 50 mM $LiClO_4$ with a constant potential	0.039–40.84	0.013	[38]
Catechol	MWCNT–NF-PMG	CP/CNT/E	5 mM PMG present in pH 1.5 M H_2SO_4 aqueous solution; CV	360.0–4350.0	29.3	[55]
L-dopa	PPy-CNT-GCE	CP/CNT/E	Aqueous solution containing 0.02 M tiron and 0.01 M Py; CV	1.0–100.0	0.10	[28]
NADH	PMG-CNT	CP/CNT/E	0.4 mM MG in 0.01 M borate buffer (pH 9.1) with 0.1 M $NaNO_3$; CV	NR	NR	[66]
Lysine	LOx/AuNP/c-MWCNT/PANI/Au, LOx/AuNP/c-MWCNT/DAB/Au	CNT/CP/E	Aniline in 1.0 M HCl and DAB (0.05 M) in PBS (pH 7.4); CV	5.0–600.0, 20.0–600.0	5.0, 600.0	[67]
Glucose	PNb-SWCNT	CP/CNT/E	PBS (pH 8.5) containing Nb; CV	100.0–8.5×10^3	0.05	[63]
	GOx/poly(2,6-DAP)/MWCNT/GC	CP/CNT/E	0.1 M HCl and 4 mM 2,6-DAP; CV	0.42–8.0×10^3	0.13	[62]
	GOx-PoPD/AuNP-ATP-	CP/CNT/E	5 mM oPD and 2 mg mL^{-1} GOx in acetate	50.0–8.85×10^3	15.0	[64]

Analyte	Electrode	Preparation pathway	Polymerization media; technique	Linear range (μM)	LOD (μM)	References
	cMWCNT		buffer (pH 5.4); CV			
	PPy/CNT /GOD	CP/CNT/E	0.5 M Py in PBS (pH 7); CV	500.0–5.0×10^4	500.0	[43]
	Au/POAP /CNT/GOD	CP/CNT/E	5.0 mM o-AP+13 mg/mL CNT+50 U/mL GOD in acetate buffer (pH 5.0)	10.0–10.0×10^3	10.0	[65]

ATP, 4-aminothiophenol; Chit, chitosan; DM, dodecylamine monolayer; LOx, lysine oxidase; NR, not reported; PABA, poly(aniline boronic acid); PAMAM, poly(amidoamine).

Table 1. Applications of CNT-CP composite electrodes

Dopamine (DA) plays an important role as a neurotransmitter in the renal, hormonal, cardiovascular, and central nervous systems, as determined in different samples by voltammetry with different kinds of CP consisting of CNT composite electrodes. MWCNT-chitosan/poly(amidoamine)/DNA nanocomposite-modified gold electrode, which was fabricated by layer-by-layer modification, was used for DA and uric acid (UA) under the coexistence of ascorbic acid (AA) [46]. The biosensor-based polymer-CNT showed strong catalytic activity toward the oxidation of DA and UA. The originally overlapped signals of UA, DA, and AA oxidation at the bare electrode was also separated into three well-defined peaks with this biosensor. The detection limits of DA and UA were found to be 0.03 and 0.07 mM, respectively. Zhang et al. [51] reported the preparation of PSS-SWCNT-modified glassy electrode composite electrode and use for detecting DA in the presence of AA. In this study, PSS can provide a good microenvironment for DA due to the negative charges of polymeric surface. The results show that the oxidation peak current is proportional to the concentration of DA in the range of 1.6×10^{-8} to 6.0×10^{-4} M. The detection limit (S/N=3) was calculated as 8.0×10^{-9} M. These values indicated that the PSS/SWCNT film-modified GCE exhibits remarkable selectivity to determine DA in the presence of AA. Another CNT-CP composite electrode, which was used for DA and UA determination, is poly(acrylic acid)-(PAA) MWCNT/GCE [56]. PAA-MWCNT composite was prepared by mixing MWCNT powder with PAA aqueous solution under sonication. There is no interference of excess AA due to AA that demonstrates no voltammetric peak at PAA-MWCNT/GCE. PAA-MWCNT/GCE can catalyze the electro-oxidation of UA, which makes it possible to detect the lower level of UA. The lowest detection limits (S/N=3) were 20 nM DA and 110 nM UA. The higher sensitivity and selectivity should be originating from the combination of MWCNT and polymer film. The DA determination in the presence of excess AA was also determined by another CNT-CP electrode, which is Prussian blue nanoparticles supported on poly(*m*-aminobenzene sulfonic acid)-SWCNT-modified edged plane pyrolytic graphite electrode (EPPGE) [57]. The detection limit was calculated as 2.8 nM for DA. The modified electrode was found to be electrochemically stable and reusable and can be used for drug analysis.

DA can be also electrochemically detected with a thin layer of *in situ* polymerized poly(aniline boronic acid) (PABA)/CNT composite and a thin layer of the highly permselective NF film [58]. The NF film was used for the elimination of the AA interference by a charge-discriminating membrane to preferentially accumulate the positively charged DA and reject the negatively charged ascorbate at the electrode surfaces in physiological pH. The detection limit (1.5 nM) is slightly higher than that without NF (0.6 nM). The CNT-CP composite increased the effective electrode surface area; therefore, the density of boronic acid groups in polymeric backbone is available for DA binding. Another CNT-CP composite electrode, which was used for DA determination, was prepared by layer-by-layer modification with dodecylamine monolayer, CNT, and tiron-doped PPy [25]. The modified electrode was sensitive to DA in the presence of excess AA. A linear response was obtained between the oxidation peak current and DA concentration in the range of 2.0×10^{-8} to 1.0×10^{-4} M with a detection limit of 3.0 nM (S/N=3). The useful lifetime of composite electrode can be estimated as 40 days. All of these values indicated that the new electrode was used for the determination of DA in real samples. The analysis of DA in the presence of AA and UA was also carried out with alternative CNT-CP composite electrode, which is polyglutamic acid (PGA) patterned CNT electrodes [59]. PGA selectively detected DA in the presence of AA and UA due to the negative charges on both AA (pKa=4.1) and UA (pKa=5.75) and positive charges on DA (pKa=8.89) at physiological pH (7.2). Hence, the negative charges on the carboxylate groups on the PGA-SWCNT film surface interact with positive charge DA. DA detection was linear in the range of 3.3 to 26.6 μM (R^2=0.994) and the limit of detection (LOD) was found as 0.38 μM. PGA-SWCNT composite electrodes are used for detecting DA in real sample. Lin et al. also studied AA, DA, and UA determination with MWCNT/PEDOT-FAD hybrid composite [31]. The electrocatalytic properties of AA, DA, and UA were improved with MWCNT/PEDOT-FAD hybrid composite compared to PEDOT, PEDOT-FAD, MWCNT/PEDOT, and PEDOT-FAD/MWCNT due to the good recognition of AA, DA, and UA based on the obvious redox peaks and current responses. With this composite electrode, the detection limit of AA, DA, and UA could be estimated as 4.0×10^{-4}, 6.0×10^{-6}, and 2.0×10^{-6} M, respectively. The proposed hybrid composite was also used for H_2O_2 determination. It will be useful in the development of multifunctional biosensors.

In addition, poly(3-methylthiophene) (P3MT)-modified GCE with NF/SWCNT film was also developed for the determination of DA in the presence of excess AA and UA [35]. The experimental results implied that the enhancement of the DA current response with composite electrode was caused by combining the advantages of P3MT, CNTs with NF. The detection limit was estimated 5.0 nM (S/N=3). Furthermore, the interferences of AA and UA were effectively reduced. This CNT-CP composite electrode could be applied to the determination of DA contents in DA hydrochloride injection and human serum. Another combination of CNT and CP for the sensitive and selective determination of DA is NF, SWCNTs, and P3MT film-modified GCE [37]. The modified electrode displays higher electrocatalytic activity towards AA, DA, and UA because of the combining unique properties of SWCNT (high specific surface area and electrocatalytic and adsorptive properties) with the cation selectivity of NF. The combination of CP and CNTs improves the sensitivity and selectivity of bare electrode towards DA determination. Among the studies given in Table 1, the lowest detection limit was found as 0.6 nM with 3-aminophenylboronic acid, and SWCNT includes ssDNA composite electrode. The cause of the high sensitivity is explained that boronic acid groups are accessible for DA binding and increasing of the effective electrode surface area due to DNA-wrapped SWCNTs.

Another biologically important compound, epinephrine (EP), which exists in the tissues and body fluids in the form of cations that control the function of the nervous system, was also determined with CNT-CP composite electrodes. One of them is poly(brilliant cresyl blue) and dihexadecyl phosphate dispersed MWCNT GCE (PBCB-MWCNTs-DHP/GCE), which was prepared by an *in situ* electropolymerization method [39]. The authors emphasized that the improved response of EP with PBCB-MWCNTs-DHP/GCE could be caused from the hydrogen bonding and p-p interactions between EP and PBCB. The experimental results indicate that the current response displays a good linear relationship with the concentration of EP in the range of 5.0×10^{-8} to 1.0×10^{-5} M with a detection limit of 1.0×10^{-8} M. Another study described the preparation of GCE modified with MWCNTs and PMG film (PMG/MWCNT/GCE) to develop a sensor for simultaneous determination of AA, EP, and UA [60]. The obtained catalytic peak current was linearly dependent on the AA, EP, and UA concentrations in the ranges of 0.4 to 100.0, 0.1 to 100.0, and 0.3 to 90.0 mM, respectively. The detection limits for AA, EP, and UA were 0.23, 0.0820, and 0.12 mM, respectively. The sensor was applied to the determination of AA, EP, and UA in real samples. In addition, conductive CNT-CP composite containing MWCNT and poly(methylene blue) (PMB) was synthesized on GCE, gold, and ITO electrodes by potentiostatic methods for EP, AA, and DA [61]. The detection limits using semiderivative square wave voltammetry in the presence of 10 mM AA are 96 mM EP and 8.53 mM DA (S/N=3). Although the sensitive determination of EP was carried out with all CP-CNT composite electrode, the best sensitivity (0.01 μM) and linearity (0.05–10.0 μM) were obtained at PBCB-modified MWCNT composite electrode.

The voltammetric determination of catechol and catechin (CAT), which are known as antioxidants, was also determined using CNT-CP composite electrodes. PMG-coated MWCNT composite electrodes [55] were used for biochemical compounds catechol and quinol. The electrocatalysis studies indicated that well-separated voltammetric peaks of catechol and quinol were obtained with a peak separation of 147 mV. The sensitivity values for catechol and quinol were calculated as 0.4 and 3.2 mA mM^{-1} cm^{-2}, respectively. Another composite electrode used for CAT determination was proposed as SWCNT/poly(hydroxymethylated-3,4-ethylenedioxythiophene) (PEDOTM) composite electrodes [38]. The CNT-CP composite electrode has a wide linear response for CAT in the concentration range between 0.039 and 40.84 μM, with a low detection limit of 0.013 μM. Also, the modified electrode showed a good reproducibility and long-term stability as well as high selectivity. The sensor was applied analysis of CAT in commercial green tea.

Glucose determination is important for clinical diagnostics, bioscience research, and food chemistry [62]. For this reason, the CNT combined with CP modified enzymatic or nonenzymatic electrodes are used for the detection of glucose. One of these electrodes that have been used for glucose determination is SWCNTs functionalized with poly(Nile blue A) composite electrode (PNb-SWCNT) fabricated by electrochemically cycling the potential of the SWCNT/GCE in Nb solution [63]. The response of the composite electrode showed a linear dependence on the concentration of glucose in the range of 0.01 to 8.5 mM with correlation coefficient of 0.997 and the detection could be estimated as 5.0 mM (S/N=3). Kamyabi et al. described another CNT-CP composite electrode, which is covalently immobilized glucose oxidase on a poly(2,6-diaminopyridine)/CNT electrode [GOx/poly(2,6-DAP)/MWCNT/glassy electrode] [62]. The reason of the markable improvement of glucose current response was explained as the synergistic effect of the high active surface area of both CP and CNT. The sensor showed two linear responses to glucose in the concentration ranging from 0.42 μM to 8.0 mM with a correlation coefficient. The composite electrode was applied to glucose determination in the biological samples. Gold nanoparticle (AuNP) functionalized MWCNT-poly(*o*-phenylenediamine) (PoPD) composite film electrode [64] was also synthesized for glucose determination. The biosensor presented a linear range from 0.05 to 8.85 mM with a sensitivity of 27.93 mA mM^{-1} cm^{-2} and a detection limit of 0.015 mM for the detection of glucose. The modified electrode was applied to the determination of glucose in human blood serum samples. Another study for glucose determination is the use of CNT-PPy composite film electrode [43]. The biosensor exhibits excellent response performance to glucose with a linear range from 0.5 to 50 mM. Additionally, the biosensor displays rapid response, high sensitivity, and long-term stability. One of the CNT-CP composite electrodes for glucose determination is glucose oxidase immobilized poly(*o*-aminophenol)/CNT/Au electrode (Au/POAP/CNT/GOD), which is electrochemically polymerized at a gold electrode surface [65]. The biosensor has a lower detection limit (0.01 mM) than Au/POAP/GOD due to the high surface area and excellent electrical conductivity.

A comparison of detection limits showing the best result for sensitive determination glucose was obtained by the use of PNb-SWCNT composite electrode (LOD=0.05 μM) and the best linearity between the analytical signal and glucose concentration found by the use of GOx/poly(2,6-DAP)/MWCNTs composite electrode (0.42–8.0×10^3 μM).

p-Acetamidophenol is an antipyretic and analgesic drug compound that has extensive applications in pharmaceutical industries. It is important to keep the drug concentration in therapeutic range in that an accurate and sensitive determination for this compound is necessary. Potentiostatic methods have been used for the synthesis of highly active, conductive biocomposite film (MWCNT-PANIFAD) on gold and screen-printed carbon electrodes. The composite electrode contains MWCNTs and PANI-FAD copolymer (PANIFAD) that are used for the *p*-acetamidophenol determination successfully [29]. Levodopa (L-dopa) is a biologically important compound that increases the DA level of the brain, which helps the treatment of Parkinson's disease. An excessive level of L-dopa in plasma causes nausea and vomiting in that the detection and its concentration determination is a necessary feature in pharmaceutical and clinical procedures, which are achieved successfully by the CNT-CP electrode prepared by the modification of GCE with MWCNT and PPy doped with tiron [28]. Nicotinamide adenine dinucleotide (NADH) is a coenzyme found in all living organisms and is involved in many important enzymatic reactions. SWCNT/PNb nanocomposite electrode and poly(methylene green)-CNT (PMG-CNT)-modified electrode were used to determine the NADH in a fast, easy, and accurate way [63, 66]. Another application field of CNT-CP are vitamins that they applied successfully to find the vitamin content of commercial juices. PEDOT methanol/ ascorbate oxidase/NF-SWCNTs (PEDOTM/AO/NF-SWCNT) were used for this purpose [36].

Amino acids are also biologically essential compounds, and their level in the human system is associated with some diseases. For instance, lysine, an essential amino acid, causes intellectual disability or behavioral problems when its concentration exceeds the therapeutic range. The most straightforward, rapid, and cost-effective way is the use of electrochemical sensing techniques as alternative among all other techniques. Many composite electrodes were prepared using AuNPs and carboxylated MWCNTs (c-MWCNT) decorated on (i) PANI and (ii) poly(1,2-diaminobenzene) (DAB) electrodes for the electrochemical detection of lysine [67]. Tryptophan (L-Trp) is another essential amino acid that is also added to food products and pharmaceutical formulations. The detection of L-Trp quantity in the brain is too important due to its toxic effect. Liu et al. suggested the electrode surface that is prepared with conductive PGA film and highly active CNTs, which determined L-Trp with high selectivity, good stability, and reproducibility [68].

Paracetamol (acetaminophen; *N*-acetyl-*p*-aminophenol) is a long established and one of the most popular and commonly used analgesic and antipyretic drugs that reduce fever and relieve cough, cold, and muscular aches, migraine headache, backache, and toothache pains [69]. Therefore, the selective and sensitive determination of paracetamol is an important research field in pharmaceutical chemistry. Barsan et al. have studied paracetamol determination with redox polymer PMG and functionalized CNTs (fCNTs). They used two configurations (both CNT/CP/E and CP/CNT/E) for electrode preparation. In the first type of electrode configuration, MG was electropolymerized on a graphite composite electrode and then combined with fCNT (fCNT/PMG/CE). The second type of composite electrode configuration was prepared by electropolymerization of MG on the fCNT covered graphite composite surface (PMG/fCNT/CE) to evaluate the performance of both surfaces on the voltammetric behavior of paracetamol. The best analytical performance for paracetamol determination was obtained at PMG/fCNT/CE using amperometric technique. To identify the electrical conductivity of the composite electrode surfaces, electrochemical impedance studies were also performed and the EIS results show that electrodes containing both PMG and fCNT were superior to those with only one component for application as sensors [70]. In another study, MWCNT-poly(thionine) (PTH) nanostructures were formed on carbon film electrodes (CFE), the polymer film being formed beneath or on top of MWCNT, to give two different sensor architectures, PTH/MWCNT/CFE or MWCNT/PTH/CFE. CV experiments and EIS results represented the faster diffusion, and easier electron transfer was observed at PTH/ MWCNT/CFE for sensing acetaminophen [71]. On the contrary, the voltammetric behavior of paracetamol was also studied with molecularly imprinted polymer film of *o*-phenylenediamine, which was combined with carboxyl functionalized MWCNTs onto a GCE surface. In this study, paracetamol was determined successfully in the range of 2.0×10^{-7} to 4.0×10^{-5} mol L $^{-1}$, with a detection limit of 5.0×10^{-8} mol L^{-1} by the help of CNT-CP composite. The recovery value of sensor ranging from 94 to 105% proves the possible application of paracetamol in real samples [72].

L-AA (vitamin C) is an essential antioxidant for human beings, with many physiological and biochemical functions. AA has been widely applied in large quantities to food products, drinks, animal feed, pharmaceutical formulations, and cosmetics due to its valuable properties, such as pH regulating and antioxidant. An abnormal AA concentration in human body fluids is associated with various diseases [73]. Therefore, a highly sensitive, accurate, and selective determination of methods for AA is needed due to the its biological importance in pharmaceutical, clinical, and food industry samples. In recent years, many research groups have developed new types of electrochemical sensors and methods for AA detection. In one of the studies, CNT-CP composites were prepared with a polyimide (PI)/aminofunctionalized MWCNT composite material for the determination of AA [74].

Sphingolipids are long-chain aliphatic amines, with C18 and small amounts of C16 to C19 dihydroxy bases are their naturally most common components. Sphingolipids have crucial roles in membrane and cell regulation as second messengers for growth; therefore, the development of the accurate quantification method of different sphingoid bases is demanded. Carballo et al. achieved the detection of long-chain aliphatic amines using the CNT-CP composite electrode. Poly(Ni-protoporphyrin) film (pNiPP) containing MWCNT was used to cover a GCE to obtain an active surface for sphingolipid detection. The hybrid material (pNiPP/MWCNT) successfully combines the permselectivity of pNiPP with the high conductivity of MWCNT [75]. DNA sensors have attracted great interest because of their application importance in medical genome research, diagnosis, and forensic science. Li et al. offered flexible MWCNT/polydimethylsiloxane (PDMS)-based electrodes that were fabricated and used for electrochemical DNA sensors. They achieved a successful determination by differential pulse voltammetry with a detection limit of 1.3×10^2 pM with good DNA sequence differentiation ability [76]. In another study, AuNP/SWCNT/PANI composite-modified disposable pencil graphite electrode (AuNP/SWCNT/PANI/PGE) prepared by CNT-CP/E pathway was used for DNA detection. PGE was immersed in a solution containing monomer (0.1 M aniline) and SWCNTs (2 mg mL^{-1}) in 0.1 $MHClO_4$ and the constant potential of 0.8 V was applied for the modification of both CP and CNT on the electrode surface. The electrode provided a simple, easy, time-effective, fast, and cheap detection scheme for DNA [77].

Xanthine (2,6-dihydroxypurine) is a purine base present in human body tissues and fluids and in other organisms. Many stimulants are derived from xanthine, including caffeine and theobromine. The determination of xanthine in blood sample and tissue is crucial for the diagnosis of gout, hyperuricemia, xanthinuria, and renal failure. In addition, in the food industry, elevated levels of hypoxanthine and xanthine are important biomarkers as a sign of meat spoilage. Dervisevic et al. offered a composite CNT-CP surface for enzyme immobilization of xanthine oxidase, which was developed by incorporating MWCNT in poly(glycidyl methacrylate) and vinylferrocene (GMA-co-VFc) copolymer film. The proposed p(GMA-co-VFc)/MWCNT coated electrodes have improved analytical performance for xanthine determination in terms of good sensitivity, selectivity, low response time (4 s), wide linear range, and low detection limit of 0.12 mM. This xanthine biosensor gave reliable results in measuring xanthine concentration in the fish meat [78].

UA (2,6,8-trihydroxypurine) is the primary end-product of purine metabolism, and its concentration must be kept at a certain level in the body. The excessive level of UA in the urine and serum resulted in diseases, such as Lesch-Nyhan, gout, hyperuricemia, cardiovascular, and kidney-related diseases. Therefore, the detection of UA is of great importance in many areas, including physiological investigations and disease diagnosis. The electrochemical accurate, selective, and sensitive determination of UA has led to the development of active electrode surfaces. For this aim, Erden et al. proposed a new UA biosensor that was constructed based on ferrocene containing polymer poly(vinylferrocene) (PVF), c-MWCNTs, and gelatin (GEL)-modified GCE. The c-MWCNT/GEL/PVF composite was characterized by SEM, CV, and EIS to express the modification of both CNT and CP contents on the surface and its activity towards UA oxidation. The proposed CNT-CP composite electrode showed high selectivity and sensitivity accompanied by the low detection limit of 2.3×10^{-8} M to UA [79]. Serotonin (5-hydroxytryptamine, 5-HT) is a neurotransmitter playing an important role in the emotion system, where low levels of 5-HT are related to several disorders, such as anxiety, depression, and migraine. Therefore, it seems necessary to develop the catalytic surface to determine the 5-HT level accurately. According to this aim, Cesarino et al. prepared composite electrode surface by the CNT-CP/E pathway. The Pt electrode was immersed in 25.0 mL of 0.2 mol L^{-1} NaCl solution containing 443 μL Py and 5.0 mg MWCNT and sonicated for 10 min with 70% amplitude. MWCNTs, PPy was successfully modified on the Pt electrode surface where 5-HT was determined selectively with an LOD value of 0.15 μmol L^{-1} [80]. In addition, 5-HT determination was also carried out with poly(3,4-ethylenedioxypyrrole)-SWCNT [P(EDOP-SWCNT)] composite electrodes [41]. P(EDOP-SWCNT)-coated GCE improved oxidation peak currents and lowered the oxidation overpotential due to a stereoporous interfacial layer CNT-CP composite electrode. Hence, the sensor demonstrated high sensitivity and selectivity for 5-HT and showed excellent stability and reproducibility. The linear response was in the range of 1.0×10^{-7} to 1.0×10^{-5} M, with a correlation coefficient of 0.998 on the anodic current. The lower detection limit was calculated as 5.0 nM. On the contrary, environmental protection is a critical concern, as polymer-CNT composites are used for the determination of pollutants, such as pesticides and phenolic compounds. Phenolic compounds are present in many wastewater streams of the oil, paint, polymer, and pharmaceutical industries, which should be determined successfully for environmental protection. With the help of CNT-CP composite electrode, which is prepared by the combination of MWCNT-PPy-horseradish peroxidase, 18 phenol derivatives were determined by amperometric technique with high sensitivity [42]. 4-Nitrophenol, another phenolic compound, was reported as a potential carcinogen and mutagen determined with GCE modified with poly(diphenylamine)/MWCNTs-β-CD (PDPA/MWCNT-β-CD) film electrode in ppb level [81]. 2-Aminophenol (2-AP) and 4-aminophenol (4-AP) are toxic industrial wastes that cause serious hazards on biological systems, so their accurate determination is important. With the help of improved characteristics of CNT-CP composite electrodes, which are prepared by electropolymerization of β-CD and L-arginine (L-Arg) on the surface of CNTs@graphene nanoribbons (CNTs@GNRs) core-shell heterostructure, 2-AP and 4-AP were determined simultaneously in nanomolar level [82]. The crucial role of highly toxic phenolic compounds such as catechol and quinol in industrial applications causes their coexistence as environmental pollutants. PMG-coated MWCNT film electrode is used for the determination of catechol and quinol with ease of fabrication and sufficient stability [55]. Pesticides are another environmental concern, which reach a destination other than their target species, including air, water, bottom sediments, and food. MWCNT/P3MT composite electrode was used for the sensitive determination of various kinds of pesticides, including herbicide (isoproturon), insecticides (voltage, cypermethrin, deltamethrin, and fenvalerate), and an acaricide (dicofol) [32]. In another study, organophosphorus pesticides that known as cholinesterase inhibitors were determined by biosensors that are prepared by the combination of acetylcholinesterase (AChE) enzyme, a CP; poly(4-[2,5-di(thiophen-2-yl)-1H-pyrrol-1-yl]benzenamine) [poly(SNS-NH_2)]; and fCNTs. The proposed biosensor offered a fast response time (6 s), a wide linear range (0.05 and 8.00 mM), and a low detection limit (0.09 mM), which is successfully used for the real sample analysis in fortified tap water samples [83].

Hydrazine is a water-soluble volatile colorless liquid that is used in various chemical industries, although it is a harmful substance for the environment. It causes serious damage to the lungs, kidneys, and the central nervous system if it is absorbed by the human system [84]. Thus, the development of an effective surface for the sensitive determination of hydrazine is required. 4,5-Dihydro-1,3-thiazol-2-ylsulfanyl-1,2-benzenediol containing sulfur and nitrogen groups was electrochemically polymerized on an MWCNT/GCE; this resulted to CNT-CP composite that has wide linear range with an LOD of 0.6 mM for hydrazine determination [85]. Persulfate is a stable and powerful oxidant that is used for the degradation of organic contaminants in groundwater and wastewater. Although persulfate has a wide range of applications, there are only several methods for its determination. One of them is the electrochemical technique that is studied with the composite electrode prepared by the combination of electropolymerized silicomolybdate (SiMO) and PEDOT on MWCNT/GCE. The SiMO-PEDOT/MWCNT electrode can effectively analyze $S_2O_8^{2-}$ due to its low overpotential, high stability, fast response, low cost, and high sensitivity with a 10^{-6} M detection limit [86]. Nitrite is a versatile chemical compound used in a wide range of applications from dye manufacture to food preservation. The potential for nitrite to be metabolized to carcinogenic *N*-nitrosamine in the stomach and the conversion of hemoglobin to methemoglobin that interferes with the oxygen transport system and its widespread presence in food products and beverages have necessitated its monitoring [87]. For this purpose, composite electrode consisting of 5-amino-1,3,4-thiadiazole-2-thiol electrochemically deposited on acid functionalized MWCNTs was used for nitrite determination in nanomolar level with amperometry [88]. In another study, poly(vinylferrocenium)/MWCNT composite-modified disposable PGE (PVF^+/MWCNT/PGE) was prepared by electropolymerization of PVF in the presence of MWCNTs with one-step electropolymerization. The prepared composite CNT-CP electrode was successfully applied to a commercial mineral water sample for nitrite determination [89].

5. Conclusion

This chapter summarized the preparation, characterization, and analytical applications of CNT-CP composite materials in electrochemistry. CNT-CP composite electrodes are prepared in three different ways, which are polymer attachment after CNT modification, CNT modification after polymerization, and CP and CNT modified together on electrode surface. The activity and synergistic effect of the prepared CNT-CP composite electrodes towards organic or inorganic compounds changed with the preparation pathway, which also affects the thickness of the polymer on bare electrode or CNT surface. CNT-CP composite electrodes were characterized by SEM, TEM, AFM, and EIS techniques to understand the morphology and prove the existence of modifiers on the prepared surfaces. Although the surface morphology of CNT-CP composite electrode surface is lightened with SEM, the more detailed image about surface in atomic size with narrow focusing areas is obtained with TEM. Another characterization tool is EIS, which is generally used to improve the understanding of multistep reactions such as diffusion impedance, charge transfer, double-layer capacitance, and solution resistance. The observing of atomic-scale images of the surface is achieved using AFM. The existing remarkable features of CNTs, such as large accessible surface area, high mechanical strength, and low electrical resistance, improved the activity of CPs, which, based on CNT-CP composite electrodes, make as great candidate for a wide range of electroanalytical and fuel cell applications.

Author details

Şükriye Ulubay Karabiberoğlu[1], Çağrı Ceylan Koçak[2] and Zekerya Dursun*

*Address all correspondence to: zekerya.dursun@ege.edu.tr

1 Ege University, İzmir, Turkey

2 Dokuz Eylül University, İzmir, Turkey

References

[1] Smith BW, Luzzi DE. Carbon nanotubes. In: Ventra MD, Evoy S, Heflin, Jr JR, editors. Introduction to Nanoscale Science and Technology: Nanostructure Science and Technology. 1st ed. Boston: Kluwer Academic; 2004. pp. 137–183.

[2] Yang J, Li X, Liu C, Ma G. Changes of structure and electrical conductivity of multi-walled carbon nanotubes film caused by 3 MeV proton irradiation. Applied Surface Science. 2015;325:235–241. DOI: 10.1016/j.apsusc.2014.11.070

[3] Martins-Júnior PA, Alcântara CE, Resende RR, Ferreira AJ. Carbon Nanotubes: Directions and perspectives in oral regenerative medicine. Journal of Dental Research. 2013;92:575–583. DOI: 10.1177/0022034513490957

[4] Vashist SK, Zheng D, Al-Rubeaan K, Luong JHT, Sheu FS. Advances in carbon nanotube based electrochemical sensors for bioanalytical applications. Biotechnology Advances. 2011;29:169–188. DOI: 10.1016/j.biotechadv.2010.10.002

[5] Seligra PG, Nuevo F, Lamanna M, Famá L. Covalent grafting of carbon nanotubes to PLA in order to improve compatibility. Composites: Part B. 2013;46:61–68. DOI: 10.1016/j.compositesb.2012.10.013

[6] Luo YL, Wei XP, Cao D, Bai RX, Xu F, Chen YS. Polystyrene-block-poly (tert-butyl methacrylate)/multiwall carbon nanotube ternary conducting polymer nanocomposites based on compatibilizers: Preparation, characterization and vapor sensing applications. Materials and Design. 2015;87:149–156. DOI: 10.1016/j.matdes.2015.08.030

[7] Díez-Pascual AM, Naffakh M, Marco C, Gómez-Fatou MA, Ellis GJ. Multiscale fiber-reinforced thermoplastic composites incorporating carbon nanotubes: A review. Current Opinion in Solid State and Materials Science. 2014;18:62–80. DOI: 10.1016/j.cossms.2013.06.003

[8] Ates, M. A review study of (bio)sensor systems based on conducting polymers. Materials Science and Engineering C. 2013;33:1853–1859. DOI: 10.1016/j.msec.2013.01.035

[9] Al-Oqla FM, Sapuan SM, Anwer T, Jawaid M, Hoque ME. Natural fiber reinforced conductive polymer composites as functional materials: A review. Synthetic Metals. 2015;206:42–54. DOI: 10.1016/j.synthmet.2015.04.014

[10] Lien TTN, Lam TD, An VTH, Hoang TV, Quang DT, Khieu, DQ, Tsukahara T, Lee YH, Kim JS. Multi-wall carbon nanotubes (MWCNTs)-doped polypyrrole DNA biosensor for label-free detection of genetically modified organisms by QCM and EIS. Talanta. 2010;80:1164–1169. DOI: 10.1016/j.talanta.2009.09.002

[11] Peng C, Jin J, Chen GZ. A comparative study on electrochemical co-deposition and capacitance of composite films of conducting polymers and carbon nanotubes. Electrochimica Acta. 2007;53:525–537. DOI: 10.1016/j.electacta.2007.07.004

[12] Spitalsky Z, Tasis D, Papagelis K, Galiotis C. Carbon nanotube–polymer composites: Chemistry, processing, mechanical and electrical properties. Progress in Polymer Science. 2010;35:357–401. DOI: 10.1016/j.progpolymsci.2009.09.003

[13] Zhang D, Kandadai M, Cech J, Roth S, Currans S. Poly(L-lactide) (PLLA)/multiwalled carbon nanotubes (MWCNT) composite: Characterization and biocompatibility evaluation. Journal of Physical Chemistry B. 2006;110:12910–12915. DOI: 10.1021/jp061628k

[14] Babaa R, Bantignie, Michel T, Fiorito S, Poncharal P, Zahab A. Carbon nanotubes reinforced PE and PEEK: fabrication and characterization. In: 20th IWPE Molecular Nanostructures; 4–11 March 2006; Austria. Kirchberg; 2006. p. 42

[15] Yang J, Yao Z, Tang C, Darvell BW, Zhang H, Pan L, Liu J, Chen Z. Growth of apatite on chitosan-multiwall carbon nanotube composite membranes. Applied Surface Science. 2009;255:8551–8555. DOI: 10.1016/j.apsusc.2009.06.013

[16] Luo YL, Wei XP, Cao D, Bai RX, Xu F, Chen YS. Polystyrene-block-poly (tert-butyl methacrylate)/multiwall carbon nanotube ternary conducting polymer nanocomposites based on compatibilizers: Preparation, characterization and vapor sensing applications. Materials and Design. 2015;87:149–156. DOI: 10.1016/j.matdes.2015.08.030

[17] Kocak S, Aslışen B. Hydrazine oxidation at gold nanoparticles and poly(bromocresol purple) carbon nanotube modified glassy carbon electrode. Sensors and Actuators B. 2014;196:610–618. DOI: 10.1016/j.snb.2014.02.061

[18] Bae HB, Ryu JH, Byun BS, Jung SH, Choi SH. Facile synthesis of novel Pt-Ru@PPy-MWNT electrocatalysts for direct methanol fuel cells. Current Applied Physics. 2010;10:44–50. DOI: 10.1016/j.cap.2009.11.039

[19] Aguirre MdC, Rivas BL, Farfal CP. Poly(3-methyltiophene)-multi walled carbon nanotubes composite electrodes. Procedia Materials Science. 2015;8:251–260. DOI: 10.1016/j.mspro.2015.04.071

[20] Yang J, Li X, Liu C, Ma G. Changes of structure and electrical conductivity of multi-walled carbon nanotubes film caused by 3 MeV proton irradiation. Applied Surface Science. 2015;325:235–241. DOI: 10.1016/j.apsusc.2014.11.070

[21] Esplandiu MJ. Electrochemistry on carbon-nanotube-modified surfaces. In: Alkire RC, Kolb DM, Lipkowski J, Ross PN, editors. Advances in Electrochemical Science and Engineering: Chemically Modified Electrodes. 1st ed. Weinheim, Germany: Wiley; 2009. pp. 117–169.

[22] Vaisman L, Wagner HD, Marom G. The role of surfactants in dispersion of carbon nanotubes. Advances in Colloid and Interface Science. 2006;128:37–46. DOI: 10.1016/j.cis.2006.11.007

[23] Barsan MM, Ghica ME, Brett CMA. Electrochemical sensors and biosensors based on redox polymer/carbon nanotube modified electrodes: A review. Analytica Chimica Acta. 2015;881:1–23. DOI: 10.1016/j.aca.2015.02.059

[24] Serafín V, Agüí L, Yáñez-Sedeñon P, Pingarrón JM. Electrochemical immunosensor for the determination of insulin-like growth factor-1 using electrodes modified with carbon nanotubes-poly(pyrrole propionic acid) hybrids. Biosensors and Bioelectronics. 2014;52:98–104. DOI: 10.1016/j.bios.2013.08.021

[25] Li J, Liu Y, Wei W, Luo S. Fabrication of tiron doped poly-pyrrole/carbon nanotubes on low resistance monolayer modified glassy carbon electrode for selective determination of dopamine. Analytical Letters. 2011;44:1226–1240. DOI: 10.1080/00032719.2010.511744

[26] Shahrokhian S, Hosseini P, Kamalzadeh Z. Investigation of the electrochemical behavior of mesalazine on the surface of a glassy carbon electrode modified with CNT/

PPY doped by 1,5-naphthalenedisulfonic acid. Electroanalysis. 2013;25:2481–2491. DOI: 10.1002/elan.201300333

[27] Qi H, Li X, Chen P, Zhang C. Electrochemical detection of DNA hybridization based on polypyrrole/ss-DNA/multi-wall carbon nanotubes paste electrode. Talanta. 2007;72:1030–1035. DOI: 10.1016/j.talanta.2006.12.032

[28] Shahrokhian S, Asadian E. Electrochemical determination of L-dopa in the presence of ascorbic acid on the surface of the glassy carbon electrode modified by a bilayer of multi-walled carbon nanotube and poly-pyrrole doped with tiron. Journal of Electroanalytical Chemistry. 2009,636:40–46. DOI: 10.1016/j.jelechem.2009.09.010

[29] Li Y, Umasankar Y, Chen SM. Polyaniline and poly(flavin adenine dinucleotide) doped multi-walled carbon nanotubes for *p*-acetamidophenol sensor. Talanta. 2009;79:486–492. DOI: 10.1016/j.talanta.2009.04.017

[30] Aguirre MdC, Rivas BL, Farfal CP. Poly(3-methyltiophene)-multi walled carbon nanotubes composite electrodes. Procedia Materials Science. 2015;8:251–260. DOI: 10.1016/j.mspro.2015.04.071

[31] Lin KC, Huang JY, Chen SM. Simultaneous determination of ascorbic acid, dopamine, uric acid and hydrogen peroxide based on co-immobilization of PEDOT and FAD using multi-walled carbon nanotubes. Analytical Methods. 2014;6:8321–8327. DOI: 10.1039/c4ay01639d

[32] Sundari PA, Manisankar P. Development of nano poly(3-methyl thiophene)/multiwalled carbon nanotubes sensor for the efficient detection of some pesticides. Journal of Brazilian Chemistry Society. 2011;22:746–755. DOI: 10.1590/S0103-50532011000400019

[33] Yang CY, Hung YT, Chen SM, Tsai TH, Lou BS, Liu X. Functionalized MWCNT with poly(3,4-ethylenedioxythiophene) hybrid film modified electrode for selectively determination of ascorbic acid and hydrazine. International Journal of Electrochemical Science. 2015;10:1128–1135.

[34] Nie T, Lu L, Bai L, Xu J, Zhang K, Zhang O, Wen Y, Wu L. Simultaneous determination of folic acid and uric acid under coexistence of L-ascorbic acid using a modified electrode based on poly(3,4-ethylenedioxythiophene) and functionalized single-walled carbon nanotubes composite. International Journal of Electrochemical Science. 2013;8:7016–7029.

[35] Wang HS, Li TH, Jia WL, Xu HY. Highly selective and sensitive determination of dopamine using a Nafion/carbon nanotubes coated poly(3-methylthiophene) modified electrode. Biosensors and Bioelectronics. 2006;22:664–669. DOI: 10.1016/j.bios.2006.02.007

[36] Wen YP, Lia D, Lu Y, He HH, Xu JK, Duan XM, Liu M. Poly(3,4-ethylenedioxythiophene methanol)/ascorbate oxidase/Nafion-single-walled carbon nanotubes biosensor for

voltammetric detection of vitamin C. Chinese Journal of Polymer Science. 2012;30:460–469. DOI: 10.1007/s10118-012-1140-4

[37] Quan DP, Tuyen DP, Lam TD, Ngo PT, Binh TNH, Viet PH. Electrochemically selective determination of dopamine in the presence of ascorbic and uric acids on the surface of the modified Nafion/single wall carbon nanotube/poly(3-methylthiophene) glassy carbon electrodes. Colloids and Surfaces B: Biointerfaces. 2011;88:764–770. DOI: 10.1016/j.colsurfb.2011.08.012

[38] Yao Y, Zhang L, Wen Y, Wang Z, Zhang H, Hu D, Xu J, Duan X. Voltammetric determination of catechin using single-walled carbon nanotubes/poly(hydroxymethylated-3,4-ethylenedioxythiophene) composite modified electrode. Ionics. 2015;21:2927–2936. DOI: 10.1007/s11581-015-1494-z

[39] Yi H, Zheng D, Hu C, Hu S. Functionalized multiwalled carbon nanotubes through *in situ* electropolymerization of brilliant cresyl blue for determination of epinephrine. Electroanalysis. 2008;20:1143–1146. DOI: 10.1002/elan.200704141

[40] Ragupathy D, Park JL, Lee SC, Kim JC, Gomathi P, Kim MK, Lee SM, Ghim HD, Rajendran A, Lee SH, Jeon KM. Electrochemical grafting of poly(2,5-dimethoxy aniline) onto multiwalled carbon nanotubes nanocomposite modified electrode and electrocatalytic oxidation of ascorbic acid. Macromolecular Research 2011;19:764–769. DOI: 10.1007/s13233-011-0802-3

[41] Kim SK, Bae SR, Ahmed MS, You JM, Jeon S. Selective determination of serotonin on poly(3,4-ethylenedioxy pyrrole)-single-walled carbon nanotube-modified glassy carbon electrodes. Bulletin of the Korean Chemical Society. 2011;32:1215–1220. DOI: 10.5012/bkcs.2011.32.4.1215

[42] Korkut S, Keskinler B, Erhan E. An amperometric biosensor based on multiwalled carbon nanotube-poly(pyrrole)-horseradish peroxidase nanobiocomposite film for determination of phenol derivatives. Talanta. 2008;76:1147–1152. DOI: 10.1016/j.talanta.2008.05.016

[43] Brânzoi V, Pilan L, Brânzoi F. Amperometric glucose biosensor based on electropolymerized carbon nanotube/polypyrrole composite film. Revue Roumaine de Chimie. 2009;54:783–789.

[44] Dong Q, Li Y, Zhu L, Ma T, Guo C. Electrocatalytic oxidation of methanol and formaldehyde on platinum-modified poly(o-methoxyaniline)-multiwalled carbon nanotube composites, International Journal of Electrochemical Science. 2013;8:8191–8200.

[45] Ajami N, Panah NB, Danaee I. Oxytetracycline nanosensor based on poly-*ortho*-aminophenol/multi-walled carbon nanotubes composite film. Iran Polymer Journal. 2014;23:121–126. DOI: 10.1007/s13726-013-0207-6

[46] Liu X, Peng Y, Qu X, Ai S, Han R, Zhu X. Multi-walled carbon nanotube-chitosan/poly(amidoamine)/DNA nanocomposite modified gold electrode for determination of

dopamine and uric acid under coexistence of ascorbic acid. Journal of Electroanalytical Chemistry. 2011;654:72–78. DOI: 10.1016/j.jelechem.2011.01.024

[47] Li Y, Wang P, Wang L, Lin X. Overoxidized polypyrrole film directed single-walled carbon nanotubes immobilization on glassy carbon electrode and its sensing applications. Biosensors and Bioelectronics. 2007;22:3120–3125. DOI: 10.1016/j.bios.2007.02.001

[48] Liu M, Wen Y, Li D, Yue R, Xu J, He H. A stable sandwich-type amperometric biosensor based on poly(3,4-ethylenedioxythiophene)–single walled carbon nanotubes/ascorbate oxidase/Nafion films for detection of L-ascorbic acid. Sensors and Actuators B. 2011;159:277–285. DOI: 10.1016/j.snb.2011.07.005

[49] Mugadza T, Nyokong T. Synthesis, characterization and application of monocarboxy-phthalocyanine-single walled carbon nanotube conjugates in electrocatalysis. Polyhedron. 2011;30:1820–1829. DOI: 10.1016/j.poly.2011.04.020

[50] Orazem ME, Tribollet B. Electrochemical Impedance Spectroscopy. Hoboken, New Jersey: Wiley; 2008. 325 pp.

[51] Zhang Y, Cai Y, Su S. Determination of dopamine in the presence of ascorbic acid by poly(styrene sulfonic acid) sodium salt/single-wall carbon nanotube film modified glassy carbon electrode. Analytical Biochemistry. 2006;350:285–291. DOI: 10.1016/j.ab.2006.01.002

[52] Karim MR, Lee CJ, Lee MS. Synthesis and characterization of conducting polythiophene/carbon nanotubes composites. Journal of Polymer Science: Part A: Polymer Chemistry. 2006;44:5283–5290. DOI: 10.1002/pola.21640

[53] Dhibar S, Bhattacharya P, Hatui G, Das CK. Transition metal doped poly(aniline-co-pyrrole)/multi-walled carbon nanotubes nanocomposite for high performance supercapacitor electrode materials. Journal of Alloys and Compounds. 2015;625:64–75. DOI: 10.1016/j.jallcom.2014.11.108

[54] Inzelt G. Conducting Polymers: A New Era in Electrochemistry. Fritz Scholz, editor. Berlin/Heidelberg: Springer-Verlag; 2008. 282 pp. DOI: 10.1007/978-3-540-75930-0

[55] Umasankar Y, Periasamy AP, Chen SM. Electrocatalysis and simultaneous determination of catechol and quinol by poly(malachite green) coated multiwalled carbon nanotube film. Analytical Biochemistry. 2011;411:71–79. DOI: 10.1016/j.ab.2010.12.002

[56] Liu A, Honma I, Zhou H. Simultaneous voltammetric detection of dopamine and uric acid at their physiological level in the presence of ascorbic acid using poly(acrylic acid)-multiwalled carbon-nanotube composite-covered glassy-carbon electrode. Biosensors and Bioelectronics. 2007;23:74–80. DOI: 10.1016/j.bios.2007.03.019

[57] Adekunle AS, Farah AM, Pillay J, Ozoemena KI, Mamba BB, Agbool BO. Electrocatalytic properties of Prussian blue nanoparticles supported on poly(m-aminobenzene-sulphonic acid)-functionalised single-walled carbon nanotubes towards the detection

of dopamine. Colloids and Surfaces B: Biointerfaces. 2012;95:186–194. DOI: 10.1016/j.colsurfb.2012.02.043

[58] Ali SR, Ma Y, Parajuli RR, Balogun Y, Lai WYC, He H. A nonoxidative sensor based on a self-doped polyaniline/carbon nanotube composite for sensitive and selective detection of the neurotransmitter dopamine. Analytical Chemistry. 2007;79:2583–2587. DOI: 10.1021/ac062068o

[59] Bui MPN, Li CA, Seong GH. Electrochemical detection of dopamine with poly-glutamic acid patterned carbon nanotube electrodes. Bio Chip Journal, 2012;6:149–156. DOI: 10.1007/s13206-012-6207-3

[60] Raoof JB, Ojani R, Baghayeri M. Fabrication of layer-by-layer deposited films containing carbon nanotubes and poly(malachite green) as a sensor for simultaneous determination of ascorbic acid epinephrine, and uric acid. Turkish Journal of Chemistry. 2013;37:36–50. DOI: 10.3906/kim-1112-54

[61] Yogeswaran U, Chen SM. Multi-walled carbon nanotubes with poly(methylene blue) composite film for the enhancement and separation of electroanalytical responses of catecholamine and ascorbic acid. Sensors and Actuators B. 2008;130:739–749. DOI: 10.1016/j.snb.2007.10.040

[62] Kamyabi MA, Hajari N, Turner APF, Tiwari A. A high-performance glucose biosensor using covalently immobilised glucose oxidase on a poly(2,6-diaminopyridine)/carbon nanotube electrode. Talanta. 2013;116:801–808. DOI: 10.1016/j.talanta.2013.07.068

[63] Du P, Wu P, Cai C. A glucose biosensor based on electrocatalytic oxidation of NADPH at single-walled carbon nanotubes functionalized with poly(Nile blue A). Journal of Electroanalytical Chemistry. 2008;624:21–26. DOI: 10.1016/j.jelechem.2008.07.031

[64] Bas SZ. A gold nanoparticle functionalized multiwalled carbon nanotube-poly(*o*-phenylenediamine) composite film for glucose biosensing applications. Analytical Methods. 2014;6:7752–7759. DOI: 10.1039/c4ay01524j

[65] Pan D, Chen J, Yao S, Tao W, Nie L. An amperometric glucose biosensor based on glucose oxidase immobilized in electropolymerized poly (*o*-aminophenol) and CNT composite film on gold electrode. Analytical Sciences. 2005;21:367–371.

[66] Li H, Worley KE, Barton SC. Quantitative analysis of bioactive NAD+ regenerated by NADH electro-oxidation. ACS Catalysis. 2012;2:2572–2576. DOI: 10.1021/cs3004598

[67] Chauhan N, Singh A, Narang J, Dahiya S, Pundir CS. Development of amperometric lysine biosensors based on Au nanoparticles/multiwalled carbon nanotubes/polymers modified Au electrodes. Analyst. 2012;137:5113–5122. DOI: 10.1039/c2an35629e

[68] Liu X, Luo L, Ding Y, Ye D. Poly-glutamic acid modified carbon nanotube-doped carbon paste electrode for sensitive detection of L-tryptophan. Bioelectrochemistry. 2011;82:38–45. DOI: 10.1016/j.bioelechem.2011.05.00

[69] Chitravathi S, Munichandraiah N. Voltammetric determination of paracetamol, tramadol and caffeine using poly(Nile blue) modified glassy carbon electrode. Journal of Electroanalytical Chemistry. 2016;764:93–103. DOI: 10.1016/j.jelechem.2016.01.021

[70] Barsan MM, Toledo CT, Brett CMA. New electrode architectures based on poly(methylene green) and functionalized carbon nanotubes: Characterization and application to detection of acetaminophen and pyridoxine. Journal of Electroanalytical Chemistry. 2015;736:8–15. DOI: 10.1016/j.jelechem.2014.10.026

[71] Ghica ME, Ferreira GM, Brett CMA. Poly(thionine)-carbon nanotube modified carbon film electrodes and application to the simultaneous determination of acetaminophen and dipyrone. Journal of Solid State Electrochemistry. 2015;19:2869–2881. DOI: 10.1007/s10008-015-2926-4

[72] Peng Y, Wu Z, Liu Z, An electrochemical sensor for paracetamol based on an electropolymerized molecularly imprinted ophenylenediamine film on a multi-walled carbon nanotube modified glassy carbon electrode. Analytical Methods. 2014;6: 5673–5681. DOI: 10.1039/c4ay00753k

[73] Pardakhty A, Ahmadzadeh S, Avazpour S, Gupta VK. Highly sensitive and efficient voltammetric determination of ascorbic acid in food and pharmaceutical samples from aqueous solutions based on nanostructure carbon paste electrode as a sensor. Journal of Molecular Liquids. 2016;216:387–391, DOI: 10.1016/j.molliq.2016.01.010

[74] Chang KC, Chu CM, Chang CH, Cheng HT, Hsu SC, Lan CC, Chen HH, Peng YY, Yeh JM. Photoisomerization of electroactive polyimide/multiwalled carbon nanotube composites on the effect of electrochemical sensing for ascorbic acid. Polymer International. 2015;64: 373–382. DOI: 10.1002/pi.4797

[75] Carballo R, Rinaldi AL, Dabas PC, Rezzano IN. A carbon nanotube/poly [Ni-(protoporphyrin IX)] composite for amperometric detection of long chain aliphatic amines. Bioelectrochemistry. 2015;104:51–57. DOI: 10.1016/j.bioelechem.2015.03.002

[76] Li J, Lee EC. Carbon nanotube/polymer composite electrodes for flexible, attachable electrochemical DNA sensors. Biosensors and Bioelectronics. 2015;71:414–419. DOI: 10.1016/j.bios.2015.04.045

[77] Eksin E, Bolat G, Kuralay F, Erdem A, Abaci S. Preparation of gold nanoparticles/single-walled carbon nanotubes/polyaniline composite-coated electrode developed for DNA detection. Polymer Bulletin. 2015;72:3135–3146. DOI 10.1007/s00289-015-1457-6

[78] Dervisevic M, Custiuc E, Çevik E, Şenel M. Construction of novel xanthine biosensor by using polymeric mediator/MWCNT nanocomposite layer for fish freshness detection. Food Chemistry. 2015;181:277–283. DOI: 10.1016/j.foodchem.2015.02.104

[79] Erden PE, Kaçar C, Öztürk F, Kılıç E. Amperometric uric acid biosensor based on poly(vinylferrocene)-gelatin-carboxylated multiwalled carbon nanotube modified glassy carbon electrode. Talanta. 2015;134:488–495. DOI: 10.1016/j.talanta.2014.11.058

[80] Cesarino I, Galesco HV, Machado SAS. Determination of serotonin on platinum electrode modified with carbon nanotubes/polypyrrole/silver nanoparticles nanohybrid. Materials Science and Engineering C. 2014;40:49–54. DOI: 10.1016/j.msec.2014.03.030

[81] Zarei K, Teymori E, Kor K. Very sensitive differential pulse adsorptive stripping voltammetric determination of 4-nitrophenol at poly (diphenylamine)/multi-walled carbon nanotube-β-cyclodextrin-modified glassy carbon electrode. International Journal of Environmental Analytical Chemistry. 2014;94:1407–1421. DOI: 10.1080/03067319.2014.962529

[82] Yi Y, Zhu G, Wu X, Wang K. Highly sensitive and simultaneous electrochemical determination of 2-aminophenol and 4-aminophenol based on poly (L-arginine)-β-cyclodextrin/carbon nanotubes@graphene nanoribbons modified electrode. Biosensors and Bioelectronics. 2016; 77:353–358. DOI: 10.1016/j.bios.2015.09.052

[83] Kesik M, Kanik FE, Turan J, Kolb M, Timur S, Bahadir M, Toppare L. An acetylcholinesterase biosensor based on a conducting polymer using multiwalled carbon nanotubes for amperometric detection of organophosphorous pesticides. Sensors and Actuators B. 2014;205:39–49. DOI: 10.1016/j.snb.2014.08.058

[84] Saengsookwaow C, Rangkupan R, Chailapakul O, Rodthongkum N. Nitrogen-doped graphene-polyvinylpyrrolidone/gold nanoparticles modified electrode as a novel hydrazine sensor. Sensors and Actuators B. 2016;227:524–532. DOI: 10.1016/j.snb.2015.12.091

[85] Fakhari AR, Ahmarb H, Hosseini H, Movahed SK. Fabrication of novel redox-active poly (4,5-dihydro-1,3-thiazol-2-ylsulfanyl-3-methyl-1,2-benzenediol)-gold nanoparticles film on MWCNTs modified electrode: Application as the electrochemical sensor for the determination of hydrazine. Sensors and Actuators B. 2015;213:82–91. DOI: 10.1016/j.snb.2015.02.015

[86] Lin KC, Wu TH, Chen SM. A highly sensitive persulfate sensor based on a hybrid nanocomposite with silicomolybdate doping poly(3,4-ethylenedioxythiophene) on multi-walled carbon nanotubes. RSC Advances. 2015;5:59946–59952. DOI: 10.1039/c5ra09009a

[87] Wang Q, Ma S, Huang H, Cao A, Li M, He L. Highly sensitive and selective spectrofluorimetric determination of nitrite in food products with a novel fluorogenic probe. Food Control. 2016;63:117–121. DOI: 10.1016/j.foodcont.2015.11.023

[88] Rajalakshmi K, John SA. Highly sensitive determination of nitrite using FMWCNTs-conducting polymer composite modified electrode. Sensors and Actuators B. 2015;215:119–124. DOI: 10.1016/j.snb.2015.03.050

[89] Kuralay F, Dumangöz M, Tunç S. Polymer/carbon nanotubes coated graphite surfaces for highly sensitive nitrite detection. Talanta. 2015;144:1133–1138, DOI: 10.1016/j.talanta.2015.07.095

Chapter 16

Carbon Nanotubes Supported Conducting Polymer Electrode for Supercapacitor

Chenzhong Yao, Bohui Wei and Yexiang Tong

Additional information is available at the end of the chapter

http://dx.doi.org/10.5772/63245

Abstract

Supercapacitors (SCs) as portable systems and electric vehicles have aroused great interest due to their high power in short term. The composite materials of carbon nanotubes and conducting polymers have been considered for SC electrodes. Carbon nanotubes (CNTs), due to their outstanding electric properties and nanoscale texture, such as large specific area, low cost, and cycle stability, exhibit a large and stable double layer capacitance. However, the pure CNTs have low specific capacitance and relatively poor energy density, which limit the commercial application for SCs. On the other hand, conducting polymers have been intensively investigated as the electrode material in SCs. Higher electrical conductivity, larger pseudo-capacitance, and faster doping/dedoping rate during charge/discharge process are their main advantages. The possible application of conducting polymers in SCs is dictated by their significant capacitance values and huge faradaic capacitance. They undergo a redox reaction to store charge in the bulk of the material and thereby increase the energy stored and reduce self-discharge. But they are not ideal materials used alone as advanced capacitors materials, because of their low mechanical strength, poor electrical conductivity and low porosity.

This chapter looks at the major conducting polymer materials, namely, polyaniline, polypyrrole, and ethylene-evinyl acetate copolymer, as well as composites of these materials with CNTs. These composites of the CNTs with these electroactive species propose a facile solution to the capacitance limitation of the CNTs and the structural defects of these electroactive species. In order to achieve low contact resistance and high surface area, surface activation through dispersion or ordered assembling of CNTs have been investigated.

Keywords: Supercapacitors, conducting polymer, electrode, electrodeposition, carbon nanotubes

1. Introduction

In recent years, much great attention has been paid to the development of flexible supercapacitors (SCs) and batteries. SCs have not only high power density, high cyclic stability and high discharge rate, comparable to dielectric capacitors, but also high energy density comparable to that of the batteries [1–4].

According to the electrode material and the operational mechanism of SCs, there are three major types of SCs. Electrochemical double-layer capacitor (EDLC) is the first type. In this system, the capacitance can be stored as a build up of charge in the electrical double-layer in the solution interface close to the surface of the carbon to balance the charge in the carbon material [5]. In charging and discharging process, carbon based electrodes undergo faradaic processes. They can provide a reversible adsorption for the ions on the high surface area [6]. Recently, many kinds of carbon materials and their composites have also been used in SCs with high energy density (155 Wh kg^{-1}) [7–11].

The second type SCs are transition metal oxides based electrochemical capacitors, which are developed based on the fast and reversible redox reactions at the surface of active materials (such as MnO_2 and RuO_2) [12–27]. A high specific capacitance can be obtained in this type of SCs, but the defect is the poor power performance due to its relatively high electrical resistance.

Conducting polymer–based SC is the third type [28–35]. The conductive polymers offer a high specific capacitance and low production cost, though the conductive polymer–based SCs usually have poor stability during cycling because of the destabilization of the polymeric backbone structure. The composite materials of carbon nanotubes and conducting polymers have been considered for SC electrodes. Carbon nanotubes, due to their outstanding electric properties and nanoscale texture, such as large specific area, low cost, and cycle stability, exhibit a large and stable double layer capacitance. However, the pure CNTs have low specific capacitance and relatively poor energy density, which limit the commercial application for SCs. On the other hand, conducting polymers have been intensively investigated as the electrode material in SCs. Higher electrical conductivity, larger pseudo-capacitance, and faster doping/dedoping rate during charge/discharge process are their main advantages [36]. The possible application of conducting polymers in SCs is dictated by their significant capacitance values and huge faradaic capacitance. They undergo a redox reaction to store charge in the bulk of the material and thereby increase the energy stored and reduce self-discharge. But they are not ideal materials used alone as advanced capacitors materials, because of their low mechanical strength, poor electrical conductivity and low porosity.

This chapter focuses on the major conducting polymer materials, namely, polyaniline (PANI), polypyrrole (PPy), and ethylene-evinyl acetate copolymer (EVA), as well as composites of these materials with carbon nanotubes. These composites of the CNTs with the electroactive species propose a facile solution to the capacitance limitation of the CNTs and the structural defects of these electroactive species. In order to achieve low contact resistance and high surface area, surface activation through dispersion or ordered assembling of CNTs have been investigated [37].

2. CNTs–PANI

2.1. Electrodeposition of CNTs–PANI

Over the past few years, various CNTs–PANI electrodes have been synthesized by many methods [38–53]. Among them, electrochemical preparation is a promising method because of its simplicity, ease of scale-up, low cost, and environmental friendliness.

In 2014, our group prepared nano PANI via cyclic voltammetry technique in acidic condition [51]. The prepared PANI has a large area with rough surface. The symmetric SCs based on the composites of PANI, CNT and poly dimethylsiloxane (PDMS) show a high capacitance and

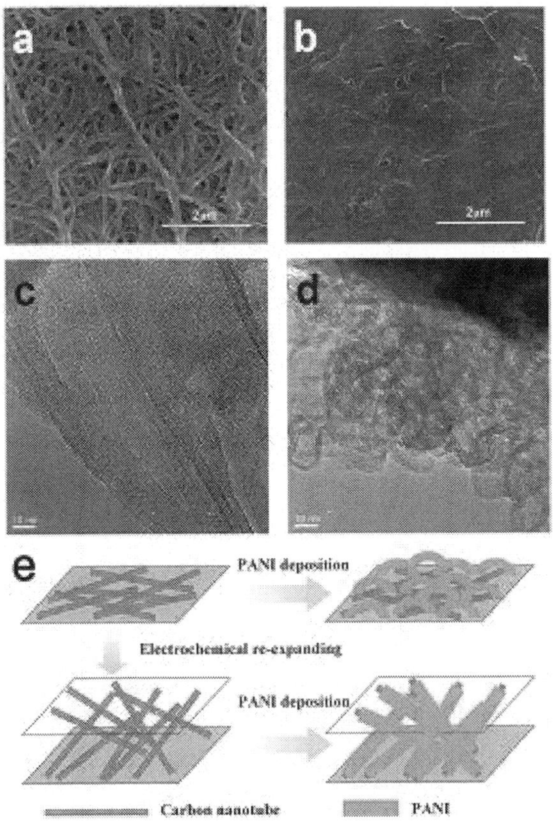

Figure 1. (a) SEM image of CNT/PANI hydrogel film; (b) SEM image of CNT/PANI film; (c) TEM image of CNT/PANI hydrogel film; (d) TEM image of CNT/PANI; (e) scheme of forming process. (Reproduced with permission from Zeng et al. [52]. The Royal Society of Chemistry). © 2015.

stretchability. Zeng et al. [52] also obtained the nano PANI on the untreated CNT hydrogel film by the similar method, as shown in **Figure 1**.

Cai et al. also prepared CNT–PANI composites by electrochemical polymerization [53]. Bare MWCNT fibers were first dipped into the electrolyte containing 1 M H_2SO_4 and 0.1 M aniline for 5 h, so the aniline monomer can be effectively infiltrated into them. Electrodeposition of aniline was mainly performed from 50 to 2000 s at a potential of 0.75 V. After polymerization, the as prepared multi-wall carbon nanotubes (MWCNTs) –PANI composite fibers were washed with deionized water and dried at room temperature for over 2 h.

Zhang et al. fabricated the MWCNTs/PANI composite films for SC electrode by an in-situ electrochemical polymerization onto stainless-steel sheet from an aniline solution containing a small content of MWCNT, and the whole process with no sonication [54]. The electrochemical polymerization was carried by the CV technique at 100 $mV \cdot s^{-1}$ between -0.2 and -1.2 V for 100 cycles in a solution of 0.5 M H_2SO_4 + 0.325 M aniline dissolved MWCNT in different mixture ratios. Three different electro-polymerization methods (potentiodynamic, potentiostatic and galvanostatic) of PANI have been studied by Tran et al. [55]. Potentiodynamic is carried out with a condition of sweeping voltage between -0.1 and 1 V at 20 $mV \cdot s^{-1}$; potentiostatic is applying a constant potential of 0.8 V; and galvanostatic is applying a constant current of 9 mA. After deposition, samples were thoroughly washed with deionized water to remove any adsorbed electrolyte and dried at 60°C for at least 2 h.

2.2. Structure and electrochemical performance of CNTs–PANI composite electrodes

Our group reported that MWCNT-based stretchable film can be formed via water surface assisted method [51]. In the research, combined with the PDMS, the composite electrodes have a uniform surface, excellent conductivity, and high stretchability. The conductivity of the MWCNT/PDMS film is 0.55 $S \cdot cm^{-1}$ with 3 wt% MWCNTs, which will increase to 4.19 $S \cdot cm^{-1}$ with 10 wt% [51]. This variation of the conductivity is due to the formation of 3D-like net structure. The 10wt% MWCNT/PDMS film also has a good stretchability, which can reach to 50% without damaging its conductivity and structure. When the content of MWCNTs is over 10%, the film is hard to be continuous. The 10wt% MWCNT/PDMS film substrate combined with PANI nanofibers show an excellent SC performance at a scan rate of 5 $mV \cdot s^{-1}$ (specific capacitance: 1023 $F \cdot g^{-1}$, areal capacitance: 481 $mF \cdot cm^{-2}$) In this work, a solid-state symmetric SC device was assembled with PANI/10wt% MWCNT/PDMS, which shows excellent stretchability and capacitance. An elastomeric polyurethane textile separator and PVA/H2SO4 gel electrolyte was shown in **Figure 2**. This device was able to deliver the highest volumetric capacitance of 2.1 $F \cdot cm^{-3}$ and the highest specific capacitance of 159 $F \cdot g^{-1}$ at 5 $mV \cdot s^{-1}$. Additionally, its maximum volumetric energy density is about 0.15 $mW \cdot h \cdot cm^{-3}$. After 500 cycles under the condition of stretching and releasing, the as-assembled devices can keep more than 95% capacitance. These results indicate that this device possesses good and stable electrochemical behavior.

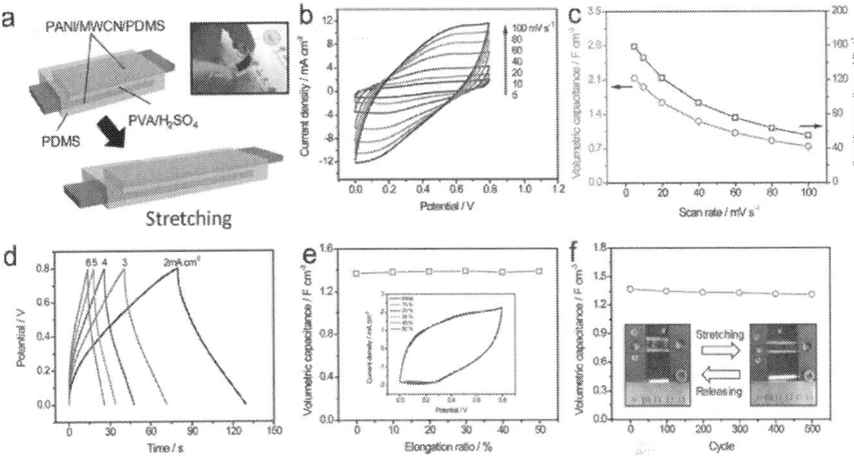

Figure 2. Schematic diagram and electrochemical capacitive curves of PANI–SSC device. (a) Schematic diagram. (b) CV curves and (c) Volumetric capacitance and specific capacitance at various scan rate. (d) Charge/discharge curves at different current densities (e) CV curves with different strain strength at 10 mV·s^{-1}. (f) Stability of under the condition of dynamic stretching and releasing. (Reproduced with permission from Yu et al. [51]). © 2014, John Wiley and Sons.

PANI coated onto individual CNTs and CNT bundles in the activated CNT film can be formed into CNT/PANI hydrogel film [52]. This nanostructure largely benefits the uniform dispersion of PANI among the CNT network scaffold, thus its PANI utilization efficiency is relatively high. In their experiment, the diffusion of electrolyte towards the electrodes was blocked. This is due to the formation of a PANI layer, which can cover the open channels of the scaffold in the film. The optimized deposition condition of PANI is 400 cycles in their study. Along with the increasing cycles, the rate performance will become poor, which may be caused by the decreasing of utilization efficiency and surface area of PANI. The capacitance of the CNT/PANI hydrogel film at 1 mA·cm^{-2} is 0.68 F·cm^{-2}, which is much higher than many flexible electrodes such as carbon films, conductive polymers and even metallic oxides. The excellent electrochemical performance may be due to the porous microstructure. This structure maintains the facile electrolyte penetration, fast proton exchange, and metallic conductivity, thus ensuring the high electrochemical performance of the flexible film electrodes. The all-solid state flexible SC is assembled by two identical CNT/PANI hydrogel film electrodes and a layer of H_2SO_4/PVA gel electrolyte, which has the sandwich structure. Specific capacitance (Csp) of the capacitor at 1 mA·cm^{-2} is 184.6 mF·cm^{-2}. Even at a current density of up to 5 mA·cm^{-2}, the Csp of this device can retain a high value of 120 mF·cm^{-2}, which indicates that the good rate performance of this SC device. The rolled-up SC is similar to that under pristine condition at 5 mA·cm^{-2} and the Csp values are almost the same. The cyclic performance of the flexible SC of the flexible SC remained almost the same after 500 cycles of charge–discharge. The electrochemical performances are shown in **Figure 3**.

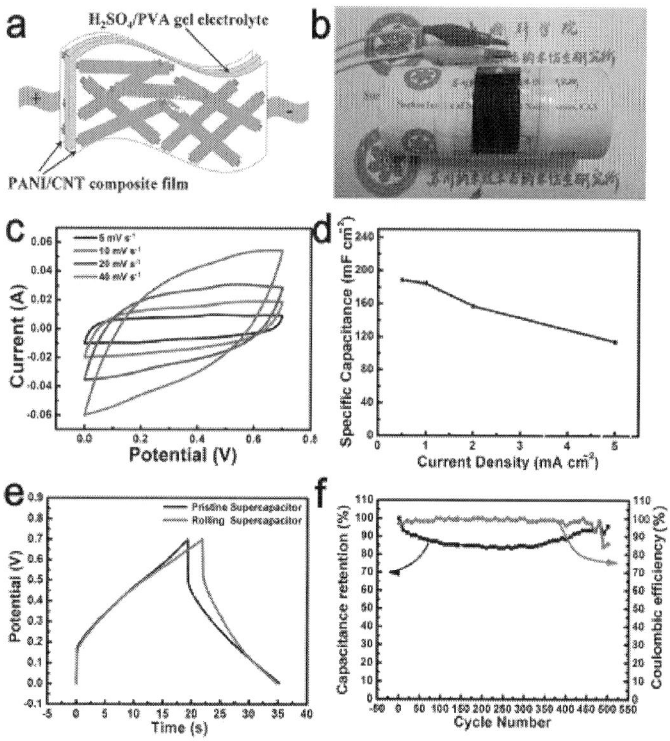

Figure 3. (a) Schematic diagram of the CNT/PANI hydrogel film; (b) the capacitor device image; (c) CV curves and (d) rate performance of the device; (e) galvanostatic charge/discharge curves at 5 mA·cm^{-2}; (f) stability of the device. (Reproduced with permission from Zeng et al. [52]. The Royal Society of Chemistry). © 2015.

MWCNTs–PANI composite fiber remained almost unchanged in diameter. The resulting MWCNTs–PANI composite fibers were flexible and can be easily bent [53]. No obvious damage to the structure was observed, and both tensile strength and electrical conductivity remained almost unchanged after bending for over a hundred cycles, as shown in **Figure 4**. PANI could be stably attached on the aligned MWCNTs possibly due to the interaction between them in the composite fiber. The aligned MWCNTs–PANI composite fibers were then coated with a layer of H_3PO_4–poly(vinyl alcohol) gel electrolyte on the surface. Two modified composite fibers were finally twisted to fabricate the SC wire. Cyclic voltammograms of MWCNTs–PANI (weight percentage of 24%) composite fiber at a scan rate of 10 mV·s^{-1} show a typical rectangular shape, which is corresponding to a double layer capacitor obtained for the bare MWCNT fiber, that can been seen in **Figure 5**. While redox peaks at 0.3–0.4 V that indicates a pseudo-capacitance derived from different oxidation states of PANI are observed for the MWCNTs–PANI composite fiber. The specific capacitances increase to the maximum value (274F·g^{-1}) with the increasing PANI weight percentage from 0 to 40%. Interestingly, further increase of PANI has decreased the specific capacitance, e.g., 245 F·g^{-1} at 70 wt%.

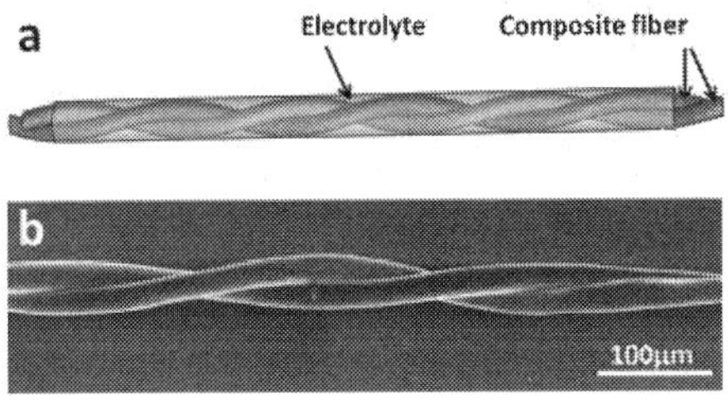

Figure 4. Two aligned MWCNT–PANI composite fibers twisted into a supercapacitor wire. (a) Schematic illustration. (b) Typical scanning electron microscopy image. (Reproduced with permission from Cai et al. [53]. The Royal Society of Chemistry). © 2013.

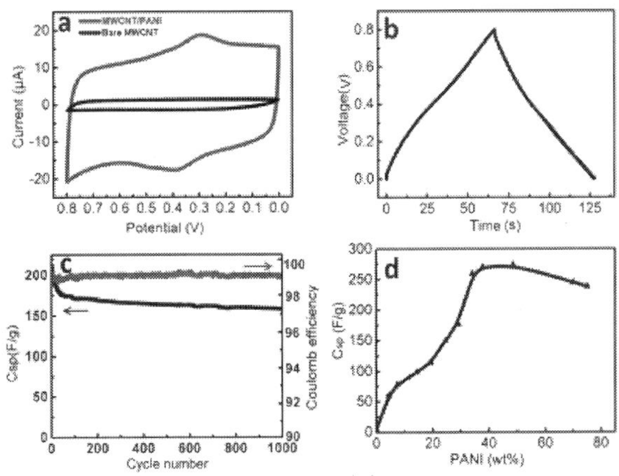

Figure 5. Electrochemical properties of the supercapacitor wires. (a) Cyclic voltammogram and (b) galvanostatic charge–discharge curves based on 24 wt% PANI; (c) dependence of specific capacitance on cycle number and (d) the PANI weight percentage on cycle number of a supercapacitor based on a PANI weight percentage of 34%. (Reproduced with permission from Cai et al. [53]. The Royal Society of Chemistry). © 2013.

Zhang et al. [54] also considered that three-dimensional network structure is conducive to increase the contacting chance between the PANI and the electrolyte, and more active sites of the composite films for faradic reaction and larger SCs than pure PANI film. The highest SCs of 500 F·g^{-1} was obtained for MWCNTs/PANI-stainless-steel sheet composite films when the content of MWCNT is only 0.8 wt%. This kind of composite electrode shows a higher specific capacitance, better power characteristic, better cyclic stability, and more promising for applications in SCs than a pure PANI film electrode.

3. CNTs–PPy

3.1. Electrodeposition of CNTs–PPy

PPy, one of the most interesting conducting polymers, have been studied in SCs due to due to its low cost, high specific capacitance, excellent flexibility and ductility [55]. Over the past few years, various CNT–PANI electrodes have been synthesized by many methods, such as *in situ* chemical polymerization, chemical oxidative polymerization and electrochemical polymerization [57–62].

Electrodeposition or electrochemical polymerizations of CNTs-PPy composites has been extensively reported [63–64]. Electropolymerization of PPy films is an attractive method that can dissolve the pyrrole monomer in a solvent, and form a film at the electrode surface by the applied anodic potential. Carbon nanotubes are suitable for electrode materials because of their good electronic and mechanical properties and high stability [65–67]. Composites of PPy and carbon nanotubes have been prepared by chemical or electrochemical oxidation and the capacitance [26, 68–70] of these composites have also been studied.

Chen et al. carried out electropolymerization of PPy in a three electrode configuration with CNT film as the working electrode, Ag/AgCl electrode as the reference electrode and a graphite rod as the counter electrode [56]. The electrolyte contains 0.1 M pyrrole and 0.05 M Na_2SO_4, and the electrodeposition was conducted at a constant current density of 1 $mA{\cdot}cm^{-2}$ for different time lengths. The fabrication process of CNT/PPy films is illustrated in **Figure 6**. PPy can be electroplated at a constant current density in pyrrole and dodecylbenzenesulfonate (NaDBS) solution in acetonitrile [71].

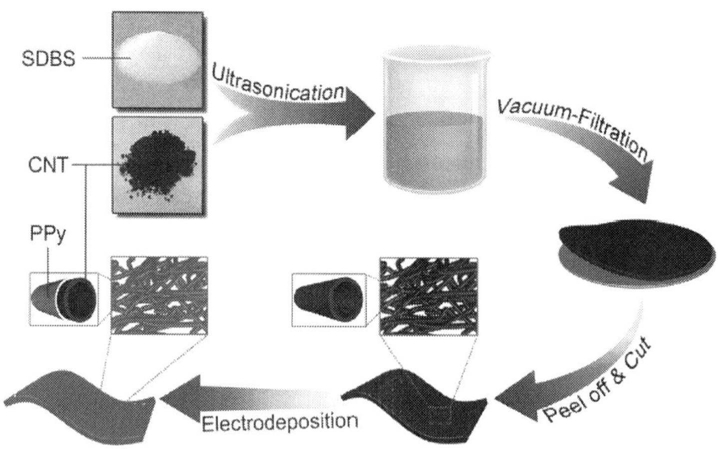

Figure 6. Schematic procedure of the paper-like CNT film and the CNT/PPy film. (Reproduced with permission from Chen et al. [56]). © 2015, Elsevier.

Pulsed potentioamperometric method was also applied by Fang et al. [72]. The MWCNTs membrane with a stainless steel (SS) ring as the current collector was used for the working electrodes. The counter electrode and reference electrode are Pt coil and a Ag/AgCl (saturated KCl) electrode, respectively. Pulsed potentioamperometric deposition of PPy was carried out in aqueous solutions containing pyrrole and 1.0 M KCl. The electrode potential was programmed with a waveform consisting of a potential step to 1.05 V (vs. Ag/AgCl).

The solution of 0.1 M pyrrole and $NaClO_4$ was also used as electrolyte for PPy deposition. In order to deposit the PPy on the CNTs evenly, the CNT films were supported by two stainless steel wire meshes [73–74]. Pretreatment of CNT is very important to improve the electrode capacitance. Generally, CNT should be immersed in 30 wt% H_2O_2 solution for 2 days and in 37 wt% HCl solution for 1 day to remove the amorphous carbon and catalyst particles.

3.2. Structure and electrochemical performance of CNTs-PPy composite electrodes

The detailed structure of the electrodeposited CNTs/PPy film analyzed by transmission electron microcopy show that a layer of amorphous PPy is well coated on the outer surface of the crystalline CNT, which is beneficial to the fast diffusion and migration of ions in the polymer, thereby improving the performance of the CNTs/PPy film [56]. The electrodes show an ultimate tensile strength of 16MPa for this film. The actual capacitance of the robust CNTs/PPy electrode is calculated to be 0.28 F·cm^{-2} at a current density of 1.4 mA·cm^{-2}, which is 5.6 times larger than that of the pure CNTs electrode (0.05 F·cm^{-2}). All-solid-state SCs assembled with the robust CNT/PPy electrodes sandwiching by PVA/H_2SO_4 solid-state electrolyte exhibit

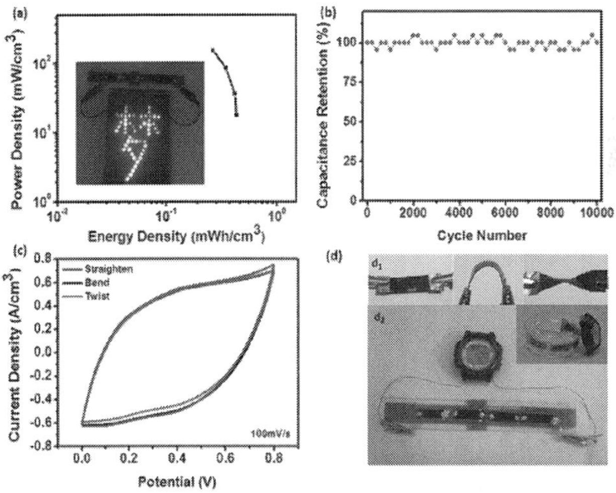

Figure 7. Electrochemical properties of SC device. (a) Ragone plots; (b) Cycle performance at 0.1 A·cm^{-3}; (c) CV curves at 100 mV·s^{-1} under different conditions; (d) Digital photos. (Reproduced with permission from Chen et al. [56]). © 2015, Elsevier.

high electrochemical performance (device volumetric capacitance of 4.9 F·cm^{-3} at 0.05 A·cm^{-3}) and a favorable rate performance (device volumetric capacitance of 3 F·cm^{-3} at 0.5 A·cm^{-3}). The values are higher than those of other solid-state SCs reported previously based on transition metal oxides and CNT [75–82]. The power density of the device achieved 0.15 W·cm^{-3}while the energy density was 0.26 mWh·cm^{-3}. Its cycling stability is excellent that the capacitance retention is larger than 95% after 10,000 cycles. The results are shown in **Figure 7**.

Xu et al. fabricated core–shell structured composite of PPy and CNTs (PPy@CNTs) by electrodeposition, which indicates the PPy is distributed evenly in the CNT fiber [73]. When the content of PPy is 51 wt%, this composite shows a maximum value of 350 F·g^{-1} and has high electrochemical stability under bending and twisting.

Vertically aligned carbon nanotubes (VACNT)-PPy nanocomposites prepared by electro polymerization shows a specific capacitance of 5 mF·cm^{-2}, which is 500% higher than that of as-grown VACNT or pure PPy film [71]. Different from the PANI, the deposition time and the thickness of PPy have no effect on the surface phenomenon. Fast redox reaction on PPy can enhance the energy density in the VACNT-PPy nanocomposite. The VACNT-PPy nanocomposite with 1 min electroplating time also has the lowest leakage current (about 5% of the device current at 0.5 V) compared to uncoated VACNTs and the 4 min electroplated VACNT-PPy nanocomposite. Life cycle testing indicates that the 4 min electrode reached its maximum capacitance after 1000 cycles. After 3000 cycles, the 4 min electroplated VACNT-PPy nanocomposite retained 92% of its maximum capacitance (capacitance at 1000 cycles). But the electrochemical capacitance maintained only 65% for the 1 min electrode over 3000 cycles.

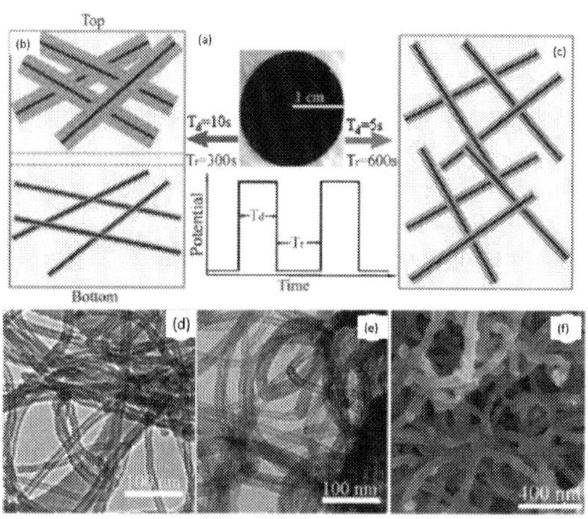

Figure 8. (a) Schematic of MWCNT film (top)and the pulsed electrodeposition condition (bottom); (b) and (c) Schematic of the PPy/MWCNTs under pulse deposition; (d) and (e) TEM image of the MWCNTs and PPy/MWCNTs; (f) SEM image of the PPy/MWCNT membrane (Reproduced with permission from Fang et al. [72]). © 2010, Elsevier.

Pulse deposition of PPy on MWCNT can improve the SCs value about 30 times [72]. In general, short deposition time allows monomers to diffuse from the bulk solution into the space between carbon nanotubes inside the membrane and thus produces more uniform PPy coating. The clogging of the space between carbon nanotubes at the membrane exterior is found to lower the SC value, as shown in **Figure 8**. PPy deposition with short pulses allows the polymer backbone to rearrange to form more ordered structure. Both the structural uniformity and high conductivity of PPy coating help to provide higher pseudo-capacitance in this application.

4. EVA-CNT

EVA is very attractive as a very promising membrane material due to its excellent mechanical property such as good softness, high flexibility and good processability. Additionally, conductive EVA-based membranes are easily obtained by combining EVA with conductive additives (CNT). In previous reports, some materials were often used for the substrates, such as carbon cloth, nickel foam, titanium foil, and et al. However, Lens wiping paper has been paid much more attention for the preparation of the conductive substrates, due to its low cost, lightweight and flexibility [83]. The EVA/40 wt% CNT paper exhibits a high conductivity of 12.1 S·cm^{-1} with good flexibility, lightweight and durability. MnO_2 was electrodeposited on the as-prepared EVA/CNTs paper via potentiostatic technique. The CV curves of 45 s-

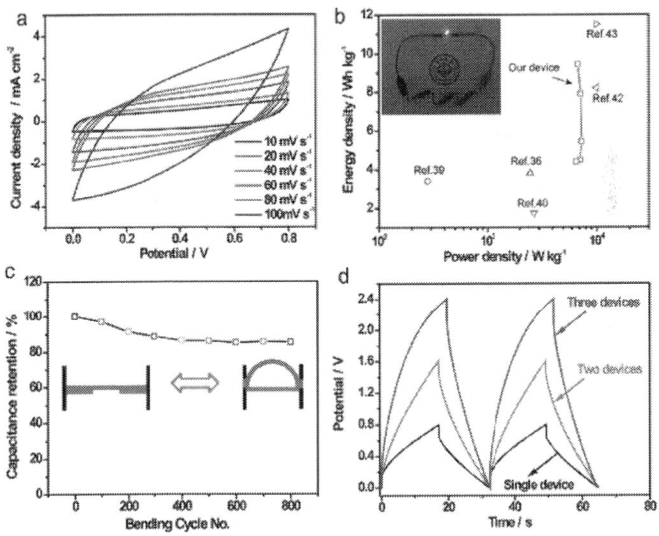

Figure 9. Electrochemical performance of 45s-MnO_2/EVA/40 wt% CNT-SC device. (a) CV curves; (b) Ragone plots; (c) capacitance retention; (d) galvanostatic charge-discharge curves (Reproduced with permission from Zhang et al. [83]). © 2014, Elsevier.

MnO_2/EVA/40 wt% CNT electrodes collected at various scan rates exhibit quasi-rectangular shapes and do not significantly change when the scan rate increases from 10 to 200 mV·s^{-1}, revealing the good capacitive behavior and high-rate capability, which demonstrates that MnO_2 is a good pseudo-capacitive electrode material for SCs. The current density of the MnO_2/EVA/40 wt% CNT electrode increases with the increasing MnO_2 deposition time from 15 to 120 s. The 120 s-MnO_2/EVA/40 wt% CNT yields a good electrochemical capacitance of 0.083 F®cm^{-2} (0.36 mg·cm^{-2}) at the scan rate of 10 mV·s^{-1}. The as-prepared electrode has better performance than the others, because the substrate of EVA/40 wt% CNT has good electrical conductivity (3.7 Ω) and rough. With increasing of the deposition amount of MnO_2, the charge-transfer resistance can increase to 22.9 Ω. If the deposition time of MnO_2 is shortened, a high capacitance will be achieved, which has been confirmed by our experiments [83], as shown in **Figure 9**.

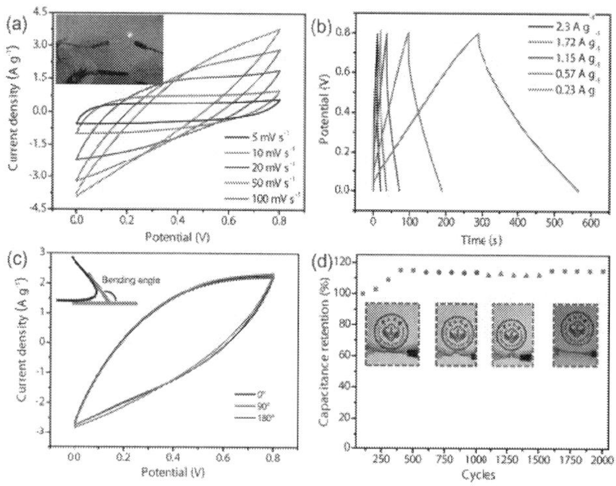

Figure 10. Electrochemical performance of MnO_2/EBC supercapacitor. (a) CV curves; (b) galvanostatic charge/discharge curves; (c) CV curves at different bending angles; (d) cycle stability at stress–recovery states. (Reproduced with permission from Zhang et al. [84]). © 2013, The Royal Society of Chemistry.

We also report, for the first time, the synthesis of conductive flexible membranes of EVA composites filled with carbon black (CB) and CNT and their application as flexible substrates in SCs [84]. We found that the prepared EVA(49.5wt%)-CB(49.5wt%)-CNT(1wt%) (denoted as ECB) showed the best flexibility and electrical conductivity of 0.17 S·cm^{-1}. To demonstrate the feasibility of the ECB substrate as a conductive substrate for flexible SCs, MnO_2 was grown on the surface of the as-prepared membrane substrate by a simple electrochemical deposition method for 20 min. The ECB substrates after being coated with a layer of MnO_2 are still well flexible, and can even be folded and twisted without any destruction. Electrochemical measurements demonstrated that the MnO_2/ECB composite exhibited a high specific capacitance of 169.5 F·g^{-1}, which is a 2- and 2.5-fold enhancement as compared to the MnO_2/FTO and

MnO_2/PETI composites. The shape of charge/discharge curves of the MnO_2/EBC electrodes is more symmetrical and their discharge time is substantially more prolonged than the other electrodes, such as the MnO_2/Ti, MnO_2/FTO, MnO_2/PETI. The specific capacitance of the MnO_2/ECB electrode increased to 111% during the first 1100 cycles and was almost constant with slight fluctuations during the subsequent 900 cycles. The power density of 10 950 $W{\cdot}kg^{-1}$ and energy density of 30.4 W $h{\cdot}kg^{-1}$ were achieved by the MnO_2/EBC electrode. By assembling two pieces of the MnO_2/EBC composite with a separator, the SC devices were assembled. The device could light a red LED (work voltage: 1.5 V) well for about 10 s after being charged at 2.5 V for 10 s [84]. The bending of the electrodes has no obvious effect on the electrochemical performance, as shown in **Figure 10**. Similarly, stress-recovery cannot affect the cycle stability obviously, with only less than 5% change. In short, structure of the electrodes will not be destroyed by the bending and stress states. So many good properties of MnO_2/ECB electrode will make it applicable in flexible SCs widely spread in the future.

5. Summary

In this chapter, we summarized recent advances in the development of composite electrode materials composed of CNTs and conductive polymer (PANI/PPy/EVA) prepared by electrodeposition. These kinds of composite electrodes can increase energy density without sacrificing their power density. Also, to avoid the use of current collectors, conductive additives and binders, the electrodes are designed to have flexibility that can reduce the total weight of the electrodes. Electrodeposition of conductive polymers onto CNT films can prevent the transmission and deposition of aniline/pyrrole molecules into the internal part of the film. However, electrodeposition often blocks electrolyte channels at the outer surface of CNT film which suffers from ECP aggregation and high resistances due to poor CNT interconnection. High loading mass of conductive polymers and cohesion strength between CNT and conductive polymers are still an important challenge.

Acknowledgements

This work was supported by the Natural Science Foundations of China (21576230, 21403306 and 21273290, 21476271),and the Natural Science Foundations of Guangdong Province (S2013030013474, 2015B010118002),

Author details

Chenzhong Yao[1], Bohui Wei[1] and Yexiang Tong[2,3*]

*Address all correspondence to: chedhx@mail.sysu.edu.cn

1 Department of Applied Chemistry, Yuncheng University, Yuncheng, China

2 MOE of the Key Laboratory of Bioinorganic and Synthetic Chemistry, KLGHEI of Environment and Energy Chemistry, The Key Lab of Low-carbon Chem & Energy Conservation of Guangdong Province, School of Chemistry and Chemical Engineering, Sun Yat-sen University, Guangzhou, China

3 Department of Chemistry, College of Science, Shantou University, Shantou, China

References

[1] B. E. Conway, Electrochemical supercapacitors: Scientific fundamentals and technological applications, Plenum Press, New York, 1999.

[2] R. Kötz, M. Carlen, Principles and applications of electrochemical capacitors, Electrochim. Acta, 2000, 45: 2483–2498.

[3] A. G. Pandolfo, A. F. Hollenkamp, Carbon properties and their role in supercapacitors, J. Power Sources, 2006, 157: 11–27.

[4] J. Li, X. Q. Cheng, A. Shashurin, M. Keidar, Review of electrochemical capacitors based on carbon nanotubes and graphene, Graphene, 2012, 1: 1–13.

[5] S. Bose, T. Kuila, A. K. Mishra, R. Rajasekar, N. H. Kim, J. H. Lee, Carbon-based nanostructured materials and their composites as supercapacitor electrodes, J. Mater. Chem., 2012, 22: 767–84.

[6] M. Inagaki, H. Konno, O. Tanaike, Carbon materials for electrochemical capacitors, J. Power Sources, 2010, 195: 7880–7903.

[7] Q. Cheng, J. Tang, J. Ma, H. Zhang, N. Shinya, L. C. Qin, Graphene and carbon nanotube composite electrodes for supercapacitors with ultra-high energy density, Phys. Chem. Chem. Phys., 2011, 13: 17615–17624.

[8] D. N. Futaba, K. Hata, T. Yamada, T. Hiraoka, Y. Hayamizu, Y. Kakudate, O. Tanaike, H. Hatori, M. Yumura, S. Iijima, Shape-engineerable and highly densely packed single-walled carbon nanotubes and their application as supercapacitor electrodes, Nat. Mater., 2006, 5: 987–994.

[9] Y. Wang, Z. Q. Shi, Y. Huang, Y. F. Ma, C. Y. Wang, M. M. Chen, Y. S. Chen, Supercapacitor devices based on graphene materials, J. Phys. Chem. C, 2009, 113: 13103–13107.

[10] M.D. Stoller, S. J. Park, Y. W. Zhu, J.H. An, R. S. Ruoff, Graphene-based ultracapacitors, Nano Lett., 2008, 8: 3498–3502.

[11] D. W. Wang, F. Li, J. P. Zhao, W. C. Ren, Z. G. Chen, J. Tan, Z. S. Wu, I. Gentle, G. Q. Lu, H. M. Cheng, Fabrication of graphene/polyaniline composite paper via *In Situ*

anodic electropolymerization for high-performance flexible electrode, ACS Nano, 2009, 3: 1745–1752.

[12] A. E. Fischer, K. A. Pettigrew, D. R. Rolison, R. M. Stroud, J. W. Long, Incorporation of homogeneous, nanoscale MnO_2 within ultraporous carbon structures via self-limiting electroless deposition:? implications for electrochemical capacitors, Nano Lett., 2007, 7: 281–286.

[13] J. K. Chang, M. T. Lee, W. T. Tsai, M. J. Deng, H. F. Cheng, I. W. Sun, Pseudocapacitive mechanism of manganese oxide in 1-ethyl-3-methylimidazolium thiocyanate ionic liquid electrolyte studied using X-ray photoelectron spectroscopy, Langmuir, 2009, 25: 11955–11960.

[14] B. Babakhani, D.G. Ivey, Anodic deposition of manganese oxide electrodes with rod-like structures for application as electrochemical capacitors, J. Power Sources, 2010, 195: 2110–2117

[15] Q. Cheng, J. Tang, J. Ma, H. Zhang, N. Shinya, L. C. Qin, Graphene and nanostructured MnO_2 composite electrodes for supercapacitors, Carbon, 2011, 49: 2917–2925.

[16] J. Yan, Z. J. Fan, T. Wei, W. Z. Qian, M. L. Zhang, F. Wei, Fast and reversible surface redox reaction of graphene–MnO_2 composites as supercapacitor electrodes, Carbon, 2010, 48: 3825–3833.

[17] L. Li, R. Li, S. Gai, S. Ding, F. He, M. Zhang, P. Yang, MnO_2 nanosheets grown on nitrogen-doped hollow carbon shells as a high-performance electrode for asymmetric supercapacitors, Chem. Eur. J., 2015, 21: 7119–7126.

[18] G. R. Li, Z. P. Feng, Y. N. Ou, D. C. Wu, R. W. Fu, Y. X. Tong, Mesoporous MnO_2/carbon aerogel composites as promising electrode materials for high-performance supercapacitors, Langmuir, 2010, 26: 2209–2213.

[19] G. P. Wang, L. Zhang, J. J. Zhang, A review of electrode materials for electrochemical supercapacitors, Chem. Soc. Rev., 2012, 41: 797–828.

[20] P. C. Chen, H. T. Chen, J. Qiu, C.W. Zhou, Inkjet printing of single-walled carbon nanotube/RuO_2 nanowire supercapacitors on cloth fabrics and flexible substrates, Nano Res., 2010, 3: 594–603.

[21] J. Xu, Q.F. Wang, X.W. Wang, Q. Y. Xiang, B. Liang, D. Chen, G. Z. Shen, Flexible asymmetric supercapacitors based upon Co_9S_8nanorod//Co_3O_4@RuO_2 nanosheet arrays on carbon cloth, ACS Nano, 2013, 7: 5453–5462.

[22] I. Ryu, M. H. Yang, H. Kwon, H. K. Park, Y.R. Do, S. B. Lee, S. Yim, Coaxial RuO_2–ITO nanopillars for transparent supercapacitor application, Langmuir, 2014, 30: 1704–1709.

[23] C. C. Hu, C. W. Wang, K. H. Chang, M. G. Chen, Anodic composite deposition of RuO_2/reduced graphene oxide/carbon nanotube for advanced supercapacitors, Nanotechnology, 2015, 26: 274004.

[24] S. Cho, A. I. Inamdar, S. M. Pawar, J. Kim, Y. Jo, J. Han, H. Kim, W. Jung, H. B. Kim, H. Im, High performance electrochemical supercapacitors based on electrodeposited RuO_2 on graphene sheet electrodes, J. Nanoelectron. Optoelectron., 2015, 10: 286–289.

[25] T. Zhai, X. H. Lu, F. X. Wang, H. Xia, Y. X. Tong, MnO_2 nanomaterials for flexible supercapacitors: performance enhancement via intrinsic and extrinsic modification, Nanoscale Horiz., 2016, 1: 109–124.

[26] X. F. Lu, G. R. Li, Y. X. Tong, A review of negative electrode materials for electrochemical supercapacitors, Sci. China Tech. Sci., 2015, 58: 1–10.

[27] X. F. Lu, Q. Li, J. X. Feng, P. P. Fang, X. H. Lu, P. Liu, G. R. Li, Y. X. Tong, The effective design and controlled synthesis of nanomaterials for supercapacitors, Sci. Sinica Chim., 2014, 44:1255–1268.

[28] J. Li, L. Cui, X. G. Zhang, Preparation and electrochemistry of one-dimensional nanostructured MnO_2/PPy composite for electrochemical capacitor, Appl. Surf. Sci., 2010, 256: 4339–4343.

[29] G. A. Snook, P. Kao, A. S. Best, Conducting-polymer-based supercapacitor devices and electrodes, J. Power Sources, 2011, 196: 1–12.

[30] I. Shown, A. Ganguly, L. C. Chen, K. H. Chen, Conducting polymer-based flexible supercapacitor, Energy Sci. Eng., 2015, 3: 2–26.

[31] R. Ramya, R. Sivasubramanian, M. V. Sangaranarayanan, Conducting polymers-based electrochemical supercapacitors—progress and prospects, Electrochim. Acta, 2013, 101: 109–129.

[32] W. Chen, C. Xia, R. B. Rakhi, H. N. Alshareef, A general approach toward enhancement of pseudocapacitive performance of conducting polymers by redox-active electrolytes, J. Power Sources, 2014, 267: 521–526.

[33] N. Kurra, R. Wang, H. N. Alshareef, All conducting polymer electrodes for asymmetric solid-state supercapacitors, J. Mater. Chem. A, 2015, 3: 7368–7374.

[34] Z. Wang, P. Tammela, P. Zhang, M. Strømme, L. Nyholm, High areal and volumetric capacity sustainable all-polymer paper-based supercapacitors, J. Mater. Chem. A, 2014, 2: 16761–16769.

[35] Y. T. Weng, H. A. Pan, N. L. Wu , G. Z. Chen, Titanium carbide nanocube core induced interfacial growth of crystalline polypyrrole/polyvinyl alcohol lamellar shell for wide-temperature range supercapacitors, J. Power Sources, 2015, 274: 1118–1125.

[36] Z. Z. Zhu, G. C. Wang, M.Q. Sun, X. W. Li, C. Z. Li, Fabrication and electrochemical characterization of polyaniline nanorods modified with sulfonated carbon nanotubes for supercapacitor applications, Electrochim. Acta, 2011, 56: 1366–1372.

[37] R. K. Sharma, L. Zhai, Multiwall carbon nanotube supported poly(3,4-ethylenedioxythiophene)/manganese oxide nanocomposite electrode for super-capacitors, Electrochim. Acta, 2009, 54: 7148–7155.

[38] R. Y. Suckeveriene, E. Zelikman, G. Mechrez, M. Narkis, Literature review: Conducting carbon nanotube/polyaniline nanocomposites. Rev. Chem. Eng., 2011; 27:15–21.

[39] M. A. Bavio, G. G. Acosta, T. Kessler, Synthesis and characterization of polyaniline and polyaniline-carbon nanotubes nanostructures for electrochemical supercapacitors, J. Power Sources, 2014, 245: 475–481

[40] G. Otrokhov, D. Pankratov, G. Shumakovich, M. Khlupova, Y. Zeifman, I. Vasil'eva, O. Morozova, A. Yaropolov, Enzymatic synthesis of polyaniline/multi-walled carbon nanotube composite with core shell structure and its electrochemical characterization for supercapacitor application, Electrochim. Acta, 2014, 123: 151–157.

[41] Q. Cheng, J. Tang, N. Shiny, L.C. Qin, Polyaniline modified graphene and carbon nanotube composite electrode for asymmetric supercapacitors of high energy density, J. Power Sources, 2013, 241: 423–428.

[42] F. Yang, M. Xu, S.J. Bao, Q.Q. Sun, MnO_2-assisted fabrication of PANI/MWCNT composite and its application as a supercapacitor, RSC Adv., 2014, 4: 33569–33573.

[43] K. Immonen, K. Nättinen, J. Sarlin, J. Hartikainen, Conductive plastics with hybrid materials. J. Appl. Polym. Sci., 2009, 114: 1494–1502.

[44] C. Jin, T.C. Nagaiah, W. Xia, B. Spliethoff, S. Wang, M. Bron, W. Schuhmannb, M. Muhler, Metal-free and electrocatalytically active nitrogen-doped carbon nanotubes synthesized by coating with polyaniline, Nanoscale 2010, 2: 981–987.

[45] V. Branzoi, F. Branzoi, L. Pilan, Electrochemical fabrication and capacitance of composite films of carbon nanotubes and polyaniline, Surf. Interface Anal., 2010, 42: 1266–1270.

[46] G. M. do Nascimento, T. B. Silva, P. Corio, M. S. Dresselhaus, Charge-transfer behaviour of polyaniline single wall carbon nanotubes nanocomposites monitored by resonance Raman spectroscopy, J. Raman. Spetros., 2011, 41: 1587–1593.

[47] A. A. Mikhaylova, E.K. Tusseeva, N. A. Mayorova, A. Y. Rychagov, Y. M. Volfkovich, A.V. Krestinin, O. A. Khazova, Single-walled carbon nanotubes and their composites with polyaniline. Structure, catalytic and capacitive properties as applied to fuel cells and supercapacitors, Electrochim. Acta, 2011, 56: 3656–3665.

[48] S. Srivastava, S. S. Sharma, S. Agrawal, S. Kumar, M. Singha, Y.K. Vijay, Study of chemiresistor type CNT doped polyaniline gas sensor, Syn. Met., 2010, 160: 529–534.

[49] M. Ding, Y. Tang, P. Gou, M. J. Reber, A. Star, Chemical sensing with polyaniline coated single-walled carbon nanotubes. Adv. Mater 2011; 23: 536–540.

[50] Y. Liao, C. Zhang, Y. Zhang, V. Strong, J. Tang, X. G. Li, K. Kalantar-zadeh, E. M. V. Hoek, K. L. Wang, R. B. Kaner, Carbon nanotube/polyaniline composite nanofibers: facile synthesis and chemosensors. Nano Lett., 2011, 11: 954–959.

[51] M. Yu, Y. Zhang, Y. Zeng, M. S. Balogun, K. Mai, Z. Zhang, X. Lu, Y. Tong, Water surface assisted synthesis of large-scale carbon nanotube film for high-performance and stretchable supercapacitors, Adv. Mater., 2014, 26: 4724–4729.

[52] S. Zeng, H. Chen, F. Cai, Y. Kang, M. Chen, Q. Li, Electrochemical fabrication of carbon nanotube/polyaniline hydrogel film for all-solid-state flexible supercapacitor with high areal capacitance, J. Mater. Chem. A, 2015, 3: 23864–23870.

[53] Z. Cai, L. Li, J. Ren, L. Qiu, H. Lin, H. Peng, Flexible, weavable and efficient microsupercapacitor wires based on polyaniline composite fibers incorporated with aligned carbon nanotubes, J. Mater. Chem. A, 2013, 1: 258–261.

[54] J. Zhang , L. B. Kong, B. Wang, Y. C. Luo, L. Kang, In-situ electrochemical polymerization of multi-walled carbon nanotube/polyaniline composite films for electrochemical supercapacitors, Synth. Met., 2009, 159: 260–266.

[55] C. Tran, R. Singhal, D. Lawrence, V. Kalra, Polyaniline-coated freestanding porous carbon nanofibers as efficient hybrid electrodes for supercapacitors, J. Power Sources, 2015, 293: 373–379.

[56] Y. Chen, L. Du, P. Yang, P. Sun, X. Yu, W. Mai, Significantly enhanced robustness and electrochemical performance of flexible carbon nanotube-based supercapacitors by electrode positing polypyrrole, J. Power Sources, 2015, 287: 68–74.

[57] K. H. An, K.K. Jeon, J. K. Heo, S.C. Lim, D. J. Bae, Y. H. Lee, High-capacitance supercapacitor using a nanocomposite electrode of single-walled carbon nanotube and polypyrrole, J. Electrochem. Soc., 2002, 149: A1058–A1062.

[58] C. Peng, S. Zhang, D. Jewell, G. Z. Chen, Carbon nanotube and conducting polymer composites for supercapacitors, Prog. Nat. Sci., 2008, 18: 777–788.

[59] S. Biswas, L. T. Drzal, Multilayered nanoarchitecture of graphene nanosheets and polypyrrole nanowires for high performance supercapacitor electrodes, Chem. Mater., 2010, 22:5667–5671.

[60] X. Lu, H. Dou, C. Yuan, S. Yang, L. Hao, F. Zhang, L. Shen, L. Zhang, X. Zhang, Polypyrrole/carbon nanotube nanocomposite enhanced the electrochemical capacitance of flexible graphene film for supercapacitors, J. Power Sources, 2012, 197: 319–324.

[61] L. Yang, Z. Shi, W. Yang, Polypyrrole directly bonded to air-plasma activated carbon nanotube as electrode materials for high-performance supercapacitor, Electrochim. Acta, 2015, 153: 76–82.

[62] Y. Yesi, I. Shown, A. Ganguly, T. T. Ngo, L. C. Chen, K. H. Chen, Directly-grown hierarchical carbon nanotube@polypyrrole core–shell hybrid for high-performance flexible supercapacitors, ChemSusChem, 2016, 9: 370–378.

[63] D. P. Dubal, S. H. Lee, J. G. Kim, W. B. Kim, C. D. Lokhande, J. Mater. Chem. Porous polypyrrole clusters prepared by electropolymerization for a high performance supercapacitors, 2012, 22: 3044–3052.

[64] A. Faye, G. Dione, M. M. Dieng, J. J. Aaron, H. Cachet, C. Cachet, J. Appl. Usefulness of a composite electrode with a carbon surface modified by electrosynthesized polypyrrole for supercapacitor applications, Electrochem., 2010, 40: 1925–1931

[65] S. Iijima, Helical microtubules of graphitic carbon, Nature, 1991, 354: 56–58.

[66] M. R. Falvo, G. J. Clary, R. M. Taylor, V. Chi, F. P. Brooks, S. Washburn, R. Superfine, Bending and buckling of carbon nanotubes under large strain, Nature, 1997, 389: 582–584.

[67] X. Lu, M. Yu, G. Wang, Y. Tong, Y. Li, Flexible solid-state supercapacitors: design, fabrication and applications, Energy Environ. Sci., 2014, 7, 2160–2181

[68] H. L. Filiatrault, G. C. Porteous, R. S. Carmichael, G. J. E. Davidson, T. B. Carmichael, "Elastomeric emissive materials: stretchable light-emitting electrochemical cells using an elastomeric emissive material," Adv. Mater., 2012, 24: 2673–2678.

[69] C. Yan, P. S. Lee, Stretchable energy storage and conversion devices, Small, 2014, 10: 3443–3460.

[70] B. Yue, C. Wang, X. Ding, G. G. Wallace, Polypyrrole coated nylon lycra fabric as stretchable electrode for supercapacitor applications, Electrochim. Acta 68 2012, 68: 18–24.

[71] R. Warren, F. Sammoura, K. S. Teh, A. Kozinda, X. Zang, L. Lin, Electrochemically synthesized and vertically aligned carbon nanotube–polypyrrole nanolayers for high energy storage devices, Sens. Actuators A, 2015, 231: 65–73.

[72] Y. Fang, J. Liu, D.J. Yu, J. P. Wicksted, K. Kalkan, C. O. Topal, B.N. Flanders, J. Wu, J. Li, Self-supported supercapacitor membranes: polypyrrole-coated carbon nanotube networks enabled by pulsed electrodeposition, J. Power Sources, 2010, 195: 674–679.

[73] R. Xu, J. Wei, F. Guo, X. Cui, T. Zhang, H. Zhu, K. Wang, D. Wu, Highly conductive, twistable and bendable polypyrrole–carbon nanotube fiber for efficient supercapacitor electrodes, RSC Adv., 2015, 5: 22015–22021.

[74] W. Zhao, S. Wang, C. Wang, S. Wu, W. Xu, M. Zou, A. Ouyang, A. Cao, Y. Li, Double polymer sheathed carbon nanotube supercapacitors show enhanced cycling stability, Nanoscale, 2016, 8: 626–633.

[75] M. F. El-Kady, V. Strong, S. Dubin, R. B. Kaner, Laser scribing of high-performance and flexible graphene-based electrochemical capacitors, Science, 2012, 335: 1326–1330.

[76] X. Xiao, T. Li, P. Yang, Y. Gao, H. Jin, W. Ni, W. Zhan, X. Zhang, Y. Cao, J. Zhong, L. Gong, W.C. Yen, W. Mai, J. Chen, K. Huo, Y.L. Chueh, Z. L. Wang, J. Zhou, Fiber-based all-solid-state flexible supercapacitors for self-powered systems, ACS Nano, 2012, 6: 9200–9206.

[77] X. Xiao, X. Peng, H. Jin, T. Li, C. Zhang, B. Gao, B. Hu, K. Huo, J. Zhou, Freestanding mesoporous VN/CNT hybrid electrodes for flexible all-solid-state supercapacitors, Adv. Mater., 2013, 25: 5091–5097.

[78] X. Lu, M. Yu, G. Wang, T. Zhai, S. Xie, Y. Ling, Y. Tong, Y. Li, H-TiO_2@MnO_2//H-TiO_2@C core–shell nanowires for high performance and flexible asymmetric supercapacitors, Adv. Mater., 2013, 25: 267–272.

[79] X. Xiao, T. Ding, L. Yuan, Y. Shen, Q. Zhong, X. Zhang, Y. Cao, B. Hu, T. Zhai, L. Gong, J. Chen, Y. Tong, J. Zhou, Z.L. Wang, WO_{3-x}/MoO_{3-x} core/shell nanowires on carbon fabric as an anode for all-solid-state asymmetric supercapacitors, Adv. Energy Mater., 2012, 2: 1328–1332.

[80] X. Xiao, T. Li, Z. Peng, H. Jin, Q. Zhong, Q. Hu, B. Yao, Q. Luo, C. Zhang, L. Gong, J. Chen, Y. Gogotsi, J. Zhou, Nano Energy, 2014, 6: 1–9.

[81] W. Wang, W. Liu, Y. Zeng, Y. Han, M. Yu, X. Lu, Y. Tong, A novel exfoliation strategy to significantly boost the energy storage capability of commercial carbon cloth, Adv. Mater., 2015, 27: 3572–3578.

[82] Y. Zeng, Y. Han, Y. Zhao, Y. Zeng, M. Yu, Y. Liu, H. Tang, Y. Tong, X. Lu, Advanced Ti-doped Fe_2O_3@PEDOT core/shell anode for high-energy asymmetric supercapacitors, Adv. Energy Mater., 2015, 5, 1402176.

[83] Z. Zhang, W. Wang, C. Li, L. Wei, X. Chen, Y. Tong, K. Mai, X. Lu, Highly conductive ethyleneevinyl acetate copolymer/carbon nanotube paper for lightweight and flexible supercapacitors, J. Power Sources, 2014, 248: 1248–1255.

[84] Z. Zhang, T. Zhai, X. Lu, M. Yu, Y. Tong, K. Mai, Conductive membranes of EVA filled with carbon black and carbon nanotubes for flexible energy-storage devices, J . Mater. Chem. A, 2013, 1: 505–509.

Chapter 17

Carbon Nanotube-Polymer Composites for Energy Storage Applications

Du Yuan, Yun Gunag Zhu and Chuankun Jia

Additional information is available at the end of the chapter

http://dx.doi.org/10.5772/63057

Abstract

Renewable energy has attracted growing attention due to energy crisis and environmental concern. The renewable power is featured by its intermittent and fluctuating nature, which requires large-scale electrical energy storage devices for dispatch and integration. Among the current energy storage technologies (e.g., pumped hydro, flywheel, compressed air, superconducting magnet, electrochemical systems), electrochemical storage technologies or batteries that reversely convert electrical energy into chemical energy demonstrate extremely great potential in the stationary and transportation applications. Scale determines the device type. Redox flow batteries (RFBs), as one of the large-scale types, are capable of complying with power station. Meanwhile, portable smart electronic devices promote the development of small-scale energy storage systems, such as Li-ion batteries and supercapacitors. With delicate device configuration, materials play critical roles on pursuing advancing performance, in terms of electrode, current collector, and separator. Carbon nanotube (CNT)/polymer composites exhibit promising potentials in the above key entities, which integrate the merits of conductivity, mechanical strength, flexibility, and cost. Therefore, this chapter is devoted to the design and application of carbon nanotube/polymer composites in different kinds of energy storage systems.

Keywords: carbon nanotube, carbon nanotube/polymer composite, energy storage, lithium battery, supercapacitor, flow battery

1. Introduction

Electricity generation from renewable sources, such as solar and wind, has been increasing in the context of growing environmental concerns and constraints of the fossil fuels and hydrocarbon reserves [1, 2]. The renewable power is featured by the intermittent and fluctuating nature, which requires large-scale electrical energy storage devices to dispatch and integrate the renewable power. Among the current storage technologies (e.g., pumped hydro, flywheel, compressed air, superconducting magnet, electrochemical systems), electrochemical storage technologies or batteries that reversely convert electrical energy into chemical energy demonstrate extremely great potential in the stationary and transportation applications. Among the electrochemical storage technologies, redox flow battery is one of the promising technologies for large-scale energy storage application. Meanwhile, portable smart electronic devices promote the development of small-scale energy storage systems, such as Li-ion batteries and supercapacitors. Key elements, such as membranes and electrodes, are crucial for their application in different batteries. Polymer/carbon nanotubes (CNTs)-composited materials have been successfully used in above-mentioned energy storage batteries. In the following sections, we introduce the application of polymer/CNT for different electrochemical storages.

2. Carbon nanotube/polymer-composited membrane for redox flow batteries

2.1. Redox flow battery

Redox flow batteries (RFBs) have widely received attention for both medium-and large-scale storage needs in the past decades. As schematically shown in **Figure 1**, a typical RFB system consists of two external reservoirs that store soluble charged/discharged electrolytes, two electrodes, and a membrane that effectively separates the anode and cathode electrolytes, while allowing the transport of non-active ions (e.g., H^+, Cl^-) to complete the reaction and a flow circulation system. In RFB, the chemical energy is converted to electrical energy (discharge) or vice versa (charge) when liquid electrolytes are pumped from storage tanks to flow-through electrodes [3, 4].

Unlike conventional batteries that store energy in electrode materials, RFB systems store energy in anolyte and catholyte solutions. As such, the most prominent advantage of RFB system is that the power and energy capacity of one RFB system can be designed separately. The power rating of the system is determined by the number of cells in the cell stack and the area of electrode, whereas the energy storage capacity is controlled by the concentration of the active redox couple species in electrolytes and the solution volume. As a result, it is possible to independently optimize the RFB's storage capacity and the power output for different applications. Both energy and power capacity of RFB can be easily increased by simply adding the volume of electrolytes and the numbers of cells in the stack, respectively. Consequently, the incremental cost of each additional energy storage capacity unit is lower than other types of battery technologies. In addition, the process of manufacturing a modular production of RFB stacks can be simplified by omitting the cell-making process, and thus, the cost would be reduced compared with solid-state batteries (e.g., Li-ion and NiMH). Furthermore, unlike practical solid-state battery, the simple design of RFB system can minimize the energy density loss resulted from negative impact of inactive materials because the volume of energy-bearing reactive electrolyte can be independently scaled up [5–7].

With the unique mechanism of the RFB system, there are some additional advantages of RFB systems over other battery systems: quick response, long cycle life, no high operational temperature, deep-discharge capacity, simple electrode reactions, low maintenance cost, safety.

2.2. History of RFB

The modern RFB system (iron-chromium RFB) was proposed in the 1970s by Lawrence Thaller at the National Aeronautics and Space Administration (NASA). Since then a number of other redox flow batteries have been invented and developed. Based on different redox couples used, these RFBs can be divided into the following: polysulfide/bromine flow battery (PSB), all vanadium redox flow battery (VRB), vanadium-bromine flow battery (VBB), vanadium-cerium flow battery (VCB), vanadium/iron redox flow battery (VIB), vanadium/air flow battery, zinc/bromine flow battery (ZBB), zinc-cerium flow battery (ZCB), iron/chromium flow battery (ICB), iron-titanium redox flow battery (ITB), soluble lead-acid battery, etc. Among these various RFBs, ZBB, ICB, VRB, and PSB have been demonstrated at a few hundred kW and even multi-MW levels for different applications [8]. Until now among these flow batteries, VRB is the most widely studied and successfully demonstrated one due to its low crossover of vanadium ions and good performances, which will be discussed in details.

2.3. Vanadium redox flow battery

The vanadium redox flow battery (VRB) proposed by Skyllas-Kazacos et al. shows great promise as a new, highly efficient, and reliable energy storage system for a wide range of applications such as solar cell and wind turbine generators, remote area power systems, emergency back-up applications, and uninterruptible power sources (UPS) [9]. The VRB employs V(II)/V(III) and V(IV)/V(V) redox couples in its anolyte and catholyte, respectively, and H_2SO_4 as the supporting electrolyte [10]. The ion-exchange membrane responsible for separation and proton transportation is crucial for a VRB. And the carbon felt usually used as the electrodes provides the electrochemically active surface for the redox reaction to take place. The electron exchange occurs at the electrodes in the aqueous phase according to the following reactions:

Cathodic side:

$$VO^{2+}+H_2O \Leftrightarrow VO_2^{+}+2H^{+}+e \quad E^0=0.99V \tag{1}$$

Anodic side:

$$V^{3+}+e \Leftrightarrow V^{2+} \qquad E^0=-0.26V \tag{2}$$

Overall reaction:

$$VO^{2+}+H_2O+V^{3+} \Leftrightarrow VO_2^{+}+2H^{+}+V^{2+} \quad E^0=1.25V \tag{3}$$

Since invention, the VRB has received extensive attention around the world, especially in Japan, Europe, the USA, and China. Though many systems have been installed in a variety of fields, the practical application of the VRB is still hindered by the high cost and low energy density. Accordingly, developing a chemically stable, low-cost, and high-quality membrane is one of the promising ways to overcome these two challenges of VRB.

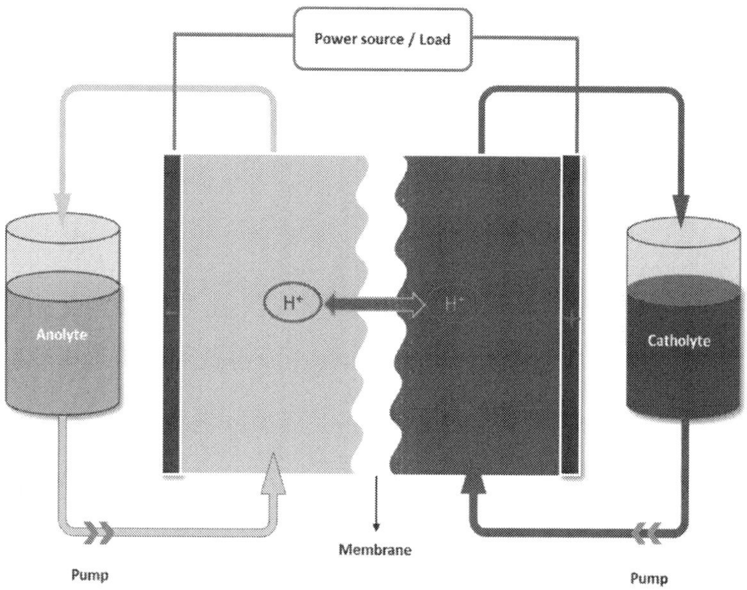

Figure 1. Schematic of vanadium redox flow battery.

2.4. Membranes for VRB application

In VRB, an ion-exchange membrane plays a critical role on separating the anode and cathode electrolytes effectively, namely, no migration or a very low permeability of vanadium ions, while allowing the transport of ions to complete the circuit during passage of current. An ideal membrane should possess high proton conductivity, good chemical stability, high tensile strength, low permeability for vanadium ions, and low cost [11]. Among ion exchange membranes, perfluorosulfonic acid (PFSA) membranes, such as Nafion (DuPont), are most commonly used due to their high chemical stabilities and proton conductivities. However, PFSA membranes usually suffer from high cost and especially high permeability of vanadium ions. To reduce the migration of V ions, many approaches such as doping of inorganic materials, surface modification, and layering have been employed to modify the Nafion membranes. Although some important progress proceeds with lower vanadium species in modified Nafion than the pristine one, the cost is still far from afforded that limits the commercialization [12, 13]. It is therefore of equivalent significance to reduce the cost. Recently, non-perfluoronated polymers such as sulfonated poly(ether ether ketone) (SPEEK) and sulfonated poly(ether sulfone) (SPES) have been received wide attention for VRB application due to their low vanadium ion permeability and low cost [14]. The ion selectivity and mechanical strength are highly dependent on the degree of sulfonation (DS). SPEEK membranes with high DS possess high ion selectivity and low areal resistance, but poor mechanical property, high vanadium ions permeability, and stability. On the other hand, SPEEK membranes with low DS exhibit good mechanical strength and stability but poor ion selectivity and high areal resistance [15]. Consequently, it is important to reduce the vanadium ions crossover and increase the mechanical strength of SPEEK membranes, while keeping their high ion selectivity for VRB. One effective approach to address this issue is to embed inorganic nanoparticles/carbon nanostructures into the SPEEK matrix such as, WO_3, graphene, rGO, and GO [16–18], which can significantly increase the mechanical strength and stability of the composite membranes with high DS SPEEK matrix.

CNT is well known as its good conductivity, high surface area, well-directed 1D conductive path, low cost, and good mechanical stability. Integrating their merits, CNT/polymer nanocomposites become promising candidates in membranes for VRB application. General strategies include co-casting and post-synthesis. Casting of CNT/polymer may require further surface modification on CNT to possess desirable functional groups/charge. Polymeric materials can also be synthesized on CNT architectures/frameworks by CVD or vacuum filtration through chemical/electrochemical polymerization.

CNT/polymer-composited membranes, such as CNT/Nafion and CNT/SPEEK, have been successfully used in fuel cell and shown not only good physical properties (e.g., mechanical strength, chemical stability, and ion selectively) but also good cell performances [19]. However, there are no reports about CNT/polymer membrane for VRB before 2015. In order to combine the good properties of CNT and SPEEK matrix and eliminate their individual disadvantages, Jia et al. firstly reported a novel SPEEK/short-carboxylic multiwalled CNT membrane for VRB application [20]. They introduced only 0.7 wt% CNT into SPEEK matrix using solution casting method. With the combination of functionalized CNTs with excellent electrocatalytic activity and the low cost of SPEEK membrane, this novel nanotube-polymer-composited membrane consequently exhibits good chemical strength, high ion selectively, and low vanadium permeability. Detailed study showed that the short-carboxylic multiwalled CNTs are well dispersed in the traditional SPEEK membrane due to the good interaction between short-carboxylic multiwalled CNTs and SPEEK matrix. In addition, the well dispersion of carbon nanotubes and good π-π interactions, electrostatic interactions, and van der Waals interactions between the aromatic groups of SPEEK polymer and the sidewall of CNTs increased the mechanical strength significantly (61.8 MPa) compared to a pure SPEEK membrane (39 MPa). Moreover, as shown in **Figure 2**, the carbon nanotubes used as an inner filler can dead or decrease the size of ion transport pathway significantly and thus restrain the vanadium permeability. Furthermore, the hydrophilic cluster formed by different –COOH, –OH groups on the side wall of CNTs will not only further decrease the ion transport pathway but also provide enriched proton transport pathways. Therefore, the permeability of vanadium ions through this composited membrane (3.22×10^{-7} cm^2 min^{-1}) is lower than that through the pure SPEEK (4.03×10^{-7} cm^2 min^{-1}) membrane and Nafion 212 membrane (8.23×10^{-7} cm^2 min^{-1}), which leads to a low self-discharge of the VRB cell and thus high Columbic efficiency. On the other hand, the inner proton transport pathway formed by various functional groups on the CNTs could enhance the proton selectivity and ion conductivity and thus enhance the voltage efficiency and energy efficiency of the VRB cell.

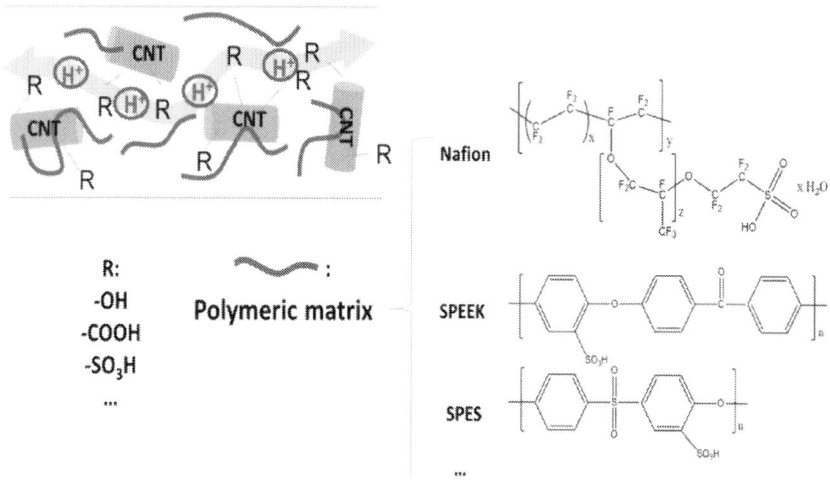

Figure 2. Mechanism of proton transport in polymer/CNT-composited membranes.

Although the cell performance of the VRB cell with SPEEK/CNT is better than those only with Nafion membranes or SPEEK membrane individually, the chemical properties of the SPEEK membrane should be further optimized. The chemical stability of the composited membranes could be optimized using stable matrixes such as Nafion, SPES, PBI, and SPEEK with low DS. In addition, the properties of polymer/CNT-composited membranes could be further controlled and optimized in different polymer matrixes via different designed R group-functionalized CNTs (e.g., CNT-OH, CNT-COOH, CNT-SO_3H, and CNT-NH_2). The advancing composited CNT/polymer membranes are always designed to pursuit the desirable performance of VRB. In addition, these polymer/CNT-composited membranes provide us suitable approaches to synthesize membranes and shine new light on novel membrane-electrode assembly for VRB.

2.5. Going beyond: toward other redox flow batteries

Non-aqueous redox flow batteries based on non-aqueous electrolytes have received widely studied recently. Although being far less developed than aqueous redox flow batteries, of the non-aqueous redox flow batteries seem more attractive due to their wider electrochemical window (>2 V) and greater energy density. However, the lithium-conducting membranes and low power density are still the main challenges for their commercialization. Therefore, the lithium-conducting polymer (Nafion, Polyvinylidene difluoride, PEEK, Polybenzimidazole, PES, SPES, and SPEEK)/functionalized CNT-composited membranes such as CNT-SO_3Li/polymer, CNT-COOLi/polymer, and CNT-OLi/polymer are expected to be promising candidates for applications in the non-aqueous redox flow lithium batteries, redox flow lithium air batteries, and redox flow lithium sulfur batteries [21–23].

3. The application of CNT-polymer compositions in flexible Li-ion batteries

Because of the environmental contamination and energy crisis, energy storage systems have attracted much attention from research institutes and industries. Lithium ion batteries (LIBs) due to their high energy densities and long lifetime are the main commercial batteries, especially for portable devices [24, 25]. Among the possible candidates of lithium storage, carbon nanotubes (CNTs) have emerged as the promising ones. Their peculiar structure and unique properties, such as high electrical conductivity and tensile strength, make them well suited as a critical component in novel anode materials for enhanced lithium storage [26]. For example, it is possible to design free-standing and flexible electrodes using CNTs, which are important parts for the flexible batteries [27]. In addition, some organic active materials could work as electrodes in lithium ion batteries, which could realize much high capacity [28]. These materials have the minimal environmental footprint, lightweight, flexibility, and chemical tenability, which could guarantee its promising potentials in flexible LIBs.

3.1. The working mechanism of LIBs

The working mechanism of lithium ion batteries is demonstrated in **Figure 3**.

1. During the charging and discharging process, the lithium ions move between the anode and cathode via the electrolyte, which is typically a lithium salt such as LiPF6 dissolved in organic solvent such as ethylene carbonate. The electrochemical reactions are expressed as reactions 1 and 2. The most common anode is graphite with a theoretical capacity of 372 mAh/g, and the cathode is $LiCoO_2$ with a theoretical capacity of 272 mAh/g [29]. The electrochemical properties of cathode and anode dramatically draw influence on the lithium storage properties. For example, lithium ions can only combine on every 2nd carbon hexagon in the graphite sheet which limits the amount of lithium ions to one for every six carbon atoms [30]. Thus, its theoretical capacity of graphite is 372 mAh/g.

$$LiMO_2 \leftrightarrow Li_{1\text{-}x}MO_2 + xLi^+ + xe^- \text{(cathode side reaction)} \quad (1)$$

$$xLi^{+} + xe^{-} + 6C \leftrightarrow Li_{x}C_{6}\text{(anode side reaction)} \quad (2)$$

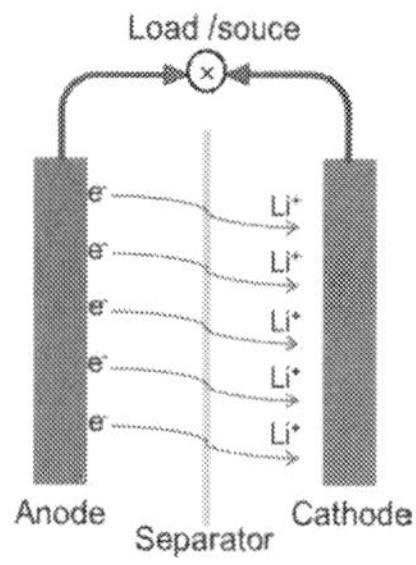

Figure 3. The basic structure of lithium ion batteries.

3.2. The significance of flexible Li-ion batteries

Flexible electronics will play more and more significant role in the future electronic markets, for example, flexible displays, sensors, batteries, etc. [31, 32]. Due to the characters of flexible devices, the corresponding energy storage systems should be flexible as well. Due to their high operation voltage, high energy density, and long life span, LIBs have been chosen as the most promising candidates for flexible electronics. Some concept and prototype flexible electronics have been demonstrated. However, there are lots of challenges to realize good flexible LIBs, for example, the package of the electrolytes and electrodes. As one key part in flexible LIBs, electrode with intrinsically flexible should be the best candidate [33].

3.3. The traditional organic polymer in LIB

Organic materials became more and more important in the LIBs due to their merits, such as renewable, designable, and flexible. various organic materials are used in LIBs, such as organic radical polymers [34–36], conducting polymers [37, 38], organosulfur compounds [39, 40], bipolar porous polymeric frameworks [41], layered compounds [42], trioxotriangulene derivates [43], tetracyano quinodimethane (TCNQ) [44], pyrene-tetraone (PYT) [45], polyimide [46], and organic carbonyl compounds. Several reported organic compounds and their basic information in energy storage systems are shown in **Table 1**. Despite the rapid development of organic-based electrode materials, some critical obstacles still exist. For example, most organic electrodes suffer from the rapid capacity fading upon cycling due to the dissolution of active organic chemicals into organic electrolytes. And the power densities of these organic electrodes are relative low due to their poor electronic conductivity [47, 48]. In order to utilize these organic materials in flexible LIBs, CNTs were employed as free-standing conductive networks and current collectors [46, 49]. Current collector is the necessary component to support electrode materials and conduct electrons in the LIBs. The common current collectors are Cu and Al with high densities and smooth surface, which leads to low energy density and poor adhesion. However, light and rough CNT-based current collectors could address these issues.

Polymers	Structures	Redox potentials (V)	Functions
Organic radical polymer (poly[4-(nitronyl nitroxyl) styrene]) [35]	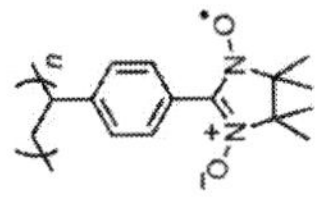	−0.61 and 0.72 V vs Ag/AgCl	Active material
Conducting polymer (poly(3,4-ethylenedioxythiophene)) [38]	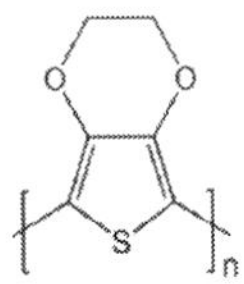	1.25 V vs Ag/AgCl	Conductive additives for active material
Organosulfur compounds (lithium polysulfidophosphates) [39]	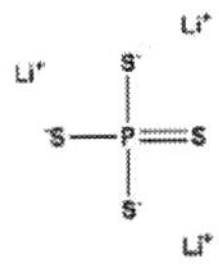	~2.2 V vs Li/Li^+	Active material
Bipolar porous polymeric frameworks (amorphous covalent triazine-based frameworks) [41]	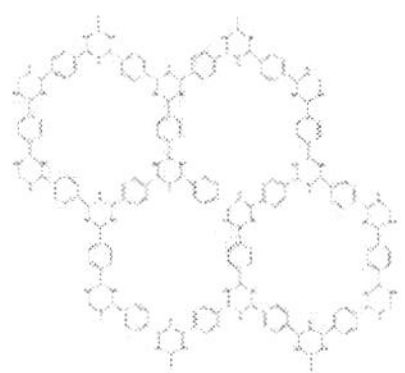	2.8 and 4.2 V vs Li/Li^+	Active material
Layered compounds (polypyrrole-iron-oxygen (PPy-Fe-O) coordination complex) [42]	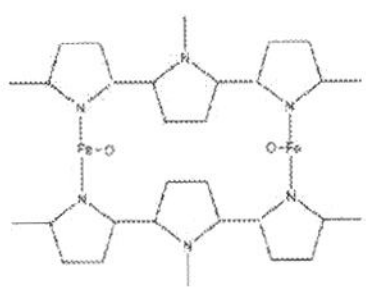	~1.5 V vs Li/Li^+	Active material
Trioxotriangulene derivates (tri-tert-butylated TOT derivative) [43]	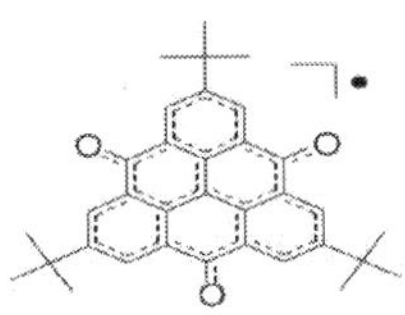	1.4 and 3.8 V vs Li/Li^+	Active material
Tetracyano quinodimethane [44]	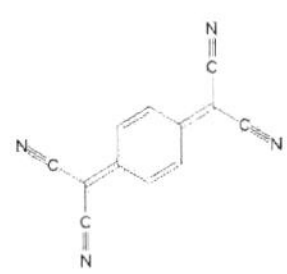	1.0 V vs. Li/Li^+	Active material

Polymers	Structures	Redox potentials (V)	Functions
Pyrene-tetraone (pyrene-4,5,9,10-tetraone) [45]	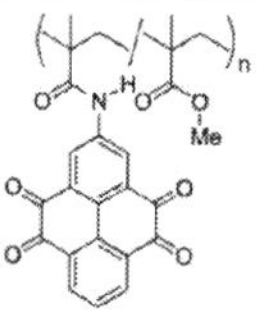	2.2 and 2.8 V vs. Li/Li$^+$	Active material
Polyimide (pyromellitic dianhydride (PMDA) and 2, 6diaminoanthraquinone (AQ) [46]	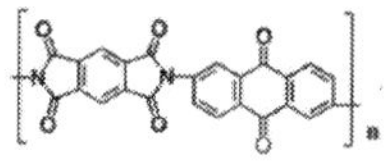	2.2 and 2.4 V vs Li/Li$^+$	Active material

Table 1. The reported several organic polymers in LIBs.

3.4. The methods of preparation of flexible CNT/polymer electrodes

Free-standing electrodes by combining CNTs and organic polymers have been widely developed towards multifunctional electrodes.

CNTs can be used as active anode materials, which are combined with conducting polymers to form as flexible electrodes. Chen et al. used aligned carbon nanotubes (ACNTs) and poly(3,4-ethylenedioxythiophene) (PEDOT) to produce flexible electrodes. In this composition, the ACNTs are the active anode materials and the PEDOT is used for enhancement of electrochemical activity and electronic conductivity of the electrode. PEDOT film was deposited onto the ACNT array using chemical vapor-phase polymerization. Cui et al. used PEDOT as conductive polymer for enhancement of electrochemical performance of Si nanowires. Electropolymerization is applied for generation of cation radicals and coupling of the radicals for chain growth. Although the previous methods achieved high electrochemical performance, the processes seem relatively complicated. Using vacuum or pressure filtration, free-standing CNT papers could also be produced [50]. This method is much facile, and the produced CNT papers also exhibit good mechanical properties.

On the other hand, CNTs can be mainly applied as the effective current collectors with organic polymers being employed as active material as either high-capacitance anode or cathode for the flexible electrodes. Lee et al. prepared a free-standing lumiflavine (LF)-single-walled carbon nanotubes (SWNTs) nanohybrids for fast and sustainable energy storage. The hybrid electrode was formed via interaction without disruption of their redox-active properties. The hybrid electrode could possess high conductivity and Li-ion accessibility. In addition, the structure could also prevent the dissolution of polymer into the electrolyte. During the preparation, the reassembly of bulk LF particles into molecular layers is significant for the improvement of stability and kinetics of Li storage in LF.

Li et al. developed one fabrication technique for SWNT/Nafion composite, which can be used as anode in Li-ion battery [51]. The battery exhibited reversible capacity of ~850 mAh/g after 15 cycles. Using this composite anode, they demonstrated the integration of electrode and separator to simplify device architecture and decrease overall weight.

Wang et al. prepared multiwalled carbon nanotubes (MWNTs) functionalized with a hyperbranched aliphatic polyester and two different poly(ethylene glycol)s by the reactions of carbonyl chloride groups on the surface of MWNTs and hydroxyl groups of polymers [52]. Due to the different electrode-electrolyte interfaces, the hyperbranched polymer-functionalized MWNT showed a significant improvement over linear polymer-functionalized MWNTs in lithium insertion/deinsertion capacity and cycle stability.

Wu et al. used the composite of polyimide derivative and SWNTs as electrodes for flexible LIBs. Before preparation of hybrid electrodes, they firstly synthesized an active polymer using a condensation polymerization of pyromellitic dianhydride (PMDA) and 2,6-diaminoanthraquinone (AQ). Then, the polymer was combined with SWNTs to form a binder-free flexible organic cathode film, which exhibited high capacity and good rate performance.

3.5. The perspective of CNT/polymer in flexible LIBs

CNTs, with high conductivity and good mechanical properties, are promising materials for flexible LIBs. On the other hand, some polymers with high conductivity can further enhance the performance of flexible LIBs by combining with CNTs. In addition, some polymers can work as active electrode materials with high specific capacities, which can also be used in flexible LIBs. The composites of CNT/polymer own the flexibility, high conductivity, high energy densities, which are suitable for flexible LIBs.

4. Carbon nanotube-polymer nanocomposites for supercapacitors

Among energy storage devices, supercapacitor possesses its unique merits of high power intensity and long lifetime. It can be categorized into electric double-layer capacitor (EDLC), pseudocapacitor, and newly evolving lithiumion capacitor [53]. With advantages on synergetic effects on electrochemical performance, CNT-polymer nanocomposites play significant roles on supercapacitors that integrate interfacial adsorption and redox reaction, hierarchical microstructure, conductivity, mechanical strength, and flexibility [54–57]. The following section focuses on current design and material synthesis of CNT-polymer nanocomposites in the pursuit of true performance matrices, that is, volumetric/areal performance.

4.1. Why supercapacitor in energy storage?

Comparing to lithium ion battery (LIB), supercapacitor enables fast/ultrasfast fully charging/discharging (in seconds) and much higher power delivery (>10 kW/kg) for shorter times [54]. It becomes an indispensable entity towards the development of sustainable and clean energy.

In supercapacitors, EDLCs represent the majority of commercially manufactured products. The electrode materials of EDLC are carbon allotropes (carbon onions, CNT, reduced graphene oxide (rGO), graphene (G), activated carbon, etc.) [58, 59]. The energy is stored at the electrolyte/carbon interface by charging the double-layer capacitance via reversible ion adsorption on the surface of carbon. In order to enhance the capacitance of carbon electrode, good electrical conductivity, high specific surface area (SSA), and electrochemical stability are the dominant factors attract great research interest.

Numerous endeavors have been devoted to increase energy density of supercapacitor. Organic electrolyte is applied to enlarge the voltage window (typically 0.8–1 V for aqueous electrolyte). On the other hand, pseudocapacitors chemically store charge via redox reaction at the surface/interface region. The electrode materials for pseudocapacitors include transition metal oxides/hydroxides and conducting polymers [e.g., polyaniline (PANI), polypyrrole (Ppy), and poly(3,4-ethlenedioxythiophene) (PEDOT), etc.]. To utilize their high capacitance and address the relative low conductivity, nanocomposites with conducing framework endowed with pseudocapacitive nanostructure are under intense study.

Most importantly, recent research focuses primarily on improving areal/volumetric energy densities of supercapacitor with mounting demand for compact and portable energy storage systems [60–62]. Though Ragone Plot provides a representative summary, the concern on the true performance matrices stems from the mass loading/packing density of active materials where the areal/volumetric properties turn to be realistic in device aspect ($C_a = C_g \times m$, $C_v = C_g \times \varrho$, where m and ϱ are the mass loading and packing density for the electrode respectively). Promising volumetric performances can be achieved in micro-/ultrathin configuration; however, they do not scale up linearly with thickness. The great challenge hence lies in the storage of large amount of energy with simultaneous promising areal/volumetric performances.

4.2. Current challenge and strategies for CNT-based electrode materials

Several merits of CNT deserve re-emphasis: its high conductivity, highly accessible outer surface, and its outstanding mechanical strength. The unique characteristics of CNT facilitate its role in supercapacitor as electrode, conducting additive, template/scaffold for nanocomposite electrode, platform for flexible/stretchable device [54–57]. However, the shortcomings are worthy of noting. The SSA of CNT is only in moderate level and is lower than activated carbon, mesoporous carbon, rGO, and G [63]. Also, its packing density needs to be improved considering areal/volumetric performances.

Recent research has explored numerous approaches to improve the performance of CNT-based electrodes from the angles of packing density, pores to enhance SSA, incorporation of redox nanostructures via surface functionalization [64–66]. The design of hierarchical CNT-based nanocomposites is now a general strategy to pursuit high performance. An integration of 1D (CNT) and 2D (GO, rGO, and G) carbon nanostructure becomes a greatly promising representative that serves further as scaffold for carbon-redox nanostructure hybrid electrodes [67–69]. Nevertheless, the central focus on areal/volumetric energy densities should not be override by the arts of material synthesis.

4.3. CNT-polymer nanocomposites for supercapacitors

Before focusing on CNT-conducting polymer nanocomposites, CNT-cellulose deserves certain lines. Victor et al. firstly demonstrated the fabrication of vertically aligned CNT-unmodified plant cellulose paper supercapacitor [70]. With however relatively low capacitance, this design started the concept of paper device with rationalization of active materials and triggered the development of ultrathin, flexible, and paper-like devices.

4.3.1. CNT-PANI

Conducting polymers generally possess high charge densities and relatively low costs. The redox reactions allow for high energy storage and reduced self-discharge; however, the power density may suffer from slow diffusion [71]. Conducting polymer-based composites with desirable pores and better conductivities hence seek better opportunities. PANI is one of the most promising candidates because of its multiple redox states, relatively high conductivity, environmental stability, and flexibility in terms of the synthetic methods and consequently the so-formed nanostructures.

From the device perspective, a binder-free structure is ideal to eliminate dead cell components, which benefits the areal/volumetric performance. Free-standing CNT-based frameworks via simple and facile ways therefore call for the demand. Meng et al. have prepared a CNT-PANI network starting from forming a free-standing CNT network [72]. Applying the in situ chemical polymerization of aniline monomer, a wholly uniform coating of ~50 to 90 nm of PANI was formed. A flexible all solid state symmetric device with ~23 wt% of PANI was further formed with H_2SO_4/PVA gel, giving a C_v of ~37 F cm^{-3} and correspondingly an E_v of ~2.4 × 10^{-3} Wh L^{-3}. To construct the flexible supercapacitor, cellulose, the most abundant renewable organic materials in nature, is one attractive substrate material. Li et al. utilized ecofriendly bacterial cellulose (BC) paper by vacuum filtration that consists of ribbon-shaped ultrafine nanofibers as substrate for the BC-CNT-PANI paper electrode [73]. BC paper endows the device with porosity, high tensile strength, high water retention, and good contact between electrodes and aqueous electrolyte due to its hydrophilicity. CNT ink was then deposited on the BC paper with strong binding via hydrogen bonding and electrostatic interaction, followed by electrodeposition of PANI. The binder-free electrode showed an appreciable C_g of 656 F g $^{-1}$ in 1 M H_2SO_4. An optimal deposition time for the PANI coating indicates the critical value of thickness beyond which the pores in the paper are filled by PANI that blocks ion diffusion towards interior. This lightweight and cost-effective supercapacitor paper is a promising candidate for stretchable/flexible device. Hyder et al. utilized an electrostatic layer-by-layer (LBL) system to synthesize CNT-PANI films [74]. PANI nanofibers (~30 to 70 nm × 0.4–3 μm) were chemically polymerized, where the nanoscale concerns the redox reactions localizing in the thin surface layer, and meanwhile, the amine or imine surface groups were present. On the other hand, CNT surface was modified with carboxylic acid groups. The pH-dependence on surface charges for both entities facilitates the behavior like weak polyelectrolytes with controllable LBL film thickness and morphology, where high PANI content (58 wt%) can be achieved. The resulted film (e.g., 1.3 μm, 0.64 g cm^{-3}) consisted of an interpenetrating nano-porous network yielding excellent electrochemical performance: a C_v of ~238 F cm^{-3} in 1 M $LiPF_6$ (comparing to ~180 F cm^{-3} for LBL CNT film) and a C_v of ~248 F cm^{-3} in 1 M H_2SO_4, and an E_v of ~220 Wh L^{-1} at 100 kW L^{-1}. Note that the mass loading of 0.15–0.23 mg cm^{-2} was relatively low that was a direct consequence of LBL process, which lead to a C_a less than 0.1 F cm^{-2}.

Besides above, multifunctional design, for example, integrating advantageous mechanical strength, is making strides. Benson et al. reported a CNT-PANI fabric by electrodeposition of PANI on the CVD-formed CNT fabric (15 μm) [75]. The strong π-π interaction between CNT (benzenoid rings) and PANI (quinoid rings) resulted in strong chemical bonding and high interfacial shear strength. The electrode with ~31 wt% PANI provided a C_a of ~0.7 F cm^{-2} and a C_v of 308 F cm^{-3}, which showed a 10-fold increase comparing to the initial CNT fabric. The synthesized nanocomposites demonstrated remarkable mechanical properties: a maximum tensile strength of 484 MPa and an elastic modulus of 19 GPa, which were comparable to CNT-epoxy composites, strong natural fibers and tissues, aerospace-grade metal alloys. This ultratough design may offer a solution to a system requires both structural and energy storage functions. Also, stretchable fiber-shaped supercapacitors find their applications in wearable device. Zhang et al. developed a superelastic (stretched by over 400% strain) symmetric supercapacitor by CNT-PANI nanocomposite sheets [24]. A prestrain method was employed to form the CNT sheet, followed by a layer of PANI coated via electropolymerization. At 50 wt% of PANI, the electrode provided a length specific capacitance of 0.9 mF cm^{-1} and a Ca of ~3 mF cm^{-2}. Most importantly, the device exhibited superior performance under dynamical test: ~71% retention after stretching at a strain of 300% for 5000 cycles, ~90% retention after bending for 5000 cycles, and ~96% retention at a high stretching speed of 30 mm s^{-1}. Very recently, a highly compressible all-solid-state supercapacitor was integrated by Niu et al., based on sponge-CNT-PANI electrodes [77]. The compression tolerance of the sponge is utilized, followed by a repeated "dipping and drying" strategy for CNT coating. Chemical polymerization of PANI was performed to form the porous hierarchical skeleton/skin architecture. The total mass loading of active materials was 6.3 mg cm^{-2}, with a contribution of 4.1 mg cm^{-2} from PANI. A layer of gel electrolyte PVA/H_2SO_4 was selected to overcome the contact between neighboring electrode materials at large compression strain. At the relaxed state, the supercapacitor showed a high C_a of 2.7 F cm^{-2} and a C_v of 3.4 F cm^{-3}, with 3% loss under 60% strain. It delivered an E_a of 0.01 Wh L^{-2} and an E_v of 0.12 Wh L^{-1}. The highly compressible supercapacitor will pave the way for advanced applications in the area of the compressible energy storage/compression-tolerant electronics.

4.3.2. CNT-Ppy

Ppy is another popular conducting polymer included in the work awarded 2000 Nobel Prize in Chemistry. A natural cellulose-Ppy all-polymer paper has been synthesized to achieve superior areal/volumetric performance of ~5.7 F cm^{-2}/236 F cm^{-3} at 20 mg cm^{-2}/290 μm, where the scaling up facilities the practical applications [78]. Besides hierarchical structure, Ppy plays a role of nitrogen source for N-doped carbon materials, as demonstrated in Chen et al.' work [79]. The carbonization at 500–1100°C of the Ppy-coated macroscopic-scale carbonaceous nanofibers leads to the N-doped carbon nanofibers, where the incorporated heteroatoms favored the performance.

4.3.3. CNT-Nafion

Nafion is a state-of-art polymer electrolyte membrane as discussed in Section 2.4. Due to its ionic conductivity, it is a promising candidate for solid electrolytes in supercapacitor. Huang et al. sprayed CNT in diluted H_2SO_4 solution on both side of a Nafion membrane to construct a flexible solid state supercapacitor, where high H^+ mobility of protonated Nafion is fully utilized [80]. In a recent work by Daniel et al., Nafion 115 solution was dip-coated on the aligned CNT forest and covered its top to form a separating layer [81]. The symmetric EDLC without additional separator was then soaked in 1 M H_2SO_4 for much improved electrochemical performance comparing to its dehydrated state. In this design, Nafion acts as both separator and ionic conductor, brings promising mechanical strength, and favors operation at high temperature up to 60°C (~2.6 times enhancement on C_g).

4.4. Outlook

There is no doubt that the 3D hierarchical CNT-polymer architectures have made striking progress. Based on such scaffolds, more emerging advancing materials/device designs will be expected. One primary concern lies in the synthesis of newly discovered conducting polymers. Towards green chemistry and renewable resources, integration of CNT with other natural polymers than cellulose, like lignin currently under intense study, may be further explored [82]. Also, the control of macropores/mesopores/micropores will still be one of the focuses. In the ultrathin configuration, commonly obtained relatively low C_a with highly appreciated C_v urges the enhancement of the areal mass loading, where combination of multiple redox nanostructures with synergic effects may play significant roles. In the flexible/stretchable configuration, scaling up of the high C_v nanocomposites is still a challenge, which is also valid for ultrathick configuration. Furthermore, the stability of CNT-conducting polymer that puts influence on cycling stability warrants in-depth research. Above all, the understanding on fundamental interface chemistry and ionic transportation governs the development of energy storage [83].

Author details

Du Yuan, Yun Gunag Zhu and Chuankun Jia*

*Address all correspondence to: msejiac@nus.edu.sg; jack2012ding@gmail.com

Department of Materials Science and Engineering, Faculty of Engineering, National University of Singapore, Singapore, Singapore

References

[1] W. Wang, Q. Luo, B. Li, X. Wei, L. Li, Z. Yang. Adv. Funct. Mater. 2013, 23, 970.

[2] W. Jiang, Z. Yan, D. Feng. Renew. Sustain. Energy Rev. 2009, 13, 2485.

[3] C. Jia, Q. Liu, C. Sun, F. Yang, S. Heald, Y. Liu, Z. Li, W. Lu, J. Xie. ACS Appl. Mater. Interface. 2014, 6, 17920.

[4] A. Weber, M. Mench, J. Meyers, P. Ross, J. Gostick, Q. Liu. J. Appl. Electrochem. 2011, 41, 1137.

[5] M. Skyllas-Kazaocs, M.H. Chakrabarti, S.A. Hajimolana, F.S. Mjalli, M. Saleem. J. Electrochem. Soc. 2011, 158, R55.

[6] X.F. Li, H.M. Zhang, Z.S. Mai, H.Z. Zhang, I. Vankelecom. Energy Environ. Sci. 2011, 4, 1147.

[7] G. Wei, C. Jia, J. Liu, C. Yan. J. Power Sources 2012, 220, 185.

[8] Z.G. Yang, J.L. Zhang, M.C.W. Kintner-Meyer, X.C. Lu, D.W. Choi, J.P. Lemmon, J. Liu, Chem. Rev. 2011, 111, 3577.

[9] M. Skyllas-Kazacos, M. Rychcik, R.G. Robins, A.G. Fane, M.A. Green. J. Electrochem. Soc. 1986, 133, 1057.

[10] M. Skyllas-Kazacos, C. Peng, M. Cheng. Electrochem. Solid State Lett. 1999, 2, 121.

[11] C. Jia, J. Liu, C. Yan. J. Power Sources. 2012, 203, 190.

[12] J. Xi, Z. Wu, X. Qiu, L. Chen. J. Power Sources. 2007, 166, 531.

[13] Q. Luo, H. Zhang, J. Chen, D. You, C. Sun, Y. Zhang. J. Membr. Sci. 2008, 325, 553.

[14] X. Ling, C. Jia, J. Liu, C. Yan. J. Membr. Sci. 2012, 415–416, 306.

[15] C. Jia, J. Liu, C. Yan. J. Power Sources. 2010, 195, 4380.

[16] W. Dai, Y. Shen, Z. Li, L. Yu, J. Xi, X. Qiu. J. Mater. Chem. A. 2014, 2, 12423.

[17] W. Dai, L. Yu, Z. Li, J. Yan, L. Liu, J. Xi, X. Qiu. Electrochim. Acta. 2014, 132, 200.

[18] N. Wang, S. Peng, H. Wang, Y. Li, S. Liu, Y. Liu. Electrochem. Commun. 2012, 17, 30.

[19] B.P. Tripathi, M. Schieda, V.K. Shahi, S.P. Nunes. J. Power Sources. 2011, 196, 911.

[20] C. Jia, Y. Cheng, X. Ling, G. Wei, J.O. Liu, C. Yan. Electrochim. Acta. 2015, 153, 44.

[21] Q. Huang, H. Li, M. Gratzel, Q. Wang. Phys. Chem. Chem. Phys. 2013, 15, 1793.

[22] C. Jia, F. Pan, Y. Zhu, Q. Huang, L. Lu, Q.Wang. Sci. Adv. 2015, 1(10), e1500886.

[23] Y. Zhu, C. Jia, J. Yang, F. Pan, Q. Huang, Q. Wang. Chem. Commun. 2015, 51, 9451.

[24] M. Armand, J.M. Tarascon. Nature. 2008, 451, 652.

[25] J.B. Goodenough, K.-S. Park. J. Am. Chem. Soc. 2013, 135, 1167.

[26] B.J. Landi, M.J. Ganter, C.D. Cress, R.A. DiLeo, R.P. Raffaelle. Energy Environ. Sci. 2009, 2, 638.

[27] J. Chen, Y. Liu, A.I. Minett, C. Lynam, J. Wang, G.G. Wallace. Chem. Mater. 2007, 19, 3595.

[28] Y. Liang, Z. Tao, J. Chen. Adv. Energy Mater. 2012, 2, 742.

[29] B. Liu, W. Xu, P. Yan, X. Sun, M.E. Bowden, J. Read, J. Qian, D. Mei, C.-M. Wang, J.-G. Zhang. Adv. Funct. Mater. 2015, 26, 605

[30] C. Wang, G. Wu, W. Li. J. Power Sources. 1998, 76, 1.

[31] J.A. Rogers, T. Someya, Y. Huang. Science. 2010, 327, 1603.

[32] G. Zhou, F. Li, H.-M. Cheng. Energy Environ. Sci. 2014, 7, 1307.

[33] H. Nishide, K. Oyaizu. Science 2008, 319, 737.

[34] K. Oyaizu, H. Nishide. Adv. Mater. 2009, 21, 2339.

[35] T. Suga, H. Ohshiro, S. Sugita, K. Oyaizu, H. Nishide. Adv. Mater. 2009, 21, 1627.

[36] T. Suga, S. Sugita, H. Ohshiro, K. Oyaizu, H. Nishide. Adv. Mater. 2011, 23, 751.

[37] P. Novák, K. Müller, K.S.V. Santhanam, O. Haas. Chem. Rev. 1997, 97, 207.

[38] Y. Yao, N. Liu, M.T. McDowell, M. Pasta, Y. Cui. Energy Environ. Sci. 2012, 5, 7927.

[39] Z. Lin, Z. Liu, W. Fu, N.J. Dudney, C. Liang. Angew. Chem. 2013, 125, 7608.

[40] W. Ai, L. Xie, Z. Du, Z. Zeng, J. Liu, H. Zhang, Y. Huang, W. Huang, T. Yu. Sci. Rep. 2013, 3.

[41] K. Sakaushi, G. Nickerl, F.M. Wisser, D. Nishio Hamane, E. Hosono, H. Zhou, S. Kaskel, J. Eckert. Angew. Chem. Int. Ed. 2012, 51, 7850.

[42] Y. Mao, Q. Kong, B. Guo, X. Fang, X. Guo, L. Shen, M. Armand, Z. Wang, L. Chen. Energy Environ. Sci. 2011, 4, 3442.

[43] Y. Morita, S. Nishida, T. Murata, M. Moriguchi, A. Ueda, M. Satoh, K. Arifuku, K. Sato, T. Takui. Nat. Mater. 2011, 10, 947.

[44] Y. Chen, S. Manzhos. Phys. Chem. Chem. Phys. 2016, 18, 1470.

[45] T. Nokami, T. Matsuo, Y. Inatomi, N. Hojo, T. Tsukagoshi, H. Yoshizawa, A. Shimizu, H. Kuramoto, K. Komae, H. Tsuyama. J. Am. Chem. Soc. 2012, 134, 19694.

[46] H. Wu, Q. Yang, Q. Meng, A. Ahmad, M. Zhang, L. Zhu, Y.G. Liu, Z. Wei. J. Mater. Chem. A, 2016, 4, 2115

[47] H. Chen, M. Armand, G. Demailly, F. Dolhem, P. Poizot, J.M. Tarascon. ChemSusChem 2008, 1, 348.

[48] D.J. Kim, S.H. Je, S. Sampath, J.W. Choi, A. Coskun. RSC Adv. 2012, 2, 7968.

[49] M. Lee, J. Hong, H. Kim, H.D. Lim, S.B. Cho, K. Kang, C.B. Park. Adv. Mater. 2014, 26, 2558.

[50] S. Hossain, Y.-K. Kim, Y. Saleh, R. Loutfy. J. Power Sources 2003, 114, 264.

[51] X. Li, F. Gittleson, M. Carmo, R.C. Sekol, A. D. Taylor. ACS Nano 2012, 6, 1347.

[52] X.H. Wang, H. Liu, Y. Jin, C.H. Chen. J. Phys. Chem. B 2006, 110, 10236.

[53] P. Simon, Y. Gogotsi, B. Dunn. Science 2014, 343, 1210.

[54] P. Simon, Y. Gogotsi. Acc. Chem. Res. 2012, 46, 1094.

[55] L. Li, Z. Wu, S. Yuan, X. Zhang. Energy Environ. Sci. 2014, 7, 2101.

[56] Z. Niu, L. Liu, L. Zhang, W. Zhou, X. Chen, S. Xie. Adv. Energy Mater. 2015, 5, 1500677.

[57] Y.G. Zhu, Y. Wang, Y. Shi, J.I. Wong, H.Y. Yang. Nano Energy. 2014, 3, 46.

[58] Z. Yu, L. Tetard, L. Zhai, J. Thomas. Energy Environ. Sci. 2015, 8, 702.

[59] F. bonaccorso, L. Colombo, G. Yu, M. Stoller, V. Tozzini, A.C. Ferrari, R.S. Ruoff, V. Pellegrini. Science. 2015, 347, 6217.

[60] Y. Gogotsi, P. Simon. Science. 2011, 34, 917.

[61] X. Yang, C. Cheng, Y. Wang, L. Qiu, D. Li. Science. 2013, 341, 534.

[62] Z. Wu, Z. Liu, K. Parvez, X. Feng, K. Mullen. Adv. Mater. 2015, 27, 3669.

[63] H. Jiang, P.S. Lee, C. Li. Energy Environ. Sci. 2013, 6, 41.

[64] D.N. Futaba, K. Hata, T. Yamada, T. Hiraoka, Y. Hayamizu, Y. Kakudate, O. Tanaike, H. Hatori, M. Yumura, S. Iijima. Nat. Mater. 2006, 5, 987.

[65] C. Choi, J.A. Lee, A.Y. Choi, Y.T. Kim, X. Lepro, M.D. Lima, R.H. Baughman, S.J. Kim. Adv. Mater. 2014, 26, 2059.

[66] H. Wang, H. Dai. Chem. Soc. Rev. 2013, 42, 3088.

[67] D. Yu, K. Goh, H. Wang, L. Wei, W. Jiang, Q. Zhang, L. Dai, Y. Chen. Nat. Nanotechnol. 2014, 9, 555.

[68] N. Jung, S. Kwon, D. Lee, D. Yoon, Y.M. Park, A. Benayad, J. Choi, J.S. Park. Adv. Mater. 2013, 25, 6854.

[69] Y.G. Zhu, Y. Wang, Y. Shi, Z.X. Huang, L. Fu, H.Y. Yang. Adv. Energy Mater. 2014, 4, 201301788.

[70] V.L. Pushparaj, M.M. Shaijumon, A. Kumar, S. Murugesan, L. Ci, R. Vajtai, R.J. Linhardt, O. Nalamasu, P.M. Ajayan. Proc. Natl. Acad. Sci. 2007, 104, 13574.

[71] Q. Zhang, E. Uchaker, S.L. Candelaria, G. Cao. Chem. Soc. Rev. 2013, 42, 3127.

[72] C. Meng, C. Liu, L. Chen, C. Hu, S. Fan. Nano Lett. 2010, 10, 4025.

[73] S. Li, D. Huang, B. Zhang, X. Xu, M. Wang, G. Yang, Y. Shen. Adv. Energy Mater. 2014, 4, 1301655.

[74] M.N. Hyder, S.W. Lee, F.C. Cebeci, D.J. Schmidt, Y. Shao-Horn, P.T. Hammond. ACS Nano. 2011, 5, 8552.

[75] J. Benson, I. Kovalenko, S. Boukhalfa, D. Lashmore, M. Sanghadasa, G. Yushin. Adv. Mater. 2013, 25, 6625.

[76] Z. Zhang, J. Deng, X. Li, Z. Yang, S. He, X. Chen, G. Guan, J. Ren, H. Peng. Adv. Mater. 2015, 27, 356.

[77] Z. Niu, W. Zhou, X. Chen, J. Chen, S. Xie. Adv. Mater. 2015, 27, 6002.

[78] Z. Wang, P. Tammela, P. Zhang, M. Stromme, L. Nyholm. J. Mater. Chem. A 2014, 2, 16761.

[79] L. Chen, X. Zhang, H. Liang, M. Kong, Q. Guan, P. Chen, Z. Wu, S. Yu. ACS Nano 2012, 6, 2092.

[80] C. Huang, P.S. Grant. Sci. Rep. 2013, 3, 2393.

[81] D.P. Cole, A.L.M. Reddy, M.G. Hahm, R. McCotter, A.H.C. Hart, R. Vajtai, P.M. Ajayan, S.P. Kama, M.L. Bundy. Adv. Energy Mater. 2014, 4, 1300844.

[82] H. Li, D. Yuan, C. Tang, S. Wang, J. Sun, Z. Li, T. Tang, F. Wang, H. Gong, C. He. Carbon 2016, 100, 151.

[83] Y. Gogotsi. ACS Nano 2014, 8, 5369.

Chapter 18

Applications of Carbon Nanotubes and Their Polymer Nanocomposites for Gas Sensors

Dr. Abdul Hakim Shah

Additional information is available at the end of the chapter

http://dx.doi.org/10.5772/63058

Abstract

Because of extensive applications in industrial and environmental monitoring, biomedicines and pharmaceutics, etc., gas sensors are being focused widely by the research community since a few decades. Generally, gas-sensing materials include semiconducting metal oxides, vapour sensitive polymer, porous silicon, etc. Based on the gas-sensing principle of adsorption/desorption of target gas molecules on the sensors' surface, significant enhancement in sensitivity could be achieved by increasing the interfacial contact between the sensors' surface and analytes (target gases). Carbon nanotubes (CNTs), due to their unique electron transport phenomenon, have proven their ability to utilize them as sensing material in conductometric gas sensors. This chapter consists of three major sections. *First section* contains studies about the fundamentals of gas sensors and definitions of technical parameters used to characterize them. *Second section* describes up-to-date structural and chemical studies of the CNTs in detail in connection with the dependence of electrical transport phenomena upon these properties. Their gas-sensing mechanism and several literature reports about such investigations are also quoted and explained in easy language. *Third section* describes CNTs-polymer nanocomposites for conductometric gas sensors, which have been described in details and a comprehensive way. Conclusions have been drawn, and references are enlisted at the end of the chapter.

Keywords: CNTs, polymers, nanocomposites, gas sensors, sensing mechanism

1. Gas sensors; review

1.1. Criterion and their physical significance

Chemical sensors (gas sensors) are being intensely focused due to their extensive applications in industries, environmental monitoring, space exploration, food technology, biomedicine and pharmaceutics. Their performance is characterized by several technical parameters such as sensitivity, lowest detection limit, response time, recovery time, selectivity and working temperature. Highly sensitive gas sensors are required for the leakage detection of explosive and toxic gases in industries. Gas sensors with quick response have strong demand in environmental monitoring and globe warming.

An effective gas sensor must have following decisive factors:

a. It must have high value of sensitivity magnitude and discrimination ability (selectivity).

b. It must have fast speed of responding.

c. It must have lower detection threshold value.

d. Its work temperature must be low for minimum power consumption.

e. It must be stable in repeating use.

In order to obtain above criterion, materials such as polymers, semiconducting metal oxides and silicon with high porosity are generally used. Taking in mind the principle of adsorption/desorption, maximization of the contact's cross-section between target gas molecules and gas-sensing material's surface is one of the major tactics to enhance the sensitivity magnitude.

Detection of different gases, particularly the volatile organic compounds (VOCs) such as ammonia, amines, ethanol and acetone, has vital significance in industries. *Ammonia* being pungent and toxic in nature is a by-product in industries such as fertilizers, refrigerant, as well as dung from agriculture, flora and fauna. Being a poisonous and explosive-nature gas, it could harm people at ppm or ppb level leakage. *Propylamine* ($C_3H_7NH_2$) is a high hazardous gas with caustic and exasperating nature; create pollution in chemical and fisheries industries. Similarly, in order to monitor alcoholic percent in the blood, *ethanol* sensing is necessary from the clinical and forensic point of view. *Acetone* is on high degree among all volatile organic compounds, which has applications as an organic solvent in the laboratories and industries. In the literature, there are recent investigations about its harmful internal as well as external effects on the health, for example, eyes' annoyance, sluggishness and sleepiness of the body nervous system when it is exposed to acetone continuously. In food technology, acetone is emitted by potatoes in their long-time storage. This shows that how much is it important to be detected timely [1, 2, 91].

The first gas sensor was designed by Seiyama in 1962, which was based on ZnO thin film with working temperature 400°C. Later on, several other material scientists contributed to this research area and their efforts took a big turn after the introduction of nanosized materials. Introduction of nanomaterials has made it more efficient to produce sensitive, selective, economic, small-sized sensing systems with low power consumption. Moreover, high surface area and hollow structure of the nanomaterials have exposed them effective for good adsorption. Therefore, nanomaterials such as carbon nanotubes (CNTs), metal oxides nanostructures (nanowires, nanofibres and nanoparticles) are under investigation for gas sensors.

1.2. Performance parameters for gas sensing

In the gas sensors research, in addition to the measurement of resistance (conductivity) or other physical properties, a set of characteristic parameters need to be evaluated for the performance prophecy of the materials [3]. A brief description of certain technical parameters as stated in the last section is presented as under.

1.2.1. Sensitivity

It is defined as 'the change of measured signal (response) of a gas sensor per unit change in analyte (target gas) concentration'. More specifically, 'it is the slope of a calibration graph'. As we know that the metal oxide-based gas sensors work on the principle of resistance change (so-called conductometric), therefore, their sensitivity is described in terms of change of resistance. Mathematically, we can write.

$$S^{n}_{red(oxd)} = R_{air(gas)} / R_{gas(air)} \tag{1}$$

$$S^{p}_{\mathrm{red(oxd)}} = R_{\mathrm{gas(air)}} / R_{\mathrm{air(gas)}} \tag{2}$$

where the left hand sides of Eqs. (1) and (2) represent sensitivities of n-type and p-type semiconducting materials towards reducing and oxidizing gases, respectively. R_{air} is the base-line resistance of the sensor in air, and R_{gas} is the value of resistance when the sensor is exposed to the target gas. Some researchers do mention the % values of the sensitivities, which could be obtained by simply multiplying the right-hand side of Eqs. (4) and (5) by 100.

1.2.2. Selectivity

Selectivity determines the selective response of the gas sensor towards a specific target gas among a group of gases, or more precisely, the particularity of the gas sensor towards a specific target gas molecule in a mixture of gases. For example, for a mixture of target gases A, B, C and D, the selectivity of the sensor towards the target gas A is given as;

$$Sel_{A} = S_{A} / \sum S_{i} \tag{3}$$

where S_A denotes sensitivity of the sensors towards target gas A, and $\sum S_i$ is the sum of sensitivities for all the gases A, B, C and D, that is $\sum S_i = S_A + S_B + S_C + S_D$. Generally, the sensitivities are taken in percentage (% response) for the calculation of selectivity.

1.2.3. Stability

It is 'the ability of a sensor to maintain its reproducible performance of response through certain cycles or a period of time'. In order to evaluate the stability, repeated response of the sensor is measured for a particular concentration level, through the certain cycles of operation or periodically through a time zone.

In order to investigate the reproducibility of the gas sensor, *coefficient of variation* (CV) is calculated for a repeated response of the sensor towards a constant value of target gas concentration through a range of cycles. Mathematically 'it is the ratio of standard deviation to the mean of response magnitude'.

$$CV = \sigma / \mu \tag{4}$$

1.2.4. Optimum working temperature

This is the value of temperature, at which the sensor possesses maximum sensitivity for a particular concentration level of the target gas. This is an important parameter regarding the performance of sensor, and it is the characteristic of material used, because every metal oxide works at particular operating temperature. Evaluation of power consumption is based on the working temperature. The higher the optimum working temperature, the higher is the power consumption.

1.2.5. Response and recovery speeds

These mean the response and recovery times of a gas sensor. Response time is defined is 'the time required to obtain 90% of the saturated value of resistance during its change upon exposure to certain target gas'. Recovery time is 'the time needed for attainment of 90% of the base-line value of resistance after the release of target gas'. Both response and recovery times determine the speed of sensor towards its cyclic operation.

1.2.6. Detection threshold or detection limit or limit of detection (LOD)

LOD is defined as 'the lowest concentration value of the target gas, which the sensor is able to detect at a particular value of operating temperature'. This is an important parameter estimating the lowest concentration extreme for a gas sensor, at which the resistance onsets changing.

1.2.7. Dynamic range

It is the range of target gas Concentrations, starting from detection limit to the highest value of limiting concentration, through which the sensor's response is measured. It evaluates the sensor's performance through an extended concentration zone.

1.2.8. Linearity

It is 'the deviation of the response calibration graph versus target gas concentration, from straight line'. It is obtained by linear fit of the response concentration curve for a metal oxide-based gas sensor, and the slope of linear fit is measured.

1.2.9. Life cycle

This is an important parameter which covers both the shelf-life as well as operating life. Operating life time is 'the period of time over which the sensor could be operated continuously'. The shelf-life is 'the maximum length of time, for which the sensor is stored after operation'.

2. Carbon nanotubes (CNTs)

2.1. Background and general classification

In 1991, during his experiment of electric arc discharge for fullerene production, Iijima [4] noticed that certain novel structures also produced which were named afterwards as carbon nanotubes. They are actually graphene sheets rolled specifically in the tube shape. Since that time, carbon nanotubes are the title of hardcore investigative research in different applications. Depending upon the range of diameters, carbon nanotubes may be classified in single-walled and multi-walled.

a. **Single-walled carbon nanotubes (SWCNTs)** consist of a bundle of string like which is a single-walled nanotube with diameter of the order of a nanometer (~1.4 nm) and inter-tube distance upto 0.3 nm approximately.

b. **Multi-walled carbon nanotubes (MWCNTs)** contain several graphene sheets rolled up in layers form, resulting in concentric tubes with increasing diameter. Due to significant Van der Wall forces, the MWCNTs are amassed together, giving typical diameter upto 10–20 nm, and their interlayer spacing upto ~0.34 nm.

More simply saying, carbon nanotubes can be considered as a macromolecule of carbon, very similar to a graphite sheet (which is used as your pencil lead) but wound in a cylindrical reel form (**Figure 1**). When graphite is coiled, the carbon arrangement takes as much strength as one hundred times of steel's and almost 2 mm long. Furthermore, these nanotubes exhibit flexibility, thermal stability and inertness chemically. They nanotubes are capable of either metallic or semi-conducting nature but it depends on the 'twirl' of the tube.

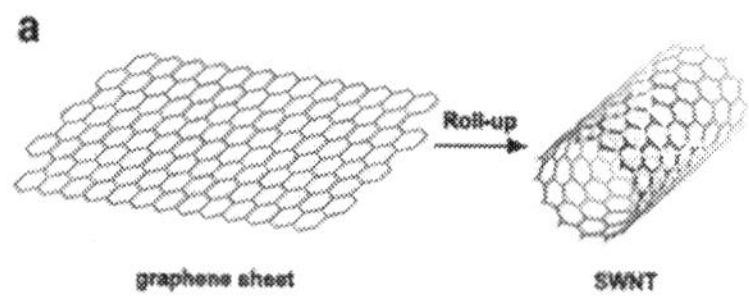

Figure 1. Roll of graphene into SWCNT [5].

Two important parameters for carbon nanotubes on which their electronic properties are supposed to be dependent are diameter and helicity (Chirality). Chirality is the degree of twist of the tube and is found by Chirality vector, denoted by C_h, given by [5]

$$C_h = n\mathbf{a} + m\mathbf{b} \tag{5}$$

where **a** and **b** are unit vectors of the hexagonal lattice. n and m are integers which denote chiral vector as (n,m) expression (**Figure 2**). Axis of nanotubes and chiral vector both are orthogonal to each other. Using the values of n and m, diameter and Chiral angle (Chirality) are calculated as following:

$$d_t = C_h / \pi \tag{6}$$

and

$$cos\theta = \frac{2n+m}{2\sqrt{n^2+m^2+nm}} \tag{7}$$

Figure 2. Schematic representation of a 2D graphene sheet illustrating lattice vectors a1 and a2, and the roll-up vector na1 + ma2 [5].

Tubes are classified into zigzag and armchair tubes, according to the different values of θ which can take the values from 0° to 30°.

a. For θ = 0°, m = 0, we have (n, 0) type tubes, which are called zigzag tubes.

http://dx.doi.org/10.5772/63058

b. For $\theta = 30°$, m = n, we have (n, n) type tubes, which are called armchair tubes.

Metallic or semiconducting nature of the tubes can also be determined by (n, m) indices. Metallic conductivity comes for the three cases.

a. Armchair tubes with indices (n, n).

b. Zigzag tubes with indices (n, 0) in which n and n–m are the multiples of 3.

All other tubes with arbitrary (n, m) values are semiconducting. For a random distribution of nanotubes, statistical analysis predicts that approximately ⅓ are metallic and ⅔ are semiconducting [6].

2.2. Electronic transport

Discovery of carbon nanotubes and their studies thereafter have opened new frontiers of science and industries. Unique geometrical features of carbon nanotubes are convincing towards facing of certain quantum mechanical problems which might be helpful in low-dimensional physics as well. In this section, we will study certain specific aspects of the electron transport in both single-walled and multi-walled CNTs which cannot be taken into account in case of mesoscopic physics generally. A new phenomena of Aharonov-Bohm will be described Which has recently been investigated in CNTs-based field effect transistor devices. The role of chemical doping/composite in case of intrinsic transport and interfacial effects will also be given a touch. Such interfacial effects have vital significance in sensing technology.

It has been reported that electronic transport properties of the nanotubes are connected with their electronic band structures which consist of multiple one-dimensional sub-bands sliced from the Dirac sliced from the Dirac dispersion cone of graphene.

2.3. Gas sensors based on CNTs

Conductometric gas sensors work on the principle of change in resistance due to the interaction with target gas molecules. It has been reported that gas sensors based on semiconducting carbon nanotube offer high sensitivity, quicker response, and better portability towards certain target gases (such as NO_2 and NH_3) as compared to conventional thin-film gas sensors [7].

2.3.1. History and working principle

After the discovery of CNTs in 1991, just like other devices, a gigantic research run was embarked on to investigate their use in gas-sensing devices as well. Their synthesis, possessions (such as electronic, mechanical and thermal) and applications in different areas are being studied deeply. Both calculative and experimental works are also being conducted for their better understanding. Researchers all over the world are in search of their performance as gas sensors such as NASA's researchers are doing for the identification of atmospheric components of various planets. When they are exposed to certain gases, variation in their different properties is detected by various methods.

2.3.2. *Device fabrication techniques based on CNTs*

Carbon nanotubes are grown through arc-discharge, laser ablation and chemical vapour deposition (CVD) methods [8]. In order to fabricate CNTs-based gas sensor device, three techniques are generally applied [9].

2.3.2.1. *Casting technique*

One way is to caste CNTs on interdigitated electrodes (IDEs). In this method, the electrodes are fabricated by photolithography technique and evaporation of Ti and Au on silicon oxide. As-grown SWCNTs are purified with acid first and then by air oxidation before they could be integrated with the IDEs [10]. The major advantage of this method is that purified nanotubes are obtained so that the impact of impurities can be minimized. These purified nanotubes are then dispersed in the dimethylformamide (DMF) and drop-deposited onto the electrode area. Finally, the DMF is evaporated such that a network of nanotubes is subsequently obtained. This device fabrication technique was developed firstly by Li et al. for SWCNTs-based gas sensor device.

2.3.2.2. *Screen printing*

This is another simple technique which is widely used for the fabrication of gas sensor device. In this method, CNTs paste is mixed with organic binders (terpineol, ethyl cellulose) and glass frits are printed on electrodes coated glass or tube. The organic binders are removed by evaporation through annealing in N_2 atmosphere. Lee at al. first used this technique to fabricate MWCNTs-based gas sensor for NO_2 gas detection.

2.3.2.3. *Dielectrophoresis (DEP)*

It is another technique which is used for the fabrication of CNTs-based sensors. It is the electrokinetic motion of dielectrically polarized materials in non-uniform electric fields which is used to manipulate CNTs for separation, orientation and positioning. This technique can establish a good electrical contact between CNTs and the electrodes. Suspension of highly purified CNTs in ethanol is ultrasonicated for some time. A glass substrate is patterned with castle-wall-typed interdigitated microelectrode so that high and low are formed periodically. This electrode is surrounded by a silicon rubber spacer in the form of a sealed chamber, and the CNTs suspension is continuously injected by a peristaltic pump. An AC voltage is applied for the DEP trapping of CNTs on the microelectrode for a definite period of time. Finally, the DEP process is stopped and ethanol is evaporated at room temperature. Suehiro et al. applied this technique for the MWCNTs-based gas sensor devices to detect various target gas vapours such as NH_3, NO_2, SO_2 and HF.

2.3.3. *Gas adsorption on carbon nanotubes; parametric analysis*

Using first-principles calculations by density function theory, adsorption of various target gases on CNTs has been studied by Peng and Cho as well as Zhao. Such an adsorption is characterized by three parameters, that is tube-molecule distance (d), adsorption energy (E_a) and charge transfer (Q) [11].

2.3.3.1. *Tube–molecule distance (d)*

It is defined as the nearest distance between atoms on the molecule and the nanotube for T site (top of a carbon atom), or the distance between the centre of the gas molecule and the centre of the carbon hexagon (carbon–carbon bond) for the B site (top of centre of the C–C bond).

2.3.3.2. *Adsorption energy (E_a)*

It is defined as the total energy gained by molecule adsorption at equilibrium distance. Mathematically,

$$E_a = E_{total\,(Tube+gas\ molecule)} - E_{(tube)} + E_{(Gas\ molecule)} \tag{8}$$

When $E_a < 0$, the reaction is exothermic and spontaneous, that is high value of adsorption energy releases more energy during the reaction. However, for $E_a > 0$, it is relatively hard to continue the reaction because of the amount of energy required.

2.3.3.3. *Charge transfer (Q)*

It denotes the total Mulliken charge number on the molecules, If Q is positive then charge transfers from molecule to tube. It has been shown computationally that large charge transfer changes the resistance (or conductance) of the tubes. Such changes are recorded by voltage variation which could be related to the gate capacitance of the device. Mathematically,

$$\Delta Q = C_g \Delta V_g = \delta\theta\left(\pi dl/\sigma\right) \tag{9}$$

Here, C_g is the nanotube capacitance, ΔV_g is the change in voltage recorded during experiment. 'd' is diameter, and 'l' is the length the CNT, σ is the molecules cross-sectional area, and δθ is the coverage of the molecules over the CNT's surface.

2.3.4. *Electronic transport dependency over molecular adsorption of target gases*

We know that generally the electronic properties of CNTs are determined by their chirality and diameter [12]. However, experimental evidences demonstrate that the electronic properties are also very sensitive to the gaseous environment around CNTs, particularly when they are exposed to oxygen. Such a variation in CNTs electronic properties are attributed towards the charge transfer between CNT and target gas molecules where the molecules act as electrons' donors or accepters. In case of no charge transfer, the resistance variation is mainly due to the variation in carriers' free lifetimes and mobility. It has therefore been concluded that the electrical response of SWNTs bundle to various target gases can either be due to the change of fermi level due to the charge transferring or by the additional scattering channel for conduction electrons in the walls of the CNT. Mathematically, this can be understood by following equation [13]:

$$S = S_o + \left(\rho_a / \rho_o \right) (S_a - S_o) \tag{10}$$

where S_o and S_a denote the thermopower contributed from the host resistivity ϱ_o, and the impurity resistivity ϱ_a associated with target gas adsorbed, respectively. This equation is also known as the Nordheim-Gorter (N-G) relation [14]. Its shows mathematically whether the gas molecules are physisorbed or chemisorbed.

2.4. Mechanism-based design classification of CNTs-based gas sensors

As we have mentioned earlier that the electronic properties of CNTs are very sensitive to the exposed environment due to the gas molecule adsorption. Physical quantities such as thermopower, resistance and density of states of single SWNT or SWNT bundles vary on exposing them to certain gases. Sensors with resistance variation on exposure to the target gases are generally termed as conductometric gas sensors. Due to the easy manufacture and portability, conductometric sensors are the most commonly used architecture in the design of CNTs gas sensors. Such sensors have further following subclasses on the basis of architecture.

2.4.1. Semiconducting CNTs FET gas sensors

We know that CNTs can be semiconducting as well as metallic. If we connect a semiconducting CNT in between two metal both sides, it forms a metal/semiconducting/metal-type device which behaves as p-type transistor. Therefore, the resistance variation of CNTs to the target gas adsorption can be recorded with field emit transistors (FETs). Compared with SWCNTs, the mechanism of MWCNTs' response to gas adsorption is more complicated due to the multilayer tube structure. However, they also show high sensitivity to specific gases experimentally [15, 16].

2.4.2. CNTs-based two terminal gas sensors

The change in resistance of CNTs after exposing them to gaseous atmosphere can also be detected by resistors with two terminals connected with dc voltage. In these kinds of gas sensors, the interdigitated electrodes are applied to provide larger sensing areas and sufficient contacts between the electrodes and the coated CNTs [17].

2.4.3. Carbon nanotubes enhanced ionization chamber for gas sensing

The major drawback of the chemical gas sensors is that gas molecules with low adsorption energy are hardly detectable with them. Therefore, another type of sensors called ionization gas sensor are applied. Such sensors are based on the finger printing ionization characteristics of the target gases. Their working is based on the gaseous molecules' ionization due to their collisions with the accelerated electrons. Practically, there is no adsorption and chemical interaction between the sensing device and target gas molecules. Hence, adsorption energy and charge transfer with the sensing material are not the desired parameters in such sensors. Detection of inert gases or gas mixtures can easily be achieved by ionization chamber, which in turn delivers better selectivity. However, such sensors are bulky in architectures, with considerable high power consumption and breakdown voltage, which are industrially functional drawbacks. Carbon nanotubes can lead to the large field enhancement factor, thereby increasing the electric field around the tips reasonably and produce corona discharge at very low voltages [18]. This has shown importance in ionization gas sensors [19–27]. Besides this, CNTs are good emitter sources of electrons because they have very sharp tip curvature and low electron escaping work function.

Another important factor that dominates the ionization characteristic is the inter-electrodes distance. According to Paschen's curve, in order to induce the breakdown phenomenon, a certain gap distance is required, and a decrease in the breakdown voltage occurs with reducing the inter-electrodes distance.

Integrated dielectric barrier discharge (DBD) mechanism is a general technique used to fabricate the novel ionization gas sensor with short-gas spacing, CNTs and DBD coating to realize low power consumption and breakdown voltage. It has been shown that the DBD layer effectively lowers the breakdown voltage and improves the reproducibility of the device [25].

2.4.4. CNTs-based gas-sensing capacitor

These sensors are based on the randomly aligned carbon nanotubes and are used for humidity sensing. The structure of such a sensor is shown in **Figure 3**.

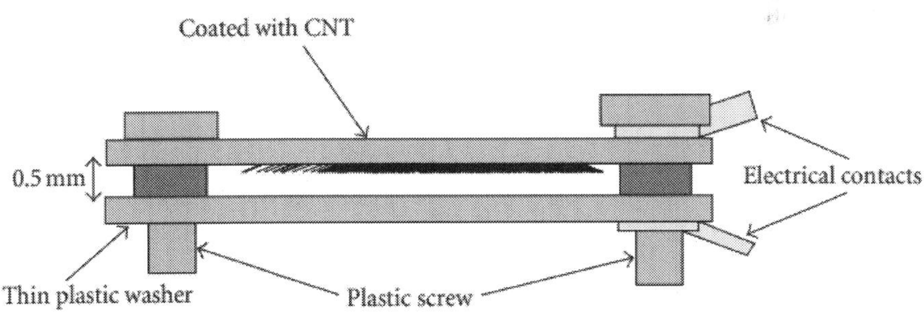

Figure 3. Parallel plate capacitive as a humidity sensor with MWNTs deposited on one plate.

Capillary condensation is responsible for the increase in the capacitance under high relative humidity. Such a condensation effect is induced by the nanopores and interstitial gaps in the random aligned CNTs structure.

2.4.5. CNTs-based resonator sensor

The working principle of these sensors is based on the change of the electrical properties (resonant frequency) of the sensing materials in the gaseous exposure. Circular disk resonators with SWCNTs coated on top of the disk have been used as resonator sensor. Resonant frequency is given by following relation;

$$f_o = 1.841c \Big/ 2\pi a\sqrt{\epsilon_r} \tag{11}$$

where 'f_0 is the resonant frequency, c is the light's speed in vacuum, a is the radius of the disk, and ε_r denotes the relative dielectric constant of the substrate'. When the CNTs-based resonator sensor is exposed to certain target gas vapours, the alternation of the dielectric constant of the circular disk with CNTs on top induces the resonant frequency shift.

2.5. Pre-functionalization of CNTs, a recent strategy

Although as-prepared/as-obtained carbon nanotubes can be used to fabricate gas sensors of any type as discussed in the last section; however, their sidewall functionalization improves their performance which opens new doors for their applications in many areas [28–30].We have studied that the electronic sensitivity of CNTs to various target gases is highly limited by the binding energy and charge transfers of the molecules with the CNTs wall. When they are functionalized, active bonds appear on the side walls of CNTs which would indeed enhance their sensing capability towards wide range of gaseous vapours [31].

It has been investigated the resistance response of chemically functionalized MWNTs (f-CNTs) to alcohol vapours with ultralow power consumption [32]. For the fabrication of the sensor, purified MWNTs are chemically functionalized in acidic media. To do so, pure CNTs (commonly, MWCNTs) are sonicated in a mixture acids. It is believed that after chemical functionalization, the polar COOH groups are attached onto the nanotubes surface. The sensors will therefore give stronger responses to the alcohol. This is due to the fact that their absorption efficiency with these volatile organic molecules will be increased due to the dipole–dipole interactions (mainly hydrogen bonding) between the COOH and the polar organic molecules. Although CNTs have been proven as a good material for sensor, yet, the pure CNTs have almost zero response to H_2 because they have very week binding energy [33]. However, when they are functionalized with materials such as Pt or Pd, CNTs may be very sensitive to hydrogen as well [34, 35]. Basic mechanism for such improvement is that the hydrogen molecules dissociate into atomic hydrogen on Pt/Pd surfaces and form PtH_2/PdH_2, which lowers the work function of Pt/Pd and causes the electron transfer from Pt/Pd to MWNTs [36–38].

The question is that how can we functionalize CNTs efficiently? Several methods are used to obtain f-CNTs. Reflux of CNTs in acidic media (HNO_3 or HCl or H_2SO_4 or their mixtures) is very common nowadays. Such a chemical acidification activates the surface bonding of CNTs which would then readily interact with the target gas vapours as discussed earlier.

2.6. Literature survey regarding CNTs-based gas sensors

Since its discovery in 1991, carbon nanotubes have been the core research topic in various areas of applications. Sufficient work has been done by others; however, there is still certain gap in some circumstances, and therefore, researchers are still in an extensive run to optimize the properties of carbon nanotubes for different industries such as cathodes for supercapacitors, batteries, and gas sensors.

In order to obtain gas sensors based on carbon nanotubes, the first step is to fabricate a good sensor device which could execute certain functional and architectural advantages such as easy in manufacturing, portable, economical, capable to be utilized in all weathers with minimum seasonal effects, consume low power during operation, high sensitive as well as selective and reproducible as well. Various methods are used to fabricate a good sensor device as described in previous sections. Earlier, several efforts in this regard have been reported in literature. Li et al. designed a conductive gas sensor by casting SWCNTs on the interdigitated electrodes (IDEs) [39]. Such a device is shown in **Figure 4**.

Lee et al. used screen-printing technique to fabricate a simple gas sensor device based on CNTs. CNT paste was mixed with MWNTs, terpineol, ethyl cellulose, and glass frits onto electrode-coated glass as the gas-sensing element for NO_2 detection. The organic binder was removed by annealing in N_2 ambient [40].

Dielectrophoresis (DEP) method has been used by Suehiro et al. [41]. During fabrication, the CNTs with high purity were ultrasonicated in ethanol for 60 min. The system to fabricate the CNTs-based gas sensor is schematically depicted in **Figure 5**.

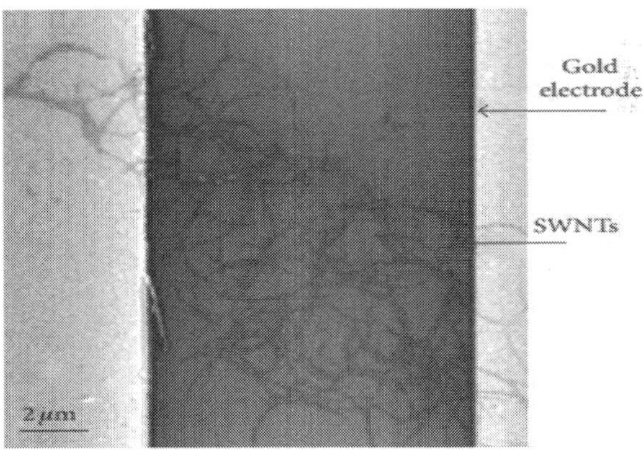

Figure 4. SEM image of SWCNTs across two gold electrodes [39].

Such DEP-fabricated CNTs gas sensors successfully detected various target gas vapours such as ammonia, nitrogen dioxide, sulphur dioxide and hydrogen fluoride.

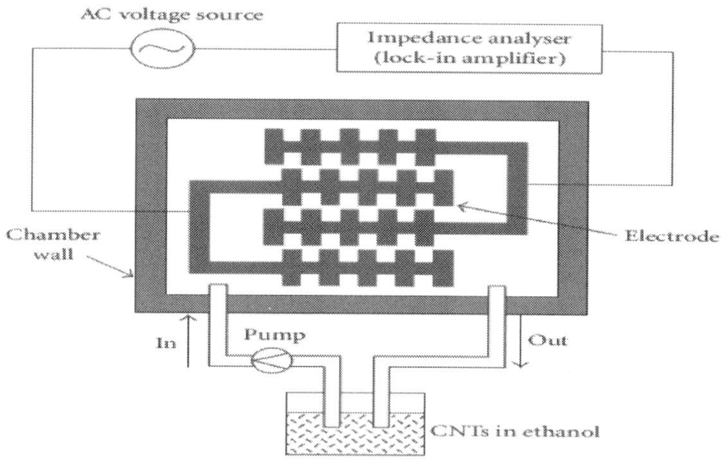

Figure 5. Schematic diagram of the experimental set up for MWCNTs-based gas sensor device by DEP method.

A three terminal gas sensor device based on vertically aligned CNTs for N_2 gas has been fabricated successfully by Huang et al. [42]. The CNTs were grown by thermal CVD on a silicon substrate at 700°C. Substrate was pre-heated and then cooled down at room temperature in N_2 ambient. A 30-nm-thick Fe layer was sputtered on the substrate as catalyst, and C_2H_2 was used as the carbon source. The as-grown CNTs mat is shown in **Figure 6**.

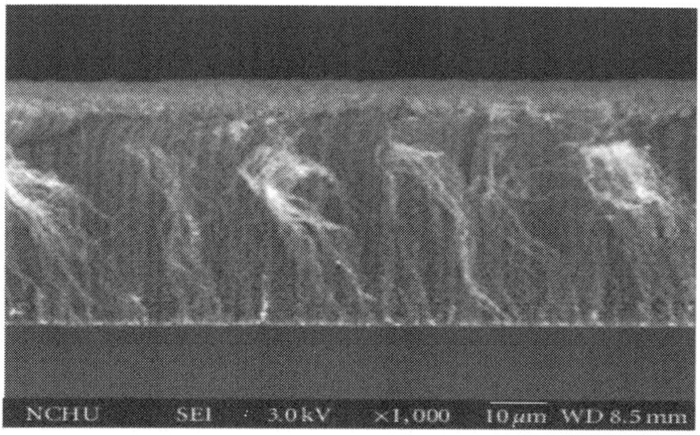

Figure 6. Vertically aligned CNTs mat (cross-sectional image) [45].

An ammonia sensor based on laterally aligned MWCNTs has been fabricated by Jang et al. [43]. A substrate of N-type heavily doped Silicon wafer with SiO_2 layer on top was used, and the

Nb electrodes were deposited and patterned on it. Such a gas sensor device with detailed steps of fabrication is depicted in **Figure 7**.

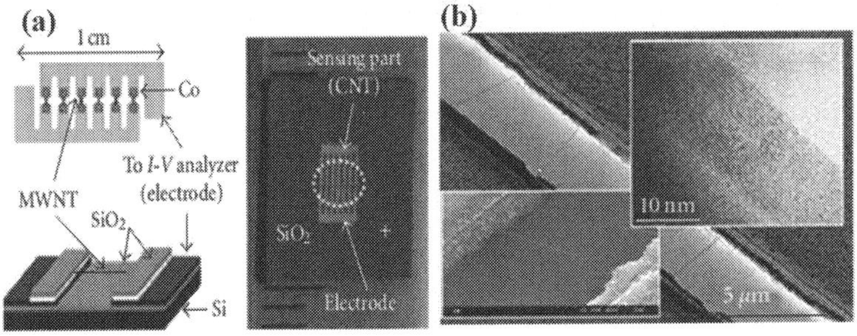

Figure 7. (a) Ammonia chemical sensor based on laterally aligned MWCNTs, (b) aligned CNTs on SiO_2 surface. TEM image of laterally grown CNTs (inset) [43].

NH_3OH and HCl sensor devices based on self-aligned and self-welded MWCNTs were prepared successfully between pre-patterned electrodes by Tabib-Azar and Xie, through a low-pressure CVD (LPCVD) method [44].

In order to study the adsorption of target gas molecules over the CNTs surface and the resultant-induced effects, Peng and Cho applied first-principles calculations using density functional theory and investigated the adsorption of NO_2 gas on SWCNTs surface and resultant changes in various parameters [11]. Binding configuration of NO_2 on semiconducting (10, 0) SWCNTs surface was obtained which showed that the NO_2 gas molecule of such a configuration was found to bind with SWNT with adsorption energy of 0.3 eV. It was also found that the NO_2 molecule has high diffusion kinetics on nanotubes surfaces. Electron density analysis was also carried out which reveals that charge transfer is induced from C atom to the NO_2 gas molecule, thus leading to p-type (hole) doping of these (10, 0) SWCNTs.

Such DFT calculations were also performed by Zhao et al. with the adsorption of molecules of various gases such as NO_2, O_2, NH_3, N_2, CO_2, CH_4, H_2O, H_2 and Ar, on both individual SWCNT and SWCNT bundles [48]. In case of individual SWNT, the data calculated for equilibrium tube–molecule distance, adsorption energy and charge transfer towards various target gas molecules on (10, 0), (17, 0) and (5, 5). SWNTs is shown in **Table 1**.

Results in **Table 1** reveal that the gaseous molecules (except for NO_2 and O_2) are charge donors with minute charge transfer (0.01–0.035 e) and weak binding energies (≤0.2 eV). These gas molecules can be identified as physisorbed. For the gases O_2 and NO_2, it is shown that both of them are charge acceptors with large values charge transfer and adsorption energies. These results are in good consistency with the experimental data [46, 47]. Furthermore, no clear dependence of adsorption on the tube size and chirality was proved.

	NO_2	O_2	H_2O	NH_3	CH_4	CO_2	H_2	N_2	Ar
(10, 0) SWNT									
d (Å)	1.93	2.32	2.69	2.99	3.17	3.20	2.81	3.23	3.32
E_a (meV)	797	509	143	149	190	97	113	164	57
Q (e)	–0.061	–0.128	0.035	0.031	0.027	0.016	0.014	0.008	0.01
Site	T	B	T	T	C	C	C	C	C
(5, 5) SWNT									
d (Å)	2.16	2.46	2.68	2.99	3.33	3.54	3.19	3.23	3.58
E_a (meV)	427	306	128	162	122	109	84	123	82
Q (e)	–0.071	–0.142	–0.033	0.033	0.022	0.014	0.016	0.011	0.011
Site	T	B	T	T	C	C	C	C	C
(17, 0) SWNT									
d (Å)	2.07	2.50	2.69	3.00	3.19	3.23	2.55	3.13	3.34
E_a (meV)	687	487	127	133	72	89	49	157	82
Q (e)	–0.089	–0.096	0.033	0.027	0.025	0.015	0.012	0.006	0.01
Site	T	B	T	T	C	C	C	C	C

The optimal adsorption sites: T (top of a carbon atom), B (top of the centre of the C–C bond), C (top of the centre of carbon hexagon).

Table 1. Data for equilibrium tube–molecule distance (d), adsorption energy (E_a) and charge transfer (Q) of various molecules on (10, 0), (17, 0) and (5, 5) individual SWNTs [45].

In case of SWCNTs bundle, following four distinct sites are specified for the adsorption of gas molecules [48, 13]:

a. External surface of the nanotubes bundle,

b. The groove formed at the contact between adjacent tubes on the outside of the bundle,

c. The interior pore of individual tubes,

d. Interstitial channel formed between three adjacent tubes within the bundle.

Adsorption of the target gas on these sites is decided by two parameters such as

a. The binding energy of the target gas molecules,

b. Availability of the sites on CNT surface.

Reason for such determination is that out of above-mentioned four sites, some sites may not adsorb the gas molecules due to mismatch of the dimension of molecule and the site diameter. The c-site (interior pore of the tube) can be accessed by molecules only in following cases [49];

a. SWCNT is uncapped,

b. SWCNT has defects on the tube walls.

Binding energies and contribution of specific surface area for the adsorption of H_2 gas molecules as calculated by Williams and Eklund [47], are shown in **Figure 8**.

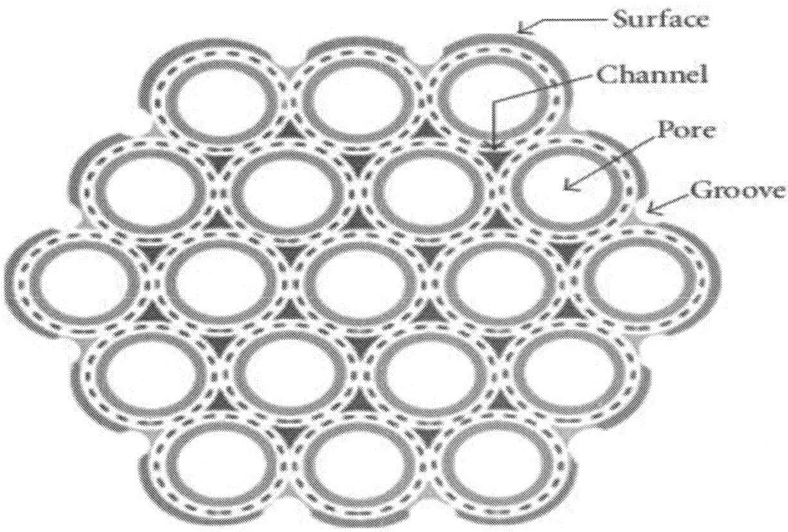

Figure 8. Schematic structure of SWNT bundle with available sites for gas adsorption [47].

The binding energy data in **Table 2** shows that in the order of increasing, following order holds good:

$$E_{B(channels)} > E_{B(grooves)} > E_{B(pores)} > E_{B(surface)}, \tag{12}$$

S. no.	Site	Binding energy (eV)	Specific surface area (m²/g)
1.	Channel	0.119	45
2.	Grooves	0.089	22
3.	Pores	0.062	783
4.	Surface	0.049	483

Table 2. Binding energies and specific surface areas of specified sites.

Specific surface areas for the pores and surface have larger values.

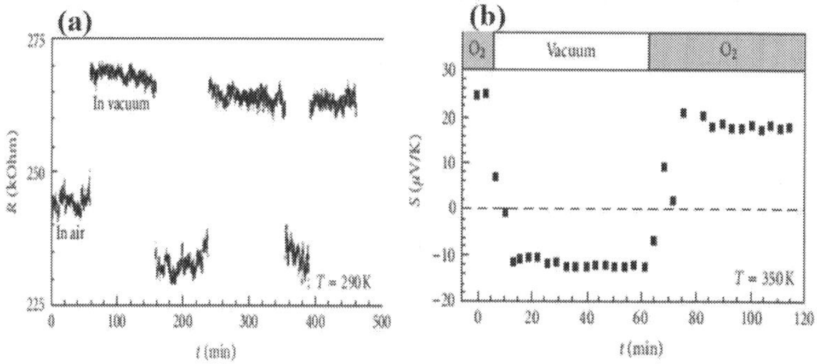

Figure 9. (a) Sensitivity of the electrical resistance 'R' of SWNT films to the gas exposure, and (b) sensitivity to environmental conditions of thermoelectric power 'S' for SWNTs. In vacuum, S is n-type, whereas in oxygen exposure, S is p-type, with a larger magnitude [46].

Several other parameters, such as electrical resistance, thermoelectric power and local density of states, for CNTs exposed to target gas molecules have also been measured. The work done by Collins et al. [46] in this regards sounds better. Using transport measurement and scanning tunnelling spectroscopy, they measurement these parameters and found that all these three parameters change reversibly even when SWCNTs are exposed to small concentration of oxygen (**Figure 9**).

Likewise, effect of degassing on these parameters for SWCNTs was studied by Sumanasekera et al. at 500 K [49] and it was found that thermoelectric power decrease slowly from 54 μV/K and reversibly approach to a constant value of −44 μV/K (**Figure 10**). Spin–lattice relaxation of all the SWCNTs was also studied using nuclear magnetic resonance (NMR) for which a dramatic increase was observed in the oxygen exposure.

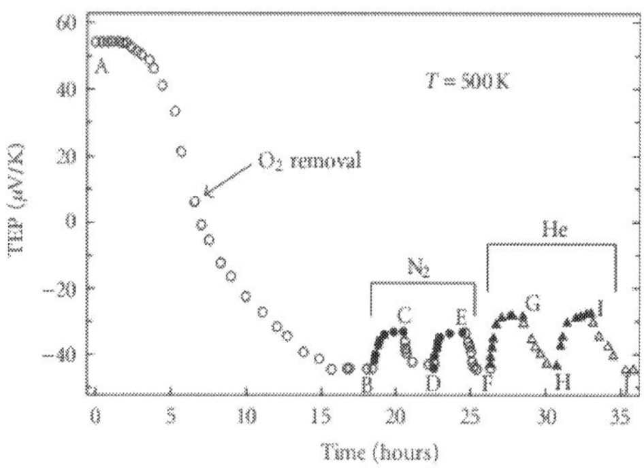

Figure 10. Thermoelectric power 'S' versus time for a mat of SWNTs at T = 500 K initially saturated with air at ambient conditions.

http://dx.doi.org/10.5772/63058

Besides oxygen, the effect of alcohol (methanol, ethanol, isopropanol and butanol), ammonia and water molecules on the electrical transport of SWNT bundles have also been studied and it was found that both thermoelectric power and resistivity (resistance) varies strongly with the alcoholic exposure [40]. All the alcoholic molecules were found to be physisorbed on the tube wall as indicated by the Nordheim-Gorter N-G equation plot (**Figure 11**).

A linear N–G plot indicates physisorption of the gaseous molecules. Significant effect of water vapours over the resistance was observed as well; however, it did not alter the thermoelectric power. Curved-type N–G plot for the ammonia and water shows the chemisorption of the adsorbed molecules.

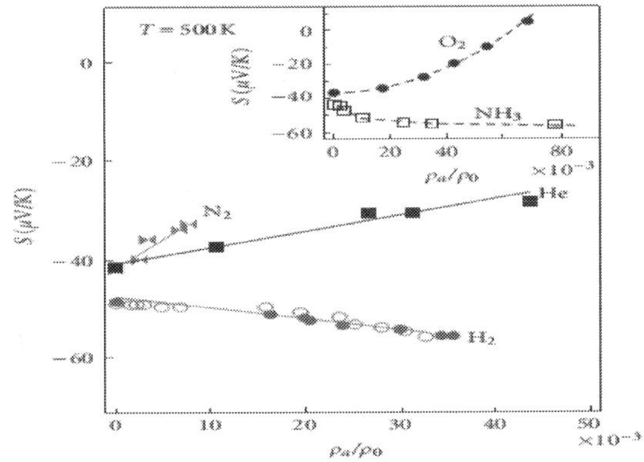

Figure 11. N–G plots (S versus ϱ) showing the effect of gas adsorption on the electrical properties of the SWCNTs bundle [49].

As far as the design and sensing properties of the CNTs-based gas sensors on the basis of sensing mechanism is concerned, several researchers took the charge in this regard. The efforts made by Kong et al. [34] are of vital significance, who fabricated a semiconducting SWCNT FET from a single SWNT for chemical sensing. The single SWCNT was grown on SiO_2/Si substrate with patterned catalyst islands, through a controlled CVD in which the polysilicon on the back side acted as gate of the transistor. Room-temperature plot of conductance versus time was recorded towards NO_2 and NH_3 (**Figure 12**). It was observed that exposing the device to 200 ppm of NO_2, a sharp increase in the conductance (about three orders of magnitude) is obtained (**Figure 12a**). The sensitivity magnitude of 100–1000 with a quicker response time of 2–10 s response indicated that the device is significantly sensitive towards NO_2 gas molecules. In 1% NH_3 exposure, a dramatic decrease in conductance along with sensitivity of 10–100 was observed which further reveals the sensitivity of the SWCNT-based FET gas sensor device towards ammonia as well (**Figure 12b**).

Another contribution by Huang et al. [41] has been reported who fabricated the gas sensor device based on the vertically aligned MWCNTs bundle with FET structure through thermal CVD and investigated change of resistance (and hence sensitivity) towards the nitrogen (N_2)

gas. P-type semiconducting behaviour was observed and resistance increased with exposure to N_2. The increasing resistance showed an electron donor role of the CNTs and decreasing the concentration of conducting holes on CNTs. The sensor showed relative fast response and the resistance returned back to its initial value in vacuum, which indicated the physisorption of the N_2 molecules.

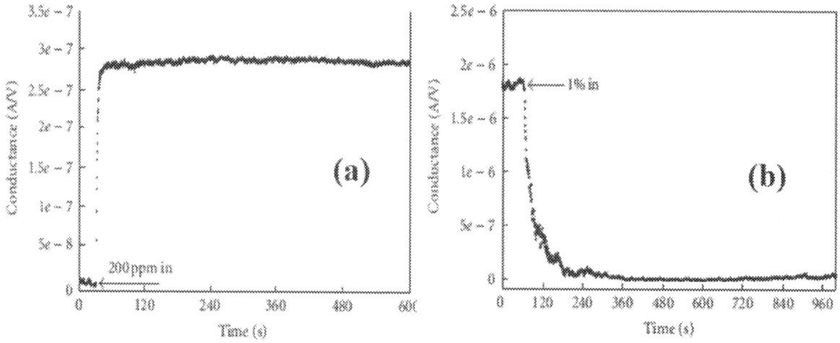

Figure 12. Conductance change of the single SWCNT-based FET sensor when exposed to (a) NO_2 and (b) NH_3 gases [34].

Gas-sensing performance of the sensor based on chemically functionalized MWCNTs has been investigated by Sin et al. [31], and he found that the sensor responded to alcohol vapours at low working temperature (with very low power consumption). The sensor also responded linearly to the alcohol vapour in the concentration range of 1–21 ppm with good selectivity to oxygen, water and alcoholic vapours. These functionalized MWCNTs deliver 0.9–9.6% higher sensitivity magnitude as compared with the un-functionalized MWCNTs.

All these literature reports reveal the feasible sensing nature of carbon nanotubes (both SWCNTs and MWCNTs) and describe their significance as sensing material. Just like the use of CNTs in other applications, a competitive research run in the field of chemical (gas) sensors has been in practice since its first discovery. However, in order to improve the stability and sensitivity effects of CNTs, recently preparation of their composites with the external materials (conducting polymers, etc.) is being intensely focused by the research community.

3. CNTs/polymer nanocomposites

3.1. Review and significance

Polymer is basically a large molecule (macromolecule), which is composed of many repeated units. Such a repeated structure delivers polymers following unique properties;

a. Low density

b. Reasonable strength

c. Good flexibility

d. Easily processable

Polymers occur in two types, that is natural and synthetic. Due to a wide range of properties, both types of polymers exhibit play vital and significant role in industries.

In spite of their numerous properties, the limited mechanical properties of polymers have restricted their use in many engineering applications. Therefore, research community is in continuous run to explore new and novel polymeric materials with improved properties. The first step in this regard was the amalgamation of different polymers to obtain new materials with enhanced properties. However, this leads to only a marginal improvement in properties and their industrial use still remained a question. Then, the strategy of using different organic and inorganic fillers was applied to improve the strength and stiffness of the polymers. The use of long fibres as fillers with the thermosetting polymers enhanced the mechanical properties. However, the processing of polymers with such kinds of fillers is extremely difficult which lacks its commercialization. Therefore, small particles/fibres reinforced composites were applied as a next strategy. Different metals' particles, carbon black and silica, etc., were used as these fillers. Advancement in mechanical properties to a desired range was obtained with the loading of these fillers to a higher level, and consequently this increased cost and they were also not easily processable. Therefore, further tactics in this regard were applied which include the loading of nanofillers at low levels and these nanofiller reinforced polymer matrix are known as polymer nanocomposites.

3.2. Polymer nanocomposite materials

As we have described earlier, those polymer nanocomposites are a new class of advanced composite materials, which have been receiving much attention on research and commercial levels. The dimensions of nanofillers incorporated in these materials are only up to a few nanometers (about 1/10,000 of human hair), due to which they offers ultra-large interfacial specific area between the nano-element and polymer matrix. Resultantly, the polymer nanocomposites display significant enhancement in toughness, thermal stability, etc., without compromising on their other properties such as stiffness or optical clarity.

3.3. Why CNTs–polymer nanocomposites

As compared with the other nanofillers, carbon nanotubes (CNTs) have been shown as the most promising nanofiller materials for polymer composites and this credit goes to their remarkable mechanical and electrical properties [50]. CNTs have been under-investigation as a nanofiller for polymer nanocomposites since the day of its first discovery. In order to take the advantages of CNTs and their extra ordinary properties in everyday life, CNT/polymer nanocomposites were explored and investigated for many purposes and these research efforts have led them favourable in many fields, which include energy sector, defence equipments, automobile industries sports, supercapacitors and gas sensors. These multi-dimensional applications of CNT/polymer nanocomposites are actually based on their flexible processing methods, ultra-high strength, light in weight and good durability. Moreover, the high value of electrical conductivity, CNT/polymer nanocomposites are usable as electrostatic discharge (ESD) and electromagnetic interference (EMI) shielding materials. However, the feasibility of CNTs to fabricate their polymer nanocomposite materials depends mainly on their homogeneous dispersion throughout the polymer matrix without losing their integrity.

In the following sections, we will discuss different fabrication methods of polymer nanocomposites and their numerous properties briefly. Then exploration of their gas-sensing properties will be described in detail. Finally, we will sort out the gas-sensing mechanisms for these nanocomposite materials in the light of literature report.

3.3.1. Preparation strategies

We have learnt that CNTs have strong tendency of aggregation due to their large surface area which provide them the ability to be bonded with the polymer matrices. Such a combination lasts until and unless large forces are applied. Such large and un-bearable forces mainly come from the vigorous mixing of the polymer which can damage the CNTs structure as well. Therefore, during their processing of CNT/polymer nanocomposites, the biggest issue is the proper dispersion of individual CNTs in the polymer matrices so that the advantages of CNTs' prominence be fully realized. Pre-functionalization of CNTs chemically has helped is their dispersion to some extent, but then the main issue is their long term stability. Following methods are in use to process the CNT/polymer nanocomposites in recent research.

a. Solution blending

b. Melt mixing

c. In situ polymerization

3.3.1.1. Solution processing (solution blending)

This is most common method used to prepare CNT/polymer nanocomposites. In this method, both CNTs and polymer are mixed in a suitable solvent. The main advantage of this method is that CNTs are mixed rigorously by dispersion. Generally, it consists of three steps such as

Dispersion of nanotubes in a suitable solvent by magnetic stirring, ultrasonication or refluxing. However, sonication for longer time can shorten the nanotubes which reduce the aspect ratio of CNTs and are unfavourable to the properties of nanocomposites.

Mixing of CNT dispersion with the polymer (either at room temperature or elevated temperature),

Final yield of nanocomposite by precipitation or casting of a film.

It is notable that both organic and aqueous medium are used to produce CNT/polymer nanocomposites [51]. To minimize the shortening of nanotubes by sonication, suitable surfactants are used to get better dispersion. Sodium dodecylbenzene sulfonate, K-30, etc., are common surfactants used for this purpose [52].

In spite of the advantages of the solution-processing technique for the better nanotube dispersion, it is industrially less favourable due to its less yield.

3.3.1.2. *Melt blending*

In order to obtain large scale production on industrial level, melt processing is a preferred option because it is relatively economical and simple to process. Generally, this method consists of two steps;

Melting of the polymer pellets to form a viscous liquid,

Apply high shear stress to disperse the CNTs in this viscous liquid.

This method has been used successfully to prepare MWCNT/polycarbonate, [50] MWCNT/nylon-6, [51] SWCNT/polypropylene [52] and SWCNT/polyimide [53] nanocomposites.

Although this is a simple technique, the major problem with it is that the application of high shear force and high temperature for melting purpose may weaken the properties of nanocomposites as well as the CNT fragmentation. Therefore, there must be an optimum value of the shear stress and temperature to get the required dispersion. To overcome these issues, several modifications have been introduced in melt blending, for example, combining the solution-melt blending techniques and then applying several cycles of melt processing [54], introduction of polymer-coated MWCNT (instead of pure MWCNT) into the melted polymer to promote compatibilization [55].

3.3.1.3. *In situ polymerization*

This technique is common nowadays, particularly to prepare of polymer grafted nanotubes where polymers are insoluble and thermally unstable. The main advantage of this method is that it enables grafting of polymer macromolecules onto the walls of CNTs. Besides this, it enhances the miscibility of CNTs with almost each polymer matrix. Depending on the desired molecular weight and molecular weight distribution of polymers, several types of polymerizations (such as chain transfer, radical, anionic and ring-opening metathesis) can be used for in situ polymerization processing. For the first time, this was applied for the preparation of PMMA/MWCNT nanocomposites [56–58]. Later on, the preparation of hydroxyl functionalized MWCNTs/acrylic acid-grafted PTT nanocomposites [59], SWCNT-reinforced polyimide nanocomposites [60], etc., have been reported by using this technique.

3.3.2. *Inter-constituents' interaction in the CNT/polymer nanocomposite*

We have studied that introduction of CNTs to the polymer matrices upgrades their different properties including electrical, mechanical, morphology and thermal properties, of the resultant nanocomposites. Best performance of the CNTs/polymer nanocomposites is generally governed by several factors, purification of CNTs to remove certain impurities such as amorphous and metallic carbon, optimal CNTs' dispersion, CNTs curvature and controlled content ratio of CNTs in the nanocomposite are of major significance in the performance of CNTs/polymers nanocomposites. Besides this, the optimal performance of the CNTs/polymer nanocomposites is also decided by the selection of suitable polymers as well. These performance parameters of the CNTs/polymer nanocomposites are generally evaluated in terms of the inter-CNTs-polymer interaction. Study and sorting out of the interaction characteristics of the nanocomposite system have, therefore, primary significance [61–63].

Interaction between different constituents of the CNTs/polymer nanocomposites consists of following three classes [63–66];

a. Interaction between nanotubes in a bundle,

b. Interaction between different shells (in case of MWCNTs),

c. Interaction between CNTs and polymer matrix.

In the CNTs/polymer nanocomposites, the constituents, that is CNTs and polymers preserve their original character both chemically and physically. However, they also deliver additional properties from combination. These added properties mainly depend on the following factors:

a. Interfacial bonding between CNT and polymer,

b. Mechanical load transfer from the polymer matrix to CNT, and

c. Interfacial strength

Various investigations have proved that in case of insufficient load transfer (e.g. micromechanical interlocking, bonds, etc.) from the polymer matrix to CNT, the resultant nanocomposite display properties less than that of expected. Therefore, strong interaction between CNTs and polymer at their interface is one of the pre-requite condition for better mechanical performance of the nanocomposite. Additionally, the effective dispersion of CNTs in the polymer matrix has also vital significance in this context. Ultrasonication and refluxing are two means for such dispersion. Interfacial chemical bonding may be obtained by pre-functionalization of CNTs which offers a strong interface and leads to effective stress (load) transfer. However, it must be noted that only non-covalent functionalization is supporting in this effect, whereas covalent functionalization of CNTs is not suitable because it may disturb the covalent bonds of graphene sheet.

In order to evaluate such an interfacial interaction, computational methods are used. Molecular dynamics (MD)-based simulation has been used frequently due to its flexibility and being more accurate. This kind of MD simulation is based on two different approaches. In the first approach, a CNT is dragged out of the nanocomposite and in the second approach, a CNT and polymer units are fetched together for studying their interaction. Results of both types of approaches are evaluated in term of some technical parameters, which are given below [67–70]:

a. **Interaction energy (Interfacial-binding energy):** This is obtained by taking the difference between potential energy of the nanocomposite and sum up of the individual ingredients (CNTs and polymers). It is denoted by E_{int} and is given by

$$E_{int.} = E_{nc} - \left(E_{CNT} + E_{polymer}\right) \tag{13}$$

If we divide $\mathbf{E_{int}}$ cross-sectional area of the CNT and normalize it by the number of polymer molecules, then we get the interfacial binding energy.

$$(B.E)_{interfacial} = \text{Norm}\left[E_{int.} / A_{CNT} \right] \quad (14)$$

b. Strain–strain curve and shear modulus: It is defined as the work done, which is required to drag out a CNT divided by the cylindrical volume of the CNT.

$$\tau_{interfacial} = W_{dragout} / \pi r^2 1 \quad (15)$$

where r is the external radius of the CNT and 'l' is its length.

3.4. CNTs/polymer composites for gas sensors

3.4.1. Working principle

After learning about the interfacial interaction in CNTs/polymer nanocomposites, we are now able to describe the device applications of these nanocomposite materials in the gas sensors. Such nanocomposite materials when synthesized for the gas-sensing systems, generally contains organic polymers as the major polymeric materials. However, there is no restriction on the use of conducting/non-conducting polymers. Both can be applied in the gas sensor system.

Conducting polymers possess semiconducting nature because of their heterocyclic compounds, which exhibit physicochemical characteristics. Such properties can lead to the reversible changes in the conductivity of the sensing layer, which can be readily observed when polar chemicals are adsorbed on their surface, even at room temperature [71]. Such a change in conductivity is based on the following two causes [72]:

a. By charge transfer between gas molecules and the polymer,

b. The swelling of polymer film.

Non-conductive polymers can also be used as gas-sensing materials. In spite of the conductivity changes, the bulk dissolution of gas into the film may change their other physical properties [73], for example inside the polymer-coated resonators (by Avramov et al. [74] and Yantchev et al. [72]). Conducting particles-insulating matrix composition is another nanocomposite used as gas-sensing materials where the polymer acts as the insulating matrix, while the dispersed conducting particles provide the conducting path for sensing [75]. Due to the adsorption of interested analyst, there are volumetric changes of the matrix polymer. CNTs, due to its one-dimensional structure and good conductivity, are ideally used as the dispersed particles in this conducting particles-insulating matrix composition for gas-sensing systems. Due to this fact, CNTs/polymer composites have been widely studied for gas sensors [76–78].

3.4.2. Sensing mechanism

Interfacial bonding between CNTs and polymer executes main contribution to the gas-sensing effect of such nanocomposites. However, on the basis of experiments, following common sensing mechanisms have been suggested.

a. Due to the occurrence of quantum mechanical tunnelling effects, conducting paths are formed inside the CNT/polymer composites and the distance between these conducting paths is such that hopping of electrons between them occurs. With the increase in distance between the adjoining carbon nanotubes, their mutual contact resistance between them increases. After the polymer nanocomposites are exposed to target gas, adsorption of organic vapours leads to the swelling of the polymer matrix that increases the volume, and hence the distance between adjoining nanotubes. Consequently, the contact resistance increases too. This entire dependency of all the parameters, particularly (the coverage of swelling and resultant electrical response) mainly depends on the polymer's solubility in the solvent. Such a mechanism was initially suggested in the literatures [79–82].

b. Another mechanism is based on the charge transfer due to the adsorption of polar organic molecules which affects the electronic properties of the semiconductor CNT surface. Such an adsorption of gaseous molecules is mainly due to the fact that the nanocomposite's surface and solvent molecule both exhibit polar nature. The attached polar functional groups to the CNT surface adsorb solvent molecules. Polarity of gaseous molecules decides the ability of the sensing response. More polar is the adsorbed molecule, quick response is there, and vice versa. Similarly, the polar groups on the CNTs' surface also increase directly with the adsorption of solvent molecules, and hence, a better response is obtained. This fact has been observed for CNT/PMMA nanocomposites with polar solvents like methanol which showed an improved response to f-CNT/PMMA composites even though they are not good solvents for PMMA [83, 84]. It was observed that the reversibility of these sensors with such polar solvents is very poor. This means that even though the sensor is re-exposed to air, the resistance value does not return to its initial value completely, that is the polar interaction between the CNT surface and these solvents is high so that desorption from the surface is slow [85].

3.5. Literature survey

Due to their exclusive and remarkable properties, CNT/polymer nanocomposites-based gas sensors are the focus of recent gas sensor research. Aim of the entire efforts is to lower the working temperature and enhance the sensing parameters such as sensitivity, response speed and selectivity. Earlier, An et al. [86] prepared SWNT/polypyrrole nanocomposite by in situ polymerization of the pyrrole and SWMTs mixture (**Figure 13**).

Gas-sensing device was fabricated through spin-casting on the already patterned interdigitated electrodes. Sensing response of this nanocomposite was investigated towards NO_2 gas which showed better sensitivity than the pure individual materials (CNTs or polypyrrole); however, the major drawback with this nanocomposite is that it degrades when it is exposed to the gas for longer time (**Figure 14**).

Figure 13. FESEM images of (a) pure Ppy, (b) pure SWNT bundles and (c) SWNT/Ppy nanocomposite. Inset: optical microscope image of the interdigitated electrodes.

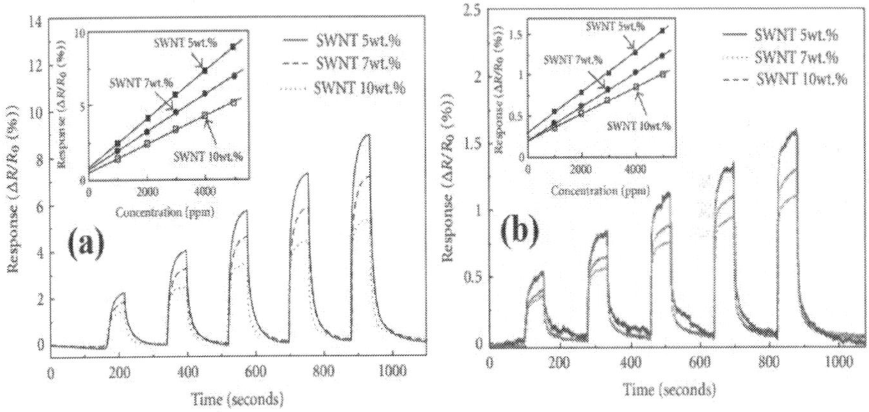

Figure 14. The change in sensitivity as a function of gas exposure time for 3000 ppm NO_2.

Then Cho et al. [87] introduced SWCNT-ethyl cellulose (SWCNT/EC) nanocomposite. This nanocomposite was prepared by solution-blending process with the use of chloroform as media. The interdigitated electrode technique was used to fabricate the gas-sensing device, and its response was investigated towards benzene and ethanol. Resistance observed almost a linear variation with the gaseous concentration, and it gave reasonable sensitivity to both of the gases (**Figure 15**).

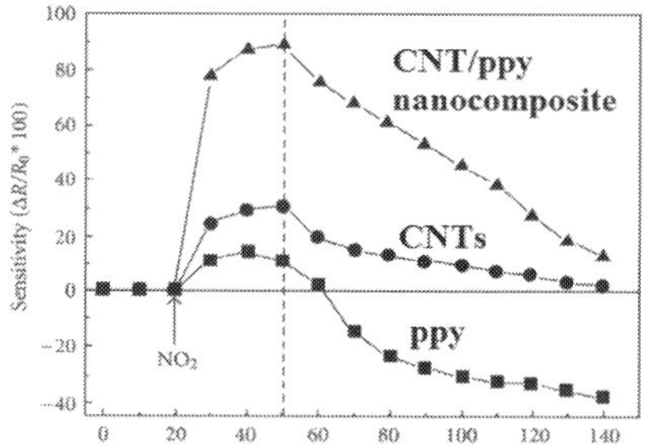

Figure 15. (a) Gas-sensing properties of SWNT–Polymer (EC) composites for 1000–5000 ppm benzene vapours, (b) gas-sensing properties of SWNT–Polymer (EC) composites for 1000–5000 ppm ethanol vapours (insets in the upper left corners show the linear relationship) [87].

Using similar idea, Philip et al. [85] prepared composite thin film of MWCNT/polymethylmethacrylate (MWCNT/PMMA) and functionalized-MWCNT/PMMA (f-MWCNTs/PMMA) nanocomposites. Their sensitivity response was investigated to a variety of gases (dichloromethane, chloroform, acetone, etc.). Both of nanocomposites response well; however, the enhancement in sensitivity and response time of f-CNT/PMMA composites was more significant (**Figure 16**).

In addition of CNT/polymer composite, the sensing characteristics of polymer-coated CNTs film (CNT as a major material) towards specific gases were also tested. Li et al. [88] studied the sensing character of this kind of structure by coating of polymer on CNTs. Chlorosulfonated polyethylene (polymer) was dissolved in tetrahydrofuran (THF) solvent, which was used as a coating solution for Cl_2 sensing, and hydroxypropyl cellulose was dissolved in the solvent chloroform as coating on SWCNTs to detect HCl. Both the nanocomposites responded well to the corresponding gases. **Table 3** tabulates familiar CNT-polymer nanocomposites, which have been investigated for gas sensors.

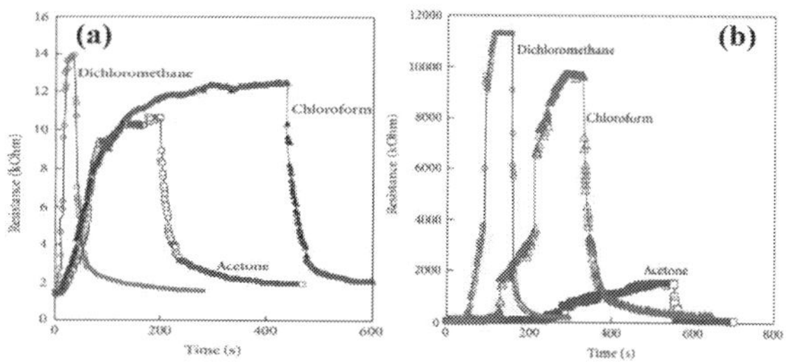

Figure 16. Response of (a) CNT/PMMA and (b) f-CNT/PMMA composites to dichloromethane, chloroform and acetone vapours [85].

Organic polymer	CNT type	Sensor configuration	Targeted analytes	Functionalization method	Detection limit	Response time (s)	Reversibility
PEI, Nafion	SWNTs	ChemFET	NO_2, NH_3	Non-covalently drop-coating	100 ppt (NO_2)	60–120	Reversible (UV)
PMMA	MWNTs	Chemiresistor	Dichloromethane, chloroform, acetone	Mixed as composite film	N/S	2–5	Reversible
PEI/Starch mixture	SWNTs	ChemFET	CO_2	Non-covalently dip-coating	500 ppm	60	Reversible
PEI	SWNTs	ChemFET	NO	Non-covalently drop-coating	5 ppb	70	Reversible
Polypyrrole	SWNTs	ChemFET	NO_2	Covalent chemical polymerization	N/S	600–1800	Reversible
Poly(3-methylthiophene)	MWNTs	Chemiresistor	CH_2Cl_2, $CHCl_3$, CCl_4, CH_4	Covalent chemical polymerization	N/S	60–120	Reversible
PABS	SWNTs	Chemiresistor	NH_3, NO_2, H_2O	Covalent chemical functionalization	20 ppb (NO_2), 100 ppb (NH_3)	600	Reversible
Poly(vinyl acetate), Polyisoprene	MWNTs	Chemiresistor	Tetrahydrofuran, ethanol, cyclohexane	Non-covalently drop-coating	N/S	1200	Reversible
Polyaniline	SWNTs	Chemiresistor	NH_3	Non-covalent electrochemical functionalization	50 ppb	600	Reversible
Polyaniline	MWNTs	Chemiresistor	Triethylamine	*In situ* polymerization (PANI–SWNT composite)	500 ppb	200–400 (up to 600 for recovery)	Reversible

Table 3. Some well-known CNT-polymer nanocomposite materials investigated for gas sensors [89].

Likewise, we had prepared polystyrene/MWCNTs nanocomposite through solvent evaporation technique and investigated improvement in its electrical properties and thermal stability, which is quite reasonable [90].

4. Conclusion

In summary, the gas sensors based on both CNTs (in single and bundle form) and CNT/ polymers nanocomposites are being focused to lower the working temperature, enhance response speed and sensitivity. However, the problem of their degradation during exposure for longer times or high temperature is still to be resolved. Efforts in this line are in continuous run. Exploration of new and novel nanocomposite materials of polymers with CNTs is increasing day by day, which will indeed sort out solution to unresolved technical and industrial issues.

Author details

Dr. Abdul Hakim Shah*

Address all correspondence to: ahakim_physics@yahoo.com

Department of Material Physics and Nanotechnology, Khushal Khan Khattak University, Karak, Pakistan

References

[1] L . Yang, H. Lin, Z. Zhang, et al. Gas sensing of tellurium-modified silicon nanowires to ammonia and propylamine, Sens. Actuators B Chem. 177:260–264, 2013.

[2] Z. Zhu, C.T. Kao, R.J. Wu. Appl. A highly sensitive ethanol sensor based on Ag@TiO_2 nanoparticles at room temperature, Surf. Sci. 320:348–355, 2014.

[3] X.H. Xia, et al. Self-supported hydrothermal synthesized hollow Co_3O_4 nanowire arrays with high supercapacitor capacitance. J. Mater. Chem. 21(25):9319–9325, 2011.

[4] S. Iijima. Helical microtubules of graphitic carbon. Nature. 354(6348):56–58, 1991.

[5] T.W. Odom, J.L. Huang, P. Kim, C.M. Lieber. Structure and electronic properties of carbon nanotubes. J. Phys. Chem. B. 104:2794–2809, 2000.

[6] DCD03 2/06/87-M/ DREA mod. 17 Dec 1997.

[7] A.H. Wei, et al. Nanotube molecular wires as chemical sensors, Science. 287:622-25, 2000.

[8] H. Dai. "Nanotube growth and characterization." In: Carbon Nanotubes: Synthesis, Structure, Properties and Applications, M.S. Dresselhaus, G. Dresselhaus, Eds., pp. 29–54, Springer, New York, NY, USA, 2000.

[9] Y. Wang, J.T.W. Yeow. A review of carbon nanotubes-based gas sensors. J. Sens. 2009, 493904 (1- 24), 2009.

[10] J. Li, Y. Lu, Q. Ye, et al. "Carbon nanotubes sensors for gas and organic vapour detection. Nano Lett. 3(7):929–933, 2003.

[11] J. Zhao, A. Buldum, J. Han, et al. Gas molecule adsorption in carbon nanotubes and nanotube bundles. Nanotechnology. 13(2):195–200, 2002.

[12] J. Han. In: Carbon Nanotubes: Science and Application, M. Meyyappan, Ed., CRC Press LLC, Boca Raton, FL, USA, 2005.

[13] C.K.W. Adu, G.U. Sumanasekera, B.K. Pradhan, et al. Carbon nanotubes: a thermoelectric nano-nose. Chem. Phys. Lett. 337(1–3):31–35, 2001.

[14] R.D. Barnard. Thermoelectricity in Metals and Alloys. John Wiley & Sons, New York, NY, USA, 1972.

[15] O.K. Varghese, P.D. Kichambre, D. Gong, et al. Gas sensing characteristics of multiwall carbon nanotubes." Sens. Actuators B. 81(1):32–41, 2001.

[16] J. Chung, K.H. Lee, J. Lee, et al. Multi-walled carbon nanotubes experiencing electrical breakdown as gas sensors. Nanotechnology. 15(11):1596–1602, 2004.

[17] T. Ueda, S. Katsuki, K. Takahashi, et al. Fabrication and characterization of carbon nanotube based high sensitive gas sensors operable at room temperature. Diam. Relat. Mater. 17(7–10):1586–1589, 2008.

[18] Z. Hou, D. Xu, B. Cai. "Ionization gas sensing in a microelectrode system with carbon nanotubes." Appl. Phys. Lett. 89(21):3, Article ID 213502, 2006.

[19] A. Modi, N. Koratkar, E. Lass, et al. Miniaturized gas ionization sensors using carbon nanotubes. Nature. 424(6945):171–174, 2003.

[20] W.A. Heer, A. Chatelain, D. Ugarte. "A carbon nanotube field-emission electron source." Science. 270(5239):1179–1180, 1995.

[21] N. Jonge, Y. Lamy, K. Schoots, et al. High brightness electron beam from a multi-walled carbon nanotube. Nature. 420(6914):393–395, 2002.

[22] S.J. Kim. Gas sensors based on Paschen's law using carbon nanotubes as electron emitters. J. Phys. D. 39(14):3026–3029, 2006.

[23] S. Kim. CNT sensors for detecting gases with low adsorption energy by ionization. Sensors. 6(5):503–513, 2006.

[24] J. Wu, H. Liu, D. Xu, et al. In: Proceedings of the 3rd IEEE International Conference on Nano/Micro Engineering and Molecular System, Sanya, China, January 2008.

[25] A. Wadhawan, R.E. Stallcup II, K.F. Stephens II, et al. Effects of O_2, Ar, and H_2 gases on the field-emission properties of single-walled and multiwalled carbon nanotubes. Appl. Phys. Lett. 79(12):1867–1869, 2001.

[26] Z. Yong, L. Junhua, L. Xin, The structure optimization of the carbon nanotube film cathode in the application of gas sensor. Sens. Actuators A. 128:278–289, 2006.

[27] J.L. Bahr, J.M. Tour. Covalent chemistry of single-wall carbon nanotubes. J. Mater. Chem. 12(7):1952–1958, 2002.

[28] A. Hirsch. "Functionalization of single-walled carbon nanotubes." Angew. Chem. Int. Ed. 41(11):1853–1859, 2002.

[29] S. Niyogi, M.A. Hamon, H. Hu, et al. Chemistry of single-walled carbon nanotubes. Acc. Chem. Res. 35(12):1105–1113, 2002.

[30] S. Peng, K. Cho. Ab initio study of doped carbon nanotube sensors. Nano Lett. 3(4): 513–517, 2003.

[31] M.L.Y. Sin, G.C.T. Chow, G.M.K. Wong, et al. Ultralow-power alcohol vapor sensors using chemically functionalized multiwalled carbon nanotubes. IEEE Trans. Nanotechnol. 6(5):571–577, 2007.

[32] I. Sayago, E. Terrado, E. Lafuente, et al. Hydrogen sensors based on carbon nanotubes thin films. Synth. Met. 148(1):15–19, 2005.

[33] S. Dag, Y. Ozturk, S. Ciraci, et al. Adsorption and dissociation of hydrogen molecules on bare and functionalized carbon nanotubes. Phys. Rev. B. 72(15):8, Article ID 155404, 2005.

[34] J. Kong, M.G. Chapline, H. Dai. Functionalized carbon nanotubes for molecular hydrogen sensors. Adv Mater. 13(18):1384–1386, 2001.

[35] A.T. Gee, B.E. Hayden, C. Mormiche, et al. The role of steps in the dynamics of hydrogen dissociation on Pt(533). J. Chem. Phys. 112(17):7660–7668, 2000.

[36] R.A. Olsen, C. Bădescu, S.C. Ying, et al. *Adsorption and diffusion on a stepped surface: atomic hydrogen on Pt(211).* J. Chem. Phys. 120(24):11852–11863, 2004.

[37] A. Mandelis, C. Christofides. Physics, Chemistry and Technology of Solid State Gas Sensor Devices, John Wiley & Sons, New York, NY, USA, 1993.

[38] J. Li, Y. Lu, Q. Ye, et al. Carbon nanotubes sensors for gas and organic vapour detection. Nano Lett. 3(7):929–933, 2003.

[39] Y.D. Lee, W.S. Cho, S.I. Moon, et al. Gas sensing properties of printed multiwalled carbon nanotubes using the field emission effect. Chem. Phys. Lett. 433(1–3):105–109, 2006.

[40] J. Suehiro, G. Zhou, M. Hara. Fabrication of a carbon nanotube-based gas sensor using dielectrophoresis and its application for ammonia detection by impedance spectroscopy. J. Phys. D. 36(21):L109–L114, 2003.

[41] C.S. Huang, B.R. Huang, Y.H. Jang et al. Three-terminal CNTs gas sensor for N_2 detection. Diam. Relat. Mater. 14(11–12):1872–1875, 2005.

[42] Y.T. Jang, S.I. Moon, J.H. Ahn, et al. A simple approach in fabricating chemical sensor using laterally grown multi-walled carbon nanotubes. Sens. Actuators B. 99(1):118–122, 2004.

[43] M.T. Azar, Y. Xie. Sensitive NH_3OH and HCl gas sensors using self-aligned and self-welded multiwalled carbon nanotubes. IEEE Sens. J. 7(10):1435–1439, 2007.

[44] S. Peng, K. Cho. Chemical control of nanotube electronics. Nanotechnology. 11(2):57–60, 2000.

[45] C. Cantalini, L. Valentini, L. Lozzi, et al. NO_2 gas sensitivity of carbon nanotubes obtained by plasma enhanced chemical vapor deposition. Sens. Actuators B. 93(1–3): 333–337, 2003.

[46] P.G. Collins, K. Bradley, M. Ishigami, et al. Extreme oxygen sensitivity of electronic properties of carbon nanotubes. Science. 287(5459):1801–1804, 2000.

[47] K.A. Williams, P.C. Eklund. Monte Carlo simulations of H_2 physisorption in finite-diameter carbon nanotube ropes. Chem. Phys. Lett. 320(3–4):352–358, 2000.

[48] G. Stan, M.W. Cole. Hydrogen adsorption in nanotubes. J. Low Temp. Phys. 110(1–2): 539–544, 1998.

[49] G.U. Sumanasekera, C.K.W. Adu, S. Fang, et al. Effects of gas adsorption and collisions on electrical transport in single-walled carbon nanotubes. Phys. Rev. Lett. 85(5):1096–1099, 2000.

[50] P. Poetschke , A.R. Bhattacharyya, A. Janke. Melt mixing of polycarbonate/multi-wall carbon nanotube composites. Compos. Interfaces. 10, 389–404, 2003.

[51] R. Bandyopadhyaya, E. NativRoth, O. Regev, et al. Stabilization of individual carbon nanotubes in aqueous solutions. Nano Lett. 2:25–28, 2002.

[52] M.F. Islam, E. Rojas, D. Bergey, et al. High weight fraction surfactant solubilization of single-wall carbon nanotubes in water. Nano Lett. 3:269–273, 2003.

[53] T. Liu, I.Y. Phang, L. Shen et al. Morphology and mechanical properties of multiwalled carbon nanotubes reinforced nylon-6 composites. Macromolecules. 37:7214–7222, 2004.

[54] A.R. Bhattacharyya, T.V. Sreekumar, T. Liu, et al. Crystallization and orientation studies in polypropylene/single wall carbon nanotube composite. Polymer. 44:2373–2377, 2003.

[55] E.J. Siochi, D.C. Working, C. Park, et al. Melt processing of SWCNT-polyimide composite fibers. Compos. Part B. 35B:439–446, 2004.

[56] R. Haggenmueller, H.H. Gommans, A.G. Rinzler. Aligned single-wall carbon nanotubes in composites by melt processing methods. Chem. Phys. Lett. 330:219–225, 2000.

[57] Z. Jin, K.P. Pramoda, S.H. Goh, et al. Poly(vinylidene fluoride)-assisted melt-blending of multi-walled carbon nanotube/poly(methyl methacrylate) composites. Mater. Res. Bull. 37:271–278, 2002.

[58] Z. Jia, Z. Wang, C. Xu, et al. Study on poly(methyl methacrylate)/carbon nanotube composites. Mater. Sci. Eng. A. A271:395–400, 1999.

[59] C.V. Santos, A.L. Hernandez, F.T. Fisher et al. Improvement of thermal and mechanical properties of carbon nanotube composites through chemical functionalization. Chem. Mater. 15:4470–4475, 2003.

[60] K.W. Putz, C.A. Mitchell, R. Krishnamoorti. Elastic modulus of single-walled carbon nanotube/poly(methyl methacrylate) composites. J. Polym. Sci. Part B Polym. Phys. 42:2286–2293, 2004.

[61] C.S. Wu. Synthesis and characterization of poly(trimethylene terephthalate) composites incorporating multi-walled carbon nanotubes. J. Appl. Polym. Sci. 114:1633–1642, 2009.

[62] C. Park, Z. Ounaies, K.A. Watson et al. Dispersion of single wall carbon nanotubes by in-situ polymerization under sonication. Chem. Phys. Lett. 364:303–308, 2002.

[63] M. Rahmat, P. Hubert. Carbon nanotube-polymer interactions in nanocomposites: a review. Compos. Sci. Technol. 72:72–84, 2011.

[64] D. Qian et al. Load transfer and deformation mechanisms in carbon nanotube-polystyrene composites. Appl. Phys. Lett. 76(20):2868, 2000.

[65] M.F. Yu, B.I. Yakobson, R.S. Ruoff. Controlled sliding and pullout of nested shells in individual multiwalled carbon nanotubes. J. Phys. Chem. B. 104(37):8764, 2000.

[66] D. Qian, W.K. Liu, R.S. Ruoff. Load transfer mechanism in carbon nanotube ropes. Compos. Sci. Technol. 63(11):1561, 2003.

[67] H. Chen et al. Influence of nanotube chirality, temperature, and chemical modification on the interfacial bonding between carbon nanotubes and polyphenylacetylene. J. Phys. Chem. C. 112(42):16514, 2008.

[68] Liao K, Li S. Interfacial characteristics of a carbon nanotube–polystyrene composite system. Appl. Phys. Lett. 79(25):4225, 2001.

[69] C. Wei. Adhesion and reinforcement in carbon nanotube polymer composite. Appl. Phys. Lett.88(9):093108, 2006.

[70] S.J.V. Frankland, V.M. Harik. Analysis of carbon nanotube pull-out from a polymer matrix. Surf. Sci. 525(1–3):103–8, 2003.

[71] P.N. Bartlett, P.B.M. Archer, S.K. Ling-Chung. Conducting polymer gas sensors—part I: fabrication and characterization. Sens. Actuators. 19(2):125–140, 1989.

[72] V.M. Yantchev, V.L. Strashilov, M. Rapp, et al. Theoretical and experimental mass-sensitivity analysis of polymer-coated SAW and STW resonators for gas sensing applications. IEEE Sens. J. 2(4):307–312, 2002.

[73] M.C. Petty, R. Casalini. Gas sensing for the 21st century: the case for organic thin films. Eng. Sci. Educ. J. 10(3):99–105, 2001.

[74] I.D. Avramov, S. Kurosawa, M. Rapp, et al. Investigations on plasma-polymer-coated SAW and STW resonators for chemical gas-sensing applications. IEEE Trans. Microw. Theory Tech. 49(4):827–837, 2001.

[75] S.M. Cho, Y.J. Kim, Y.S. Kim, et al. The application of carbon nanotube-polymer composite as gas sensing m. Proc. IEEE Sens. 2:701–704, 2004.

[76] Y. Wanna, N. Srisukhumbowornchai, A. Tuantranont, et al. The effect of carbon nanotube dispersion on CO gas sensing characteristics of polyaniline gas sensor. J. Nanosci. Nanotechnol. 6(12):3893–3896, 2006.

[77] L. Valentini, V. Bavastrello, E. Stura, et al. Sensors for inorganic vapor detection based on carbon nanotubes and poly(o-anisidine) nanocomposite material. Chem. Phys. Lett. 383(5–6):617–622, 2004.

[78] M. Wienecke, M.C. Bunescu, M. Pietrzak, et al. PTFE membrane electrodes with increased sensitivity for gas sensor applications. Synth. Met. 138(1–2):165–171, 2003.

[79] L. Valentinia, J. Biagiottia, J.M. Kennya et al. Morphological characterization of single-walled carbon nanotubes-PP composites. Compos. Sci. Technol. 63:1149–53, 2003.

[80] B. Safadi, R. Andrews, E.A. Grulke. Multiwalled carbon nanotube polymer composites: synthesis and characterization of thin films. J. Appl. Polym. Sci. 84:2660–9, 2002.

[81] H.W. Goh, S.H. Goh, G.Q. Xu, et al. Dynamic mechanical behavior of in situ functionalized multi-walled carbon nanotubes/phenoxy resin composite. Chem. Phys. Lett. 373:277–83, 2003.

[82] A. Allaouia, S. Baiab, H.M. Chengb, et al. Mechanical and electrical properties of a MWNT/epoxy composite. Compos. Sci. Technol. 62:1993–8, 2002.

[83] T. Someya, J. Small, P. Kim, et al. Alcohol vapor sensors based on single-walled carbon nanotube field effect transistors, Nano Lett. 3:877–81, 2003.

[84] G.R. Hutchison, M.A. Ratner, T.J. Marks, et al. Adsorption of polar molecules on a molecular surface, J. Phys. Chem. B. 105:2881–4, 2001.

[85] B. Philip, J.K. Abraham, A. Chandrasekhar. Carbon nanotube/PMMA composite thin films for gas-sensing applications. Smart Mater. Struct. 12:935–939, 2003.

[86] K.H. An, S.Y. Jeong, H.R.H. Wang, et al. Enhanced sensitivity of a gas sensor incorporating single-walled carbon nanotube–polypyrrole nanocomposites. Adv. Mater. 16(12):1005–1009, 2004.

[87] S.M. Cho, Y.J. Kim, Y.S. Kim, et al. The application of carbon nanotube–polymer composite as gas sensing materials." Proc. IEEE Sens. 2:701–704, 2004.

[88] J. Li, Y. Lu, M. Meyyappan. Nano chemical sensors with polymer-coated carbon nanotubes. IEEE Sens. J. 6(5):1047–1051, 2006.

[89] T. Zhang, S. Mubeen, N.V Myung, M.A Deshusses. Recent progress in carbon nanotube-based gas sensors. Nanotechnology. 19:332001 (14 pp), 2008.

[90] A.H. Shah, T.Z. Rizvi. Improvement in electrical and thermal behavior of polystyrene/multiwalled carbon nanotubes nanocomposites. Measurement. 46(4):1541–1550, 2013.

[91] A.H. Shah, Y. Liu, G.S. Zakharova, et al. "Synthesis of vanadium pentoxide nanoneedles by physical vapour deposition and their highly sensitive behavior towards acetone at room temperature." RSC Adv. 5:23489–23497, 2015.